Infectious Disease Ecology

Infectious Disease Ecology

THE EFFECTS OF ECOSYSTEMS ON DISEASE AND OF DISEASE ON ECOSYSTEMS

Edited by

Richard S. Ostfeld,
Felicia Keesing,
and Valerie T. Eviner

PRINCETON UNIVERSITY PRESS   PRINCETON AND OXFORD

Published by Princeton University Press, 41 William Street, Princeton,
New Jersey 08540

In the United Kingdom: Princeton University Press, 3 Market Place,
Woodstock, Oxfordshire OX20 1SY

Library of Congress Cataloging-in-Publication Data

Infectious disease ecology: the effects of ecosystems on disease and of disease on
ecosystems / edited by Richard S. Ostfeld, Felicia Keesing, and Valerie T. Eviner.
p. cm.
Includes index.
ISBN 978-0-691-12484-1 ((hardcover) : alk. paper)
ISBN 978-0-691-12485-8 ((pbk.) : alk. paper)
1. Ecosystem health.   2. Host-parasite relationship—Environmental
aspects.   3. Communicable diseases in animals—Environmental aspects.
I. Ostfeld, Richard, 1954–   II. Keesing, Felicia.   III. Eviner, Valerie T.
QH541.15.E265I54   2008
571.9—dc22   2007028039

British Library Cataloging-in-Publication Data is available

This book has been composed in Sabon

Printed on acid-free paper. ∞

press.princeton.edu

Printed in the United States of America

3   5   7   9   10   8   6   4   2

CONTENTS

PART III Management and Applications

PART IV Concluding Comments: Frontiers in the Ecology of Infectious Diseases

ACKNOWLEDGMENTS

THIS BOOK IS A PRODUCT of the 11th Cary Conference, held in May 2005 at the Institute of Ecosystem Studies in Millbrook, New York, Cary Conferences have been held biennially since 1985 and are designed to bring together a diverse group of scientists to discuss broad topics of importance to the field of ecology. The 2005 conference focused on the ecology of infectious diseases, particularly the effects of ecosystems, broadly defined, on infectious disease dynamics, the effects of infectious diseases on ecosystems, and the potential for ecology to play a useful role in predicting and managing infectious disease outbreaks. The full list of past Cary Conferences and the books they have generated can be found on the Web site of the Institute of Ecosystem Studies (www. ecostudies.org).

The focus and the structure of this conference arose from numerous discussions with colleagues. In particular, the members of our steering committee—Alan Berkowitz, Charles Canham, Kathryn Cottingham, Peter Daszak, Andrew Dobson, Robert Holt, Gene Likens, Michael Pace, Alison Power, O. J. Reichman, and Lee Talbot—helped us shape the conference from the earliest days of its planning with constructive and invaluable suggestions. The entire scientific staff at IES provided helpful input on several occasions. Organizers of recent Cary Conferences, especially Charlie Canham, Gary Lovett, and Alan Berkowitz, were also extremely helpful in answering the great number of logistical questions we generated.

We received financial support for the conference through grants from the National Science Foundation, the Doris Duke Charitable Foundation, the United States Department of Agriculture, the National Oceanic and Atmospheric Association, Dutchess County, New York, and the IES. The IES also provided a great deal of logistical support. Without the generous support of all of these institutions, neither the conference nor this book would have been possible.

Coordinating a gathering of scientists from around the world and then housing and feeding them for three days is no easy task. Our conference coordinator, Claudia Rosen, did a superb job. We and all the conference participants appreciated her organization, good humor, and constant energy. During the conference itself, she worked with a team of graduate students—Brian Allan, Dustin Brisson, Julia Butzler, Jacob Griffin, Doug McCauley, Katie Prager, Lynn Schnurr, and Kirsten Schwarz—who coordinated transportation and many other logistics for conference attendees. We are extremely grateful to them for their help, and for also managing to be energetic, enthusiastic, and valuable participants in the conference itself at the same time. Several other members of the IES staff deserve special mention. Tanya Rios did her usual outstanding job preparing conference materials and grant proposals. Marie Smith guided us through the grant process and maintained her sense of humor and her professionalism long after most others would have lost them. Matt Gillespie and Lynn Sticker provided excellent help with managing and handling manuscripts.

We would particularly like to thank the authors of the chapters in this volume. Not only have they put tremendous effort into providing a synthesis of critical issues in disease ecology, but they have done so quickly, responding without apparent animosity to our demands of repeatedly short deadlines coupled with difficult writing assignments. We would also like to thank all of the attendees of the Cary Conference. This book is a reflection not only of its authors but also of the attendees' eager engagement in discussions across disciplines. The ideas and excitement the group generated have greatly enriched our perspectives and writings.

Finally, Robert Kirk and Sam Elworthy of Princeton University Press have been enthusiastic about this project since its conception.

Richard S. Ostfeld
Felicia Keesing
Valerie T. Eviner

LIST OF CONTRIBUTORS

Sonia Altizer, Institute of Ecology, University of Georgia,
Athens, GA 30602, USA   179
Michael Begon, School of Biological Sciences, University
of Liverpool, Biosciences Building, Liverpool,
L69 7ZB UK   12
Alan R. Berkowitz, Institute of Ecosystem Studies, Box AB, 65
Sharon Turnpike, Millbrook, NY 12545, USA   448
Ottar N. Bjornstad, Pennsylvania State University, 501 ASI
Building, University Park, PA 16801, USA   179
Carol Brewer, Division of Biological Sciences, University of
Montana, Missoula, MT 59812, USA   284
James M. Brown, Biology Department, University of New
Mexico, Albuquerque, NM 87131, USA   223
Jeremy J. Burdon, CSIRO Plant Industry, GPO Box 1600,
Canberra ACT 2601, Australia   179
Carla E. Cáceres, Department of Animal Biology, University
of Illinois at Urbana-Champaign, Urbana, IL 61801, USA   223
Stephen R. Carpenter, Center for Limnology, University of
Wisconsin, Madison, WI 53706-1492, USA.   71
Isabella M. Cattadori, Center for Infectious Disease Dynam-
ics, Pennsylvania State University, University Park, PA
16802, USA   347
F. Stuart Chapin III, Department of Biology and Wildlife,
Institute of Arctic Biology, University of Alaska, 129
Arctic Health, West Ridge, Fairbanks, AK 99775, USA   284
Jonathan M. Chase, Department of Biology, Box 1137, Wash-
ington University, St. Louis, MO 63130, USA   223
James E. Childs, Department of Epidemiology and Public
Health, Yale School of Medicine, 60 College Station, PO
Box 208034, New Haven, CT 06520-8034, USA   441
Keith Clay, Department of Biology, Indiana University, Jordan
Hall 159, Bloomington, IN 47405, USA   145
Sharon K. Collinge, University of Colorado, Dept of EBIO and
ENVS, 334 UCB, Boulder, CO, 80309-0334 USA   129
Patricia Ann Conrad, VM:PMI, 1 Shields Avenue, University
of California, Davis, CA 95616, USA   448
C. M. Cox, The Land Institute, Salina, KS 67401, USA   368

Jennifer Rudgers, Department of Biology, Indiana University,
  Jordan Hall 159, Bloomington, IN 47405, USA   145
Dave Strayer, Institute of Ecosystem Studies, Box AB,
  65 Sharon Turnpike, Millbrook, NY 12545, USA   179
Kathryn P. Sutherland, Department of Biology, Rollins
  College, 1000 Holt Avenue, Winter Park, FL 32789, USA   387
Lee M. Talbot, Environmental Science & Policy, George
  Mason University, 3048 David J. King Hall, 4400
  University Drive, Fairfax, VA 22030, USA   284
Peter Holmes Thrall, Centre for Plant Biodiversity Research,
  CSIRO Plant Industry, GPO Box 1600, Canberra ACT
  2601, Australia   179
Tammy Tinther, Department of Biology, Jordan Hall 159,
  Indiana University, Bloomington, IN 47405, USA   145
Maria Uriarte, Department of Ecology, Evolution, and
  Environmental Biology, Columbia University, 1021
  Schermerhorn Extension, 1200 Amsterdam Ave.,
  New York, NY 10027, USA   179
Daniel A. Vasco, Institute of Ecology, University of Georgia,
  Athens, GA 30602-2202, USA   48
Jessica R. Ward, Department of Ecology & Evolutionary
  Biology, Cornell University, E451 Corson Hall, Tower
  Road, Ithaca, NY 14853, USA   304
Margaret A. Waterman, Biology Department, Southeast
  Missouri State University, MS 6200, Cape Giradeau, MO
  63701, USA   448
Helen J. Wearing, Institute of Ecology, University of Georgia,
  Athens, GA 30602-2202, USA   48
Bruce A. Wilcox, Leahi Hospital, 3675 Kilauea Avenue, 3rd
  Floor, Honolulu, HI 96816, USA   284

INTRODUCTION

Felicia Keesing, Richard S. Ostfeld, and Valerie T. Eviner

PATHOGENS ARE UBIQUITOUS. We are all familiar with the cold viruses that give us sniffles, coughs, and aches, and with the more frightening pathogens that cause diseases such as AIDS, malaria, and tuberculosis, which kill millions of people each year. But pathogens affect much more than our own health. Farmers struggle with fungi that attack their crops, managers of endangered species worry about the potential impact of an epidemic on the fragile populations under their steward-ship, and tourists find their favorite snorkeling destinations devastated by coral bleaching diseases. Because of concerns like these, enormous quantities of energy and resources are deployed each year in the diag-nosis and treatment of infectious diseases of humans, nonhuman ani-mals, and plants. Yet annihilation of harmful pathogens is an unrealistic goal in most cases, and many other pathogens play critical positive roles in ecosystems, from recycling nutrients to increasing biological diversity. We are just beginning to recognize the degree to which patho-gens, and the diseases they cause, are embedded within ecological systems.

Infectious diseases necessarily involve interactions among at least two species, the pathogen and the host species it infects. For many patho-gens, such as the virus that causes avian flu, the fungus that causes soy-bean rust, and the protist that causes African sleeping sickness, more than one species can serve as a host. And many pathogens are transmit-ted from host to host by at least one species of vector, such as a mos-quito, an aphid, or a tick. Understanding the dynamics of any particular disease system, then, involves understanding at best a simple but more often a complex system of interactions among the organisms most di-rectly involved in disease transmission. Ecologists would seem to be nat-ural allies of a suite of health specialists, including epidemiologists, physicians, veterinarians, and agricultural scientists.

With notable exceptions, however, ecologists have not traditionally studied infectious diseases or have considered disease outbreaks as dis-turbances rather than inherent parts of the ecosystem. Similarly, most biomedical scientists have not considered the broader ecological contexts of disease. But the need to integrate these disciplines has become increas-ingly apparent in the past several years as we face a surge of emerging or reemerging infections, including West Nile virus encephalitis, sudden oak death, severe acute respiratory syndrome (SARS), monkeypox, and new types of avian influenza. Typically, a newly emerging infectious disease is recognized from a cluster of mysterious disease cases arising in a host population, followed by the elimination of well-known pathogens as potential causes, and finally the identification of a new pathogen, or perhaps of an old one outside its known range. Remedial action is then undertaken to prevent further spread of the disease. When the pathogen is specialized—that is, largely restricted to one host species—and is transmitted directly between individuals, the standard public health arsenal to battle disease—quarantine, vaccination, emergency public education—is usually effective. SARS is a recent example. However, when the pathogen is more generalized, infecting multiple host species, including asymptomatic reservoirs, or when it is transmitted indirectly, such as by vectors or through environmental contact, then remedial action is much more problematic, and outbreaks may be followed by poorly contained spread, often with devastating consequences. Recent examples of poorly contained outbreaks include West Nile virus in humans, horses, and wild birds; sudden oak death in live oaks and tanoaks; Ebola virus in humans and apes; Lyme disease in humans; hantavirus pulmonary syndrome in humans; and various transmissible spongiform encephalopathies in livestock, wildlife, and humans. Failure of the standard biomedical arsenal in the cases of some human, nonhuman animal, and plant diseases may be largely a consequence of the ecological complexity involved in the evolution, transmission, and maintenance of these pathogens in nature.

Other lines of evidence also suggest that a more ecological perspective would greatly enhance our understanding and management of diseases. For example, more than 75% of emerging human pathogens are zoonotic (Taylor et al. 2001), that is, they are transmitted to humans from other animals. This observation suggests that a focus on the ecological interactions between wildlife hosts and zoonotic pathogens would be fruitful. Climate change has been associated with an increase in the frequency, distribution, and severity of many infectious diseases worldwide (Harvell et al. 2002), demonstrating ecological impacts on pathogen dynamics. And the rapid spread throughout Eurasia and Africa of the H5N1 strain of avian flu virus highlights just how much we need to know about bird migration patterns to develop appropriate management strategies for a potential human pandemic (Olsen et al. 2006). Some recent studies attest to the ability of ecological approaches to inform disease prevention and management. For example, the number of Lyme disease cases can be predicted almost two years in advance simply by monitoring annual acorn production (Ostfeld et al. 2006), allowing early, targeted public warnings. As another example, planting a diversity of rice strains rather than a monoculture has been shown to increase yields and reduce rates of infection with fungal rice blast in China (Zhu et al. 2000). More and more case studies like these are being published every year.

There are, in our view, two pressing needs if we are to improve our ability to predict the occurrence, dynamics, and consequences of infectious diseases. The first and most obvious need is to forge stronger alliances between ecologists and the traditional infectious disease specialists. Biomedical, veterinary, and agricultural scientists are well equipped to track infectious disease in populations and to treat and attempt to prevent disease in individual patients or populations. The power of these disciplines to improve the quality of life for people and other animals is enormous. Nevertheless, these disciplines often are not well equipped to anticipate disease outbreaks or to track the consequences of diseases beyond direct effects on victim populations. We see a strong role for ecologists in both these endeavors. Assembling the conceptual expertise is only part of the challenge, however: disease biologists from all disciplines need to work together more effectively to integrate knowledge of the functioning of ecological systems with knowledge of pathogens, cells, tissues, and immune systems and to develop effective management strategies based on this integration.

The second need is to identify the general ecological principles that underlie the dynamics of disease systems. Case studies now exist in sufficient number to allow the vigorous pursuit of conceptual syntheses. Such syntheses would provide a crucial unification of many disparate diseases and provide guidance for researchers attacking new disease systems. We know, for example, that some hosts are more efficient at transmitting particular pathogens than others, but what generalities, if any, can be made about the role of host diversity in disease transmission? Habitat fragmentation has been shown to affect the transmission of malaria in Brazil (Vittor et al. 2006), Lyme disease in New York and New England (Allan et al. 2003; Brownstein et al. 2005), and hantavirus in Panama, but do we know enough to be able to predict the impact of habitat fragmentation—and perhaps other forms of habitat alteration—on other diseases? Are some types of pathogens more likely than others to affect ecosystem functioning, and similarly, are some ecosystems more vulnerable to the impacts of pathogens? Under what conditions do infectious diseases alter the functioning of ecological systems in desirable ways, by, for example, increasing the cycling rates of nutrients or increasing biological diversity?

This book attempts both to develop conceptual frameworks and to more fully integrate ecology with traditional disease biology. We have invited outstanding scientists and educators to provide conceptual syntheses of their areas of expertise. We have organized these efforts into three main sections. The first focuses on the effects of ecosystems, in the broadest sense, on infectious diseases, the second on the effects of infectious diseases on ecosystems, and the third on management and applications using these ideas. In an effort to foster the developing dialogue among scientific specialties, we have included contributions from ecologists, biomedical scientists, agricultural scientists, and veterinarians.

The contributions in this book exhibit a three-pronged conceptual approach. They articulate the generalities emerging from the increasing number of case studies appearing in the scientific literature, raise specific questions to guide future studies, and demonstrate both the challenge and the potential for ecologists and other disease biologists to work together. Owing to increased interest, particularly on the part of young scientists, and to rising levels of funding, opportunities to study the ecology of infectious disease are increasing. From higher agricultural yields to more diverse animal communities to reduced human suffering and mortality, we have much to gain from the marriage of ecology and disease biology.

LITERATURE CITED

Allan, B. F., F. Keesing, and R. S. Ostfeld. 2003. Effects of habitat fragmentation on Lyme disease risk. Conservation Biology 17:267–72.

Brownstein, J. S., D. K. Skelly, T. R. Holford, and D. Fish. 2005. Forest fragmentation predicts local scale heterogeneity of Lyme disease risk. Oecologia 146:469–75.

Harvell, C. D., C. E. Mitchell, J. R. Ward, S. Altizer, A. P. Dobson, R. S. Ostfeld, and M. D. Samuel. 2002. Climate warming and disease risks for terrestrial and marine biota. Science 296:2158–62.

Olsen, B., Vincent J. Munster, Anders Wallensten, Jonas Waldenström, Albert D. M. E. Osterhaus, and Ron A. M. Fouchier. 2006. Global patterns of influenza A virus in wild birds. Science 312:384–88.

Ostfeld, R. S., C. D. Canham, K. Oggenfuss, R. J. Winchcombe, and F. Keesing. 2006. Climate, deer, rodents, and acorns as determinants of variation in Lyme-disease risk. PLoS Biology 4(6):e145.

Taylor, L. H., S. M. Latham, and M. E. Woolhouse. 2001. Risk factors for human disease emergence. Philosophical Transactions of the Royal Society of London. Series B, Biological Sciences 356:983–89.

Vittor, A. Y., R. H. Gilman, J. Tielsch, G. Glass, T. Shields, W. S. Lozano, V. Pinedo-Cancino, and J. Patz. 2006. The effect of deforestation on the human rate of *Anopheles darlingi*, the primary vector of Falciparum malaria in the

Peruvian Amazon. American Journal of Tropical Medicine and Hygiene 74:3–11.

Zhu, Y., H. Chen, J. Fan, Y. Wang, Y. Li, J. Chen, J. Fan, S. Yang, L. Hu, H. Leung, T. W. Mew, P. S. Teng, Z. Wang, and C. C. Mundt. 2000. Genetic diversity and disease control in rice. Nature 406:718–22.

PART I

Effects of Ecosystems on Disease

Introduction

Felicia Keesing

FOR MANY, MANY YEARS, people have recognized connections between the environment and outbreaks of infectious disease. For example, in the fourth century B.C., Hippocrates developed a "miasmatic" theory that linked malaria transmission to the environment, a link that was strengthened in the 1700s by Giovanni Lancisi, physician to the pope, who recognized that malaria transmission decreased after drainage of wetlands. Lancisi even postulated that mosquitoes might be involved in transmission (Cook and Webb 2000). Two centuries later, in the early 1900s, another environmental connection was proposed for malaria by an Italian public health worker, who suggested that the presence of nonhuman hosts for mosquitoes might divert mosquito meals away from humans, thus reducing malaria transmission (zooprophylaxis, reviewed by Service 1991).

More recently, there has been a proliferation of examples of environmental effects on the transmission of a wide range of infectious diseases. For example, periods of heavy rainfall have been correlated with outbreaks of Rift Valley fever in Kenya (Linthicum et al. 1999), eutrophication has been linked to parasitic diseases of amphibians (Johnson and Chase 2004), and the local risk of Lyme disease has been predicted from acorn production (Ostfeld et al. 2001). These examples, and many others like them, underscore the relevance of connections between ecosystems and disease. More important, they suggest that it might be possible to develop a general framework for predicting environmental effects on disease dynamics. For example, the outbreaks of Rift Valley fever, amphibian diseases, and Lyme disease are all caused by increases in the availability of resources (e.g., rainfall, nutrients, food) that influence host behavior, abundance, and diversity, consequently influencing disease transmission.

The dynamics of infectious diseases are most immediately affected by host, vector, and pathogen diversity, abundance, and behavior. But hosts, vectors, and pathogens are embedded within ecological communities, ecosystems, and landscapes, and it is thus not surprising that processes on these larger scales influence disease dynamics (figure I.1). The contributors to this first section of the book have attempted to move beyond case studies by synthesizing current understanding of a different component of this conceptual framework.

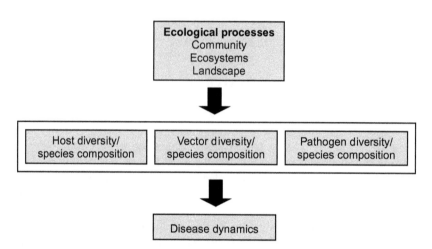

Figure I.1. Conceptual model of effects of ecosystems on disease dynamics.

In chapter 1, Michael Begon explores how host diversity could influence disease dynamics, examining under what conditions host diversity should increase versus decrease disease risk. Alison Power and Alexander Flecker in chapter 2 focus on patterns in the diversity of vectors of plant diseases by first exploring types of vector diversity, then using a comprehensive data set to describe the characteristics of vector range and host range for a suite of plant pathogens. Pejman Rohani and his co-authors in chapter 3 tackle the effects of pathogen diversity on disease dynamics by developing a mathematical framework for understanding the consequences for one pathogen of quarantining individuals infected with another. Pieter Johnson and Stephen Carpenter then ask, in chapter 4, how eutrophication influences different types of pathogens and parasites, and what factors contribute to variation in these effects. In chapter 5, Hamish McCallum investigates current understanding of the effects of landscape structure on disease dynamics, reviewing both a range of modeling efforts that incorporate spatial dynamics of transmission and a set of examples in which landscape structure either increases or decreases disease transmission.

LITERATURE CITED

Cook, G. C., and A. J. Webb. 2000. Perceptions of malaria transmission before Ross' discovery in 1897. Postgraduate Medical Journal 76:738–40.

Johnson, P., and J. Chase. 2004. Parasites in the food web: Linking amphibian malformations and aquatic eutrophication. Ecology Letters 7:521–26.

Linthicum, K .J., A. Anyamba, C. J. Tucker, P. W. Kelley, M .F. Myers, and C. J. Peters. 1999. Climate and satellite indicators to forecast Rift Valley fever epidemics in Kenya. Science 285:397–400.

Ostfeld, R.S., E.M. Schauber, C.D. Canham, F. Keesing, C.G. Jones, and J.O. Wolff. 2001. Effects of acorn production and mouse abundance on abundance and *Borrelia burgdorferi* infection prevalence of nymphal *Ixodes scapularis* ticks. Vector-Borne and Zoonotic Diseases 1:55–63.

Service, M.W. 1991. Agricultural development and arthropod-borne diseases: A review. Revista de Saude Pública 25:167–78.

CHAPTER ONE
Effects of Host Diversity on Disease Dynamics

Michael Begon

SUMMARY

THIS CHAPTER REVIEW INSIGHTS THAT HAVE BEEN GAINED from studies that explicitly acknowledge the contribution of multiple hosts to the dynamics of pathogens. I start from a simple, deterministic viewpoint, distinguishing the contrasting possible effects that a multiplicity of hosts can have on pathogen dynamics, either favoring pathogen persistence or high pathogen abundance (amplification) or reducing these (including especially the dilution effect). A review of the limited available data suggests that although the theoretical possibility of pathogen amplification by multiple hosts is clear, it is not inevitable, in part because transmission between species is rarer than has sometimes been imagined. It is emphasized that it may often be difficult to separate a dilution effect from a simple density effect: their outcomes (reduced pathogen abundance when host diversity is greater) may be the same, but the underlying biological mechanism, which is taken here to define the dilution effect, is quite different. Some empirical evidence for a dilution effect exists, but even for the most studied system, Lyme disease, there seems little empirical basis for distinguishing between a dilution effect and an effect of mouse density, and therefore little evidence either, in spite of its plausibility, for a dilution effect. I then ask briefly what further possibilities may arise with a shift to a (more realistic) perspective in which key processes are stochastic, dynamics are not necessarily equilibrial, and hosts exist as metapopulations. Finally, I consider even more briefly the possible evolutionary consequences of pathogen dynamics driven by a multiplicity of hosts.

INTRODUCTION

The injustice, throughout most of the twentieth century, of a Cinderella role for parasites and pathogens at the ecological Grand Ball has been largely rectified more recently, at least from the point of view of theoretical advances (Hudson et al. 2002). However, most work has been done on single parasites attacking single hosts, whereas in reality, most parasites are capable of affecting multiple hosts and most hosts support multiple pathogens. In this chapter I review what additional insights have been gained from studies that explicitly acknowledge the contribution of multiple hosts to the dynamics of pathogens. This is also a topic with several important applied contexts, since it covers zoonotic infections (infections circulating in wildlife but transmissible to humans) and infections of domesticated species (plant and animal) transmissible to wildlife, and vice versa.

It is also important to say what will not be reviewed here. When more than one host is attacked by a single species of pathogen, the dynamics of the hosts themselves may also be substantially affected, most notably through the process of apparent competition (Holt 1977; Hudson and Greenman 1998). Neither apparent competition nor other effects on host dynamics are discussed in this chapter. Furthermore, Keesing et al. (2006) have recently sought to broaden perspectives by including all aspects of species diversity, whether or not those species are hosts, in their discussion of the determinants of pathogen dynamics. For example, predators or competitors of a host that are not hosts themselves may affect host abundance or behavior and in this way alter pathogen dynamics. These effects, too, are mostly excluded here, though the interaction between infection and competition when species are both hosts and competitors is be discussed, as are the difficulties that may be encountered in distinguishing the effects of other species as hosts and as competitors.

Throughout this review, details of theoretical studies are kept to a minimum, focusing on the results of those studies, and expressing them, where possible and appropriate, as simple figurative illustrations. This is done in part because those theoretical results are mostly uncontroversial and can therefore be taken at face value, but also to leave more space for empirical studies, which I take, by virtue of their rarity, to be more valuable.

I start from a simple, deterministic viewpoint, distinguishing the contrasting possible effects that a multiplicity of hosts can have on pathogen dynamics, either favoring pathogen persistence or high pathogen abundance (amplification) or reducing these (including especially the dilution effect). (Studies of helminths and other macroparasites are not reviewed, but most of the conclusions drawn, in a qualitative sense at least, are equally applicable in that context.) I then ask briefly what further possibilities may arise with a shift to a (more realistic) perspective in which key processes are stochastic, dynamics are not necessarily equilibrial, and hosts exist as metapopulations. Finally, I consider even more briefly the possible evolutionary consequences of pathogen dynamics driven by a multiplicity of hosts.

A DETERMINISTIC FRAMEWORK

An exposition of the possible consequences for pathogen dynamics of moving from one to several host species can be given simply and effectively, in qualitative terms, by looking at the move from one to two host species (figure 1.1; Holt et al. 2003). Formally, figure 1.1 considers pathogens with direct transmission, but the qualitative patterns generated are equally applicable to vector-borne infections and those with free-living infective stages, that is, with a life-cycle stage capable of persisting outside either the host or a vector, such as a spore. We begin (figure 1.1a) with an illustration of the fundamental concept of a critical abundance threshold for a single host species below which there are insufficient susceptible hosts for the infection to invade and spread (the basic reproductive ratio, $R_0 < 1$) and above which invasion and spread are possible ($R_0 > 1$; e.g., Anderson and May 1991).

The consequences for pathogen dynamics of a second host species can be demonstrated simply by plotting the joint threshold curve for the two hosts in the species 1 abundance/species 2 abundance plane (Begon et al. 1992; Bowers and Turner 1997; Holt et al. 2003). The equation for this curve is given by

$$\left(\frac{S_1 - S_{1,T}}{S_1}\right)\left(\frac{S_2 - S_{2,T}}{S_2}\right) = \frac{\beta_{12}\beta_{21}}{\beta_{11}\beta_{22}}, \qquad (1)$$

where S_i is the abundance of susceptible individuals of host species i and $S_{i,T}$ is the threshold abundance for that species, and β_{ij} is the transmission coefficient from species i to species j. Thus, the left-hand side of equation (1) represents the combined threshold deficit (the extent to which the two species between them fail to provide sufficient susceptible hosts to sustain the pathogen) and the right-hand side is the ratio of inter- to intraspecific pathogen transmission. The curve itself intersects the two axes at the respective threshold abundances, and it is clear that if the abundance of either one of the hosts exceeds its threshold, then the pathogen can persist. But even if both host abundances are below their respective thresholds, as long as their joint abundance is above and to the right of the curve, there will be sufficient interspecific transmission for the pathogen to persist (e.g., figures 1.1c and d).

The shape of the joint threshold curve is given by the value of $\frac{\beta_{12}\beta_{21}}{\beta_{11}\beta_{22}}$. If this is zero (interspecific transmission in at least one of the cases is nonexistent), then the curve is rectilinear (figure 1.1b): the pathogen will persist in the presence of the two hosts species only if it would persist in the presence of at least one of them alone. This simple case is nonetheless

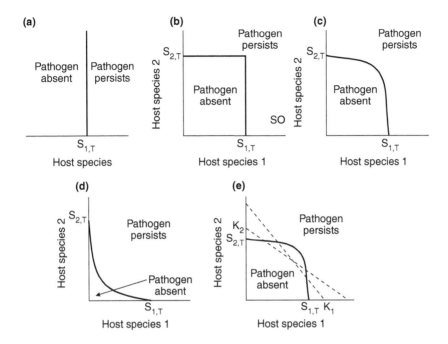

Figure 1.1. Critical threshold abundances and joint threshold curves. (a) The critical threshold abundance of a single host species, $S_{1,T}$, above which a pathogen persists and below which it cannot persist and is therefore typically absent. (b) The joint threshold curve for two host species, cutting the axes at their respective critical threshold abundances, $S_{1,T}$, $S_{2,T}$, within which the pathogen persists and beyond which it cannot persist and is therefore typically absent. In this case there is no interspecific transmission and the curve is therefore rectilinear. SO represents a joint abundance typical of spillover dynamics. (c) Similar to b, except that a small amount of interspecific transmission gives rise to an upwardly convex joint abundance curve. (d) Similar again, except that because interspecific transmission rates exceed, overall, intraspecific ones, the curve is concave. (e) Similar to d, but with the inclusion of Lotka-Volterra interspecific competition zero isoclines, at the intersection of which the two species equilibrate. (After Bowers and Turner 1997; Holt et al. 2003.)

informative, since it captures what has sometimes been called spillover dynamics (SO in figure 1.1b; Daszak et al. 2000; Power and Mitchell 2004). Here, one host is capable alone of supporting the pathogen, and a second host makes no additional contribution to this because there is no transmission from the second host to the first. But pathogens nonetheless continually spill over from the first host to the second such that this second host is either continually, regularly or at least intermittently infected. Many zoonotic infections come into this category, where humans, as the second host, are a dead end in terms of transmission; examples include rabies, bubonic plague, and hantavirus infection.

For even very small positive values of $\frac{\beta_{12}\beta_{21}}{\beta_{11}\beta_{22}}$ (interspecific transmission in both directions; figure 1.1c), the two hosts may combine to allow pathogen persistence at host abundances that would not, alone, support the pathogen; and if interspecific transmission is sufficiently effective (in either or both directions) for $\frac{\beta_{12}\beta_{21}}{\beta_{11}\beta_{22}}$ to exceed 1 (figure 1.1d), then the pathogen can persist at host abundances far below the respective thresholds.

Of course, figures 1.1a–d describe the potential of two host species to combine in favoring the persistence of a pathogen, but whether or not they do so depends on the abundances actually exhibited by those species. If the two host species are not also competitors, then the most straightforward assumption is that both hosts are at their respective carrying capacities, K_1 and K_2. The consequence of there being more than one host for the persistence of the pathogen will then depend on whether the point (K_1, K_2) is above or below the joint threshold curve in the phase plane. But if the hosts also compete, then, as Bowers and Turner (1997) describe, outcomes may be predicted by combining the joint threshold curve with conventional Lotka-Volterra competition isoclines. Figure 1.1e, for example, illustrates how two host species, both of which are capable of supporting the pathogen when alone, may fail to support it when both are present because interspecific competition drives them to a joint abundance below the joint threshold curve. That is, additional host species may, through *competition*, reduce the abundance of a focal host species and hence make pathogen persistence less likely. This is not, though, a dilution effect (see below).

The abundance and dynamics of a pathogen, as opposed to whether or not it persists, are not so easy to predict from the simple diagrams in figures 1.1b–e (nor from the mathematical models that those figures summarize). For example, a pathogen that relies for its persistence on just one host species may still exhibit surges of abundance as a result of epidemics in the other host population. (This is discussed more fully below.) Notwithstanding these complications, pathogens tend to be more abundant the farther their hosts are above their abundance thresholds, and the shapes of the joint threshold curves in figures 1.1c–e tend to place host populations farther above thresholds than they would otherwise be. Hence, the message from these models is that interspecific transmission may allow multiple hosts to have a combinatorial effect on shared pathogens, making them both more likely to persist and more abundant than they would otherwise be (an amplification effect), even when individual host species are not capable themselves of supporting the pathogen, but only if competition among the hosts does not depress their joint abundance below the relevant joint threshold curve.

LITTLE EVIDENCE FOR AN AMPLIFICATION EFFECT
OR ITS PREREQUISITES

Evidence for such an amplification effect could come either from a detectable increase in pathogen abundance, or probability of occurrence, with host richness in a community, or indirectly, from significant levels of interspecific transmission, since this is a necessary (though not of course sufficient) condition for the effect. In practice, though, in the few empirical studies that have been carried out, the tendency has been to seek a decrease in pathogen abundance with host richness, and there appears to be no clear evidence of an increase, though Ostfeld and Keesing (2000a) report that in the case of Lyme disease, disease risk to humans, and hence perhaps the abundance of the pathogen, increases with the species richness of bird hosts within the community. Power and Mitchell (2004), examining a generalist plant pathogen, barley yellow dwarf virus (BYDV), in experimentally manipulated plant communities, did find that pathogen prevalence was highest in the richest communities, but this was entirely driven by a single, most competent host species, *Avena fatua*—a case of a community function (pathogen abundance) being a reflection of community composition rather than community richness.

Evidence from transmission rates themselves is inevitably hard to come by, since these rates, even within a species, are difficult to quantify. What little evidence there is, however, again calls into question the importance, in practice, of the amplification effect. Begon et al. (1999) used longitudinal data from marked, repeatedly captured individuals of two woodland rodent host species, the bank vole, *Clethrionomys glareolus*, and the wood mouse, *Apodemus sylvaticus*, that co-occur and are both infected by cowpox virus to estimate transmission coefficients both within and between species. Within-species coefficients were successful in accounting for the observed dynamics of cowpox virus infection within the respective populations, but the estimates for between-species transmission were close to zero. Recently, Carslake et al. (2005) complemented this population-level approach by estimating, within the same system, the spatial and temporal scales over which infectious individuals posed a risk of infection to other individuals, of their own or the other species. The results are summarized in figure 1.2. In both species, the risk was

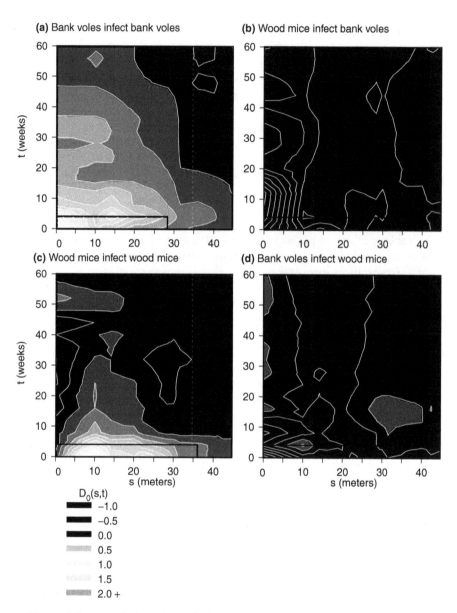

Figure 1.2. Cumulative plots of the space-time interaction, from k-function analysis, among cowpox cases within and between two host species, bank voles and wood mice. Significant space-time interaction indicates a risk of transmission from infectious to susceptible individuals at the scale indicated. Solid black lines represent a hypothesized scale of transmission, determined spatially as the mean home range diameter and temporally as the infectious period. Within the species there was significant interaction at the hypothesized scales. Between species there was no significant interaction at any scale, indicating an effective absence of interspecific transmission. (After Carslake et al. 2005.)

concentrated in the infectious period on the temporal scale and within a typical "home range" on the spatial scale. But between the species, risk appeared, again, to be negligible. Despite their close co-occurrence, these species seem to act as effectively independent reservoirs of cowpox virus infection.

Thus, while the theoretical possibility of pathogen amplification by multiple hosts is clear, it is equally clearly not inevitable, in part perhaps because transmission between species is rarer than has sometimes been imagined. For directly transmitted pathogens of animals, at least, this is not surprising. Transmission is an essentially behavioral process, and behavioral patterns within species are rarely repeated between species.

Abortive Transmission: The Dilution Effect

Joint threshold curves do not necessarily have a negative slope (see figures 1.1b–e). Rather, the addition of a second host species may, through effects on transmission pathways, make a pathogen less abundant, or less likely to persist, than in the presence of a competent host alone (figure 1.3a). This was pointed out first for pathogens with free-living infective stages (Begon and Bowers 1994), but it can also occur with directly transmitted and vector-borne infections; in the latter case it is commonly referred to as the dilution effect (Norman et al. 1999; Ostfeld and Keesing 2000a,b). In each case, though, the process is in essence the same. In the presence of the second species, a transmission event that might previously have linked an infectious and a susceptible individual of the original competent host species instead links infectious individuals to incompetent susceptibles that generate far fewer new infections than a competent host would. These largely abortive transmission events are wasted, and the second host species therefore serves only to dilute the transmission process on which the generation of new pathogens depends.

Starting with directly transmitted infections, Dobson (2004) in particular demonstrates that a dilution effect occurs far more readily if the usual default assumption of density-dependent transmission (Begon et al. 2002) is replaced by one of frequency dependence. This is interesting, coinciding as it does with an increasingly widespread questioning of the universality of density-dependent transmission (Begon et al. 1999; Fenton et al. 2002). But it is also important to note that Dobson's argument relies on there being frequency-dependent transmission *overall*: a fixed frequency of *total* contacts, within and between species. For directly transmitted animal parasites, while a case can be made for the plausibility of frequency-dependent transmission within a species (transmission

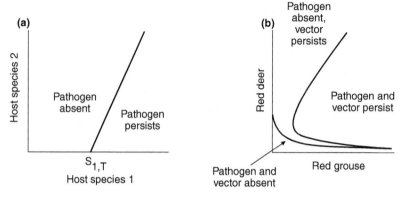

Figure 1.3. Joint threshold curves, similar to figure 1.1, but appropriate to a di-lution effect. (a) A basic dilution effect curve: the second species "wastes" trans-mission opportunities, making it less likely for a pathogen to persist or be abundant in its presence than in its absence. (b) A specific example of the curve, following Norman et al. (1999) and Gilbert et al. (2001). The pathogen is loup-ing ill virus, and the vector is the tick, *Ixodes ricinus*. Red grouse do not sup-port adult ticks and therefore cannot support ticks alone. But only red grouse become viremic and can pass on the infection effectively. Hence, bites of in-fected ticks on deer are wasted. For further details, see text.

arising through social contacts that occur at a fixed frequency), this case is far less plausible between host species (and overall).

On the other hand, with plant pathogens, direct transmission often occurs between adjacent (static) individuals, and a case can indeed be made that each individual has a fixed frequency of total contacts. Thus, the results of Mitchell et al. (2002, 2003) with respect to fungal, largely species-specific foliar pathogens of plants are interesting. As host diver-sity decreased, the pathogen load of individual host species increased. The authors themselves emphasize the role of an increase in host density in the absence of other species, increasing opportunities for transmis-sion (a move from right to left in figure 1.1e as a result of competition). But the results might also be interpreted as evidence for a dilution effect with directly transmitted pathogens, where the total frequency of con-tacts is approximately constant (as envisaged by Dobson) and hosts other than the natural host have very low competence. As I discuss shortly, it may often be difficult to separate a dilution effect (figure 1.3a) from a simple density effect (figure 1.1e): their outcomes (reduced patho-gen abundance when host diversity is greater) may be the same, but the underlying biological mechanism, which is taken here to define the dilution effect, is quite different.

For a dilution effect with directly transmitted pathogens, then, the critical contact rate (which must decline as host diversity increases for the dilution effect to apply) is that between infectious hosts and susceptible competent hosts. With free-living infectious particles, the critical rate is between these particles and susceptible competent hosts. Hence, a dilution effect is entirely plausible as long as "consumption" by the hosts plays a major role in the dynamics of the particles (incompetent hosts remove large numbers of particles from the pool). In practice, however, these dynamics are probably usually a balance largely between production by competent hosts and natural mortality or attrition, which would make a dilution effect unlikely. There seems to be no evidence of such an effect, or its absence, for infections with free-living stages.

The Dilution Effect for Vector-Borne Infections

The dilution effect has been most thoroughly studied and discussed for vector-borne infections, especially Lyme disease (tick-borne *Borrelia burgdorferi* infection) in the United States (Ostfeld and Keesing 2000a,b; Schmidt and Ostfeld 2001). For these infections it is not possible to talk about a single critical contact rate since there are two: between infectious vectors and susceptible, competent hosts and between infectious hosts and susceptible, competent vectors. Thus, the dilution effect requires not only that vector contacts be wasted on less competent host species but also that there *not* be a sufficiently large compensatory increase in vector abundance to overcome this, arising from these same incompetent hosts.

These dual effects can also be illustrated with joint threshold curves. Figure 1.3b does so, based on the parameterization of a model inspired by tick-borne louping ill virus infection in a viremic host, red grouse (*Lagopus lagopus scoticus*), coexisting with red deer (*Cervus elaphus*), which are nonviremic (Gilbert et al. 2001; Norman et al. 1999; see also Laurenson et al. 2003, Rosà et al. 2003). All life stages of the ticks feed on deer, but adult ticks do not feed on grouse, which cannot therefore support the ticks alone. Hence, the joint threshold curve for the ticks does not intersect the grouse axis. That aside, the ticks are amplified by the presence of a second species in the manner of figure 1.1d. But the effects on the virus, and the consequent curve, are more complex. At high grouse abundances there are sufficient competent hosts to support the virus, and at low grouse abundances there are too few. But at intermediate grouse abundances, as deer abundance increases, there are four zones in the phase space. First, there are too few deer to support the ticks. Then there are sufficient deer to support the ticks but too few ticks to support the virus. Then the grouse and deer combine to support sufficient ticks to support the virus, at grouse abundances at which the grouse alone could support neither ticks nor virus: the deer "rescue" the virus infection, even though they themselves are incompetent hosts for the virus. Finally, at the highest deer densities, the dilution effect sets in: too many tick bites are wasted on these nonviremic hosts, and the virus fails to persist.

This model clearly illustrates the mechanics and the potential of the dilution effect in vector-borne infections. Unfortunately, the only grouse-deer site for which Gilbert et al. could estimate parameters placed it just on the virus-persists side of the threshold curve (estimated seroprevalence in young grouse 0.5%), whereas the observed seroprevalence was zero (consistent, in principle, with parameter values far from the curve). This discrepancy is small (indeed impressive) but nonetheless fails formally to provide support for the model's accuracy.

Direct evidence consistent with a dilution effect comes from a study of flea-borne infections of *Bartonella* species (hemoparasitic bacteria) in bank voles and wood mice in Ireland (Telfer et al. 2005). Bank voles were introduced into Ireland only during the twentieth century and are still spreading northeastward. Hence, there is an invasion front behind which bank voles coexist with wood mice (the only native rodent species in Ireland) but ahead of which wood mice live alone. In Ireland, *Bartonella* has been detected in wood mice but not in bank voles. Thus, dilution effects in this natural experiment should give rise to lower prevalences of *Bartonella* in wood mice behind the invasion front than ahead of it, which is precisely what was found for the two common species of *Bartonella*, *B. taylorii* and *B. birtlesii*. On the other hand, once again, prevalence may simply be lower behind the invasion front because wood mouse abundance is lower in the presence of bank voles, and lower host abundance gives rise to lower pathogen abundance (figure 1.1e rather than figure 1.3a). In both *Bartonella* species, however, the best generalized linear model had *Bartonella* prevalence declining with two factors, bank vole density and the prevalence of the other *Bartonella* species, but the inclusion of wood mouse density did not improve the model significantly. The dilution effect is supported, but a host density effect is not.

Kosoy et al. (1997) also looked at *Bartonella* infections in rodent-host communities of different composition and diversity, and their results were proposed by Ostfeld and Keesing (2000b) as being consistent with the dilution effect: prevalence highest in the least diverse communities. However, these data amalgamate a variety of different *Bartonella* species in a variety of hosts, and there are insufficient observations to detect whether the prevalence of any given *Bartonella* species in any given host declined with diversity. Hence, in the absence of density estimates, it is again equally plausible that prevalence was high at low host diversities because host density was higher there.

THE DILUTION EFFECT AND LYME DISEASE

Lyme disease is caused by infection with the spirochete, *Borrelia burgdorferi*, transmitted by *Ixodes* ticks between a variety of vertebrate hosts in North America (mainly mammals and some birds) and transmissible to humans mostly when bitten by infected, questing tick nymphs. Infected humans may show clinical symptoms, including fever and arthritis, but they are dead-end hosts, effectively incapable of onward transmission. There are related *Borrelia* infections elsewhere in the world. In eastern North America, where most work has been carried out, the white-footed mouse, *Peromyscus leucopus*, is highly efficient at infecting feeding ticks and is widely regarded as the principal reservoir of the pathogen. A number of other species have a lower or much lower instantaneous ability to infect a vector—what many Lyme disease workers refer to as reservoir competence. Inspired by this system, Ostfeld and Keesing (2000b) describe four attributes they consider necessary for the dilution effect and applicable to the Lyme disease system: (1) the vector must be a generalist, biting a range of host species; (2) a significant proportion of infected vectors must acquire infection from infectious hosts, rather than transovarially from infected vectors of the previous generation; (3) there must be variation between host species in reservoir competence; and (4) there must be a positive correlation between reservoir competence and numerical dominance in the community such that the most competent species tend to be present in both species-poor and species-rich communities.

Studies of the dilution effect in the Lyme disease system have focused most of their attention on the risk of infection to humans. Hence, a key parameter has been the density of infectious nymphs (DIN), although in the context of pathogen abundance overall, this parameter is clearly only part of just one of the critical contact rates. Calculating DIN requires estimates of both nymphal density and infection prevalence in nymphs (NIP), and in the absence of data on the former, some studies have limited themselves to NIP. In fact, studies of the dilution effect in the Lyme disease system have been concerned largely with the effect of dilution on NIP and the risk to humans rather than on the persistence or dynamics of the pathogen (Ostfeld and Keesing 2000b).

Two recent studies have provided further evidence of a dilution effect in the Lyme disease system. LoGiudice et al. (2003), in what they themselves describe as "the most complete accounting of the distribution of vector meals and infection probabilities across Ixodid tick hosts ever collected," estimated, for a variety of vertebrate species, the typical body burden of ticks, the proportion of engorged (fully fed) tick larvae that molted successfully into nymphs, and the proportion of these infected with the Lyme disease spirochete. From these, they further estimated NIP for a variety of model host communities containing the white-footed mouse, to which further (less competent) hosts were added. They found that in these models, NIP did indeed decline with increasing host diversity, and they argue that similar trends would be found for DIN, though no attempt was made to estimate this. Allan et al. (2003) utilized the fact that in larger forest fragments, rodent diversity typically increases and the density of white-footed mice declines. They found, further, that as forest fragment size increased, NIP and DIN both declined. Although the decline in NIP occurred across the whole range of fragment sizes, however, the relationship for DIN was driven exclusively by four small fragments (ca. 1 ha), because the density of nymphs themselves was higher in these populations, which is interesting in light of the observation that the effect of fragment size on diversity and mouse density is itself most intense in fragments of less than 2 ha (Krohne and Hoch, 1999).

Overall, therefore, the evidence seems clearest for the Lyme disease system that NIP is highest in the least diverse communities (dominated by white-footed mice), and the evidence suggests that DIN and the risk to humans are also highest in these communities. However, there seems little empirical basis for distinguishing between a dilution effect and an effect of mouse density as the basis for these relationships, and therefore little evidence either, in spite of its plausibility, for there being a dilution effect in the dynamics of *B. burgdorferi* in this system.

TRANSIENT DYNAMICS AND METAPOPULATIONS

It is no doubt realistic to acknowledge that many populations spend much of their time far from any equilibrium, recovering from the latest disaster or coming back down to earth after the latest bonanza, and that many populations are in fact metapopulations, comprising patches that may or may not be occupied at any given time. But deterministic approximations that ignore these realities may often capture the essence of a system's dynamics, and in the present context it is difficult to see what difference they would make to dilution effects, other than perhaps to make (stochastic) extinction of a pathogen more likely once a dilution effect had reduced it to low abundance. Views on the nature and importance of amplification effects, however, may be altered drastically.

To see why, it is necessary first to revisit the idea of reservoir competence. Workers researching Lyme disease have used the term to mean effectiveness in passing on an infection (in practice, the proportion of vectors feeding on a host that become infected). But a broader perspective, and the one taken here, describes a competent reservoir host as one capable of sustaining a pathogen ($R_0 > 1$, long term), and a host may be highly effective at passing on infection but still not come into this category. This arises because of the dependence of R_0 on the abundance of susceptible hosts, and its dependence, too, on the average life expectancy of an infectious host. Thus, a host may be highly effective at passing on infection while it is alive, but a combination of a short infectious life and an inability to replace infected hosts faster than they are lost (through death or recovery) may prevent a host from sustaining the infection. Nonetheless, prior to the eventual extinction of the infection within the host (or even of the host itself), there may be a substantial epidemic of the infection.

In the simplified context of figure 1.1, the combination of a host that can and a host that cannot sustain an infection alone falls naturally into the category of spillover, where pathogen dynamics in both hosts are driven by the dynamics in the competent host. But if we think instead of a number of host populations of both species as subpopulations in an overall metapopulation of the pathogen, bearing in mind especially that subpopulations of different species may be sympatric, then the respective roles of the two host species are not so straightforward. Now, attention is focused not so much on infection dynamics within a subpopulation as on the extinction and colonization (invasion) rates of whole subpopulations. For the competent host species, deterministic extinction rates are, by definition, zero, but stochastic extinction rates may not be negligible if some subpopulations are small. Moreover, for the incompetent host species, all subpopulations are, again by definition, doomed to extinction unless rescued by invasion, but these subpopulations may nonetheless spawn large numbers of potential invaders of other subpopulations during the course of epidemic outbreaks.

One probable example of this is sylvatic plague (*Yersinia pestis* infection) in the southwestern United States (Gage and Kosoy 2005), where a distinction is drawn between enzootic and epizootic hosts. Enzootic hosts (deer mice, *Peromyscus* spp., and voles, *Microtus* spp.) are little affected by plague but sustain infection long term; that is, they are competent hosts according to the definition used here. Epizootic hosts (various species of prairie dogs, *Cynomys* spp., chipmunks, *Tamias* spp., ground squirrels, *Spermophilus* spp., and wood rats, *Neotoma* spp.) could not sustain the bacterium alone but may be subject to extensive diebacks from infection in which large numbers of infectious individuals are generated. While the enzootic hosts are usually credited with the long-term maintenance of sylvatic plague, the role of the many epizootic hosts should not be underestimated, not only in transmitting infection among themselves but also in driving plague dynamics in the system as a whole. The alternative terminology among plague workers that distinguishes "maintenance hosts" and "amplifying hosts" has much to recommend it.

This perspective suggests in turn that it would be fruitful to explore epidemiological models in which R_0 was calculated not for infectious individuals but for infectious subpopulations (see, e.g., Matthews et al. 2003), and in which the multiplicity of hosts was acknowledged, such that R_0 was computed from a whole matrix of interactions between subpopulations (see, e.g., Dobson 2004, following Diekmann et al. 1990). Infections in subpopulations of amplifying hosts would have a high probability of going extinct themselves, but also a high probability of invading (transmitting infection to) other subpopulations of all host types. The relative contributions of maintenance and amplifying hosts to an overall value of R_0 is not easy to predict.

Evolution in Multiple-Host Systems

Finally, acknowledging a multiplicity of hosts may also complicate the evolutionary dynamics of pathogens, if this introduces conflicting selective pressures on the pathogen to which compromise solutions must be found. Woolhouse et al. (2001), for example, have argued that for zoonotic human infections in which humans are dead-end hosts, pathogenicity may be especially high, because virulence is usually constrained by trade-offs with transmissibility, and in dead-end hosts none of these constraints can apply. However, such an argument seems to ignore the likelihood that precisely because humans are dead-end hosts, no productive evolution will occur within them, and the argument seems, moreover, to assume that pathogens want to be pathogenic but are usually constrained.

Boots et al. (2004) describe a spatially explicit simulation model of pathogen evolution in which two modes of virulence may be stable, depending on the spatial or social structure of the host: high virulence, in which interactions are mostly local because this prevents recovered, immune individuals from inhibiting spread, and low virulence, where there is more global mixing because this is associated, through a trade-off, with high transmissibility. Assuming for the sake of argument that the mechanics of the model are reasonable, its results can be applied in two ways here. First, the model suggests that when a pathogen jumps from one host species to another (or simply persists in two species with little transmission between them), then its virulence and its dynamics may be very different in the two if their social structures are sufficiently different. Second, the model may be applied to the metapopulation scenario described above, with high virulence being favored in some subpopulations and low virulence favored in others.

This highlights what is perhaps the key issue in the evolution of pathogens in multispecies systems: Under what circumstances does all the important evolution take place in the main reservoir host, with the biology in other hosts being essentially a chance by-product of that? Under what circumstances is evolution driven toward a strategy adapted to be optimum with repeated switches between maintenance and amplifying host species? This section therefore concludes on a theme that has been pervasive throughout: there are many more model predictions and plausible possibilities than there are definitive answers based on sound evidence.

LITERATURE CITED

Allan, B. F., F. Keesing, and R. S. Ostfeld. 2003. Effects of habitat fragmentation on Lyme disease risk. Conservation Biology 17:267–72.
Anderson, R. M., and R. M. May. 1991. Infectious Diseases of Humans. Oxford: Oxford University Press.
Begon, M., M. Bennett, R. G. Bowers, N. P. French, S. M. Hazel, and J. Turner. 2002. A clarification of transmission terms in host-microparasite models: Numbers, densities, and areas. Epidemiology and Infection 129:147–53.
Begon, M., and R. G. Bowers. 1994. Host–host pathogen models and microbial pest control: The effect of host self-regulation. Journal of Theoretical Biology 169:275–87.
Begon, M., R. G. Bowers, N. Kadianakis, and D. E. Hodgkinson. 1992. Disease and community structure: The importance of host self-regulation in a host–pathogen model. American Naturalist 139:1131–50.
Begon, M., S. M. Hazel, D. Baxby, K. Bown, R. Cavanagh, J. Chantrey, T. Jones, and M. Bennett. 1999. Transmission dynamics of a zoonotic pathogen within and between wildlife host species. Proceedings of the Royal Society of London. Series B, Biological Sciences 266:1939–45.
Boots, M., P. J. Hudson, and A. Sasaki. 2004 Large shifts in pathogen virulence relate to shifts in host population structure. Science 303:842–44.
Bowers, R. G., and J. Turner. 1997. Community structure and the interplay between interspecific infection and competition. Journal of Theoretical Biology 187:95–109.

Carslake, D., M. Bennett, S. Hazel, S. Telfer, and M. Begon. 2005. Inference of cowpox virus transmission rates between wild rodent host classes using space-time interaction. Proceedings of the Royal Society of London. Series B, Biological Sciences (in review).

Daszak, P., A. A. Cunningham, and A. D. Hyatt. 2000. Emerging infectious diseases of wildlife: Threats to biodiversity and human health. Science 287: 443–49.

Diekmann, O., J. A. P. Heesterbeek, and J. A. J. Metz. 1990. On the definition and the computation of the basic reproduction ratio R_0 in models for infectious diseases in heterogeneous populations. Journal of Mathematical Biology 28:365–82.

Dobson, A. P. 2004. Population dynamics of pathogens with multiple host species. American Naturalist 164:S64–78.

Fenton, A., J. P. Fairbairn, R. Norman, and P. J. Hudson. 2002. Parasite transmission: Reconciling theory and reality. Journal of Animal Ecology 71:893–905.

Gage, K. L., and M. Y. Kosoy. 2005. Natural history of plague: Perspectives from more than a century of research. Annual Review of Entomology 50:505–28.

Gilbert, L., R. Norman, K. Laurenson, H. W. Reid, and P. J. Hudson. 2001. Disease persistence and apparent competition in a three-host community: An empirical and analytical study of large-scale, wild populations. Journal of Animal Ecology 70:1053–61.

Holt, R. D. 1977. Predation, apparent competition, and structure of prey communities. Theoretical Population Biology 12:197–229.

Holt, R. D., A. P. Dobson, M. Begon, R. G. Bowers, and E. Schauber. 2003. Parasite establishment and persistence in multi-host-species systems. Ecology Letters 6:837–42.

Hudson, P., and J. Greenman. 1998. Competition mediated by parasites: Biological and theoretical progress. Trends in Ecology & Evolution 13: 387–90.

Hudson, P. J., A. Rizzoli, B. T. Grenfell, H. Heesterbeek, and A. P. Dobson. (Eds.). 2002. The Ecology of Wildlife Diseases. Oxford: Oxford University Press.

Keesing, F., R. D. Holt, and R. S. Ostfeld. 2006. Effects of species diversity on disease risk: Modes of the dilution effect in disease ecology. Ecology Letters 9:485–98.

Kosoy, M. Y., R. L. Regnery, T. Tzianabos, E. L. Marston, D. C. Jones, D. Green, G. O. Maupin, J. A. Olson, and J. E. Childs. 1997. Distribution, diversity, and host specificity of *Bartonella* in rodents from the southeastern United States. American Journal of Tropical Medicine and Hygiene 57:578–88.

Krohne, D. T., and G. A. Hoch. 1999 Demography of *Peromyscus leucopus* populations on habitat patches: The role of dispersal. Canadian Journal of Zoology 77:1247–53.

Laurenson, M. K., R. A. Norman, L. Gilbert, H. W. Reid, and P. J. Hudson. 2003. Identifying disease reservoirs in complex systems: Mountain hares as reservoirs of ticks and louping-ill virus, pathogens of red grouse. Journal of Animal Ecology 72:177–85.

LoGiudice, K., R. S. Ostfeld, K. A. Schmidt, and F. Keesing. 2003. The ecology of infectious disease: Effects of host diversity and community composition on Lyme disease risk. Proceedings of the National Academy of Sciences 100:567–71.

Matthews, L., D. T. Haydon, D. J. Shaw, M. E. Chase-Topping, M. J. Keeling, and M. E. J. Woolhouse. 2003. Neighbourhood control policies and the spread of infectious diseases. Proceedings of the Royal Society of London. Series, Biological Science B 270:1659–66.

Mitchell, C. E., P. B. Reich, D. Tilman, and J. V. Groth. 2003. Effects of elevated CO_2, nitrogen deposition, and decreased species diversity on foliar fungal plant disease. Global Change Biology 9:438–51.

Mitchell, C. E., D. Tilman, and J. V. Groth. 2002. Effects of grassland species diversity, abundance, and composition on foliar fungal disease. Ecology 83:1713–26.

Norman, R., R. G. Bowers, M. Begon, and P. J. Hudson. 1999. Persistence of tick-borne virus in the presence of multiple host species: Tick reservoirs and parasite-mediated competition. Journal of Theoretical Biology 200:111–18.

Ostfeld, R. S., and F. Keesing. 2000a. Biodiversity and disease risk: The case of Lyme disease. Conservation Biology 14:722–28.

Ostfeld, R. S., and F. Keesing. 2000b. The function of biodiversity in the ecology of vector-borne zoonotic diseases. Canadian Journal of Zoology 78:2061–78.

Power, A. G., and C. E. Mitchell. 2004. Pathogen spillover in disease epidemics. American Naturalist 164:S79–89.

Rosà, R., A. Pugliese, R. Norman, and P. J. Hudson. 2003. Thresholds for disease persistence in models for tick-borne infections including non-viraemic transmission, extended feeding and tick aggregation. Journal of Theoretical Biology 224:359–76.

Schmidt, K. A., and R. S. Ostfeld. 2001. Biodiversity and the dilution effect in disease ecology. Ecology 82:609–19.

Telfer, S., K. J. Bown, R. Sekules, M. Begon, T. Hayden, and R. Birtles. 2005. Disruption of a host-parasite system following introduction of an exotic species. Parasitology 131:1–8.

Woolhouse, M. E. J., L. H. Taylor, and D. T. Haydon. 2001. Population biology of multihost pathogens. Science 292:1109–11.

CHAPTER TWO
The Role of Vector Diversity in Disease Dynamics

Alison G. Power and Alexander S. Flecker

SUMMARY

ALTHOUGH MANY IMPORTANT DISEASESOF HUMANS, animals, and plants are transmitted by vectors, we know little about the role of vector diversity in disease dynamics. In this chapter we explore two ways in which vector diversity may affect the epidemiology of insect-transmitted plant viruses: through effects on pathogen host range and through effects on transmission rates. Our analysis of a large database of plant viruses indicated that most viruses have a low diversity of vectors and a large diversity of hosts. Nearly 60% percent of vector-transmitted plant viruses have a single known vector species, whereas less than 10% have a single known host plant. Patterns of vector diversity varied significantly according to vector taxa (insects, fungi, nematodes, mites), virus family, and viral genome (DNA, RNA). DNA viruses were more likely than RNA viruses to be transmitted by a single species of vector, but there was no difference in the probability of infecting more than one host. Vector-transmitted viruses, DNA viruses, and viruses with multiple hosts were most likely to be designated as emerging plant viruses. These patterns suggest the potential importance of vector and host diversity for disease dynamics. Despite the prevalence of single vectors for plant viruses, some important groups of viruses have multiple vectors, as well as multiple hosts. We recommend increased attention to both experimental studies and theoretical explorations of the ecological consequences of vector and host diversity.

INTRODUCTION

A number of important emerging diseases of humans are transmitted by vectors, including West Nile virus and Lyme disease, yet the role of vector diversity in pathogen epidemiology and disease dynamics is not well understood. Many disease agents, particularly viruses, are strongly dependent on the vectors that transmit them among hosts, and vectors can shape the temporal and spatial pattern of pathogen spread. Vector diversity has the potential to affect disease dynamics in two major ways: through effects on pathogen host range and through effects on transmission rates. For most vectored pathogens, increases in vector diversity strongly influence the realized host range of the pathogen by providing additional host colonization opportunities for the pathogen. An increase in the number of species of effective vectors, each with a unique feeding range, is likely to expand the viral host range. In such cases, the feeding range of the vectors may largely determine the host range of the pathogen. The second impact of vector diversity operates through transmission rate. Differences in vector diversity may result in functional differences in transmission that have strong effects on disease dynamics.

In this chapter we address both effects of vector diversity, using examples from the epidemiology of plant viruses. We begin by exploring the implications of vector diversity and describing patterns of vector diversity for plant viruses, then address the role of vector diversity in disease dynamics.

DIMENSIONS OF VECTOR DIVERSITY

Both species richness and species identity are critical components of vector diversity in plant virus systems. In the context of disease dynamics, it may be less important to focus on vector taxonomic diversity per se than on the functional diversity that may accompany taxonomic diversity. Functional diversity can take many forms. First, taxonomy can dictate transmission mechanism, and therefore species identity is likely to play an important role in the effects of vector diversity. For insect-transmitted pathogens, transmission mechanisms include nonpersistent or stylet-borne, semipersistent or foregut-borne, persistent-circulative, and persistent-propagative (Gray and Banerjee 1999; Nault 1997). Stylet-borne viruses are carried on the mouthparts of vectors and are called nonpersistent because they are retained by the vector for no more than a few hours and are lost once a vector has fed on a host. Foregut-borne viruses enter the foregut of the vector and are semipersistent in their vectors. Semipersistent viruses are retained by the vector for a few days and can be transmitted to multiple plants. Circulative, persistent viruses pass through the insect gut into the hemolymph and then into the salivary glands via highly specific transport mechanisms. They can be transmitted repeatedly to new plants and are usually retained for the lifetime of the vector. Propagative viruses are circulative viruses that replicate in the insect vector as well as in the plant host, and can sometimes be transmitted from parent vector to offspring by transovarial transmission.

It should be noted that in vector terminology, *persistence* refers to the period that the vector retains the pathogen and is capable of transmitting it, and this period can have significant consequences for rates and patterns of disease spread (Jeger et al. 1998). Typically, an individual vector species transmits viruses using only one mechanism, with the notable exception of aphids, some of which can transmit viruses nonpersistently and persistently, including by circulative and propagative transmission. An individual virus uses only one insect transmission mechanism (Nault 1997), although some some stylet-borne viruses are also transmitted by seed.

Other dimensions of vector functional diversity, such as variation in life history and behavior, may also have important effects on disease spread. The seasonal phenology of vector emergence from overwintering stages varies among vector species and can strongly shape pathogen dynamics. In addition, univoltine vectors (with a single generation per year) have the potential to influence disease spread very differently from multivoltine vectors. Although many important vectors of plant viruses, including aphids, whiteflies, and some leafhoppers, are multivoltine, some vector species are not. Although there are no comprehensive analyses of the role of diversity in vector phenology on the epidemiology of plant viruses, the dynamics of some well-studied human diseases appear to depend strongly on diversity in vector life history. For example, mosquito vectors of West Nile virus include both univoltine and multivoltine species (Crans 2004), and both may play important roles in the spread of this disease.

Diversity within a vector species may also have significant implications for virus transmission (Power and Gray 1995). The transmission efficiency of vectors can vary among genotypes, among developmental stages, and among morphological stages. For example, it has long been known that different genotypes or clones within a vector species may differ dramatically in their ability to transmit particular viruses (e.g., Björling and Ossiannilsson 1958; Sohi and Swenson 1964). Immature stages may be more or less efficient than adults in virus acquisition and transmission. For some insect-transmitted viruses, nymphs may be more efficient vectors than adults (e.g., Halstead and Gill 1971).

Diversity among vectors in host preference, feeding behavior, aggregation behavior, and movement behavior all lead to different patterns of pathogen transmission. Diversity in host plant preference is common among vector species that transmit the same virus. Moreover, seasonal variation in host plant preferences of vectors can be a critical factor driving epidemiology (Purcell 1976, 1979). As noted earlier, the feeding range of the vectors can govern the host range of the pathogen. It has been shown that the host range of plant viruses can be dramatically increased by the expansion of the feeding range of their insect vectors (e.g., Gold-bach and Peters 1994; Harrison and Robinson 1999). The importance of host preferences of different vectors has similarly been implicated in the dynamics of West Nile virus in the United States (Fonseca, Keygho-badi, Malcolm, Mehmet, et al., 2004; Fonseca, Keyghobadi, Malcolm, Schaffner, et al., 2004).

In addition to variation among vector species in the host plants they prefer to feed on, many studies have demonstrated differences among insect vector species in feeding patterns, including the type of tissue probed by the vector, the number of "tasting" probes, and the length of the feeding period (e.g., Calderon and Backus 1992; Margaritopoulos et al. 2005). These differences in species-specific vector feeding behavior can have significant effects on the probability of virus transmission (e.g., Chen et al. 1997; Collar et al. 1997; Palacios et al. 2002). Similarly, ag-gregation behavior and movement behavior are species-level characteris-tics (Hayek and Dahlsten 1988) that influence virus transmission. Previous research has shown that movement rates of vector insects can determine pathogen spread independently from vector abundance (e.g., Power 1987, 1991, 1992).

An example of the importance of vector diversity for virus transmis-sion can be found in the barley yellow dwarf virus/cereal yellow dwarf virus (BYDV) complex. This complex includes BYDV-GPV, BYDV-MAV, BYDV-PAV, BYDV-RGV, BYDV-RMV BYDV-SGV, and CYDV-RPV, all of which are transmitted in a persistent-circulative fashion by aphids. At least twenty-five species of grass-feeding aphids have been re-ported to be vectors of BYDVs, but fewer than ten species play an im-portant role in the worldwide spread of the viruses in the BYDV complex (Halbert and Voegtlin 1995).

These aphids have distinct though overlapping host ranges that allow the colonization of more than 150 grass species by the viruses in the complex (D'Arcy 1995; Halbert and Voegtlin 1995). Even closely related vector species, such as *Rhopalosiphum maidis* and *R. padi,* have very different preferences for, and performance on, various grass species, in-cluding major crops such as corn, wheat, and oats. Their efficiency at transmitting different BYDVs when given adequate feeding periods var-ies from no transmission to nearly 100% transmission (Power and Gray 1995). Moreover, there are significant differences in the movement be-havior of different vector species, even within the same plant commu-nity (Power 1991; A.G. Power, unpublished data). The loss of one or two of these primary vector species from the disease complex would be expected to have significant consequences for disease spread.

Each vector species transmits a different subset of BYDVs at varying efficiencies, and there can be considerable variation in transmission

efficiency within an aphid species (Power and Gray 1995). Recent studies have demonstrated considerable variation in the ability of aphid
populations from different geographic areas to transmit various BYDVs
(e.g., Bencharki et al. 2000; Gray and Banerjee 1998; Gray et al. 2002).
In studies with aphids from a single region, Dedryver et al. (2005)
found significant variability in BYDV transmission among both clones
and genotypes of the English grain aphid, *Sitobion avenae*, and showed
that the heritability of transmission ability was relatively high. Vectors
may also vary in their efficiency as vectors during different life stages.
For example, nymphal stages of the greenbug, *Schizaphis graminum*,
often acquire and transmit BYDVs more efficiently than adults of this
species (Halstead and Gill 1971), although this pattern also varies with
aphid clone (Gray et al. 2002).

Despite our longstanding recognition of the importance of vector diversity in some disease systems such as the BYDV system, most information is observational. To our knowledge, there have been no experimental
studies that directly test the importance of multiple vectors for disease
spread in any system. Direct empirical tests of the role of vector diversity
could contribute substantially to our understanding of the factors driving pathogen dynamics in nature.

Patterns of Vector Diversity

Although we know that some plant viruses, such as BYDV, have more
than one species of vector, it is useful to consider how common multiple
vectors are and whether the number of vectors varies with vector taxonomy, virus taxonomy, or virus genome structure. Previously, we examined overall patterns of host and vector diversity for plant viruses (Power
and Flecker 2003), using the VIDE database on plant viruses (Brunt et
al. 1996). This database, which is updated regularly, is the most current
listing of plant virus species and contains 1,673 total viruses, including
910 unique virus species. The mechanism of transmission is known for
only 59% of the viruses in the database; of these, 89% are transmitted
by vectors. Of the 474 vectored viruses, 87% are transmitted by insects,
6% by fungi, 4% by nematodes, and 2% by mites. In a previous study
(Power and Flecker 2003), we analyzed the host range and vector range
of all 910 unique viruses in the database, at the level of species, genus,
and family of host and vector. One caveat in interpreting our results is
that taxonomic relationships among viruses and species determinations
are still somewhat in flux for some virus groups.

Our analysis indicated that plant viruses tend to have few vectors and
many hosts, whether the data were analyzed at the level of species, genus,

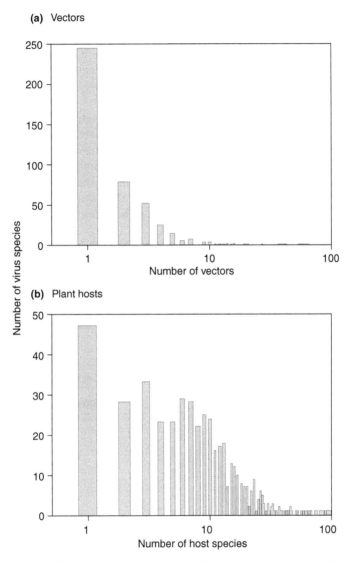

Figure 2.1. Number of plant virus species with (a) single versus multiple vectors (*n* = 447) and (b) single versus multiple hosts (*n* = 475). Data from VIDE database on plant viruses (Brunt et al. 1996).

or family (Power and Flecker 2003). Overall, we found that most viruses have a large host range and a very narrow range of vectors. While nearly 60% percent of vector-transmitted viruses have a single known vector species (figure 2.1a), less than 10% have a single known host plant (figure 2.1b). Viruses with a narrow host range are always transmitted by only one or a few vector species; viruses with many vectors never have a narrow host range.

This striking pattern suggests that the association with the vector rather than the host is the major constraint to disease spread for many plant viruses. This is consistent with molecular analyses, which suggest that areas of the viral genome responsible for regulating the specificity of insect transmission are more highly conserved than those for host infection (Power 2000). These patterns imply that attaining access to new hosts is more important than overcoming the defenses of these hosts in determining rates of virus spread.

Vector diversity varies by vector taxa (figure 2.2). Viruses transmitted by insects constitute 87% ($n = 414$) of vector-borne viruses, and these drive the pattern already described, with most viruses having one

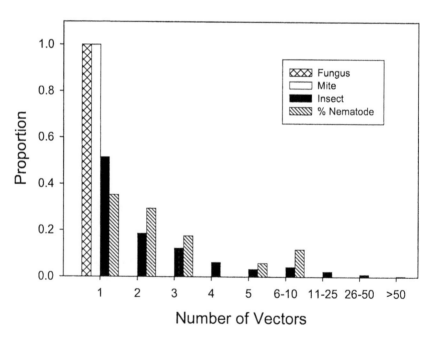

Figure 2.2. Proportion of plant virus species with single versus multiple vectors for different vector taxa: fungi ($n = 25$), mites ($n = 9$), insects ($n = 396$), and nematodes ($n = 17$).

or a few vectors. Viruses transmitted by fungal vectors (6%; $n = 26$) or mite vectors (2%; $n = 10$) have even lower vector diversity than insect-transmitted viruses, in that each is transmitted by a single species of vector. Nematode-transmitted viruses (4%; $n = 19$) are somewhat less likely to be vector specialists, with 50.0% having a single vector species.

We also examined the relationship between genome type and vector diversity. Interestingly, our analysis suggests that DNA viruses are particularly likely to have low vector diversity. Of those DNA viruses in our database that were transmitted by vectors, 83.8% were transmitted by a single species of vector, whereas only 49.1% of RNA viruses had one vector species ($\chi^2 = 28.0$, $P < 0.0001$; figure 2.3). This suggests that a jump to a new vector species is relatively improbable for most DNA viruses but may be somewhat easier for RNA viruses. Woolhouse et al. (2005) have argued that RNA viruses are most likely to jump between host species, because they typically have larger host ranges and higher mutation rates than DNA viruses. Mutation rates are estimated to be 300 times higher in RNA viruses than DNA viruses, and higher mutation rates should lead to greater genetic variation in the population and

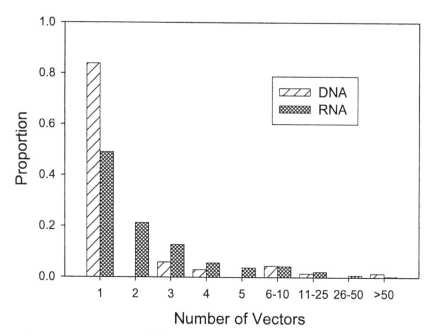

Figure 2.3. Proportion of plant virus species with single versus multiple vectors for DNA ($n = 68$) and RNA ($n = 375$) viruses.

therefore a greater likelihood of large host ranges (Woolhouse et al. 2001).

Among directly transmitted viruses infecting humans, RNA viruses are more likely than DNA viruses to infect other animal hosts (67% vs. 36%, respectively; Woolhouse et al. 2001). In contrast, we found no significant difference in the proportions of DNA versus RNA plant viruses that infect more than one host ($P \gg 0.05$). However, higher mutation rates in RNA viruses might also result in a greater probability of adapting to more than one species of vector, which may partially explain our finding that RNA viruses were more likely to have multiple vector species. A jump to a new vector species could have large impacts on transmission and disease dynamics. The epidemiology of Venezuelan equine encephalitis virus (VEEV) appears to be an example of a vector jump resulting in huge impacts on pathogen dynamics. Although the normal vectors of VEEV are seven closely related *Culex* mosquitoes, during outbreaks most of the transmission is carried out by species in other mosquito genera (Weaver and Barrett 2004). This expansion in vector range results from rapid viral adaptation to these new mosquito vectors (Brault et al. 2004).

Patterns of vector diversity also differ significantly among virus families ($\chi^2 = 98.16$, $P < 0.0001$). Some families, such as the insect-transmitted Geminiviridae, are composed largely of viruses that are exclusively transmitted by a single group of vectors using a single transmission mechanism and have extremely low vector diversity. In fact, most geminiviruses are transmitted by a single vector species. In contrast, some families contain viruses transmitted by several vector taxa, sometimes with different transmission mechanisms. For example, species of aphids, mites, and fungi are all capable of transmitting viruses in the Potyviridae using a transmission mechanism distinct to that vector group.

EMERGING PATHOGENS

The recent emphasis on understanding the characteristics of emerging pathogens led us to build on our earlier analysis by focusing on emerging plant viruses. Anderson et al. (2004) pointed out that viruses constitute the largest taxonomic group of emerging plant pathogens. Between 1996 and 2002, 47% of outbreaks of plant pathogens reported to the International Society for Infectious Diseases were caused by viruses (Anderson et al. 2004), about the same as the proportion of human diseases outbreaks (44%) or wildlife disease outbreaks (43%) caused by viruses (Dobson and Foufopoulos 2001; Taylor et al. 2001). We used two

TABLE 2.1
Emerging plant viruses identified by Madden (2001) and Anderson et al. (2004)

Virus Species	Genus	Family	Genome	Transmission
Arabis mosaic	Nepovirus	Comoviridae	RNA	Vector
Banana bunchy top	Nanavirus		DNA	Vector
Bean golden mosaic	Bigeminivirus	Geminiviridae	DNA	Vector
Beet yellow	Closterovirus		RNA	Vector
Cassava African mosaic	Bigeminivirus	Geminiviridae	DNA	Vector
Plum pox	Potyvirus	Potyviridae	RNA	Vector
Potato Andean latent	Tymovirus		RNA	Vector
Potato Andean mottle	Comovirus	Comoviridae	RNA	Unknown
Potato Y	Potyvirus	Potyviridae	RNA	Vector
Rice stripe necrosis	Furovirus		RNA	Vector
Rice yellow mottle	Sobemovirus		RNA	Vector
Soybean dwarf	Luteovirus		RNA	Vector
Tomato black ring	Nepovirus	Comoviridae	RNA	Vector
Tomato yellow leaf curl	Bigeminivirus	Geminiviridae	DNA	Vector

recent reviews of emerging plant pathogens to identify important emerging plant viruses (table 2.1; Anderson et al. 2004; Madden 2001). We then examined several characteristics of the 14 plant viruses that have been identified as emerging threats. We asked whether these emerging viruses were more likely than viruses not identified as emerging (1) to be RNA or DNA viruses, (2) to be transmitted by vectors, and (3) to be transmitted by multiple vectors.

Our results suggest that DNA plant viruses are over three times more likely than RNA plant viruses to be designated as emerging threats, relative to their proportional representation in the complete database of plant viruses ($\chi^2 = 4.72$, $P = 0.030$). Viruses transmitted by vectors are also significantly more likely to be designated as emerging threats than are directly transmitted viruses ($\chi^2 = 9.27$, $P < 0.01$). In fact, all of the emerging viruses on our list for which the transmission mechanism is known are transmitted by vectors. More than 80% of these viruses are transmitted by insects. Consistent with theoretical predictions (Daszak et al. 2000; Woolhouse et al. 2001), viruses with multiple hosts are somewhat more likely to be designated as emerging threats than are host specialists ($\chi^2 = 3.47$, $P = 0.062$). Logistic regression of threat against number of host species indicates a very significant relationship between threat and the number of hosts ($\chi^2 = 13.27$, $P < 0.001$). There is no evidence, however, that this is true for vectors; that is, viruses with multiple vectors are not more likely than viruses with a single vector species to be designated as emerging threats ($P > 0.05$). This is consistent with the pattern described above, since emerging viruses are likely to be DNA viruses and DNA viruses are likely to be vector specialists.

These results are somewhat contrary to the findings of recent analyses of viruses infectious to humans. Taylor et al. (2001) found that emerging viruses of humans were more likely to be directly transmitted than vector-borne. However, one obvious explanation for this difference is that humans are mobile, whereas plants are not. For sedentary organisms like plants, transmission that requires direct contact among hosts may not allow the rapid increase in infections that characterizes emerging pathogens.

Although emerging plant viruses were not more likely to have multiple vectors, the vectors of these viruses are likely to be host generalists, like the viruses themselves. The apparent role of generalist vectors in facilitating the emergence of plant viruses is consistent with the pattern that has been recognized in human diseases. Molyneux (2003) pointed out that generalist vectors play an important role in emerging human diseases by moving pathogens from animal reservoirs to humans. Such generalist vectors often share a number of characteristics: broad geographic distribution; species complexes or species groups; a range of hosts, including both animals and humans; efficient vectors, with high vectoral capacity; and no transovarial transmission from parent to offspring (Molyneux 2003). Most of these characteristics are similar for many important generalist vectors of plant viruses, including broad geographic distributions, species complexes, broad host ranges that may include both wild and domesticated plants, efficient transmission patterns, and low probability of transovarial transmission.

Vectors and Disease Dynamics

The role of vectors has only rarely been addressed in epidemiological models of disease. Beginning with the pioneering mathematical models of Ross (1910) and Kermack and McKendrick (1927), classic epidemiological theory has tended to focus on the simplest disease systems: single host species infected by a single, directly transmitted pathogen species. Although some of the earliest theory was designed to examine malaria dynamics (Ross 1910), vector populations were typically subsumed in the transmission parameter rather than included as explicit variables. This simplification is still common today (e.g., Anderson and May 1991), even as the importance of vectored pathogens is increasingly recognized. In the past decade or so, there have been a number of attempts to introduce more complexity into the classic models, by incorporating multiple pathogens (Dobson and Roberts 1994; Roberts and Dobson 1995) or multiple hosts (e.g., Dobson and Foufopoulos 2001), but relatively few have incorporated any vector complexity (but see Antonovics et al. 1995).

In contrast, the more descriptive, system-specific approaches typically used in plant disease models often incorporate vectors, although they have rarely been applied to multiple pathogens (e.g., Madden et al. 1983) or multiple hosts. Plant disease models that include vectors have been used for evaluating the potential of various management strategies for controlling the spread of plant pathogens (Holt et al. 1999; Jeger and Chan 1995). In addition, various aspects of vector population dynamics and behavior have been explored.

A number of recent plant disease models have focused on the response of vectors to infection of their host plant. Any effects on vector fitness that lead to higher rates of population growth on infected plants may have significant impacts on rates of pathogen spread (Holt et al. 1997; Zhang et al. 2000). In a model of the whitefly-transmitted African cassava mosaic virus based on substantial field data (Holt et al. 1997), rates of spread were very sensitive both to vector population dynamics (abundance, birth rate, mortality) and to virus transmission rates (inoculation and acquisition). Interestingly, when virus infection of hosts led to increased vector fecundity, spatial aggregation of vectors was promoted (Zhang et al. 2000). These models predict that vector aggregation should have a dual effect. Within the infected crop, it should reduce the effective contact rate between vector and host and thus lead to lower disease incidence than would be predicted with no aggregation. On the other hand, it has the potential to cause increased emigration rates of infective vectors to other hosts in the area (Zhang et al. 2000).

Plant disease models have predicted that vector preference behavior will have significant impacts on rates of pathogen spread. According to conventional wisdom, vector preference for infected hosts over healthy hosts should promote disease spread (Irwin and Thresh 1990). However, the results of recent epidemiological models suggest that the reverse should often be true (McElhany et al. 1995; Real et al. 1992). That is, a preference for healthy plants should lead to greater rates of pathogen spread, since an infective vector is less likely to "waste" a visit to a plant that is already infected. However, the effect of vector preference depends on the frequency of infected plants in the population, as well as whether the transmission system is persistent, nonpersistent, or semipersistent. Models predict that pathogens with even moderate persistence are likely to have higher rates of spread by healthy-preferring vectors at most disease frequencies (McElhany et al. 1995).

Many pathogens of humans, animals, and plants have either multiple hosts or multiple vectors, or both (Power and Flecker 2003; Woolhouse et al. 2001). In multiple-host systems, host species vary in key epidemiological traits such as resistance, tolerance, and attractiveness to vectors (Daszak et al. 2000; LoGiudice et al. 2003; Woolhouse et al. 2001), and clearly vectors vary in epidemiologically relevant traits as well. Vector species vary in host range, life history, transmission efficiency, and behavior. This results in highly heterogeneous and asymmetrical rates of pathogen transmission within and between different host species. Although the recent emphasis on pathogens with multiple hosts has resulted in a flourishing theoretical literature on the dynamics of multihost pathogens of animals (e.g., Dobson 2004; Dobson and Foufopoulos 2001; Holt et al. 2003), theory that explicitly treats multiple vectors is conspicuously absent (Jeger et al. 2004). The only attempts at incorporating more than one vector species in disease models utilize simulation models (e.g., Irwin et al. 2000) or numerical solutions to traditional SIR models (Chan and Jeger 1994). These system-specific approaches do not yet provide us with broad predictions about the importance of vector diversity for pathogen dynamics.

CONCLUSIONS

Our understanding of the influence of vector diversity on pathogen epidemiology and disease dynamics is rudimentary, but the potential for a strong effect of vector diversity is high. In this chapter we have addressed two categories of effects of vector diversity on disease dynamics, effects on pathogen host range and effects on pathogen transmission rates. We have explored the implications of these effects for the epidemiology of plant viruses. We conclude the following: (1) changes in vector diversity are likely to affect host range, because host preferences vary between vector species, and the feeding range of the vector often determines the host range of the pathogen; and (2) changes in vector diversity will result in functional changes in transmission that have strong effects on disease dynamics.

In our analysis of patterns of vector diversity for plant viruses, we found that plant viruses tend to have a low diversity of vectors and a high diversity of hosts, and the largest proportion of viruses fall into the category of host generalist/vector specialist. A majority of vector-transmitted viruses have a single known vector species, whereas relatively few viruses have a single known host plant. We also found that patterns of vector diversity varied significantly according to vector taxa (insects, fungi, nematodes, mites), virus family, and viral genome (DNA, RNA). It is interesting that DNA viruses were much more likely than RNA viruses to be transmitted by a single species of vector, suggesting that a jump to a new vector species may be more likely for RNA viruses. Contrary to our expectations based on animal viruses, we did not detect any difference in the proportions of DNA and RNA viruses that infect more than one host.

Our examination of emerging plant viruses indicated that DNA viruses were more likely to be identified as emerging threats than were RNA viruses, relative to overall frequency of the two types of viruses. We also found that viruses transmitted by vectors were more likely to be designated as emerging threats than were directly transmitted viruses, a finding that emphasizes the importance of examining vector diversity. As has been found for animal viruses, plant viruses with multiple hosts were more likely to be designated as emerging threats than were host specialists.

These patterns of vector diversity suggest the potential importance of vector diversity for disease dynamics while simultaneously highlighting the lack of both experimental studies and theoretical explorations of vector diversity. Epidemiological theory has largely ignored the potential effects of multiple vectors on disease dynamics. A few system-specific models of plant viruses incorporate vector activity and behavior related to virus transmission, and these models suggest that vector diversity may be an important determinant of the rate and extent of virus spread. However, explicit treatment of the influence of multiple vectors on disease is still lacking. The development of a body of theory treating multiple vectors would make an important contribution to our understanding of disease epidemiology and could stimulate focused experimental research on this topic in biotic communities.

Literature Cited

Anderson, P. K., A. A. Cunningham, N. G. Patel, F. J. Morales, P. R. Epstein, and P. Daszak. 2004. Emerging infectious diseases of plants: Pathogen pollution, climate change and agrotechnology drivers. Trends in Ecology & Evolution 19:535–44.

Anderson, R. M., and R. M. May. 1991. Infectious Diseases of Humans. Dynamics and Control. Oxford: Oxford University Press.

Antonovics, J., Y. Iwasa, and M. P. Hassell. 1995. A generalized model of parasitoid, venereal, and vector-based transmission processes. American Naturalist 145:661–75.

Bencharki, B., M. El Yamani, and D. Zaoui. 2000. Assessment of transmission ability of barley yellow dwarf virus-PAV isolates by different populations of

Rhopalosiphum padi and *Sitobion avenae*. European Journal of Plant Pathology 106:455–64.

Björling, K., and F. Ossiannilsson. 1958. Investigations on individual variations in the virus transmitting ability of different aphid species. Handlingar II 14:1–13.

Brault, A.C., S. A. Langevin, R. A. Bowen, N.A. Panella, B. J. Biggerstaff, B. R. Miller, and N. Komar. 2004. Differential virulence of West Nile strains for American crows. Emerging Infectious Diseases 10:2161–68.

Brunt, A., K. Crabtree, M. Dallwitz, A. Gibbs, and L. Watson. 1996. Viruses of Plants: Descriptions and Lists from the VIDE Database. Wallingford, UK: C.A.B. International.

Calderon, J. D., and E. A. Backus. 1992. Comparison of the probing behaviors of *Empoasca-fabae* and E-*Kraemeri* (Homoptera, Cicadellidae) on resistant and susceptible cultivars of common beans. Journal of Economic Entomology 85:88–99.

Chan, M.-S., and Jeger, M. J. 1994. An analytical model of plant virus disease dynamics with roguing and replanting. Journal of Applied Ecology 31: 413–27.

Chen, J. Q., B. Martin, Y. Rahbe, and A. Fereres. 1997. Early intracellular punctures by two aphid species on near-isogenic melon lines with and without the virus aphid transmission (Vat) resistance gene. European Journal of Plant Pathology 103:521–36.

Collar, J. L., C. Avilla, and A. Fereres. 1997. New correlations between aphid stylet paths and nonpersistent virus transmission. Environmental Entomology 26:537–44.

Crans, W. J. 2004. A classification system for mosquito life cycles: Life cycle types for mosquitoes of the northeastern United States. Journal of Vector Ecology 29:1–10.

D'Arcy, C. J. 1995. Symptomatology and host range of barley yellow dwarf. *In* Barley Yellow Dwarf: 40 Years of Progress, ed. C. J. D'Arcy and P. A. Burnett, 9–28. St. Paul, MN: American Phytopathological Society.

Daszak, P., A. A. Cunningham, and A. D. Hyatt. 2000. Emerging infectious diseases of wildlife: Threats to biodiversity and human health. Science 287:443–49.

Dedryver, C. A., G. Riault, S. Tanguy, J. F. Le Gallic, M. Trottet, and E. Jacquot. 2005. Intraspecific variation and inheritance of BYDV-PAV transmission in the aphid *Sitobion avenae*. European Journal of Plant Pathology 111:341–54.

Dobson, A. 2004. Population dynamics of pathogens with multiple host species. American Naturalist 164:S64–78.

Dobson, A., and J. Foufopoulos. 2001. Emerging infectious pathogens of wildlife. Philosophical Transactions of the Royal Society of London. Series B, Biological Sciences 356:1001–12.

Dobson, A. P., and M. G. Roberts. 1994. The population dynamics of parasitic helminth communities. Parasitology 109(suppl.):S97–108.

Fonseca, D. M., N. Keyghobadi, C. A. Malcolm, C. Mehmet, F. Schaffner, M. Mogi, R. C. Fleischer, and R. C. Wilkerson. 2004. Emerging vectors in the *Culex pipiens* complex. Science 303:1535–38.

Fonseca, D. M., N. Keyghobadi, C. A. Malcolm, F. Schaffner, M. Mogi, R. C. Fleischer, and R. C. Wilkerson. 2004. Outbreak of West Nile virus in North America: Response. Science 306:1473–75.

Goldbach, R., and D. Peters. 1994. Possible causes of the emergence of tospovirus diseases. Seminars in Virology 5:113–20.

Gray, S. M., and N. Banerjee. 1999. Mechanisms of arthropod transmission of plant and animal viruses. Microbiology and Molecular Biology Reviews 63:128–42.

Gray, S. M., D. M. Smith, L. Barbierri, and J. Burd. 2002. Virus transmission phenotype is correlated with host adaptation among genetically diverse populations of the aphid *Schizaphis graminum*. Phytopathology 92:970–75.

Halbert, S. and D. Voetgtlin. 1995. Biology and taxonomy of vectors of barley yellow dwarf virus. *In* Barley Yellow Dwarf: 40 Years of Progress, ed. C. J. D'Arcy and P. A. Burnett, 217–58. St. Paul, MN: American Phytopathological Society.

Halstead, B. E., and C. C. Gill. 1971. Transmission of barley yellow dwarf virus by different stages of greenbug. Phytopathology 61:749–51.

Harrison, B. D., and D. J. Robinson. 1999. Natural genomic and antigenic variation in whitefly-transmitted geminiviruses (Begomoviruses). Annual Review of Phytopathology 37:369–98.

Hajek, A. E., and D. L. Dahlsten. 1988. Distribution and dynamics of aphid (Homoptera, Drepanosiphidae) populations on *Betula pendula* in Northern California. Hilgardia 56:1–8.

Holt, J. K., J. Colvin, and V. Muniyappa. 1999. Identifying control strategies for tomato leaf curl virus disease using an epidemiological model. Journal of Applied Ecology 36:625–33.

Holt, J. K., M. J. Jeger, J. M. Thresh, and G. W. Otim-Nape. 1997. An epidemiological model incorporating vector population dynamics applied to African cassava mosaic virus disease. Journal of Applied Ecology 34:793–806.

Holt, R. D., A. P. Dobson, M. Begon, R. G. Bowers, and E. M. Schauber. 2003. Parasite establishment in host communities. Ecology Letters 6:837–42.

Irwin, M. E., W. G. Ruesink, S. A. Isard, and G. E. Kampmeier. 2000. Mitigating epidemics caused by non-persistently transmitted aphid-borne viruses: The role of the plant environment. Virus Research 71:185–211.

Irwin, M. E., and J. M. Thresh. 1990. Epidemiology of barley yellow dwarf virus: A study in ecological complexity. Annual Review of Phytopathology 28:393–424.

Jeger, M. J., and M. S. Chan. 1995. Theoretical aspects of epidemics: Uses of theoretical models to make strategic management decisions. Canadian Journal of Plant Pathology 17:109–14.

Jeger, M. J., J. Holt, F. van Den Bosch, and L.V. Madden. 2004. Epidemiology of insect-transmitted plant viruses: Modelling disease dynamics and control interventions. Physiological Entomology 29:291–304.

Jeger, M. J., F. van Den Bosch, L.V. Madden, and J. Holt. 1998. A model for analysing plant-virus transmission characteristics and epidemic development. IMA Journal of Mathematics Applied in Medicine and Biology, 15:1–18.

Kermack, W. O., and A. G. McKendrick. 1927. A contribution to the mathematical theory of epidemics. Proceedings of the Royal Society of London. Series A, Physical Sciences 115:700–721.

LoGiudice, K., R. S. Ostfeld, K. A. Schmidt, and F. Keesing. 2003. The ecology of infectious disease: Effects of host diversity and community composition on Lyme disease risk. Proceedings of the National Academy of Sciences of the United States of America 100:567–71.

Madden, L. V. 2001. What are the nonindigenous plant pathogens that threaten U.S. crops and forests? APSnet feature story. St. Paul, MN; American Phytopathological Society. Available: http://www.apsnet.org/online/feature/exotic/.

Madden, L.V., J. K. Knoke, and R. Louie. 1983. The statistical relationship between aphid trap catches and maize dwarf mosaic virus inoculation pressure. In Plant Virus Epidemiology, ed. R. T. Plumb and J. M. Thresh, 159–68. Oxford: Blackwell.

Margaritopoulos, J. T., C. Tsourapas, M. Tzortzi, O. M. Kanavaki, and J. A. Tsitsipis. 2005. Host selection by winged colonizers within the *Myzus persicae* group: A contribution towards understanding host specialization. Ecological Entomology 30:406–18.

McElhany, J. P., L. A. Real, and A. G. Power. 1995. Disease spread, spatial dynamics, and vector preference for diseased hosts: A study of barley yellow dwarf virus. Ecology 76:444–57.

Molyneux, D. H. 2003. Common themes in changing vector-borne disease scenarios. Transactions of the Royal Society of Tropical Medicine and Hygiene. 97:129–32.

Nault, L. R. 1997. Arthropod transmission of plant viruses: A new synthesis. Annals of the Entomological Society of America 90:521–41.

Palacios, I., M. Drucker, S. Blanc, S. Leite, A. Moreno, and A. Fereres. 2002. Cauliflower mosaic virus is preferentially acquired from the phloem by its aphid vectors. Journal of General Virology 83:3163–71.

Power, A. G. 1987. Plant community diversity, herbivore movement, and an insect-transmitted disease of maize. Ecology 68:1658–69.

Power, A. G. 1991. Virus spread and vector dynamics in genetically diverse plant populations. Ecology 72:232–41.

Power, A. G. 1992. Host plant dispersion, leafhopper movement and disease transmission. Ecological Entomology 17:63–68.

Power, A. G. 2000. Insect transmission of plant viruses: A constraint on virus variability. Current Opinion in Plant Biology 3:335–39.

Power, A. G., and A. S. Flecker. 2003. Virus specificity in disease systems: Are species redundant? In The Importance of Species: Perspectives on Expendability and Triage, ed. P. Kareiva and S. A. Levin. Princeton, NJ: Princeton University Press.

Power, A. G., and S. M. Gray. 1995. Aphid transmission of barley yellow dwarf viruses: Interactions between viruses, vectors and host plants. In Barley Yellow Dwarf: 40 Years of Progress, ed. C. J. D'Arcy and P. A. Burnett, 259–89. St. Paul, MN: American Phytopathogical Society Press.

Purcell, A. H. 1976. Seasonal changes in host plant preference of blue-green sharpshooter *Hordnia circellata* (Homoptera: Cicadellidae). Pan-Pacific Entomologist 52:33–37.

Purcell, A. H. 1979. Leafhopper vectors of xylem-borne plant pathogens. *In* Leafhoppers, Vectors and Plant Disease Agents, ed. K. Maramorosch and K. F. Harris, 603–25. New York: Academic Press.

Real, L. A., E. A. Marshall, B. M. Roche. 1992. Individual behavior and pollination ecology: Implications for the spread of sexually transmitted plant diseases. *In* Individual-Based Models and Approaches in Ecology, ed. D. L. DeAngelis and L. J. Gross, 525. New York: Chapman and Hall.

Roberts, M. G., and A. P. Dobson. 1995. The population dynamics of communities of parasitic helminths. Mathematical Biosciences 126:191–214.

Ross, R. 1910. The Prevention of Malaria. London: J. Murray.

Sohi, S. S., and K. G. Swenson. 1964. Pea aphid biotypes differing in bean yellow mosaic virus transmission. Entomological Experiments and Applications 7:9–14.

Taylor, L. H., S. M. Latham, and M. E. J. Woolhouse. 2001. Risk factors for human disease emergence. Philosophical Transactions of the Royal Society of London. Series B, Biological Sciences 356:983–89.

Weaver, S. C., and A. D. T. Barrett. 2004. Transmission cycles, host range, evolution and emergence of arboviral disease. Nature Reviews 2:789–801.

Woolhouse, M. E. J., D. T. Haydon, and R. Antia. 2005. Emerging pathogens: The epidemiology and evolution of species jumps. Trends in Ecology & Evolution 20:238–44.

Woolhouse, M. E. J., L. H. Taylor, and D. T. Haydon. 2001. Population biology of multihost pathogens. Science 292:1109–11.

Zhang, X.-S., J. Holt, and J. Colvin. 2000. A general model of plant-virus disease infection incorporating vector aggregation. Plant Pathology 49:435–44.

CHAPTER THREE

Understanding Host-Multipathogen Systems: Modeling the Interaction between Ecology and Immunology

Pejman Rohani, Helen J. Wearing, Daniel A. Vasco, and Yunxin Huang

SUMMARY

WE PRESENT A NEW MATHEMATICAL FRAMEWORK for exploring the ecological and immunological interactions between multiple infectious diseases. The mechanism underlying the ecological interaction is the modulation of susceptible numbers for one pathogen as a result of isolation or mortality following infection with a competitor. Immunological interactions are assumed to result from immunosuppression or cross-immunity to co-circulating pathogens, both during and after infection. This model is briefly examined to explore the consequences of these factors for the coexistence of multiple infectious diseases. We show that strong competition among pathogens reduces the region of coexistence, while substantial immunosuppression acts to facilitate pathogen community persistence. The dynamics of this model in the presence of seasonal changes in contact rates are presented and compared with historical case notification data for measles and whooping cough. We finish by highlighting how such a mathematical framework may be used to systematically investigate the role played by alternative competing mechanisms in shaping the observed dynamics of multipathogen systems.

BACKGROUND

Infectious diseases have become an increasingly important and high-profile public health issue, in large part as a result of the emergence of new pathogens (Daszak et al. 2000; Dobson and Foufopoulos 2001; Lipsitch et al. 2003), the continued persistence and resurgence of older infectious diseases (Keeling and Gilligan 2000; Orenstein et al. 2004), and concerns over possible deliberate exposure (Halloran et al. 2002). Indeed, the World Health Organization's Global Burden of Disease project estimated that in 2000, more than 10 million deaths worldwide were due to infectious and parasitic diseases. Understanding the precise mechanisms underlying disease dynamics, spread, and evolution, therefore, has never been of greater importance.

To achieve detailed understanding for a particular pathogen, epidemiologists routinely study aspects of the causative etiological agent (be it a virus, bacterium, fungus, or protozoan) and typically assume no interaction with other pathogens. A good example of such an approach would be the study of measles, which some have argued has become the *C. elegans* of large-scale epidemiological dynamics (Grenfell et al. 2001). Many decades of research, coupled with extensive long-term data, have resulted in a deep understanding of measles epidemiology, with a large body of work to explain its observed epidemics (Bartlett 1957; Bjornstad et al. 2002; Bolker and Grenfell 1996; Ellner et al. 1998; Ferguson et al. 1996; Rohani et al. 1999; Schenzle 1984; Soper 1929) and explore the most effective eradication programs (Earn et al. 1998; Hethcote 1988; Nokes and Swinton 1997).

These popular approaches to epidemiology may, however, represent an oversimplification of disease communities and may be ignoring some key interactions. In a nutshell, the understanding obtained from studying (for example) only the measles virus and its interaction with humans may paint only part of the true picture. In recent years, these single-host, single-pathogen approaches have been extended to incorporate multiple hosts (see chapter 1, this volume; Dobson 2004; Gog et al. 2002; Greenman and Hudson 1999) and multiple pathogens (Ferguson et al. 2003; Gupta et al. 1998). These studies of "community epidemiology" can be broadly categorized according to the scale of interest. At the antigenic and cellular scale, studies have typically explored the immunological interaction between pathogens as a result of coinfection within a host (Garcia-Garcia et al. 2003; Kirschner 1999; May and Nowak 1994). An especially exciting area of recent research in this field is bacterial interference, a process in which competing autochthonous microorganisms block adhesion events and prevent infection by pathogenic bacteria (Reid et al. 2001). This mechanism, which has also been studied in the context of viral infections, is considered by some to have greater public health potential than vaccines because it relies on the competitive exclusion of pathogens and does not require host immune stimulation (Huovinen 2001; Reid et al. 2001; Tano et al. 2002). There are, however, obvious concerns surrounding the prophylactic administration of live organisms. At the ecological level, shared pathogens have been demonstrated to be influential in shaping extinction dynamics by causing apparent competition between species (Holt 1977; Tomkins et al. 2001). The area that has received much attention has been the dynamics of pathogens with well-established antigenic polymorphism, such as influenza, malaria, adenoviruses, poliovirus, cholera, and dengue

(Earn et al. 2002; Ferguson et al. 2003; Gupta et al. 1998; Koelle et al. 2005; Read and Taylor 2001). In such systems, it is empirically documented that different strains fluctuate out of phase with each other. This observation is summed up by Bang (1975): "If a significant proportion of a population is not immune to a given agent, the presence of that agent as an epidemic will tend to suppress the appearance of other agents of a similar nature with which interference may occur." He went on to propose that epidemiological interference may account for spatial asynchrony in outbreaks of adenoviruses (1, 2, and 5) and the temporal asynchrony in poliomyelitis (3, 2, and 1) epidemics observed in West Bengal. More recent work in this area has highlighted the significance of cross-immunity between strains that may shape the coexistence, dynamics, and evolution of strains (Dietz 1979; Elveback et al. 1968; Gog and Grenfell 2002; Kamo and Sasaki 2002; Koelle et al. 2005; White et al. 1998).

Until recently, however, the possibility that epidemics of unrelated pathogens might interact has been ignored, despite the suggestion of its likelihood in historical epidemiological literature. For example, in his classic 1894 book, *A History of Epidemics in Britain*, the learned medical historian Charles Creighton commented that "again, the great measles epidemic of 1808 in Glasgow was indeed followed by many deaths from whooping-cough in 1809. Whatever correspondence or relation there may be between measles and whooping-cough, (and it has been remarked by many in the ordinary way of experience), it eludes the method of statistics." Creighton clearly envisaged a strong interaction between these infectious diseases, though he was unclear on the underlying mechanism and had no mathematical framework for its exploration.

One obvious candidate mechanism would be immunity-mediated interaction. Consider the words of James S. Laing, the resident physician of Aberdeen City Hospital, who in 1902 stated, "Most writers assert that there is an intimate association between epidemics of measles and epidemics of whooping-cough, and that an epidemic of the former disease strongly predisposes to the subsequent development of the latter." This reflects the widely recognized fact that after infection with measles, the immune system is suppressed for a period of time, during which an individual may be more susceptible to colonization by other (particularly bacterial) pathogens. It is interesting to note, however, that after studying the average time between successive measles and whooping cough epidemics in case notification data for Aberdeen (figure 3.1), Laing and Hay (1902) concluded, "It would thus appear as if whooping-cough rather paved the way for measles than measles for whooping cough"!

In 1998, Rohani and colleagues proposed an additional *ecological* mechanism that may also contribute to interaction—specifically, interference—among unrelated acute infectious diseases. This possible in-

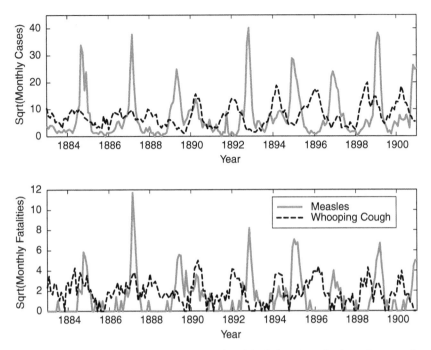

Figure 3.1. Long-term patterns in measles (black lines) and whooping cough (gray lines) epidemics in Aberdeen from 1883 to 1901 (data from Laing and Hay 1902). Panel a shows monthly case notifications, while monthly case fatalities are shown in b. The time series for both infections exhibit a strong biennial component, with a striking phase difference between measles and whooping cough outbreaks.

terference was proposed to arise from the temporary or permanent removal of potential hosts from the susceptible population for one pathogen following an acute infection by one of its direct competitors. The primary mechanism for this removal is the convalescence period, during which individuals are in isolation and hence unavailable to contract "competing" pathogens. As documented by Emerson (1937), following measles infection, children in the major U.S. cities in the 1920s and 1930s were isolated for an average of almost ten days, while isolation after an episode of whooping cough lasted nearly four weeks. Modern infection management practices similarly result in the isolation of infected children for one week after measles infection and two weeks for whooping cough (Nelson et al. 2001). In addition to the possible dynamical consequences of enforced convalescence, in conditions under which infected individuals may suffer death as a result of infection, removal from the susceptible pool can become permanent, and the interaction between pathogens is predicted to become stronger. Dynamically, this is very similar to the effects of cross-immunity in strain polymorphic systems, whereby individuals previously infected with one strain may have partial protection against infection by other competing strains (Kamo and Sasaki 2002). The significant prediction of the model of Rohani et al. (1998) was that the epidemics of competing infections would be temporally segregated, with major outbreaks out of phase with each other (as alluded to by Creighton [1894]).

Empirical support for interference effects is provided by case fatality data for measles and whooping cough for fifteen European cities in the pre- and post-World War I years (Rohani et al. 2003). In this era, infected individuals were likely to be as a result of complications following infection, typically secondary viral and bacterial lung infections (pneumonia). In sixteen cities, the population demographic characteristics (notably the per capita birth rates) were conducive to biennial outbreaks. The epidemics of measles and whooping cough were statistically significantly out of phase with each other in fifteen of these sixteen cities (Rohani et al. 2003)—that is to say, epidemic years for these infections did not coincide. This work suggests, therefore, that when disease prevalence is very high and is associated with significant mortality, as remains the case in many developing nations, it may be impossible to fully understand epidemic patterns by studying pathogens in isolation.

Although the patterns revealed in these historical data are consistent with model prediction of disease interference, there remains a need for systematic study of the different possible routes of interaction between different infectious diseases (or strains of the same disease) and their dynamical consequences. In this chapter, we aim to further develop theory on the interaction between infectious diseases. We present a novel general model for examining systems with multiple pathogens. For illustration purposes, our model analyses are focused on measles and whooping cough, but the proposed framework is flexible and may be applied to strain polymorphic as well as to other unrelated diseases. The key ingredient of the formalism we develop is the simultaneous inclusion of immunologically determined components (immunosuppression and cross-immunity) and ecological factors (isolation and infection-induced mortality).

THE TWO-DISEASE MODEL

Much of the influential epidemiological theory has been based on the SEIR (susceptible, exposed, infected, recovered) paradigm (Anderson and May 1991; Dietz 1976; Keeling and Rohani 2007). In this framework, individuals are categorized according to infection status: the infection-naïve are thought of as susceptible, upon infection they become exposed, and once latency is over they are infected and proceed to transmit the pathogen. After successfully overcoming the infection, individuals are considered recovered and immune for life. This picture does not take into account the possible influence that infection by one pathogen may exert on the community of pathogens competing for the same hosts. A case in point is childhood infections: children may contract a number of infectious diseaes, such as measles, pertussis, mumps, chickenpox, or rubella. Because any child infected with, for example, measles is unavailable to contract any other infectious disease for a period of time (perhaps up to two weeks), we may wonder what effect an outbreak of measles would have on the dynamics of the other candidate infections.

An important step in developing an understanding of the dynamical interaction between multiple infections has been to develop a novel, conceptually simple, mathematical framework that incorporates two pathogens. This work follows in the footsteps of a distinguished and significant body of work dealing with multipathogen interactions, focusing on infections such as influenza, malaria, or dengue, in which genetic diversity is well established (Andreasen et al. 1997; Dietz 1979; Earn et al. 2002; Ferguson et al. 2003; Gilbert et al. 1998; Gog and Grenfell 2002; Gomes et al. 2002; Gupta et al. 1998; Kamo and Sasaki 2002; Taylor et al. 1997). In developing the model, we envisage a simplified natural history of infection for each disease:

- All newborns are fully susceptible to both infections.

- Upon infection, a susceptible individual enters the exposed (infected but not yet infectious) class and has a probability of contracting the "competing" disease simultaneously (represented by the cross-immunity parameter ϕ_i, where $i = 1, 2$).

- After the latent period, the individual becomes infectious but is not yet symptomatic and still has a defined probability (ϕ_i, $i = 1, 2$) of becoming coinfected with the other disease.

- Typically, when symptoms appear, the disease is diagnosed and the individual is sent home to convalesce for an average period, given by $1/\delta_i$ ($i = 1, 2$). During convalescence, the competing infection may be contracted, with the transmission rate additionally modulated by the parameter ξ_i ($i = 1, 2$), which may represent isolation or temporary cross-immunity (if less than 1) or temporary immunosuppression (if greater than 1).

- Depending on the disease, host age, and host condition (typically nutritional status), infection may be fatal owing to complications (such as pneumonia and encephalitis, in the case of measles and pertussis). This is represented by per capita infection-induced mortality probabilities ρ_i ($i = 1, 2$).

- Upon complete recovery, the individual is assumed immune to the infection (disease 1) and reactivates susceptibility to disease 2, if previously not exposed to it. At this stage, we introduce the term χ_i to explore the implications of longlasting immunosuppression ($\chi_i > 1$) or cross-immunity ($\chi_i < 1$) for the transmission rate of disease j following infection with disease i.

The mathematical representation of these assumptions is presented in the appendix to this chapter. The key strength of this framework is its flexibility, allowing us to establish unambiguously the dynamical role played by each of the features of the model. For example, as demonstrated rigorously in the appendix, by removing all immune-mediated interaction between infections (i.e., $\phi_i = \chi_i = \xi_i = 1$, $i = 1, 2$) and ignoring ecological considerations ($\rho_i = 0$, $i = 1, 2$), we can strictly decouple the dynamics of the two infections; the model contains two pathogens with entirely independent transmission dynamics. It is also straightforward to extend the model to incorporate vector transmission in order to better understand the serotype dynamics of dengue, for instance (Wearing and Rohani 2006).

MODEL PREDICTIONS

One intuitively obvious possible consequence of interaction among infections is reduced abundance. Surprisingly, however, detailed equilibrium analyses have demonstrated that disease interference does not manifest itself by significantly altering infection prevalence; changes in model parameters such as the convalescence period translate into negligible changes in the number of infectives of either infection (Huang and Rohani 2005). Perhaps more surprisingly, epidemiological interference exerts little influence on the coexistence likelihood of pathogens. Defining the basic reproductive ratio of each infection as $R_0^j = \beta_j \sigma_j / (\sigma_j + \mu)(\gamma_j + \mu)$ ($j = 1, 2$), it is straightforward to show that coexistence requires $R_0^j > 1$ and

$$R_0^j > \frac{R_0^i}{1 + \alpha_i(R_0^i - 1)}, \tag{1}$$

where a_i is a convenient grouping of parameters and is given by

$$a_i = \frac{1}{\sigma_i + \mu}\left\{\phi\mu + \frac{\sigma_i}{\gamma_i + \mu}\left(\phi\mu + \frac{\gamma_i}{\delta_i + \mu}(\xi\phi\mu + \chi(1 - \rho_i)\delta_i)\right)\right\}, \tag{2}$$

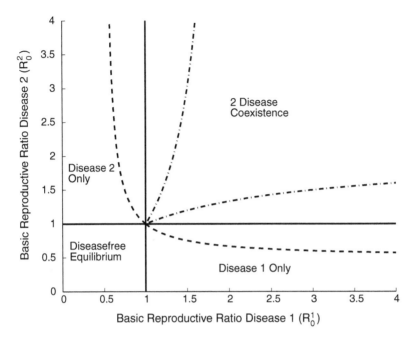

Figure 3.2. The coexistence of two infectious diseases can be affected by immunosuppression and disease interference. In the absence of immunemediated interactions ($\phi_i = \chi_i = 1, i = 1, 2$), large levels of disease-induced mortality (50%: dot-dashed line) can cause the region of two-disease coexistence to shrink somewhat. In contrast, strong levels of permanent immunosuppression ($\phi_i = 1, \chi_i = 2, \rho_i = 0, i = 1, 2$) can expand the coexistence domain. Model parameters were $\mu = 0.02$, $\frac{1}{\sigma_1} = \frac{1}{\sigma_2} = 8$ days, $\frac{1}{\gamma_1} = 5$ days, $\frac{1}{\gamma_2} = 14$ days, $\xi = 1$, $\frac{1}{\delta_1} = 7$ days and $\frac{1}{\delta_2} = 14$ days. (From Vasco et al. 2007.)

$i, j = 1, 2, j \neq i$. The diseases are assumed to have symmetrical values of ϕ, χ, and ξ (details provided in Vasco et al. 2007). In figure 3.2, we explore the conditions for disease coexistence in this model. In the absence of pathogen-induced mortality ($\rho_1 = \rho_2 = 0$), the isolation period alone has little effect on the stable two-disease equilibrium, with the coexistence criterion effectively reducing to R_0^1, $R_0^2 > 1$, since inequality (1) is always satisfied (assuming no immune effects, $\chi = 1$). It is only after we assume a 50% (dash-dotted line) probability of death following infection that the region of endemic two-disease coexistence shrinks slightly. On the other hand, if we ignore ecological factors (such as quarantining and pathogen virulence), immunosuppression resulting from one infection can facilitate the invasion and persistence of the competing disease even if the invading infection has R_0 less than 1 (dashed line).

To identify the dynamical consequences of disease interaction, Vasco et al. (2007) carried out a systematic comparison of the predicted equilibrium dynamics of the general two-disease model (described in the appendix to this chapter) with the dynamics of each infection in isolation (when $\phi = \chi = \xi = 1$). Similar to the underlying single-disease model, in the absence of seasonality in the transmission rate, the two-infection system demonstrates damped oscillations for much of the parameter space. It is possible to show, however, that when coinfection probability (ϕ) is very small, relatively large levels of permanent immunosuppression (χ) can destabilize the equilibrium (via a Hopf bifurcation), giving rise to large-amplitude cycles. The precise extent of this effect is determined by the assumed isolation periods and birth rate (Vasco et al. 2007). High birth rates make destabilization less likely, while increased convalescence periods increase the possibility.

While equilibrium studies of system dynamics are illuminating for general multipathogen interactions, an important ingredient for the specific study of childhood infections is seasonal variation in transmission rates due to the school calendar. It is well established that such external forcing can have dramatic effects on measles dynamics (Dietz 1976), with a low amplitude of seasonality generating annual epidemics, while greater levels of forcing can produce biennial and longer-term dynamics (Schwartz and Smith 1983). The dynamics of whooping cough, on the other hand, are rigidly annual, irrespective of changes in the seasonal amplitude (Rohani et al. 2002). When both infections are included in the seasonally forced two-disease model, the epidemics of measles are largely unaffected, while the pattern of whooping cough outbreaks mimics those of measles exactly (Rohani et al. 1998). The explanation for this observation lies in the primary factors that determine an infection's R_0, namely, the transmission rate (β) and the infectious period ($\frac{1}{\gamma}$). Therefore, depending on the precise combination of these traits, infections respond differentially to seasonal variation in contact rates. In general, one of the predictions of the work on two-disease models is that given $R_0^1 = R_0^2$, the bifurcation structure of the model is dictated by the infection with the higher transmission rate—in this case, measles (Huang and Rohani 2005). The most plausible explanation is that at the start of the epidemic calendar (early autumn), when there is a substantial influx of susceptibles in the young school cohort, a higher transmission rate permits an infection to get established first, and its pattern of epidemics sets the template for the competitor.

Here we extend these previous studies by examining the dynamics of the seasonal two-disease model (see the appendix to this chapter) as the coinfection probability (ϕ) and the strength of immunosuppression (χ) are varied. The results of this approach are presented in figure 3.3. The main (central) panel shows how the measles outbreak period changes as ϕ and χ

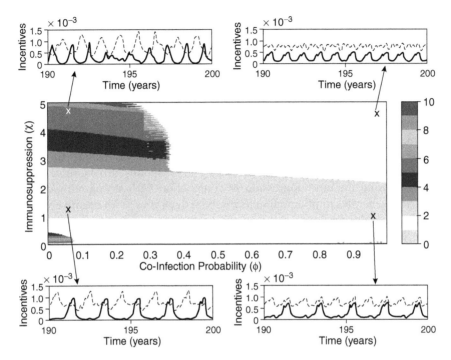

Figure 3.3. The dynamical implications of varying the probability of coinfection ($\phi_1 = \phi_2 = \phi$) and immunosuppression/cross-immunity ($\chi_1 = \chi_2 = \chi$) in a seasonally forced model. The middle panel shows the period of oscillations (in years) observed for disease 1 (measles) as the control parameters are varied. The color coding is explained in the key to the right of the panel. The top and bottom panels represent time series for measles (black lines) and whooping cough (gray lines) in the regions of parameter space marked by crosses. Model parameters are $\mu = 0.02$, $b_1 = 0.25$, $\xi = 1$, $\frac{1}{\sigma_1} = \frac{1}{\sigma_2} = 8$ days, $\frac{1}{\gamma_1} = 5$ days, $\frac{1}{\gamma_2} = 14$ days, $\frac{1}{\delta_1} = 7$ days $\frac{1}{\delta_2} = 14$ days, $\bar{b}_1 = 1{,}250$ per year and $\bar{b}_2 = 446$ per year. The dynamics were obtained by numerical integration.

are varied. For much of the parameter space (when $\phi > 0.4$), measles epidemics are either annual or biennial. When ϕ is small and coinfection is unlikely, however, changes in immunosuppression levels can give rise to bifurcations, with epidemic patterns that have periods ranging from one to ten years. Of note, χ is clearly the major determinant of dynamics. This figure can be useful in examining the combination of ecological and immunological traits that gives rise to dynamics consistent with data. As shown in figure 3.1, measles and whooping cough epidemics in Aberdeen were both biennial and clearly out of phase with one another. The dynamics summarized in figure 3.3 show, for example, that when infectious diseases are strictly independent ($\chi = \phi = 1$), measles epidemics are biennial, whereas whooping cough exhibits annual cycles (bottom right panel), in contrast to empirical findings. As the competitive strength of the interaction is increased (ϕ smaller), whooping cough epidemics also become biennial, but, of importance, their major outbreaks are out of phase with measles epidemics (figure 3.3, bottom left panel), as seen in figure 3.1. In the presence of coinfection ($\phi = 1$), increasing permanent immunosuppression leads to annual outbreaks of both infections (figure 3.3, top right panel). This effect is relatively straightforward to explain, since contracting one infection essentially "primes" individuals for contraction of the other infection, resulting in seasonally driven annual cycles. Strong immunosuppression in conjunction with strong competitive effects generates multiennial quasiperiodic dynamics, with negatively correlated epidemics.

The take-home message from this analysis is that the parameters χ and ϕ both have substantial dynamical consequences, though in subtly different ways. The permanent immunosuppression (or cross-immunity) factor χ strongly affects epidemic periods, while the coinfection parameter ϕ is largely responsible for generating negative correlation between the outbreaks of the two infections (Vasco et al. 2007).

WHEN WOULD INFECTIONS INTERFERE?

Throughout the model formulation and analyses of data, we have placed heavy emphasis on the study of measles and whooping cough. Children, however, are typically exposed to many more infectious diseases, such as mumps, rubella, or chickenpox. Would we expect *all* of these infections to dynamically interact? By extension, ideally, would any modeling work need to include all of these infections? We believe the answer to this question is likely to be no. Using simple homogeneous models without age structure, it has been demonstrated that dynamical interference effects are most pronounced between infections with a similar basic reproductive ratio (Dietz 1979; Rohani et al. 1998, 2003; Huang and Rohani 2005). A complete understanding of this issue will, however, need age dependence in contact rates to be taken into account. This is because the interference concept relies on "competition" for resources (hosts) between pathogens. For the dynamical effects of this competitive interaction to be noticeable, the pathogens should be infecting largely the same cohort of hosts. Hence, the extent of interference effects is likely to be determined by the relative distributions of age at infection. One way of examining this issue is by studying the mean age at infection, which is dictated by the transmission potential of the disease or its basic repro ductive ratio, R_0. For *SIR*-type diseases, the mean age at infection (A) is approximately the life expectancy of the host $\left(\frac{1}{\mu}\right)$ divided by R_0 (more precisely, $A \sim \frac{1}{(\mu(R_0-1))}$ (Anderson and May 1991; Keeling and Rohani 2007). Table 3.1 shows that, of the potential childhood infections, measles and whooping cough have very similar mean ages at infection (approximately four to five years; Anderson and May 1982). Hence, these diseases are likely to have been strongly "competing" for children in the same age cohorts, while their interaction with the other childhood infections is likely to have been less intense. This logic suggests that infections such as rubella and chickenpox may also be good potential candidates for the study of interference. We are currently examining this issue using the two-disease model with age-specific transmission (Huang and Rohani 2006).

CONCLUSIONS

Understanding the ecology of infectious diseases has become an increasingly important endeavour. Many of the important and high-profile infections, such as influenza, malaria, and dengue, have well-established

TABLE 3.1

Historical estimates of the mean age at infection for a number of childhood diseases in the twentieth century

Disease	Time Period	Mean Age at Infection (yr)	R_0
Measles	1944–1979	4.4–5.6	13.7–18.0
	1912–1928	5.3	12.5
Whooping cough	1944–1978	4.1–4.9	14.3–17.1
	1908–1917	4.9	12.2
Chickenpox	1913–1917	6.7	9.0
	1918–1921	7.1	8.5
Mumps	1943	9.9	7.1
	1912–1916	13.9	4.3
Rubella	1972	10.5	6.7
	1979	11.6	6.0
Poliomyelitis	1960	11.2	6.2
	1955	11.9	5.9

Note: Data from Anderson and May (1982).

antigenic polymorphism. It is generally acknowledged that taking into account immune-mediated interactions among strains is necessary for explaining the ecological and evolutionary dynamics of these infections (Elveback et al. 1968; Ferguson et al. 2003; Gog and Grenfell 2002; Levin et al. 2004). In this chapter, we have reviewed some recent work that has proposed an ecological mechanism for possible interaction among antigenically distinct infections (Rohani et al. 1998, 2003). Specifically, we have put forward a general mathematical and conceptual framework within which issues pertaining to immunological and ecological scale interactions may be explored.

Recent work on the epidemiological dynamics of dengue fever provides an illustration of how this framework may be used in a systematic way to investigate alternative competing hypotheses about the mechanisms responsible for observed patterns of disease. Dengue is a mosquito-borne flavivirus of varying clinical severity that is composed of four antigenically distinct but related serotypes. Long-term surveillance in hyperendemic regions suggests that individual dengue serotypes fluctuate out of phase with each other, while aggregated dengue data exhibit predominantly annual outbreaks together with a less pronounced, two- to four-year cyclic signature (Cummings et al. 2004; Nisalak et al. 2003). Mathematical models of dengue epidemiology have previously been developed to explain these empirical observations. These models can be broadly split into two kinds. The first kind attempts to include as much ecological detail as possible, incorporating a complex array of external environmental factors (Focks et al. 1995); the second is reductionist and focuses on the prevailing immunological hypothesis of antibody-dependent enhancement (ADE), whereby infection with one serotype increases an individual's susceptibility to (or mortality from) infection with another (Ferguson et al. 1999; Kawaguchi et al. 2003). Using the mathematical framework laid out in this chapter and extending it to incorporate a vector population, we were able to take an intermediate route and explore both ecological and immunological mechanisms in a tractable manner (Wearing and Rohani 2006). Our results suggest that ADE may not be as important to explain dengue epidemics as is currently thought. Specifically, in the presence of seasonal variation in the vector population, the key immune-mediated interaction that is necessary to explain the observed dynamics of endemic dengue incidence is the well-documented temporary period of cross-protection. This finding is important for two reasons. First, if ADE is indeed the primary mechanism generating dengue dynamics, then in our model, this gives rise to rapid disease extinction unless the host population size exceeds 10^7. Second, the levels of vaccination required to eradicate serotypes would clearly be different, depending on which factors dominate.

The near-term research agenda for exploring multipathogen interaction will likely involve a number of key issues. Developing a broad understanding of the dynamical consequences of different factors is obviously important. This will ultimately provide some insights into the kinds of patterns that may result from different modes of pathogen interaction. One of the important outstanding issues is *when* disease dynamics influence each other. The use of age-structured, multipathogen models is likely to be informative in generating theory concerning the necessary ingredients for disease interference or facilitation. Exploration of model dynamics will also permit the establishment of rigorous signatures of interaction, with the associated statistical methodologies that may be used to interrogate disease data.

An exciting potential area of application of the generalized model we have proposed in this chapter is the study of a variety of multistrain pathogens whose epidemiological dynamics are intrinsically coupled to evolutionary processes (e.g., dengue, cholera, meningitis, poliovirus, echoviruses). The next step toward a greater understanding of pathogen diversity and dynamics is merging the information contained in both epidemiological and genetic data, an approach recently referred to as phylodynamics (Grenfell et al. 2004). A cogent statistical problem in this area involves the simultaneous estimation of epidemiological and genetic parameters using ecological and genetic time series. Recent advances in population genetics, in particular coalescent theory, enable us to use sequence information sampled from rapidly evolving pathogens over time to infer ecological and genetic parameters (Drummond et al. 2002; Emerson et al. 2001; Vasco et al. 2001). For multistrain pathogens, selective pressures arising from the host immune system can potentially drive the evolutionary outcome and interact with the stochastic forces of mutation, genetic drift, and recombination to determine the final set of sampled sequences and infecteds. The precise form of cross-immunity, enhancement, or immunosuppression will determine how pathogen populations eventually become structured into different antigenic strains and persist or replace each other through time. These considerations imply that incorporating basic evolutionary mechanisms into the two-disease model presented here will have substantial conceptual consequences for our understanding of the population dynamics of epidemics.

Finally, the potential public health implications of disease interference and facilitation remain largely unexamined. Intuitively, one may expect that interference or enhancement between infections may be informative in designing successful vaccination programmes. For instance, if substantial interference between two diseases is well established, then this information can be usefully deployed to derive optimal vaccine pulses.

According to theory, successful vaccination requires the reduction of susceptible numbers in the population below some critical threshold, N/R_0 (Anderson and May 1991; Kermack and McKendrick 1927; Keeling and Rohani 2007). Therefore, violent epidemics of one disease, together with the associated reduction in susceptible persons following the quarantining of all those infected during the outbreak, can, in theory, be used to time immunization pulses so that eradication may occur with fewer units of vaccine used than predicted by single-disease models (P. Rohani, unpublished data). Additionally, we may also expect interference or enhancement effects to be relevant when contemplating vaccination using multiple vaccines (such as the measles-mumps-rubella and the diphtheria-tetanus-pertussis triple vaccines).

ACKNOWLEDGMENTS

Research was funded in part by grants from the National Institutes of Health, the National Science Foundation, and the Ellison Medical Foundation. We thank Matt Bonds, Natalia Mantilla-Beniers, and Marc Choisy for discussion.

APPENDIX

In this appendix, we describe the mathematical equations used in the two-disease model. The basic approach is similar to status-based (rather than history-based) models:

$$\frac{dS_0}{dt} = \nu N - (\lambda_1 + \lambda_2)\frac{S_0}{N} - \mu S_0 \tag{3}$$

$$\frac{dE_1}{dt} = \lambda_1 \frac{S_0}{N} - \phi_2\lambda_2 \frac{E_1}{N} - (\sigma_1 + \mu)E_1 \tag{4}$$

$$\frac{dI_1}{dt} = \sigma_1 E_1 - \phi_2\lambda_2 \frac{I_1}{N} - (\gamma_1 + \mu)I_1 \tag{5}$$

$$\frac{dE_2}{dt} = \lambda_2 \frac{S_0}{N} - \phi_1\lambda_1 \frac{E_2}{N} - (\sigma_2 + \mu)E_2 \tag{6}$$

$$\frac{dI_2}{dt} = \sigma_2 E_2 - \phi_1\lambda_1 \frac{I_2}{N} - (\gamma_2 + \mu)I_2 \tag{7}$$

$$\frac{dC_1}{dt} = \gamma_1 I_1 - \xi_2\phi_2\lambda_2 \frac{C_1}{N} - (\delta_1 + \mu)C_1 \tag{8}$$

$$\frac{dC_2}{dt} = \gamma_2 I_2 - \xi_1\phi_1\lambda_1 \frac{C_2}{N} - (\delta_2 + \mu)C_2 \tag{9}$$

$$\frac{dS_1}{dt} = (1 - \rho_1)\delta_1 C_1 - \chi_2\lambda_2 \frac{S_1}{N} - \mu S_1 \tag{10}$$

$$\frac{dS_2}{dt} = (1 - \rho_2)\delta_2 C_2 - \chi_1\lambda_1 \frac{S_2}{N} - \mu S_2 \tag{11}$$

$$\frac{dS_{12}}{dt} = (1 - \rho_1)(1 - \rho_2)\left(\lambda_2 \frac{\phi_2 E_1 + \phi_2 I_1 + \xi_2\phi_2 C_1}{N} + \lambda_1 \frac{\phi_1 E_2 + \phi_1 I_2 + \xi_1\phi_1 C_2}{N} \right)$$
$$+ (1 - \psi_2\rho_2)\chi_2\lambda_2 \frac{S_1}{N} + (1 - \psi_1\rho_1)\chi_1\lambda_1 \frac{S_2}{N} - \mu S_{12} \tag{12}$$

$$\frac{d\varepsilon_1}{dt} = \lambda_1 \frac{S_0}{N} + \phi_1\lambda_1 \frac{E_2}{N} + \phi_1\lambda_1 \frac{I_2}{N} + \xi_1\phi_1\lambda_1 \frac{C_2}{N} + \chi_1\lambda_1 \frac{S_2}{N} - (\sigma_1 + \mu)\varepsilon_1 \tag{13}$$

$$\frac{d\varepsilon_2}{dt} = \lambda_2 \frac{S_0}{N} + \phi_2\lambda_2 \frac{E_1}{N} + \phi_2\lambda_2 \frac{I_1}{N} + \xi_2\phi_2\lambda_2 \frac{C_1}{N} + \chi_2\lambda_2 \frac{S_1}{N} - (\sigma_2 + \mu)\varepsilon_2 \tag{14}$$

$$\frac{d\lambda_1}{dt} = \beta_1\sigma_1\varepsilon_1 - (\gamma_1 + \mu)\lambda_1 \tag{15}$$

$$\frac{d\lambda_2}{dt} = \beta_2\sigma_2\varepsilon_2 - (\gamma_2 + \mu)\lambda_2, \tag{16}$$

TABLE 3.2
Description of model parameters

Parameter	Epidemiological Description	Typical Range
v	Host per capita birth rate	0–1 per year
μ	Host per capita death rate	0–1 per year
β_i	Transmission rate	100–2,000 per year
$1/\sigma_i$	Latent period	1–2 weeks
$1/\gamma_i$	Infectious period	1–3 weeks
$1/\delta_i$	Isolation period	1–4 weeks
ρ_i	Probability of infection-induced mortality	0–1
ϕ_i	Coinfection probability	0–1
ξ_i	Temporary immunosuppression/ cross-immunity	≥ 0
χ_i	Permanent immunosuppression/ cross-immunity	≥ 0
ψ_i	Differential infection-induced mortality	0–1

Note: Subscripts refer to disease i ($i = 1, 2$).

where all those susceptible to both infections denoted by S_0. The variables E_i, I_i, and C_i ($i = 1, 2$) represent those currently exposed, infectious, or convalescing (respectively) after infection with disease i, with no previous exposure to any infection. The terms S_i ($i = 1, 2$) represent all individuals who are only susceptible to infection j ($j \neq i$) following recovery from i. For bookkeeping purposes, we let ε_i and λ_i/β_i represent individuals latent and infectious with disease i ($i = 1, 2$). Additionally, S_{12} are all those no longer susceptible to either infection and may include those who are still exposed or infectious with one or both diseases (i.e. also in ε_1, ε_2, λ_1 or λ_2). The total population size (N) is the sum of the first ten variables only ($N = S_0 + S_{12} + \sum_{i=1}^{2}(E_i + I_i + C_i + S_i)$). The model's parameters are explained in table 3.2.

It is straightforward to demonstrate that diseases 1 and 2 can be easily decoupled within this framework. Assume there is no disease-induced mortality for either disease ($\rho_1 = \rho_2 = 0$) and no immune-mediated interaction ($\chi_i = \phi_i = 1$, $i = 1, 2$). Then, if we let $Z_i = E_i + I_i + C_i + S_i + S_0$, $i = 1, 2$, our equations can be rewritten as:

$$\frac{dS_0}{dt} = vN - (\lambda_1 + \lambda_2)\frac{S_0}{N} - \mu S_0$$

$$\frac{dZ_1}{dt} = vN - \lambda_2 \frac{Z_1}{N} - \mu Z_1$$

$$\frac{dZ_2}{dt} = vN - \lambda_1 \frac{Z_2}{N} - \mu Z_2$$

TABLE 3.3
Timings of the major school holidays when *Term* = −1 (during all other times
Term = +1)

Holiday	Model Days	Calendar Dates
Christmas	356–6	21 December–6 January
Easter	100–115	10–25 April
Summer	200–251	19 July–8 September
Autumn Half-Term	300–307	27 October–3 November

Note: The autumn half-term break is included as this is the only short holiday that has an identifiable signature in the England and Wales data.

$$\frac{dS_{12}}{dt} = \lambda_2 \frac{Z_1 - S_0}{N} + \lambda_1 \frac{Z_2 - S_0}{N} - \mu S_{12}$$

$$\frac{d\varepsilon_1}{dt} = \lambda_1 \frac{Z_2}{N} - (\sigma_1 + \mu)\varepsilon_1$$

$$\frac{d\varepsilon_2}{dt} = \lambda_2 \frac{Z_1}{N} - (\sigma_2 + \mu)\varepsilon_2$$

$$\frac{d\lambda_1}{dt} = \beta_1 \sigma_1 \varepsilon_1 - (\gamma_1 + \mu)\lambda_1$$

$$\frac{d\lambda_2}{dt} = \beta_2 \sigma_2 \varepsilon_2 - (\gamma_2 + \mu)\lambda_2.$$

These equations represent a decoupled system with two dynamically distinct infections $(Z_1, \varepsilon_2, \lambda_2)$ and $(Z_2, \varepsilon_1, \lambda_1)$.

SEASONALITY

Following the classic work of Schenzle (1984), the transmission rate in this model is assumed to be high during school terms and low at other times. In this manner, the equation describing the transmission rate for disease i can be rewritten as follows:

$$\beta_i(t) = \bar{\beta}_i(1 + b_1 Term(t)), \tag{17}$$

where $Term(t)$ is +1 during school term and −1 at other times. The parameter $\bar{\beta}_i$ represents the baseline (or mean) transmission rate. We use the parameter b_1 to represent the amplitude of seasonality. The historical dates of school terms in England and Wales are presented in table 3.3.

REFERENCES

Anderson, R. M., and R. M. May. 1982. Directly transmitted infectious diseases: Control by vaccination. Science 215:1053–60.

Anderson, R. M., and R. M. May. 1991. Infectious Diseases of Humans. Oxford: Oxford University Press.

Andreasen, V., J. Lin, and S. A. Levin. 1997. The dynamics of co-circulating influenza strains conferring partial cross-immunity. Journal of Mathematical Biology 35:825–42.

Bang, F. B. 1975. Epidemiological interference. International Journal of Epidemiology 4:337–42.

Bartlett, M. S. 1957. Measles periodicity and community size. Journal of the Royal Statistical Society 1:48–59.

Bjornstad, O. N., B. F. Finkenstadt, and B. T. Grenfell. 2002. Dynamics of measles epidemics: Estimating scaling of transmission rates using a time-series SIR model. Ecological Monographs 72:169–84.

Bolker, B., and B. Grenfell. 1996. Impact of vaccination on the spatial correlation and persistence of measles dynamics. Proceedings of the National Academy of Sciences of the United States of America 93:12648–53.

Creighton, C. 1894. A History of Epidemics in Britain. Cambridge: Cambridge University Press.

Cummings, D. A. T., R. A. Irlzarry, N. E. Huang, T. P. Endy, A. Nisalak, K. Ungchusak, and D. S. Burke. 2004. Travelling waves in the occurrence of dengue harmorrhagic fever in Thailand. Nature 427:344–47.

Daszak, P., A. A. Cunningham, and A. D. Hyatt. 2000. Emerging infectious diseases of wildlife: Threats to biodiversity and human health. Science 287:443–49.

Dietz, K. 1976. The incidence of infectious diseases under the influence of seasonal fluctuations. Lecture Notes in Biomathematics 11:1–15.

Dietz, K. 1979. Epidemiologic interference of virus populations. Journal of Mathematical Biology 8:291–300.

Dobson, A. 2004. Population dynamics of pathogens with multiple host species. American Naturalist 164:S64–78.

Dobson, A., and J. Foufopoulos. 2001. Emerging infectious pathogens of wildlife. Philosophical Transactions of the Royal Society of London. Series B, Biological Sciences 356:1001–12.

Drummond, A. J., G. K. Nicholls, A. G. Rodrigo, and W. Solomon. 2002. Estimating mutation parameters, population history and genealogy simultaneously from temporally spaced sequence data. Genetics 161:1307–20.

Earn, D. J. D., J. Dushoff, and S. A. Levin. 2002. Ecology and evolution of the flu. Trends in Ecology & Evolution 17:334–40.

Earn, D. J. D., P. Rohani, and B. T. Grenfell. 1998. Persistence, chaos and synchrony in ecology and epidemiology. Proceedings of the Royal Society of London. Series B, Biological Sciences 265:7–10.

Ellner, S. P., B. A. Bailey, G. V. Bobashev, A. R. Gallant, B. T. Grenfell, and D. W. Nychka. 1998. Noise and nonlinearity in measles epidemics: Combining mechanistic and statistical approaches to population modeling. American Naturalist 151:425–40.

Elveback, L., E. Ackerman, G. Young, and J. P. Fox. 1968. A stochastic model for competition between viral agents in the presence of interference. American Journal of Epidemiology 87:373–84.

Emerson, B. C., E. Paradis, and C. Thebaud. 2001. Revealing the demographic histories of species using DNA sequences. Trends in Ecology & Evolution 16: 707–16.

Emerson, H. 1937. Measles and whooping cough. American Journal of Public Health 27:1–153.

Ferguson, N. M., A. P. Galvani, and R. M. Bush. 2003. Ecological and immunological determinants of influenza evolution. Nature 422:428–33.

Ferguson, N. M., S. Gupta, and R. M. Anderson. 1999. The effect of antibody-dependent enhancement on the transmission dynamics and persistence of multiple-strain pathogens. Proceedings of the National Academy of Sciences of the United States of America 96:790–94.

Ferguson, N. M., D. J. Nokes, and R. M. Anderson. 1996. Dynamical complexity in age-structured models of the transmission of measles virus. Mathematical Biosciences 138:101–30.

Focks, D. A., E. Daniels, D. G. Haile, and J. E. Keesling. 1995. A simulation model of the epidemiology of urban dengue fever: Literature analysis, model development, preliminary validation, and samples of simulation results. American Journal of Tropical Medicine and Hygiene 54:489–506.

Garcia-Garcia, J. A., J. A. Mira, J. Fernandez-Rivera, A. J. Ramos, J. Vargas, J. Macias, and J. A. Pineda. 2003. Influence of hepatitis C virus coinfection on failure of HIV-infected patients receiving highly active antiretroviral therapy to achieve normal serum beta-2 microglobulin levels. European Journal of Clinical Microbiology and Infectious Diseases 22:194–96.

Gilbert, S. C., M. Plebanski, S. Gupta, J. Morris, M. Cox, M. Aidoo, D. Kwiatkowski, et al. 1998. Association of malaria parasite population structure, HLA, and immunological antagonism. Science 279:1173–77.

Gog, J. R., and B. T. Grenfell. 2002. Dynamics and selection of many-strain pathogens. Proceedings of the National Academy of Sciences of the United States of America 99:17209–14.

Gog, J. R., R. B. Woodroffe, and J. A. Swinton. 2002. Disease in endangered metapopulations: The importance of alternative hosts. Proceedings of the Royal Society of London. Series B, Biological Sciences 269:671–76.

Gomes, M. G. M., G. F. Medley, and D. J. Nokes. 2002. On the determinants of population structure in antigenically diverse pathogens. Proceedings of the Royal Society of London. Series B, Biological Sciences 269:227–33.

Greenman, J. V., and P. J. Hudson. 1999. Multihost, multiparasite systems: An application of bifurcation theory. IMA Journal of Mathematics Applied in Medicine and Biology 16:333–67.

Grenfell, B. T., O. N. Bjornstad, and J. Kappey. 2001. Travelling waves and spatial hierarchies in measle epidemics. Nature 414:716–23.

Grenfell, B. T., O. G. Pybus, J. R. Gog, J. L. N. Wood, J. M. Daly, J. A. Mumford, and E. C. Holmes. 2004. Unifying the epidiological and evolutionary dynamics of pathogens. Science 303:327–32.

Gupta, S., N. M. Ferguson, and R. M. Anderson. 1998. Chaos, persistence and evolution of strain structure in antigenically diverse infectious agents. Science 280:912–15.

Halloran, M. E., I. Longini, A. Nizam, and Y. Yang. 2002. Containing bioterrorist smallpox. Science 298:1428–32.

Hethcote, H. 1988. Optimal ages of vaccination for measles. Mathematical Biosciences 89:29–52.

Holt, R. D. 1977. Predation, apparent competition, and structure of prey communities. Theoretical Population Biology 12:197–229.

Huang, Y., and P. Rohani. 2005. The dynamical implications of disease interference: Correlations and coexistence. Theoretical Population Biology 68:205–15.

Huang, Y., and P. Rohani. 2006. Age-structured effects determine interference between childhood infections. Proceedings of the Royal Society of London. Series B, Biological Sciences 273:1229–37.

Huovinen, P. 2001. Bacteriotherapy: The time has come. British Medical Journal 323:353–54.

Kamo, M., and A. Sasaki. 2002. The effects of cross-immunity and seasonal forcing in a multi strain epidemic model. Physica D 165:228–41.

Kawaguchi, I., A. Sasaki, and M. Boots. 2003. Why are dengue virus serotypes so distantly related? Enhancent and limiting serotype similarity between dengue virus strains.

Proceedings of the Royal Society of London. Series B, Biological Sciences 270: 2241–47.

Keeling, M. J., and C. A. Gilligan. 2000. Bubonic plague: A metapopulation model of a zoonosis. Proceedings of the Royal Society of London Series. Series B, Biological Sciences 267:2219–30.

Keeling, M. J., and P. Rohani. 2007. Modelling Infectious Diseases. Princeton, NJ: Princeton University Press.

Kermack, W. O., and A. G. McKendrick. 1927. A contribution to the mathematical theory of epidemics. Proceedings of the Royal Society of London. Series A 115:700–21.

Kirschner, D. 1999. Dynamics of co-infection with M-tuberculosis and HIV-1. Theoretical Population Biology 55:94–109.

Koelle, K., X. Rod, M. Pascual, M. Yunus, and G. Mostafa. 2005. Refractory periods and climate forcing in cholera dynamics. Nature 436:696–700.

Laing, J. S., and M. Hay. 1902. Whooping-cough: Its prevalence and mortality in Aberdeen. Public Health 14:584–98.

Levin, S. A., J. Dushoff, and J. B. Plotkin. 2004. Evolution and persistence of influenza A and other diseases. Journal of Mathematical Biology 188:17–28.

Lipsitch, M., T. Cohen, B. Cooper, J. M. Robins, S. Ma, L. James, G. Gopalakrishna, et al. 2003. Transmission dynamics and control of severe acute respiratory syndrome. Science 300:1966–70.

May, R., and M. Nowak. 1994. Superinfection, metapopulation dynamics, and the evolution of diversity. Journal of Theoretical Biology 170:95–114.

Nelson, K. E., C. M. Williams, and N. M. H. Graham. 2001. Infectious Disease Epidemiology: Theory and Practice. New York: Aspen Publishers.

Nisalak, A., T. P. Endy, S. Nimmannitya, S. Kalayanarooj, U. Thisayakorn, R. M. Scott, D. S. Burke, et al. 2003. Serotype-specific dengue virus circulation and dengue disease in Bangkok, Thailand, from 1973 to 1999. American Journal of Tropical Medicine and Hygiene 68:191–202.

Nokes, D. J., and J. Swinton. 1997. Vaccination in pulses: A strategy for global eradication of measles and polio? Trends in Microbiology 5:14–19.

Orenstein, W. A., M. J. Papania, and M. E. Wharton. 2004. Measles elimination in the United States. Journal of Infectious Diseases 189:S1–S3.

Read, A. F., and L. H. Taylor. 2001. The ecology of genetically diverse infections. Science 292:1099–102.

Reid, G., J. Howard, and B. S. Gan. 2001. Can bacterial interference precent infection? Trends in Microbiology 9:424–28.

Rohani, P., D. J. D. Earn, B. F. Finkenstadt, and B. T. Grenfell. 1998. Population dynamic interference among childhood Diseases. Proceedings of the Royal Society of London. Series B, Biological Sciences 265:2033–41.

Rohani, P., D. J. D. Earn, and B. T. Grenfell. 1999. Opposite patterns of synchrony in sympatric disease metapopulations. Science 286:968–71.

Rohani, P., C. J. Green, N. B. Mantilla-Beniers, and B. T. Grenfell. 2003. Ecological interference between fatal diseases. Nature 422:885–88.

Rohani, P., M. J. Keeling, and B. T. Grenfell. 2002. The interplay between determinism and stochasticity in childhood diseases. American Naturalist 159: 469–81.

Schenzle, D. 1984. An age-structured model of pre- and post-vaccination measles transmission. IMA Journal of Mathematics Applied in Medicine and Biology 1:169–91.

Schwartz, I. B., and H. L. Smith. 1983. Infinite subharmonic bifurcation in an SEIR epidemic model. Journal of Mathematical Biology 18:233–53.

Soper, H. E. 1929. The interpretation of periodicity in disease prevalence. Journal of Royal Statistical Society 92:34–73.

Tano, K., E. G. Hakansson, S. E. Holm, and S. Hellstrom. 2002. Bacterial interference between pathogens in otitis media and alpha-Haemolytic streptococci in an in vitro model. Acta Otolaryngology 122:78–85.

Taylor, L. H., D. Walliker, and A. F. Read. 1997. Mixed-genotype infections of malaria parasites: Within-host dynamics and transmission success of competing clones. Proceedings of the Royal Society of London. Series B, Biological Sciences 264:927–35.

Tomkins, D. M., J. V. Greenman, and P. J. Hudson. 2001. Differential impact of a shared natode parasite on two gamebird hosts: Implications for apparent competition. Parasitology 122:187–93.

Vasco, D. A., K. A. Crandall, and Y.-X. Fu. 2001. Molecular population genetics: Coalescent methods based on summary statistics. In "Computational and Evolutionary Analysis of HIV Molecular Sequences, ed. A. Rodrigo and J. Learn. Norwell, MA: Kluwer.

Vasco, D. A., H. J. Wearing, and P. Rohani. 2007. Tracking the dynamics of pathogen interactions: Modeling ecological and immune-mediated processes in a two-pathogen single-host system. Journal of Theoretical Biology 245: 9–25.

Wearing, H. J. and P. Rohani. 2006. Ecological and immunological determinants of dengue epidemics. Proceedings of the National Academy of Sciences of the United States of America 103:11802–807.

White, L. J., M. J. Cox, and G. F. Medley. 1998. Cross immunity and vaccination against multiple microparasite strains. IMA Journal of Mathematics Applied in Medicine and Biology 15:211–33.

CHAPTER FOUR

Influence of Eutrophication on Disease in Aquatic Ecosystems: Patterns, Processes, and Predictions

Pieter T. J. Johnson and Stephen R. Carpenter

SUMMARY

HABITAT ALTERATION AND DISEASE EMERGENCE are among the most pressing environmental concerns facing society. Aquatic ecosystems present a nexus of these issues. In this review, we explore interactions between a particularly widespread form of anthropogenic change, aquatic eutrophication, and the incidence of disease. Our goal was to examine broad-scale patterns in the types of parasites and pathogens favored under eutrophic conditions and how these patterns vary with environment, degree of eutrophication, and type of disease. We considered the consequences of eutrophication on macroparasitic, microparasitic, and noninfectious diseases of humans and wildlife in freshwater and marine ecosystems. We found that eutrophication has diverse effects on disease that depend on the type of pathogen, host species and condition, attributes of the aquatic system, and the degree of eutrophication. Eutrophication can alter disease dynamics by changing host density, host distribution, infection resistance, pathogen virulence, or toxicity, or by causing disease directly. Although low to moderate levels of eutrophication can increase species richness and parasite abundance, higher levels often lead to a decline of parasite species richness. However, even as parasite richness declines, pathology and disease sometimes become more severe. Eutrophication can elevate host stress, leading to increased infection, pathology, and mortality. Eutrophication also causes pronounced shifts in the types of parasites and pathogens in aquatic environments, favoring generalist or opportunistic parasites with direct or simple life cycles. Collectively, these pathogens may be particularly dangerous because they can continue to cause mortality even as their hosts decline, potentially leading to sustained epidemics or extirpations. Because nutrient loading will almost certainly become more severe and widespread in the coming decades, eutrophication will continue to be an important factor in the etiology of human and wildlife diseases. We emphasize the importance of studies integrating experiments and ecological modeling to identify mechanisms and feedbacks in the interactions between nutrient loading and host-pathogen dynamics.

INTRODUCTION

Widespread habitat alteration and the increased emergence of human and wildlife diseases are two pressing and interrelated environmental concerns facing society (Millennium Ecosystem Assessment 2005; National Research Council 2001). Aquatic ecosystems present a nexus of habitat alteration and disease emergence. Because of the diverse and productive nature of freshwater and marine systems, the necessity of water for the survival of human and wildlife populations, and the importance of water as a transmission medium, aquatic environments function as foci for disease epidemics. However, aquatic ecosystems are also subject to numerous physical, chemical, and biological changes as a result of anthropogenic activity. Eutrophication is one of the most pervasive forms of habitat alteration worldwide, often leading to dramatic changes in species diversity, biogeochemistry, and ecosystem services (Smith 1998; World Water Council 2003). Consequently, there is good reason to expect that eutrophication has important effects on the both the incidence and the severity of diseases affecting humans as well as wildlife.

In this review, we employ an empirical approach to explore interactions between aquatic eutrophication and disease. Our goal is not to review all cases but rather to examine the broad-scale patterns in what types of parasites and pathogens are favored under eutrophic conditions. We consider the consequences of eutrophication on macroparasitic, microparasitic, and noninfectious diseases in aquatic ecosystems. In light of their parallels, we do not differentiate between marine and freshwater systems, and interweave examples from each. Finally, we explore connections among eutrophication, disease, and human health, and highlight pressing areas for future investigation.

CAUSES AND CONSEQUENCES OF AQUATIC EUTROPHICATION

Eutrophication is caused by elevated levels of nutrients (particularly phosphorus and nitrogen) in aquatic systems and is associated with increased primary production, an increased importance of pelagic producers relative to benthic ones, decreased water transparency, and oxygen depletion (Bennett et al. 2001; Smith 1998). Adverse effects of eutrophication include toxic algal blooms, fish kills and deterioration of fisheries, increased costs of water treatment for municipal or industrial use, and loss of recreational amenities (Postel and Carpenter 1997). The degree of eutrophication varies with the magnitude of nutrient enrichment and other characteristics of the water body such as morphometry, hydraulic retention time, food web structure, and chemical variables (Carpenter et al. 2001; Kalff 2002; Smith 1998). Mild to moderate levels of productivity are associated with increases in species richness for phytoplankton, zooplankton, and fishes (Dodson et al. 2000). However, with more severe eutrophication, increased variability and a general species impoverishment is often observed (Dodson et al. 2000).

Eutrophication of lakes, rivers, and coastal oceans is a pervasive and growing problem worldwide (Millennium Ecosystem Assessment 2005). The increased use of agricultural fertilizers in the mid-twentieth century intensified eutrophication. Although global fertilizer use has stabilized since 1990, there is a considerable variation among nations, and eutrophication continues to be an expanding problem, particularly among developing countries (Millennium Ecosystem Assessment 2005; National Research Council 1992). Untreated sewage is a substantial nutrient source in some developing regions of the world, while diffuse (nonpoint) flows of nutrients are major inputs in developed countries and increasingly important in the developing world (Millennium Ecosystem Assessment 2005). Nonpoint sources of nutrients include synthetic fertilizers, manure from intensive livestock operations, and erosion runoff (Bennett et al. 2001; Carpenter et al. 1998; Howarth 2004). Atmospheric deposition of nitrogen oxides (derived primarily from fossil fuel combustion) and ammonia also contribute to eutrophication, particularly in marine and estuarine environments (Paerl 1997; Schlesinger 1997).

Eutrophication has proved to be a surprisingly persistent problem with few easy solutions. Although this is partially attributable to continued inputs of nutrients to aquatic ecosystems, characteristics of the phosphorus cycle also sustain eutrophication as a persistent state (Carpenter 2003). Phosphorus cycles slowly from soil to surface water, and overfertilized agricultural soils contain enough phosphorus to stabilize eutrophication for more than a century (Bennett et al. 2001). In freshwater and marine systems, phosphorus is retained in sediments and can be rapidly recycled to maintain eutrophication (Carpenter 2003; Van Capellen and Ingall 1994). Attempts to restore eutrophic lakes have often been thwarted by internal recycling of phosphorus (Carpenter 2003), demonstrating that a reduction in nutrient inputs may be insufficient to reverse eutrophication.

EUTROPHICATION AND DISEASE

For a simple theoretical host-parasite system, eutrophication functions to enhance the food resources and carrying capacity of a given host population (see Lafferty and Holt 2003). Resulting increases in host growth and population density might be expected to promote parasites of that species in at least two ways. First, a larger host population is more likely to meet or exceed the necessary transmission threshold for a given parasite to invade (Lafferty and Holt 2003), and second, better-fed hosts can facilitate increased growth and spread of their parasites (e.g., Pulkkinen and Ebert 2004; Smith et al. 2005). Added food resources may also decrease parasite-induced mortality, thereby increasing the period of time over which infected hosts can spread the parasite or the rate of parasite production. Although these patterns may hold for simple systems (e.g., marine viruses and bacteria; Middelboe et al. 2002), host-parasite dynamics are often considerably more complex, owing to the additional effects of eutrophication (direct and indirect), complex food webs, time lags, and variable modes of parasite transmission (e.g., Hall et al. 2005).

The effects of environmental stress on parasites and disease are diverse and depend critically on the type of pathogen, host community, and environmental conditions, as well as on the type and dose of a given stressor (MacKenzie et al. 1995; Lafferty and Holt 2003; Williams and MacKenzie 2003; Marcogliese 2004, 2005). Predicting the consequences of eutrophication on disease is difficult because they will vary with the host-parasite system, features of the aquatic environment, and degree of eutrophication. Some diseases may grow more severe while others may decline or disappear. The literature offers contradictory information on the effects of eutrophication on parasites. Lafferty (1997) reported that eutrophication was associated with increases in parasite infection in six of seven studied cases. Others, however, have noted substantial declines in parasite species richness and infection with ongoing eutrophication (e.g., Reimer 1995). In part, this is a consequence of the dual identity of eutrophication: at low levels, increased inputs of nutrients promote primary and secondary production in aquatic systems, effectively "fertilizing" the environment. At more extreme levels, however, declining water clarity and quality, widespread hypoxia, and toxic algal blooms make eutrophication behave more like a typical stressor. Moreover, the influence of eutrophication on a given pathogen may be mediated by the features of a given aquatic system (e.g., depth, flushing rate, and substrate) or by the host species involved (e.g., hypoxia-tolerant species) (Zander and Reimer 2002). The challenge therefore becomes identifying what types of parasite-host-environment combinations are likely to promote pathogenesis with increasing eutrophication. This requires an understanding of the types of disease-causing agents, which we review briefly in box 4.1.

MACROPARASITES AND EUTROPHICATION

The effects of eutrophication on fish macroparasites (particularly helminths) have been the subject of occasional study for more than fifty years (e.g., Wisniewski 1958). Investigative approaches have included regional comparisons of fish helminths between eutrophic and oligotrophic systems (Moser and Cowen 1991; Muzzall 1999; Valtonen et al. 1997, 2003), intensive, multihost examinations of the parasite fauna of particular systems (Esch 1971; Wisniewski 1958; Wootten 1973), and long-term monitoring of select systems exposed to variable nutrient-loading rates in both marine and freshwater environments (Kennedy et al. 2001; Marcogliese et al. 1990; Stromberg and Crites 1975; Zander and Reimer 2002). Interpretation of the results of these studies is often difficult because (1) the influence of eutrophication is confounded or correlated with other stressors (e.g., chemical contamination, salinity, food web changes), (2) the degree of eutrophication is rarely measured

---

Box 4.1 Parasites are often divided into *microparasites* (bacteria, viruses, some protozoans) and *macroparasites* (helminths, crustaceans, etc.) based on differences in size and the ability of the former group to replicate directly within the host (Anderson and May 1978). Many macroparasites have a complex life cycle involving multiple host species, often a combination of invertebrate intermediate hosts (e.g., mollusks, crustaceans) and vertebrate definitive hosts (e.g., fishes, birds, mammals) that are connected through predation or free-living infectious stages. Common examples of macroparasites include acanthocephalans (thorny-headed worms), trematodes (flatworms), nematodes (roundworms), and cestodes (tapeworms). Microparasites, such as bacteria, viruses, and some fungi, generally have direct life cycles and can replicate directly within a given host. *Obligate parasites* are organisms that can survive only in close association with their hosts (although many have brief free-living stages), whereas *facultative parasites* may be opportunistically parasitic or free-living. A *disease* is considered any deviation from a state of health and may be caused by infectious or noninfectious agents. A *pathogen* is the agent that causes or induces disease. A parasite may or may not cause disease, depending on the host species or condition, environment, and parasite involved. In addition to parasites, a variety of noninfectious agents can cause disease, including chemicals or toxins, errors in development, and trauma.

explicitly, making cross-system comparisons difficult, and (3) there is a lack of experiments, precluding identification of the mechanistic connections between eutrophication and disease.

Nevertheless, some general trends have emerged. In marine and freshwater systems, low to moderate levels of eutrophication are associated with increases in (1) generalist parasites with multiple, alternative host species, (2) overall parasite abundance and species richness, (3) parasite species that utilize benthic rather than pelagic intermediate hosts, and (4) parasites that use fishes as intermediate rather than definitive hosts, completing their life cycle in birds or mammals (allogenic over autogenic parasites) (Esch 1971; Galli et al. 1998, 2001; Marcogliese 2001; Zander and Reimer 2002). At least three mechanisms have been advanced to account for these changes. Increases in primary and secondary production associated with eutrophication can facilitate higher densities of intermediate host species (particularly benthic herbivores), thereby supporting concurrent increases in parasites that depend on those hosts (Zander et al. 2002). These increases in production can cause colonization by fish-eating birds and mammals, leading to increases in parasite species richness, particularly of allogenic species (Kennedy et al. 2001; Zander et al. 2002). Finally, increased food resources can enhance the ability of a host to tolerate infection, promoting host survival and prolonging the infectious period. A longer infectious period can facilitate increased parasite transmission and higher infection prevalence (Pulkkinen and Ebert 2004).

Some of the best examples of these patterns come from the Baltic Sea, which has become increasingly eutrophic since the early twentieth century owing to agricultural runoff. In long-term studies of the macroparasite community, Zander and Reimer (2002) reported "massive" increases in the abundance of generalist parasites during early periods of eutrophication. Hosts in more eutrophic areas exhibited higher levels of infection prevalence and parasite abundance (Zander et al. 2002). Increases were most pronounced in parasite species that utilized benthic herbivores (e.g., crustaceans and mollusks) as intermediate hosts owing to increases in benthic algal production (see also Galli et al. 1998, 2001). Eutrophic sites exhibited substantial increases in digenean infection of gobies owing to elevated densities of snail intermediate hosts (*Hydrobia* spp.), which could exceed 40,000 m^{-2} in eutrophic areas (Zander et al. 2002). Because of increased benthic production and high densities of crustacean intermediate hosts, these areas were also attractive to bird hosts and considered "epidemic centers" for transmission, with more than 85% of examined birds infected (Zander and Reimer 2002).

Eutrophication can also influence host distribution and infection dynamics. Declines in oxygen availability resulting from increased decomposition and respiration will affect the distribution of oxygen-sensitive

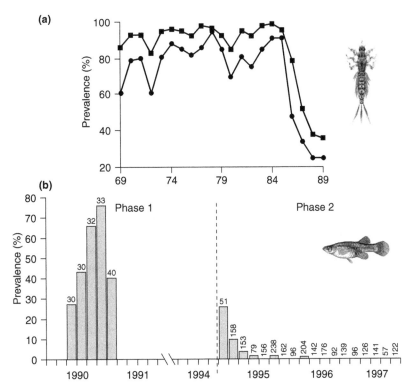

Figure 4.1. (a) Prevalence of *Crepidostomum cooperi* metacercariae in male (closed squares) and female (open squares) mayfly subimagoes (*Hexagenia limbata*) in Gull Lake, Michigan. A sewer system completed in 1984 is thought to have increased oxygen availability in the lake and allowed the mayflies to move into deeper water where they were not exposed to trematode cercariae. (From Marcogliese et al. 1990, with permission.) (b) Prevalence of mosquitofish (*Gambusia holbrooki*) infected with *Eustrongylides ignotus* at a site in north-central Florida historically subjected to sewage treatment plan inputs (Phase 1). Vertical broken line indicates time at which sewage input diverted (Phase 2, November 1994). Sample sizes for each date are listed above the bars (absence of number indicates no samples collected). (From Coyner et al. 2003.)

intermediate hosts. Between 1957 and 1972, Stromberg and Crites (1975) reported dramatic increases in a copepod-transmitted nematode, *Cathallanus oxycephalus*, in perch from Lake Erie as oxygen-sensitive mayflies declined and fish shifted their diets toward zooplankton. During this same period, copepods more than doubled in density. Similarly, Marcogliese et al. (1990) suggested that hypoxia associated with eutrophication of Gull Lake, Michigan, caused *Hexagenia* mayflies to shift their distribution toward shallower, more oxygen-rich waters, bringing them in closer proximity to sphaeriid clams—the first intermediate hosts of the trematode *Crepidostomum cooperi* (Esch et al. 1986). As a result, mayflies exhibited a fourfold increase in infection intensity between 1969 and 1976. With installation of an improved sewage system in 1984, however, eutrophication was reduced and infection in mayflies declined (figure 4.1a; Marcogliese 2005).

As eutrophication becomes increasingly severe, however, macroparasite species richness often declines (Galli et al. 1998, 2001; Marcogliese 2001; Zander and Reimer 2002). These declines are linked to decreased survival of free-living parasitic stages or to concurrent losses of intermediate hosts with increasing anoxia, sulfide levels, and light limitation. Nevertheless, some parasites thrive under these conditions. In particular, ectoparasites and generalist parasites with direct or simple (e.g., two host) life cycles may achieve high infection levels (Dušek et al. 1998; Zander and Reimer 2002), particularly in conjunction with reduced oxygen availability. Monogenetic trematodes, which typically are ectoparasitic on the gills and skin of fishes, can accelerate asphyxiation in oxygen-starved fish. Kuperman et al. (2001) hypothesized that two species of monogeneans parasitic on *Tilapia* were an important contributing factor to frequent die-offs of juvenile fish in the Salton Sea, a heavily eutrophic system with chronic oxygen depletion. Similar patterns have been noted for copepods that are ectoparasitic on fishes (Möller 1987).

Parasites that utilize intermediate hosts tolerant of low oxygen conditions can also thrive in strongly eutrophic systems. Often these hosts are planktonic or hyperbenthic species (Marcogliese 2004). Certain pulmonate snails (e.g., *Planorbella* spp.), for example, can withstand hypoxic conditions and exhibit increased growth and more frequent reproduction in eutrophic wetlands (Chase 2003a, b). These snails are intermediate hosts for the digenetic trematode *Ribeiroia ondatrae*, which causes severe limb malformations and elevated mortality in amphibian populations (Johnson and Sutherland 2003). Johnson and Chase (2004) found that periphyton production and planorbid snail biomass were positively correlated with *Ribeiroia* infection intensity and malformations among eleven species of amphibians. Similarly, with the eutrophication of Lake Constance, Hartmann and Nümann (1977) observed consistent increases in the frequency and severity of *Diplostomum* infections in yellow perch. In 1966, more than 50 tons of perch died in association with infection. The authors attributed the changes in infection to increases in macrophyte cover and snail density with elevated nutrient inputs.

Tubificid worms are another common pathogen host that can withstand extremely low levels of oxygen and promote disease. During increased nutrient loading of Lake Constance between 1950 and 1976, for example, tubificid worms increased sixfold (Hartmann and Nü-mann 1977). These worms function as important intermediate hosts for nematode parasites of fishes and birds that can cause severe disease. Infection by nematodes in the genus *Eustrongylides* (eustrongylidosis) can cause up to 80% nestling mortality in wading birds (Coyner et al. 2003). In comparisons of eutrophic and oligotrophic regions of the Great Lakes, Muzzall (1999) reported an increased prevalence and intensity of *Eustrongylides tubifex* in yellow perch from the eutrophic Saginaw Bay, which he linked to higher densities of tubificid worms and water birds. Infection was negatively correlated to perch growth and body condition. Similarly, Weisberg et al. (1986) found that *Eustrongylides* infections in mummichogs from Chesapeake Bay were greatest near sewage treatment plants. Again, infection was correlated with high densities of tubificid hosts (*Limnodrilus* sp.) and bird activity. In a particularly compelling example, Coyner et al. (2003) investigated the levels of *Eustrongylides ignotus* infection from mosquitofish in north-central Florida. Infection prevalence and intensity, tubificid density, and levels of phosphorus, nitrogen, and chlorophyll-*a* each were positively correlated with proximity to a sewage treatment plant. After sewage was diverted, infection levels in mosquitofish declined from 54% in 1990 to 0% in 1998 (figure 4.1b). The authors reported corresponding improvements in water quality during this same period.

EUTROPHICATION AND MICROPARASITES

Considerably less is known regarding the influence of eutrophication on microparasitic diseases, and we are aware of no comprehensive surveys or long-term monitoring efforts focused on the effects of eutrophication on microparasites. There are, however, numerous examples of associations between increased nutrient loading and outbreaks of microparasitic diseases. As noted for macroparasites, many of these instances involve generalist (or opportunistic) parasites with direct or simple life cycles (Khan and Thulin 1991; Overstreet 1993). Most examples involve hosts suffering from eutrophication-related stress. A variety of opportunistic protozoans and fungi colonize the gills and skin of fishes, sometimes leading to severe fish kills when combined with environmental hypoxia (Kuperman and Matey 1999; Kuperman et al. 2002). In Alabama, Overstreet (1993) observed mass mortality in spot (*Leiostomus xanthurus*) associated with heavy gill infestations by a dinoflagellate (*Amyloodinium ocellatum*). More than 40,000 dead or dying fishes were collected, and the author hypothesized that the increased infection and pathology was linked to nutrient loading and low oxygen conditions. Similarly, the colonial stalked ciliate *Epistylis* has been linked to high mortality and hemorrhagic lesions on the skin of fishes from ponds, reservoirs, and estuaries (Overstreet and Howse 1977). Infection was positively correlated with high organic content. Similar patterns were noted for oomycete infections of Atlantic menhadens in Florida (Overstreet 1988).

Interactions between marine viruses and bacteria offer some intriguing insights into the influence of trophic status on infection dynamics. Viruses are extremely abundant in aquatic environments and are thought to have dramatic effects on biogeochemical cycling, energy transfer, and host mortality (e.g., Suttle 2005; Middelboe, chapter 11, this volume). Limited evidence suggests that marine viral abundance increases with nutrient availability (Middelboe et al. 2002). In more productive environments, the frequency of visibly infected bacteria is greater than in oligotrophic environments, likely because of increased host abundance and parasite production (more viruses produced per infected cell) (Weinbauer et al. 1993). These relationships are highly variable, however, and more information is needed, particularly from highly eutrophic areas (Suttle 2005).

Nutrient enrichment can also increase transmission by providing supplementary resources directly to a pathogen. In coastal oceans and estuaries, for example, increased nutrient inputs have been linked to outbreaks of disease in coral reefs (Richmond 1993; Sutherland et al. 2004). Kim and Harvell (2002) found that the severity of a fungal pathogen (*Aspergillus*) on gorgonian sea corals in Florida was positively correlated with nitrogen concentrations and reduced water clarity. Similarly, Kuta and Richardson (2002), in a study of 190 sites on twelve patches of coral reef in the Florida Keys, found that black band disease (BBD) was linked to increased nitrite levels (see also Kaczmarsky et al. 2005). In one of the few experimental studies on eutrophication and disease, Bruno et al. (2003) conducted in situ experiments to evaluate how nutrient additions influenced the pathology of two coral diseases, aspergillosis and yellow band disease. They found that supplementary nutrients nearly doubled the infection severity and tissue loss caused by both pathogens (figure 4.2). Voss and Richardson (2005) reported that additions of nitrogen and phosphorus doubled the rate of BBD progression and coral tissue loss in laboratory and field experiments. Thus, while anthropogenic eutrophication is not necessary for disease transmission, it can exacerbate the severity of infections and resulting pathology, possibly by providing parasites with supplementary resources.

Finally, eutrophication influences microparasitic diseases through changes in intermediate host densities. As observed with macroparasites, these intermediate hosts are typically invertebrates resistant to low oxy-

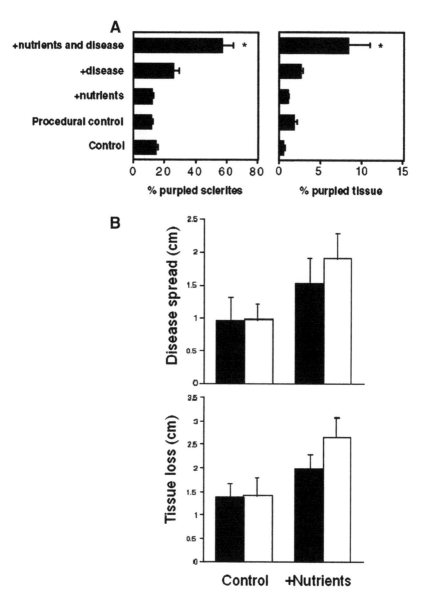

Figure 4.2. (a) Individual and combined effects of in situ nutrient additions and a fungal disease (aspergillosis) on two measures of pathology in Caribbean sea fans. (b) Influence of nutrient additions on (top) yellow band disease advancement and (bottom) host tissue loss in two species of Caribbean reef-building corals (black bars = *Montastraea franksii*, open bars = *M. annularis*). (Reprinted from Bruno et al. 2003. Copyright Blackwell Publishing Ltd.)

gen levels. Myxozoan parasites that utilize tubificid worms and fish hosts are one example. Marcogliese and Cone (2001) found the highest levels of myxozoan infection in fish directly downstream of Montreal. Fecal coliform counts were also higher in this area relative to sites upstream, and the authors attributed the heightened infection to elevated densities of tubificid hosts. Whirling disease, the myxozoan (*Myxobolus cerebralis*) disease that has caused major economic losses to the trout fisheries of Colorado and Montana, has also been linked to nutrient enrichment (Duffield et al. 1999; Thompson et al. 1999). Hatchery ponds are often highly productive environments with high densities of tubificid worms, creating conditions that have fostered disease outbreaks that sometimes spread to wild populations (Markiw 1992). As a final example, epidemics of avian cholera, which can cause enormous die-offs in numerous bird species, have been associated with increased nutrient levels. The bacterium responsible for infection, *Pasteurella multiocidia*, becomes concentrated in maggots, which are subsequently ingested or inhaled by foraging birds. With increased organic matter and warm temperatures, hypoxia-tolerant maggots can achieve large populations. Coupled with high densities of migratory birds, these conditions can lead to epidemic mortality, and outbreaks of avian cholera have increased significantly in the United States since the 1970s (Botzler 1991; Friend 2002; Holmes 1996).

EUTROPHICATION AND NONINFECTIOUS DISEASE

In addition to links between parasites and eutrophication, excess nutrient runoff is associated with a variety of noninfectious causes of morbidity and mortality. The most obvious and direct of these is hypoxia and subsequent suffocation: decomposition of algae following blooms frequently leads to oxygen depletion, which can cause massive fish kills in poorly flushed lakes and coastal estuaries (Burkholder and Glasgow 1997; Overstreet and Howse 1977; Smith 1998). In some cases, however, algal blooms can themselves cause disease. In freshwater systems, the connection between nutrients (particularly phosphorus) and toxic, often nitrogen-fixing cyanobacteria is well established (Mohammed et al. 2003 Pitois et al. 2001; Smith 1998). Various cyanobacteria (e.g., *Anabaena*, *Aphanizomenon*, *Cylindrospermopsis*) produce potent neuro- and hepatotoxins that can have acute, lethal, or chronic effects when ingested (Carmichael 2001; Landsberg 2002; Pitois et al. 2001; Smayda 1997). Noxious cyanobacterial blooms cause mortality and morbidity in livestock, wildlife, dogs, and, on rare occasions, humans (Mohammed et al. 2003; Pitois et al. 2001). In 2000, a Wisconsin youth died after ingesting water from a golf course pond. His death was attributed to liver poisoning from ingestion of toxins produced by *Anabaena*.

In marine systems, there is growing recognition of a worldwide increase in the frequency, diversity, and persistence of harmful algal blooms (Hallegraeff 1993; Paerl 1997). These blooms can be directly toxic or may become accumulated in shellfish or finfish, leading to a number of human and wildlife diseases. Toxins produced by certain diatoms and dinoflagellates have caused die-offs of both wild and cultured fishes and shellfishes (Burkholder 1998; Hallegraeff 1993; Horner et al. 1997). Blooms of *Pseudo-nitzchia australis* produce domoic acid, a potent neurotoxin that can become concentrated in planktivorous anchovies and shellfish. Over the past decade, domoic acid poisoning associated with *P. australis* blooms has caused die-offs of sea lions and pelicans in western North America (Lefebvre et al. 1999). Similarly, outbreaks of the dinoflagellate *Pfiesteria* spp. have been associated with massive die-offs of estuarine fish along the Atlantic coast (Burkholder 1998; Vogelbein et al. 2002).

Although less well established than in freshwater systems, blooms of toxic marine algae are associated with eutrophic conditions, particularly in poorly flushed or closed systems (e.g., estuaries and inland seas) (Anderson 1997; Horner et al. 1997). In Hong Kong Harbor, an eightfold increase in red tides between 1976 and 1986 was correlated with a sixfold increase in the human population and a 2.5-fold increase in nutrients (Hallegraeff 1993). Similarly, in Japan's Seto Inland Sea, a sevenfold increase in red tides was correlated with a 2.5-fold increase in chemical oxygen demand owing to inputs of untreated sewage and domestic runoff. The most severe bloom in 1972 caused the death of more than 14 million cultured yellowtail fish (*Seriola quinquiradiata*) (Hallegraeff 1993). However, following elimination of phosphate detergents and the introduction of secondary sewage treatment, red tides declined substantially. Similarly, Burkholder and Glasgow (1997) found that the density of nontoxic zoospores of *Pfiesteria* was positively correlated with ambient levels of total phosphorus in North Carolina estuaries. In laboratory experiments, inorganic phosphorus directly stimulated the release of toxic zoospores, leading the authors to suggest a role for eutrophication in precipitating *Pfiesteria*-associated fish die-offs (Burkholder and Glasgow 1997, 2001).

The mechanisms through which eutrophication affects marine algal blooms can involve direct or indirect pathways. Increased inputs of nutrients can promote harmful algal blooms and their toxicity through changes in elemental ratios. Specifically, declines in the Si:P ratio (as occurs with increasing inputs of phosphorus) can favor nuisance dinoflagellates over siliceous diatoms (Paerl 1997; Smith 1998). Hallegraeff

(1993) cited dramatic declines in Si:P ratios from the Black Sea, North Sea, and the coast of Australia as a factor contributing to increasing harmful algal blooms in these regions. Shifts in elemental ratios may select for increased toxicity, even among normally nontoxic alga (e.g., *Chrysochromulina*; Hallegraeff 1993). Increased nitrogen loading can promote domoic acid production in *Pseudo-nitzschia* via changes in the ratios of nitrogen, phosphorus, and silicon (Paerl 1997). Nutrient inputs can also promote blooms of harmful microalgae indirectly. Coastal eutrophication is commonly associated with increases in fleshy macroalgae, which function as a substrate for benthic dinoflagellates that cause ciguatera finfish poisoning (CFP; Hallegraeff 1993). Through effects on macroalgae, increased nutrient loading may have indirectly contributed to major increases in CFP cases among humans in French Polynesia between 1960 and 1984 (more than 24,000 cases) (Hallegraeff 1993).

Although most examples of noninfectious disease are direct, avian botulism emphasizes the potential for direct and indirect interactions among pathogens, hosts, and eutrophication. Avian botulism type C, one of the most important diseases of migratory birds worldwide, is caused by ingestion of a bacterially produced toxin (botulinum). Conditions favoring bacterial growth and toxin production include warm temperatures, reduced oxygen, and an abundance of decaying organic matter, all of which are also associated with eutrophication. Birds consume toxin-containing bacteria through ingestion of maggots (fleshfly or blowfly) or fish (Chattopadhyay et al. 2003; Friend 2002). In California's Salton Sea, which has become increasingly eutrophic due to heavy inputs of agricultural runoff, intense algal blooms frequently lead to anoxic conditions and major fish kills (Riedel et al. 2002). Anoxic conditions also promote growth of botulinum bacteria, and migratory birds that consume dead and dying fish frequently suffer from botulism poisoning (Friend 2002). Avian botulism type C has caused enormous die-offs of waterfowl at the Salton Sea, including several threatened or endangered species. Avian mortality from botulism and other diseases has become more frequent and more severe in recent years, and Friend (2002) estimated that, since 1989, more than 300,000 birds have died around the Salton Sea—more than in all the previous eighty-two years combined.

Eutrophication and Human Health

We have focused primarily on diseases of wildlife, but eutrophication can also have profound direct and indirect consequences for human health and disease. Increased nutrient loading in freshwater and marine systems can promote macroparasitic, microparasitic, and noninfectious diseases in human populations. We include brief examples of each for illustration.

MACROPARASITES

In parts of Russia, France, the United States, and Canada, lake eutrophication has been associated with sometimes dramatic increases in cercarial dermatitis ("swimmer's itch"), a condition caused by invasion of human skin by trematode cercariae. These parasites normally infect birds, and their penetration of human skin induces a strong immune response leading to the destruction of the invading parasite and severe skin irritation. Increases in human infections have been associated with elevated densities of snail hosts (commonly *Lymnaea* spp.) through nutrient-induced increases in benthic production and snail habitat (Beer and German 1993; de Gentile et al.1997; Scheele et al. 1999; D. Marcogliese, personal communication).

MICROPARASITES

Eutrophication may influence malaria dynamics through its effects on vector habitat. In Belize, increased nutrient runoff associated with the replacement of forest habitats with fertilized agriculture favors dominance by tall, densely growing macrophytes (e.g., *Typha* spp.) in surrounding wetlands. These environments lead to a proportional increase in the mosquito *Anopheles vestitipennis*, which is a superior malaria vector owing to its predilection for feeding on humans (figure 4.3; Rejmánková et al. 2006). There is also limited evidence linking outbreaks of cholera to cultural eutrophication of coastal waters. *Vibrio cholerae* bacteria become concentrated in fishes, shellfishes, and in biofilms on the surface of crustacean zooplankton (Colwell 1996; Epstein 1993). In Bangladesh, blooms of phytoplankton caused by increased nutrient availability and warmer temperatures promote zooplankton, which are temporally correlated with annual outbreaks of human cholera (Colwell and Huq 2001; Cottingham et al. 2003; Epstein 1993).

NONINFECTIOUS DISEASES

In freshwater, toxins produced by cyanobacteria can cause allergic skin reactions and gastrointestinal illnesses, and have been linked to liver cancer pathogenesis (Carmichael 2001). These toxins are not removed by conventional water treatment methods (Carmichael 2001; Pitois et al. 2001). In marine systems, harmful algal blooms are associated with numerous human illnesses, including amnesic, paralytic, neurological, and diarrheic shellfish poisoning and ciguatera finfish poisoning (Burkholder 1998; Hallegraeff 1993; Paerl 1997). Hoagland et al. (2002) estimated that the United States has more than 60,000 toxic exposures annually,

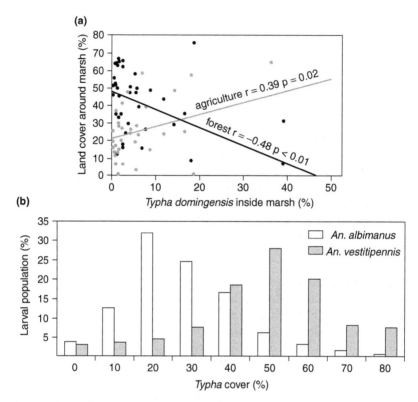

Figure 4.3. Malaria vector patterns as a function of land use and aquatic vegetation patterns in Belize. (a) Correlations between land use type and the percentage of *Typha* vegetation in surrounding marshlands. (b) Relative dominance
of two mosquito vectors as related to the percentage of *Typha* in a given marsh.
With increasing agricultural runoff and the resulting increase in *Typha* cover
within wetlands, the more efficient malaria vector (*Anopheles vestitpennis*) becomes proportionally more abundant. (Reprinted from Rejmánková et al. 2005,
with permission. Copyright Oxford University Press.)

with up to 6,500 deaths. Costs associated with public health, shellfish re-
calls, and decreased tourism approach U.S. $50 million annually. Finally,
increased nutrient loading may itself pose a direct threat to human health
(Hubbard et al. 2004). Elevated nitrates in drinking water can cause reproductive problems, methemoglobinemia (blue-baby syndrome) and
heightened cancer risk (Hubbard et al. 2004; Townsend et al. 2003; Wu
et al. 1999). Owing to fertilizer application and runoff from livestock production (particularly from swine and poultry), an estimated 10%–20% of

U.S. groundwater sources exceed the safe drinking water level for nitrates (Townsend et al. 2003).

FEEDBACKS AND THE FUTURE

Numerous diseases affecting both humans and wildlife have increased in incidence or severity in recent decades, frequently resulting from changes in the ecological interactions among a pathogen, its hosts, and the environment in which they co-occur (Daszak et al. 2000). The importance of incorporating ecology into the study of parasites and emerging diseases has been emphasized with increasing urgency in recent years. In their synthesis of the *Grand Challenges in Environmental Sciences*, the National Research Council (2001) listed infectious disease as one of the eight most pressing environmental issues, advocating a systems-level approach to understanding disease emergence. Correspondingly, in aquatic ecosystems, eutrophication is considered one of the greatest threats to global freshwater resources (Smith 1998; World Water Council 2003). Because human and wildlife populations depend on freshwater for survival, and many diseases are transmitted via waterborne stages, interactions between pathogens and aquatic eutrophication are an important frontier for understanding current and forecasting future disease epidemics.

Eutrophication will almost certainly become more widespread and more severe in the near future. Although awareness and technological innovations have slowed the problem in some regions, ongoing patterns of atmospheric deposition of reactive nitrogen, losses of wetland and riparian areas, the increasing use of fertilizers in developing nations, growing livestock populations, and urbanization all suggest that eutrophication will continue to expand (Millennium Ecosystem Assessment 2005). Moreover, even if the contributing drivers are reversed, eutrophication tends to be a persistent condition. Phosphorus levels decline only slowly in overfertilized soils. Some aquatic systems, especially lakes and reservoirs, recycle phosphorus effectively and can thereby remain eutrophic even when phosphorus inputs are reduced. Therefore, we predict that eutrophication will remain an important factor in the etiology of aquatic diseases for decades to come.

Examples in this review illustrate how eutrophication can influence disease, but we know very little about feedbacks in the other direction: can epidemic disease in aquatic systems cause or exacerbate eutrophication? Mass mortality and the resulting decomposition of fish, algae, and birds have the capacity to accelerate nutrient recycling or introduce new nutrients into aquatic environments, potentially contributing to further eutrophication and additional epidemics. For example, Glibert et al. (2005) found that, following the death and decomposition of 3,000 tons of mullet in Kuwait Bay from anoxia and bacterial infection, the resulting nutrient release fueled a secondary toxic algal bloom, leading to further die-offs in fishes. Similarly, epidemic mortality in waterfowl could promote eutrophication and further outbreaks of avian botulism. Positive feedback loops such as this one could have important implications for wildlife management and disease control.

Conclusions

Eutrophication has diverse effects on disease that depend on the type of pathogen, host species and condition, attributes of the aquatic system, and the degree of eutrophication. Our brief review reveals similarities among the responses of macroparasitic, microparasitic, and noninfectious diseases of humans and wildlife to nutrient enrichment. Eutrophication can influence disease dynamics by changing the density of suitable hosts, shifting the distribution of hosts, altering the resistance of species to infection, changing host or vector habitats, providing pathogens with supplementary resources, selecting for increased virulence or toxicity, or by causing disease directly (figure 4.4). Low to moderate levels of eutrophication in marine and freshwater systems can increase species richness and infection prevalence (particularly for macroparasites) (figure 4.5a). These changes are hypothesized to result from increases in host densities (with elevated primary and secondary benthic production), colonization by mobile carnivores (e.g., fishes, birds, mammals), and an increased ability of hosts to tolerate—and therefore spread—infection. At higher levels of eutrophication, parasite species richness often declines, possibly resulting from loss of intermediate host species and specialist parasites with decreasing oxygen and light availability (figure 4.5a and b). Infection prevalence may increase or decrease, depending on the parasite species involved.

However, even as parasite species richness declines, pathology and disease may become more severe under these conditions, for several reasons (figure 4.5c). First, increases in hypoxia, light limitation, and sulfide levels associated with eutrophication can increase stress in host populations. Generalist or opportunistic pathogens (e.g., monogenetic trematodes, ciliates, bacteria, fungi) with simple life cycles can cause severe pathology and elevated mortality in stressed hosts. Second, declines in species richness and evenness associated with eutrophication may be accompanied by increases in the abundance of select, highly tolerant taxa (e.g., some snails, tubificid worms). When these organisms func-

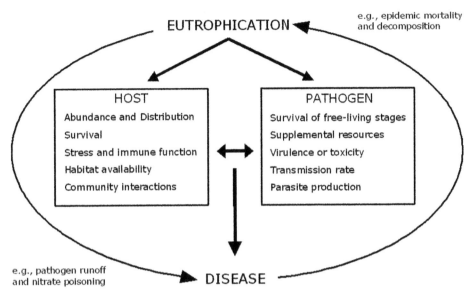

Figure 4.4. Pathways through which eutrophication can affect host-parasite interactions and, in turn, disease. Presented are examples of how eutrophication can affect attributes of host and pathogen (boxes) or cause disease directly (outside arrow). Epidemic disease can influence eutrophication through mass die-offs and subsequent decomposition of organic matter (outside arrow). The degree of eutrophication, type of pathogen, and characteristics of the aquatic system will determine whether effects are positive, negative or neutral.

tion as intermediate, alternative, or reservoir hosts for pathogens, infections in secondary or definitive vertebrate hosts can increase, potentially facilitating epidemics. Finally, eutrophic environments may select for increased virulence and toxicity among certain pathogens. Limited evidence suggests that changes in elemental ratios (nitrogen, phosphorus, silicon) resulting from anthropogenic nutrient inputs enhances or selects for toxicity in some algal species. Increased growth rates and densities associated with eutrophic conditions may also create environments conducive to pathogen evolution. Fish farms and coastal aquaculture, for example, which often combine increased nutrient inputs (via fish meal) and high host densities, are considered the single most important factor in the emergence and redistribution of fish diseases and have sometimes favored the evolution of novel, more virulent pathogens (e.g., salmon anemia virus) that can spread to wild stocks (Dobson and Foufopolous 2001; Naylor et al. 2000; Stewart 1991).

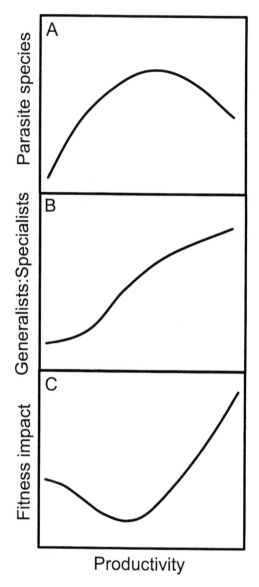

Figure 4.5. Hypothesized relationships between productivity and disease. (a) Parasite diversity. Parasite species richness is expected to exhibit a quadratic relation to eutrophication, as reported for other taxa. (b) Ratio of generalist to specialist parasites. Generalist parasites (those with multiple, alternative hosts) are favored relative to highly specialized parasites with increasing eutrophication because of increasing environmental variability. (c) Fitness impacts on infected hosts. Initial increases in resource availability may enhance a host's ability to tolerate (or resist) infection. At high levels of eutrophication, the combined effects of environmental stress and invasion by opportunistic pathogens increase the fitness consequence of infection.

Eutrophication also causes pronounced shifts in the types of parasites and pathogens in aquatic environments. In marine and freshwater environments, increased nutrient loading appears to shift the community away from parasites with narrow host requirements (i.e., specialists), complex life cycles, or intermediate hosts sensitive to reductions in light or oxygen availability. Instead, these conditions favor generalist or opportunistic parasites with direct or simple life cycles that are less dependent on the dynamics of any one host species (figure 4.5b). These parasites are expected to be more tolerant of the increased environmental variability that can accompany eutrophication. Many of these are ectoparasitic monogeneans, protozoans, fungi, or crustaceans that, often in combination with hypoxic conditions, cause increases in host mortality and morbidity (figure 4.5c). Noninfectious diseases such as harmful algal blooms or botulism poisoning represent the extreme example, in that the dynamics of the pathogen are completely divorced from the dynamics of the focal species exhibiting disease, preventing negative feedbacks from increased host mortality. Collectively, these pathogens may be particularly dangerous because they can continue to cause mortality even as their hosts decline, leading to sustained epidemics or extirpations.

Exceptions to these patterns are numerous, and there is much we do not understand regarding the relationships between eutrophication and disease. In particular, few studies have directly measured the degree of eutrophication in relation to disease, and the thresholds demarcating "light" versus "severe" eutrophication for pathogens are ill-defined. Factors that mediate or exacerbate the influence of eutrophication on diseases are poorly understood, particularly the importance of system attributes, predator community, pathogen type, and host condition. Studies that have examined the effects of eutrophication on parasites are often confounded by species introductions, chemical contamination, food web changes, and shifts in water chemistry. As a result, despite widespread associations between disease and eutrophication, the mechanisms through which eutrophication influences disease dynamics remain largely conjectural. Ecosystem-scale experiments combined with mechanistic modeling studies are needed to disentangle these interactions.

Relationships among eutrophication, species diversity, and disease also remain largely speculative. Although eutrophication is associated with a decline in the diversity of certain taxa (e.g., Dodson et al. 2000), the consequences of these changes for parasite transmission and disease are difficult to generalize. For some systems, declines in host species diversity are associated with increases in infection prevalence and disease severity (e.g., Clay et al., chapter 7, this volume; Mitchell et al. 2002;

Ostfeld and Keesing 2000; Begon, chapter 1, this volume), but these patterns are likely to vary considerably among pathogens as a function of host specificity, mode of transmission, and the responses of non-host groups. Too few epidemiological studies have examined the influence of community structure on transmission or the corresponding effects of disease on ecological communities. Understanding the reciprocal interactions between community structure and disease epidemics is a priority for subsequent investigation.

ACKNOWLEDGMENTS

We thank Elena Bennett, Julia Butzler, Jon Chase, Kathy Cottingham, Kevin Lafferty, Matthias Middelboe, Jake Vander Zanden, and especially David Marcogliese for comments and suggestions helpful in shaping this manuscript. Eliska Rejmánková generously shared the results of ongoing research on malaria dynamics in Belize.

LITERATURE CITED

Anderson, D. M. 1997. Bloom dynamics of toxic *Alexandrium* species in the northeastern US. Limnology and Oceanography 42:1009–22.

Anderson, R. M., and R. M. May. 1978. Regulation and stability of host-parasite population interactions: I. Regulatory processes. Journal of Animal Ecology 47:219–47.

Beer, S. A., and S. M. German. 1993. Ecological prerequisites in the worsening of the cercariosis situation in the cities of Russia (Moscow region as an example). Parazitologiia 27:441–49.

Begon, M. 2006. Effects of host diversity on disease dynamics. *In* Ecology of Infectious Disease: Interactions Between Diseases and Ecosystems, ed. V. Eviner, F. Keesing, and R. Ostfeld. Princeton, NJ: Princeton University Press.

Bennett, E. M., S. R. Carpenter, and N. F. Caraco. 2001. Human impact on erodable phosphorus and eutrophication: A global perspective. BioScience 51:227–34.

Botzler, R. G. 1991. Epizootiology of avian cholera in wildfowl. Journal of Wildlife Diseases 27:367–95.

Bruno, J. F., L. E. Petes, C. D. Harvell, and A. Hettinger. 2003. Nutrient enrichment can increase the severity of coral diseases. Ecology Letters 6:1056–61.

Burkholder, J. M. 1998. Implications of harmful microalgae and heterotrophic dinoflagellates in management of sustainable marine fisheries. Ecological Applications 8:S37–S62.

Burkholder, J. M., and H. B. Glasgow. 1997. *Pfiesteria piscicida* and other *Pfiesteria*-like dinoflagellates: Behavior, impacts, and environmental controls. Limnology and Oceanography 42:1052–75.

Burkholder, J. M., and H. B. Glasgow. 2001. History of toxic *Pfiesteria* in North Carolina estuaries from 1991 to the present. BioScience 51:827–41.

Carmichael, W. W. 2001. Health effects of toxin-producing cyanobacterial, the cyanoHABS. Human and Ecological Risk Assessment 7:1393–407.

Carpenter, S. R. 2003. Regime Shifts in Lake Ecosystems: Pattern and Variation. Volume 15 in the Excellence in Ecology Series. Oldendorf/Luhe, Germany: Ecology Institute.

Carpenter, S. R., N. F. Caraco, D. L. Correll, R. W. Howarth, A. N. Sharpley, and V. H. Smith. 1998. Nonpoint pollution of surface waters with phosphorus and nitrogen. Ecological Applications 8:559–68.

Carpenter, S. R., J. J. Cole, J. R. Hodgson, J. F. Kitchell, M. L. Pace, D. Bade, K. L. Cottingham, T. E. Essington, J. N. Houser, and D. E. Schindler. 2001. Trophic cascades, nutrients and lake productivity: Whole-lake experiments. Ecological Monographs 71:163–186.

Chase, J. M. 2003a. Strong and weak trophic cascades along a productivity gradient. Oikos 101:187–95.

Chase, J. M. 2003b. Experimental evidence for alternative stable equilibria in a benthic pond food web. Ecology Letters 6:733–41.

Chattopadhyay, J., P. D. N. Srinivasu, and N. Bairagi. 2003. Pelicans at risk in Salton Sea: An eco-epidemiological model. II. Ecological Modelling 167:199–211.

Colwell, R. R. 1996. Global climate change and infectious disease: The cholera paradigm. Science 274:2025–31.

Colwell, R., and A. Huq. 2001. Marine ecosystems and cholera. Hydrobiologia 460:141–45.

Cottingham, K. L., D. A. Chiavelli, and R. K. Taylor. 2003. Environmental microbe and human pathogen: The ecology and microbiology of *Vibrio cholerae*. Frontiers in Ecology and the Environment 1:80–86.

Coyner, D. F., M. G. Spalding, and D. J. Forrester. 2003. Influence of treated sewage on infections of *Eustrongylides ingnotus* (Nematoda: Dioctophymatoidea) in eastern mosquitofish (*Gambusia holbrooki*) in an urban watershed. Comparative Parasitology 70:205–10.

Daszak, P., A. A. Cunningham, and A. D. Hyatt. 2000. Emerging infectious diseases of wildlife: Threats to biodiversity and human health. Science 287:443–49.

de Gentile, L., H. Picot, P. Bourdeau, R. Bardet, A. Kerjan, M. Piriou, A. Le Guennic, C. Bayssade-Dufour, D. Chabasse, and K. E. Mott. 1997. Cercarial dermatitis in Europe: A new public health problem? Bulletin of the World Health Organization 74:159–63.

Dobson, A., and J. Foufopoulos. 2001. Emerging infectious pathogens of wildlife. Philosophical Transactions of the Royal Society, Series B 356:1001–12.

Dodson, S. I., S. E. Arnott, and K. L. Cottingham. 2000. The relationship in lake communities between primary productivity and species richness. Ecology 81:2662–79.

Duffield, J. W., D. A. Patterson, C. J. Neher, and J. B. Loomis. 1999. Economic consequences of whirling disease in Montana and Colorado trout fisheries. *In* Proceedings of the Whirling Disease Symposium, Research and Management Perspectives, Missoula, MN.

Dušek, L., M. Gelnar, and Š. Šebelová. 1998. Biodiversity of parasites in a freshwater environment with respect to pollution: Metazoan parasites of chub (*Leuciscus cephalus* L.) as a model for statistical evaluation. International Journal for Parasitology 28:1555–71.

Epstein, P. R. 1993. Algal blooms in the spread and persistence of cholera. BioSystems 31:209–21.

Esch, G. W. 1971. Impact of ecological succession on the parasite fauna in centrarchids from oligotrophic and eutrophic ecosystems. American Midland Naturalist 86:160–68.

Esch, G. W., T. C. Hazen, D. J. Marcogliese, T. M. Goater, and A. E. Crews. 1986. A long-term study on the population biology of *Crepidostomum cooperi* (Trematoda: Allocreadidae) in the burrowing mayfly, *Hexagenia limbata* (Ephemeroptera). American Midland Naturalist 116:304–14.

Friend, M. 2002. Avian disease at the Salton Sea. Hydrobiologia 473:293–306.

Galli, P., G. Crose, L. Mariniello, M. Ortis, and S. D'Amelio. 2001. Water quality as a determinant of the composition of fish parasite communities. Hydrobiologia 452:173–79.

Galli, P., L. Mariniello, G. Crosa, M. Ortis, A. Occhipinti Ambrogi, and S. D'Amelio. 1998. Populations of *Acanthocephalus anguillae* and *Pomophorhynchus laevis* in rivers with different pollution levels. Journal of Helminthology 72:331–35.

Glibert, P. M., J. J. Evans, P. H. Klesius, C. A. Shoemaker and J. A. Alexander. 2005. Comparison of two fish-kill events involving human bacterial pathogens: Influence of environmental stressors and harmful algae. Abstract presented at the annual meeting of the American Society of Limnology and Oceanography, Salt Lake City, UT.

Hall, S. R., M. A. Duffy, and C. E. Caceres. 2005. Selective predation and productivity jointly drive complex behavior in host-parasite systems. American Naturalist 165:70–81.

Hallegraeff, G. M. 1993. A review of harmful algal blooms and their apparent global increase. Phycologia 32:79–99.

Hartmann, J., and W. Nümann. 1977. Percids of Lake Constance, a lake undergoing eutrophication. Journal of the Fisheries Resources Board of Canada 34:1670–77.

Hoagland, P., D, M. Anderson, Y. Kaoru, and A. W. White. 2002. The economic effects of harmful algal blooms in the Unites States: Estimates, assessment issues, and information needs. Estuaries 25:819–37.

Holmes, J. C. 1996. Parasites as threats to biodiversity in shrinking ecosystems. Biodiversity and Conservation 5:975–83.

Horner, R. A., D. L. Garrison, and F. G. Plumley. 1997. Harmful algal blooms and red tide problems on the US west coast. Limnology and Oceanography 42:1076–88.

Howarth, R.W. 2004. Human acceleration of the nitrogen cycle: Drivers, consequences and steps toward solutions. Water Science and Technology 49:7–13.

Hubbard, R. K., J. M. Sheridan, R. Lowrance, D. D. Bosch, and G. Vellidis. 2004. Fate of nitrogen from agriculture in the southeastern coastal plain. Journal of Soil and Water Conservation 59:72–86.

Johnson, P. T. J., and J. M. Chase. 2004. Parasites in the food web: Linking amphibian malformations and aquatic eutrophication. Ecology Letters 7:521–26

Johnson, P. T. J., and D. R. Sutherland. 2003. Amphibian deformities and Ribeiroia infection: An emerging helminthiasis. Trends in Parasitology 19:332–35.

Kaczmarsky, L. T., M. Draud, and E. H. Williams. 2005. Is there a relationship between proximity to sewage effluent and the prevalence of coral diseases? Caribbean Journal of Science 41:124–37.

Kalff, J. 2002. Limnology. Upper Saddle River, NJ: Prentice Hall.

Kennedy, C. R., P. C. Shears, and J. A. Shears. 2001. Long-term dynamics of Ligula intestinalis and roach Rutilus rutilus: A study of three epizootic cycles over thirty-one years. Parasitology 123:257–69.

Khan, R. A., and J. Thulin. 1991. Influence of pollution on parasites of aquatic animals. Advances in Parasitology 30:201–38.

Kim, K., and C. D. Harvell. 2002. Aspergillosis of sea fan corals: Disease dynamics in the Florida Keys. In The Everglades Florida Bay and Coral Reefs of the Florida Keys: An Ecosystem Sourcebook, ed. Porter J. W. and K. G. Porter, 813–24. New York: CRC Press.

Kuperman, B. I., and V. E. Matey. 1999. Massive infestation by Amyloodinium ocellatum (Dinoflagellida) of fish in a highly saline lake, Salton Sea, California, USA. Diseases of Aquatic Organisms 39:65–73.

Kuperman, B. I., V. E. Matey, and S. B. Barlow. 2002. Flagellate Cryptobia branchialis (Bodonida: Kinetoplastida), ectoparasite of tilapia from the Salton Sea. Hydrobiologia 473:93–102.

Kuperman, B. I., V. E. Matey, and S. H. Hurlbert. 2001. Parasites of fish from the Salton Sea, California, USA. Hydrobiologia 466:195–208.

Kuta, K. G., and L. L. Richardson. 2002. Ecological aspects of black band disease of corals: Relationships between disease incidence and environmental factors. Coral Reefs 21:393–98.

Lafferty, K. D. 1997. Environmental parasitology: What can parasites tell us about human impacts on the environment? Parasitology Today 13:251–55.

Lafferty, K. D., and R. D. Holt. 2003. How should environmental stress affect the population dynamics of disease? Ecology Letters 6:654–64.

Landsberg, J. H. 2002. The effects of harmful algal blooms on aquatic organisms. Reviews in Fisheries Science 10:113–390.

Lefebvre, K. A., C. L. Powell, M. Busman, G. J. Doucette, P. D. R. Moeller, J. B. Silver, P. E. Miller, M. P. Hughes, S. Singaram, M. W. Silver, and R. S. Tjeerdema. 1999. Detection of domoic acid in northern anchovies and California sea lions associated with an unusual mortality event. Natural Toxins 7:85–92.

MacKenzie, K., H. H. Williams, B. Williams, B., A. H. McVicar, and R. Siddall. 1995. Parasites as indicators of water quality and the potential use of helminth transmission in marine pollution studies. Advances in Parasitology 35:85–144.

Marcogliese, D. J. 2001. Implications of climate change for parasitism of animals in the aquatic environment. Canadian Journal of Zoology 79:1331–52.

Marcogliese, D. J. 2004. Parasites: Small players with crucial roles in the ecological theater. EcoHealth 1:151–64.

Margogliese, D. J. 2005. Parasites of the superorganism: Are they indicators of ecosystem health? International Journal for Parasitology 35:705–16.

Marcogliese, D. J., and D. K. Cone. 2001. Myxozoan communities parasitizing *Notropis hudsonius* (Cyprinidae) at selected localities on the St. Lawrence River, Quebec: Possible effects of urban effluents. Journal of Parasitology 87:951–56.

Marcogliese, D. J., T. M. Goater, and G. W. Esch. 1990. *Crepidostomum cooperi* (Allocreadidae) in the burrowing mayfly, *Hexagenia limbata* (Ephemeroptera) related to the trophic status of a lake. American Midland Naturalist 124:309–17.

Markiw, M. E. 1992. Salmonid Whirling Disease. United States Department of the Interior, Fish and Wildlife Service, Fish and Wildlife Leaflet 17. Washington, DC: U.S. Department of the Interior.

Middelboe, M., T. J. Nielsen, and P. K. Bjørnsen. 2002. Viral and bacterial production in the North Water: In situ measurements, batch culture experiments and characterization of a virus-host system. Deep-Sea Research II 49:5063–79.

Millennium Ecosystem Assessment. 2005. Synthesis Report. Washington, DC: Island Press. Available: http://www.MAweb.org.

Mitchell, C. E., D. Tilman, and J. V. Groth. 2002. Effects of grassland plant species diversity, abundance, and composition on foliar fungal disease. Ecology 83:1713–26.

Mohammed, Z. A., W. W. Carmicheal, and A. A. Hussein. 2003. Estimation of microcystins in the freshwater fish *Oreochromis niloticus* in an Egyptian fish farm containing *Microcystis* bloom. Environmental Toxicology 18:137–41.

Möller, H. 1987. Pollution and parasitism in the aquatic environment. International Journal for Parasitology 17:353–61.

Moser, M., and R. K. Cowen. 1991. The effects of periodic eutrophication on parasitism and stock identification of *Trematomus bernacchii* (Pisces: Nototheniidae) in McMurdo Sound, Antarctica. Journal of Parasitology 77: 551–56.

Muzzall, P. M. 1999. Nematode parasites of yellow perch, *Perca flavescens*, from the Laurentian Great Lakes. Journal of the Helminthological Society of Washington 66:115–22.

National Research Council. 1992. Restoration of Aquatic Ecosystems: Science, Technology and Public Policy. Washington, D.C.: National Academy Press.

National Research Council. 2001. Grand Challenges in Environmental Sciences. Washington, DC: National Academy Press. Available: http://www.nap.edu/books/0309072549/html/.

Naylor, R. L., R. J. Goldburg, J. H. Primavera, N. Kautsky, M. C. M. Beveridge, J. Clay, C. Folke, J. Lubchenco, H. Mooney, and M. Troell. 2000. Effect of aquaculture on world fish supplies. Nature 405:1017–24.

Ostfeld, R., and F. Keesing. 2000. The function of biodiversity in the ecology of vector-borne zoonotic diseases. Canadian Journal of Zoology 78:2061–78.

Overstreet, R. M. 1988. Aquatic pollution problems, southeastern U.S. coasts: Histopathological indicators. Aquatic Toxicology 11:213–39.

Overstreet, R. M. 1993. Parasitic diseases of fishes and their relationship with toxicants and other environmental factors. *In* Pathobiology of Marine and Estuarine Organisms, ed. J. A. Couch and J. W. Fournie, 111–56. Boca Raton, FL: CRC Press.

Overstreet, R. M., and H. D. Howse. 1977. Some parasites and diseases of estuarine fishes in polluted habitats of Mississippi. Annals of the New York Academy of Sciences 298:427–62.

Paerl, H. W. 1997. Coastal eutrophication and harmful algal blooms: Importance of atmospheric deposition and groundwater as "new" nitrogen and other nutrient sources. Limnology and Oceanography 42:1154–65.

Pitois, S., M. H. Jackson, and B. J. B. Wood. 2001. Sources of the eutrophication problems associated with toxic algae: An overview. Journal of Environmental Health 64:25–32.

Postel, S., and S. R. Carpenter. 1997. Freshwater ecosystem services. *In* Nature's Services, ed. G. Daily, 195–214. Washington, DC: Island Press.

Pulkkinen, K., and D. Ebert. 2004. Host starvation decreases parasite load and mean host size in experimental populations. Ecology 85:823–33.

Reimer, L. W. 1995. Parasites especially of piscean hosts as indicators of the eutrophication in the Baltic Sea. Applied Parasitology 36:124–35.

Rejmánková, E., J. Grieco, N. Achee, P. Masuoka, K. Pope, D. Roberts, and R. M. Higashi. 2006. Freshwater community interactions and malaria. *In* Disease Ecology: Community Structure and Pathogen Dynamics, ed. S. Collinge and C. Ray, 90–104. Oxford: Oxford University Press.

Richmond, R. H. 1993. Coral reefs: Present problems and future concerns resulting from anthropogenic disturbance. American Zoologist 33:524–36.

Riedel, R., L. Caskey, and B. A. Costa-Pierce. 2002. Fish biology and fisheries ecology of the Salton Sea. Hydrobiologia 473:229–44.

Scheele, C. E. H., R. C. Lathrop, D. W. Marshall, E. L. Decker, D. B. Lewis, S. D. Snyder. 1999. A Survey of Swimmer's Itch-Causing Cercariae and Their Intermediate Snail Host Species in Devil's Lake, Wisconsin. Madison, WI: Wisconsin Department of Natural Resources.

Schlesinger, W.H. 1997. Biogeochemistry: An Analysis of Global Change, 2nd ed. New York: Academic Press.

Smayda, T.J. 1997. Harmful algal blooms: Their ecophysiology and general relevance to phytoplankton blooms in the sea. Limnology and Oceanography 42:1137–53.

Smith, V. H. 1998. Cultural eutrophication of inland, estuarine, and coastal waters. *In* Successes, Limitations and Frontiers of Ecosystem Science, ed. M. L. Pace and P. M. Groffman, 7–49. New York: Springer-Verlag.

Smith, V. H., T. P. Jones II, and M. S. Smith. 2005. Host nutrition and infectious disease: An ecological view. Frontiers in Ecology and the Environment 3:268–74.

Stewart, J. E. 1991. Introductions as factors in diseases of fish and aquatic invertebrates. Canadian Journal of Fisheries and Aquatic Science 45(Suppl. 1):110–17.

Stromberg, P. C., and J. L. Crites. 1975. An analysis of the changes in the preva-
    lence of *Camallanus oxycephalus* (Nematoda: Camallanidae) in western
    Lake Erie. Ohio Journal of Science 75:1–6.
Sutherland, K. P., J. W. Porter, and C. Torres. 2004. Disease and immunity in
    Caribbean and Indo-Pacific zooxanthellate corals. Marine Ecology Progress
    Series 266:273–302.
Suttle, C. A. 2005. Viruses in the sea. Nature 437:356–61.
Thompson, K. G., R. B. Nehring, D. C. Bowden, and T. Wygant. 1999. Field
    exposure of seven species or subspecies of salmonids to *Myxobolus cerebralis*
    in the Colorado River, Middle Park, Colorado. Journal of Aquatic Animal
    Health 11:312–29.
Townsend, A. R., R. W. Howarth, F. A. Bazzaz, M. S. Booth, C. C. Cleveland,
    S. K. Collinge, A. . Dobson, P. R. Epstein, D. R. Keeney, M. A. Mallin,
    C. A. Rogers, P. Wayne, and A. H. Wolfe. 2003. Human health effects of a
    changing global nitrogen cycle. Frontiers in Ecology and the Environment
    1:240–46.
Valtonen, E. T., J. C. Holmes, J. Aronen, and I. Rauthalahti. 2003. Parasite
    communities as indicators of recovery from pollution: Parasites of roach (*Ru-
    tilus rutilus*) and perch (*Perca fluviatilis*) in Central Finland. Parasitology
    126:S43–52.
Valtonen, E. T., J. C. Holmes, and M. Koskivaara. 1997. Eutrophication, pollu-
    tion and fragmentation: Effects on parasite communities in roach (*Rutilus
    rutilus*) and perch (*Perca fluviatilis*) in four lakes in central Finland. Cana-
    dian Journal of Fisheries and Aquatic Sciences 54:572–85.
Van Capellen, P., and E. D. Ingall. 1994. Benthic phosphorus regeneration, net
    primary production, and ocean anoxia: A model of the coupled marine bio-
    geochemical cycles of carbon and phosphorus. Paleoceanography 9:677–92.
Vogelbein, W. K., V. J. Lovko, J. D. Shields, K. S. Reece, L. W. Haas, and C. C.
    Walker. 2002. *Pfiesteria shumwayae* kills fish by micropredation not exo-
    toxin secretion. Nature 418:967–70.
Voss, J. D., and L. L. Richardson. 2005. Nutrient enrichment impacts severity
    and rate of progression in black band disease of corals. Abstract presented at
    the annual meeting of the American Society of Limnology and Oceanogra-
    phy, Salt Lake City, UT.
Weinbauer, M. G., D. Fuks, and P. Peduzzi. 1993. Distribution of viruses and
    dissolved DNA along a coastal trophic gradient in the Northern Adriatic Sea.
    Applied and Environmental Microbiology 59:4074–82.
Weisberg, S. B., R. P. Morin, E. A. Ross, and M. F. Hirshfield. 1986. *Eustron-
    gylides* (Nematoda) infection in mummichogs and other fishes of the Chesa-
    peake Bay Region. Transactions of the American Fisheries Society 115:776–83.
Williams, H. H., and K. MacKenzie. 2003. Marine parasites as pollution indi-
    cators: An update. Parasitology 126:S27–S41.
Wisniewiski, W. L. 1958. Characterization of the parasitofauna of a eutrophic
    lake. Acta Parasitologica Polonica 6:1–64.
Wootten, R. 1973. The metazoan parasite-fauna of fish from Hanningfield Res-
    ervoir, Essex in relation to features of the habitat and host populations. Jour-
    nal of Zoology 171:323–31.

World Water Council. 2003. World Water Action: Making Water Flow for All. London: Earthscan Publishers.

Wu, C., C. Maurer, Y. Wang, S. Z. Xue, and D. L. Davis. 1999. Water pollution and human health in China. Environmental Health Perspectives 107:251–56.

Zander, C. D., Ö. Koçogluk, M. Skroblies, and U. Strohbach. 2002. Parasite populations and communities from the shallow littoral of the Orther Bight (Fehmarn, SW Baltic Sea). Parasitology Research 88:734–44.

Zander, C. D., and L. W. Reimer. 2002. Parasitism at the ecosystem level in the Baltic Sea. Parasitology 124:S119–35.

CHAPTER FIVE
Landscape Structure, Disturbance, and Disease Dynamics

Hamish McCallum

SUMMARY

LANDSCAPE STRUCTURE INFLUENCES population density and movement patterns of hosts, vectors, and transmission stages. It is therefore self-evident that landscape will influence the dynamics of infectious diseases. Most classic models of infectious disease dynamics do not, however, explicitly include any spatial structure. Its inclusion has major implications for understanding both invasibility and persistence, because in a spatial context, an infected host is more likely to be near another infected host than is the case in a homogeneously mixing system. Understanding landscape is critical in predicting the quantitative spread of infectious disease, because landscape features can be either barriers or conduits to disease spread. Anthropogenic changes in landscape, particularly habitat destruction and fragmentation, have led to disease emergence in a number of situations. No general conclusions can be drawn about the influence of metapopulation parameters such as patch size and isolation on disease dynamics, although an important conclusion is that such effects are often not monotonic. Finally, I discuss how landscape manipulation can be used to manage pathogen threats. In particular, pathogens are an important consideration in the design of nature reserve systems and the maintenance of corridors. Although corridors can increase disease impacts, concern about disease threats should not prevent corridors being established.

INTRODUCTION

At present, three great anthropogenic changes to natural ecosystems are occurring: habitat destruction, global warming, and the introduction of alien species. All of these changes may have profound influences on disease dynamics. Two of them—habitat destruction and introductions—are essentially processes having to do with changes in connectivity within and between populations. The consequences of these changes for infectious disease are the topic of this chapter. Paradoxically, these anthropogenic changes are diametrically opposite in terms of their effects on connectivity. Habitat destruction and fragmentation act essentially to reduce the extent of connectivity between subpopulations. Increasing global trade, which is the root cause of most accidental introductions of alien species, including parasites and pathogens, causes increased connectivity between populations on different continents and in different regions. Nevertheless, the central problem for disease dynamics is to understand how changes in the degree of connectedness of individual hosts, different subpopulations of hosts, or communities of hosts influence the way in which diseases behave.

There is an extensive literature on the effects of the spatial structure of human populations on epidemiology, particularly in relation to childhood diseases such as measles (Bolker and Grenfell 1995; Grenfell and Bolker 1998; Grenfell et al. 2001; Xia et al. 2004). This work has led to some major insights into spatial structure and disease dynamics, largely because high-quality time series of data are available for major human diseases. In this chapter, I use such work where it provides insight into some of the fundamental issues concerning spatial structure and disease dynamics. However, I concentrate on changes to natural landscapes and their effects on disease dynamics, particularly in wildlife. Because the majority of the emerging diseases of humans are zoonotic in origin (Daszak et al. 2000; Kruse et al. 2004), much of this has direct bearing on human disease, although it does not form the main focus of this chapter.

It is self-evident that landscape structure affects the dynamics of infectious disease. Transmission requires contact between infectious hosts or infective stages and susceptible individuals, and because landscape affects population density and the movement of hosts, vectors, or transmission stages, it will obviously affect disease dynamics. Some interesting questions are the following:

- How does landscape structure affect qualitative aspects of disease dynamics?

- How can knowledge of landscape structure be used to improve quantitative predictions about disease spread and persistence?

- How have anthropogenic disturbances and changes to landscapes influenced disease dynamics?

- How can landscape structure be manipulated for control of infectious diseases and their impacts on host populations?

Figure 5.1 is a modification of a conceptual model, originally developed for forest disease outbreaks, of the interconnections between landscape structure and host pathogen systems. With forest pathogens, the host-pathogen system may have a substantial effect on landscape structure if large-scale tree die-offs occur (Holdenrieder et al. 2004). Wildlife pathogens are much less likely to affect landscape structure, although if a pathogen causes very high levels of mortality in a dominant herbivore,

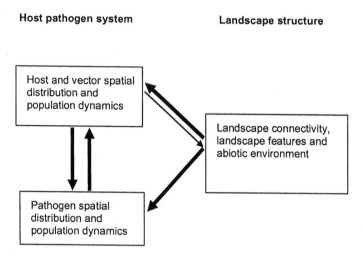

Figure 5.1. Conceptual model of the relationships between host-pathogen systems and landscape structure. The thickness of the arrows represents the likely frequency or intensity of the linkages in wildlife disease systems. (Modified from Holdenrieder et al. 2004.)

landscape-level consequences are possible. For example, following its introduction in 1995, rabbit hemorrhagic disease has reduced rabbit populations by up to 85% in some areas of semi-arid southern Australia, with a concomitant reduction in grazing pressure. However, the nature of the impact depends on subtleties of the seasonal timing of the disease and grazing impact. The disease primarily causes mortality among young animals in the breeding season (the austral spring), so that rabbit numbers are depressed by up to 95% in spring, but only by perhaps 60% in summer. This means that the reduction in grazing on annuals is much greater than the reduction in grazing on more slowly growing perennial woody plants, with a corresponding change in the plant community structure (Cooke and Fenner 2002; Mutze et al. 2002). Several other chapters in this book discuss the influence of pathogens on ecosystem and community processes, which may lead to landscape-scale effects (see Collinge et al., chapter 6, Eviner and Likens, chapter 12, and Chapin et al., chapter 13, this volume).

Landscape Structure and Qualitative Dynamics

Standard models of host-parasite interactions (Anderson and May 1978, 1979, 1991; May and Anderson 1978, 1979) do not include space explicitly. Transmission is usually assumed to be either frequency dependent (in which case the number of hosts contacted by a susceptible host per unit of time is constant, irrespective of host population size, so that it is the proportion of contacts with infected hosts that is important) or density dependent (in which case the number of host contacts depends on the density of infected hosts). The corresponding transmission terms can be written as βSI for density dependence or $\beta SI/N$ for frequency dependence, where β is a transmission parameter, S represents susceptible hosts, I represents infectious hosts, and N represents all hosts. Whether S, I, and N represent numbers in a given local population or densities per unit area in some local neighborhood differs with different models, which has been a source of confusion in the literature (McCallum et al. 2001). The implicit assumption for both density and frequency dependence, whether density or numbers are used, is that each individual in the local population has an equal probability of encountering every other individual. In all but the smallest populations or neighborhoods, this is not a sensible assumption: individuals contact others that are close to them more often than those that are farther away. Nevertheless, in some cases, the dynamics may be represented adequately by assuming that the disease spreads as if infected and susceptible hosts mixed homogeneously. This is known as the mean field assumption. The fundamental question in understanding spatial structure is to establish conditions under which the mean field assumption is appropriate, and to determine the qualitative effects of departure from that assumption.

At the opposite extreme from homogeneous mixing are lattice models, in which hosts occupy fixed points on a lattice and transmission usually is assumed to occur only between immediate neighbors (Rhodes and Anderson 1997; White et al. 1996). These models have been particularly influential in studying plant disease (Jeger 2000). Between these extremes of completely mixing hosts and purely local interactions is a vast range of models in which fully mixing subpopulations are linked by migration (Rodriguez and Torres-Sorando 2001), models in which contact rates are assumed to be heterogeneous (Woolhouse et al. 1997), and network models (Olinky and Stone 2004; Watts et al. 2005; Xia et al. 2004).

Invasion and Spatial Structure

Whether a pathogen can invade a population of fully susceptible hosts is a central question in epidemiology. Invasibility can be quantified by R_0, the basic reproductive number, which is the average number of secondary cases produced per initial infection. In purely deterministic systems, if $R_0 > 1$, the pathogen can invade, whereas if $R_0 < 1$ it will die out, although there may be a small number of secondary cases. In stochastic

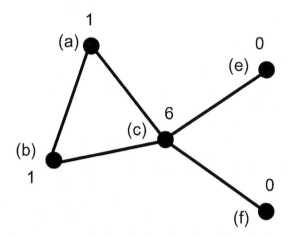

Figure 5.2. The clustering coefficient in networks. The numbers on each vertex of this network show the number of triples with that vertex at the center. For example, vertex c is the center of the triples {acb}, {ace}, {ecf}, {bcf}, {bce}, and {acf}. The network contains one closed triangle, {acb}. Thus the clustering coefficient is $C = \dfrac{3 \times 1}{8}$ (After Newman 2003.)

systems, the pathogen will not invade if $R_0 < 1$, but epidemics have a finite probability of failing even if $R_0 > 1$ (Dietz 1993) because the probability p of an epidemic occurring is approximately

$$P = 1 - \frac{1}{R_0}. \tag{1}$$

The consequences of spatial subdivision for R_0 have been modeled in classic metapopulations, lattice models, and network models (Meyers et al. 2005). As might be expected, whether transmission is assumed to be density dependent or frequency dependent has a major influence on the consequences of spatial subdivision in metapopulation models with homogeneous mixing within patches. If frequency dependence is assumed, in a model with all patches equally linked, R_0 is entirely invariant to spatial structure (Hagenaars et al. 2004), whereas if transmission is density dependent, then R_0 has a component due to intrapatch transmission, augmented by interpatch transmission (Park et al. 2001). Of note, with density-dependent transmission, the patch structure will always reduce R_0 relative to a system of the same number of individuals that mix completely.

Using a network model, Keeling (1999) showed that local spatial structure has a major effect on disease invasion. In the very early stages of invasion, rather than being surrounded by a sea of susceptibles, infected hosts rapidly become surrounded by other already infected hosts, diminishing the reproductive rate of the disease relative to that of a homogeneously mixing population (Keeling 1999). R_0 is also dependent on the structure of the network. One of the important parameters that describes the properties of a network is the clustering coefficient, which is three times the ratio of closed triangles relative to the total number of triple connections (figure 5.2). It can be thought of as the probability that two individuals that are neighbors of some other individual are also neighbors (Newman 2003). Keeling (1999) shows, using a frequency-dependent contact function, that R_0 is reduced at high values of the clustering coefficient, particularly if the overall number of connections is low. Intuitively, closed triangles mean that many contacts are "wasted" for disease transmission because they are likely to be with an already infected host.

Persistence and Spatial Structure

The effect of spatial structure on disease persistence is rather less obvious than the effect on invasion. In a purely deterministic system with input of new susceptible individuals by birth or immigration, the criteria for invasion and persistence are identical (Anderson and May 1986). In any stochastic model of any finite population, eventual extinction is inevitable. It therefore makes sense to consider the probability distribution of the time until extinction, or the probability of persistence over some specified time period. Spatial structure and connectivity can have major effects on these probabilities, and the criteria or thresholds for persistence over some long time horizon may be very different from the criteria for invasion (Park et al. 2001).

The specifics of these criteria will depend strongly on the details of the stochastic model and the type of spatial structure—metapopulation, lattice, or network—being modeled. One very important general result, however, is that pathogen persistence tends to be longest at intermediate levels of connectivity (Bolker and Grenfell 1996; Swinton et al. 2002). This is particularly the case with pathogens with a high R_0 within subpopulations, patches, or clusters. After the first introduction, there is an initial epidemic peak within the patch, followed by a deep trough in the number of infected hosts, because the pool of susceptible individuals has been locally exhausted. In this trough, local extinction of the pathogen is very likely. In poorly connected metapopulations, a local epidemic is thus likely to go extinct before transmission to another patch or cluster occurs, leading to global extinction. In very well-connected metapopulations, interpatch transmission occurs early in the epidemic, leading to a synchronized epidemic peak over the entire metapopulation, with global extinction being likely in the subsequent synchronized epidemic trough. At intermediate levels of connectivity, however, transmission to another patch is a relatively rare event that will occur over a wide range of stages of within-patch epidemics, leading to desynchronized epidemics in each patch. Although extinction on any given patch in the first postepidemic trough is highly likely, other patches will be at different stages of the epidemic cycle and will therefore be able to trigger epidemics in patches from which the pathogen has disappeared (Hagenaars et al. 2004; Keeling and Grenfell 1997; Park et al. 2001, 2002).

Spatial structure influences host as well as pathogen extinction. Pathogens that affect their host's reproductive rate can drive the host population to extinction deterministically in a lattice-based model, whereas this is not possible in a nonspatially structured model (Boots and Sasaki 2002; Sato et al. 1994). In these models, both transmission and recruitment are possible only to neighboring points in the lattice. A host infected by a pathogen that does not greatly increase mortality remains in its lattice position, infecting its immediate neighbors and hence the surrounding neighborhood. When mortality does occur, the vacant lattice position is then surrounded with infected hosts, which can recruit to that position only at a very low rate. Deterministic host extinction is thus most likely in pathogens that greatly affect fecundity but have only a minor effect on mortality. As de Castro and Bolker (2005) have pointed out in a recent review, the general effect of spatial subdivision is to generate a form of frequency-dependent transmission that facilitates parasite-induced extinction.

Macroparasites

Most discussion of parasitic disease and spatial structure has been based on microparasites. Macroparasites such as helminths have a quite different population biology: it is necessary to keep track of the frequency distribution of parasites among hosts, and infections are often endemic rather than epidemic (Anderson and May 1978). One aspect that is of particular importance for spatial dynamics is that, because macroparasites must leave their host at some point in their life cycle, there is frequently a relatively long life history stage that lives outside the particular host of interest. This is obviously the case for parasites with complex indirect life cycles, but is true even in "directly transmitted" macroparasites, such as many gastrointestinal nematodes. Eggs passed out through the gut must survive and develop into larvae in the external environment before they can become infective. The effect of either a free-living infective stage or an intermediate host is that the production of an infective stage from an infected host and the subsequent infection of a susceptible host are separated in time. Given that hosts are mobile, this also means that the infected host and the one to which infection is transferred may be separated in space (Morgan et al. 2004). The need for infectious stages to survive outside their primary host means that landscape characteristics such as microclimate that cause differences in survival of infective stages or that cause aggregation in hosts can determine whether and at what intensity the macroparasite persists. Potentially, global information system (GIS) methods can be used to identify areas of high disease risk (Pfeiffer and Hugh-Jones 2002).

LANDSCAPE STRUCTURE, DISEASE SPREAD, AND PERSISTENCE

Most models of the spread of infectious disease consider a uniform environment, even if they do not assume homogeneous mixing within it. In addition to the general qualitative processes discussed earlier, heterogeneous environments can have two different effects on disease spread and persistence. First, the rate of movement of hosts or vectors and connectivity varies through a variable landscape, with consequences for the movement of the epidemic front and whether invasion can occur. Second, population density and survival rates of hosts, vectors, and infective stages vary across a heterogeneous landscape. In the public health and veterinary literature, there is an emerging field of landscape epidemiology (De La Rocque et al. 2004; Kitron 1998; Ostfeld et al. 2005; Pfeiffer and Hugh-Jones 2002). This consists basically of using georeferenced climate data and GISs to model predicted distributions of vectors and reservoir hosts, and hence to produce maps of the risk of epidemics.

Case Studies of Disease Spread in Landscapes

Epizootics of a novel disease are a relatively uncommon event, although there is some evidence that they are occurring more frequently (Daszak et al. 2000). Some wildlife epizootics that have been reasonably well studied are rabies among raccoons in the northeastern United States, mycoplasmal conjunctivitis in house finches in the United States, West Nile virus in the United States, and myxomatosis and rabbit hemorrhagic disease in Australia. This section briefly reviews the impact of landscape structure on the spread of each of these diseases.

Rabies entered raccoon populations in western Connecticut in 1991 and has since spread across the state. Analysis and modeling of empirical data from rabies cases at a township level shows the importance of a heterogeneous landscape in affecting disease spread. Rivers were of critical importance in limiting disease spread, but the role of raccoon population density was unclear (Smith et al. 2002). This is thus a case in which the effect of the landscape on host movement (rivers are an effective barrier to raccoon movement) affects disease spread more than does the effect of the landscape on host density.

In contrast, the initial spread of myxomatosis among rabbits in Australia in 1950–51 was almost exclusively along river systems, in particular the Murray-Darling system (Ratcliffe et al. 1952). Because the primary vectors of myxomatosis in Australia are mosquitoes, in this situation the effect of the landscape on vector density was the primary determinant of pathogen spread. Rabbit hemorrhagic disease, spreading among the same rabbit populations from an identical point source almost fifty years later, showed a rather different dependence on landscape (Kovaliski 1998). The initial spread in Australia was extremely rapid: more than 50 km per week (Cooke and Fenner 2002), with occasional long-distance dispersal events of up to 300 km. There was no obvious dependence on river systems. Disease impact was greatest in arid or semi-arid regions and lower in humid regions and those with low rabbit densities (Kovaliski 1998; Mutze et al. 2002; Story et al. 2004). In Australia, the disease appears to spread locally by contact, augmented by occasional long-distance dispersal by flies (Cooke 2002). Here, landscape features affecting local host density were important and local vector distribution was less important.

West Nile virus and house finch mycoplasmal conjunctivitis are two avian pathogens that have recently spread across the United States from the northeast (Hochachka and Dhondt 2000; USGS 2005). Although there certainly are important small-scale landscape effects on prevalence and spread, in both cases the broad pattern of spread is consistent with the simplest possible diffusion models of invasion. These models predict that range expansion (or equivalently, the square root of area occupied) should be a linear function of the time since introduction (van den Bosch et al. 1992).

These case studies emphasize the simple point made previously: landscape structure has a major influence on disease spread because it influences both movement patterns and density of hosts and disease vectors. Landscape features are obviously more likely to act as barriers to directly transmitted diseases such as rabies than to pathogens with airborne vectors. For vector-transmitted diseases, landscape features that influence vector density or dispersal are likely to be the main determinants of the rate of epizootic spread.

Landscape, Persistence, and Prevalence

Among the vast number of metrics used to describe landscapes, patch size and patch isolation are the most widely used, as classic metapopulation theory suggests that they are the fundamental properties that determine metapopulation persistence (Hanski 1997). There is a massive literature on empirical studies of patch occupancy and population density as a function of these parameters in free-living organisms. Relatively few studies, however, have looked at the effect of these landscape parameters on levels of parasite infection.

In many respects, pathogen epizootics in a spatially structured host population might be expected to behave as a classic metapopulation, with patch-specific colonization and extinction parameters similar to those described by Hanski (1994) in his incidence function approach. The process of initiation of an epizootic on a patch of susceptible individuals would be expected to be a function of the distance of the patch from currently infected patches and the sizes of those patches, subject to a minimum patch size necessary to the host population size to exceed the threshold for disease introduction. The dependence of pathogen extinction on patch parameters is less straightforward. In standard metapopulation models, it is usually considered to be a random process, primarily a function of patch size, with the possibility of a "rescue effect" (Harding and McNamara 2002) caused by recolonization from adjacent patches. In a host-pathogen system, extinction of the pathogen may occur for two different reasons, with very different scaling with host population size. First, the pathogen may become extinct because the host population has become extinct. It is generally considered that pathogens will usually be driven to extinction before their hosts (McCallum and Dobson 1995), but this generalization will only apply if it is a single-host, single-pathogen system and transmission is density dependent. Nevertheless, stochastic host extinction is possible even in this case if the host population is relatively small. Pathogens can easily drive a highly susceptible host to extinction if a reservoir host exists that can maintain high levels of infection, even as the susceptible species declines to extinction (Gog et al. 2002; McCallum and Dobson 2002). However, identifying reservoirs is not straightforward, either conceptually or practically (Haydon et al. 2002). Even pathogens restricted to a single host can result in host extinction if transmission is frequency dependent, because there is no minimum threshold host density (McCallum et al. 2001).

The pathogen may also become extinct by exhausting the pool of susceptible hosts. In human disease, the concept of stochastic fadeout has become central to understanding the dynamics of infection in heterogeneous metapopulations (Bolker and Grenfell 1995; Grenfell and Bolker 1998; Keeling and Grenfell 1997). Simple unstructured deterministic models predict converging oscillations in prevalence following the introduction of a disease into a fully susceptible population with recurrent births. However, as discussed earlier, the proportion of infected hosts following a peak may drop so low with a highly infectious disease that in a finite population, stochastic disappearance of infection is almost inevitable. Only large cities with populations in excess of 250,000–400,000 (Bartlett 1957; Grenfell and Harwood 1997) have large enough populations to maintain the disease indefinitely. Persistence of the disease in the metapopulation relies on asynchrony in the timing of peaks in the subpopulations. Whether similar dynamics are common in wildlife disease is unknown, although the behavior of morbillivirus in North Sea seal populations is consistent with the idea (Swinton et al. 1998).

The key distinction between standard metapopulation models and epidemics in spatially structured systems is that the former can be considered to be more or less Markovian. That is, the transition probabilities depend on the present state of the system and not its history. This is not the case with disease models. An epidemic has a characteristic time course, and the probability of disease extinction in a particular time interval is a function of how long the disease has been present in that patch. Similarly, following fadeout, a patch will be refractory to further disease invasion until the proportion of susceptible hosts has built up sufficiently.

With these general principles in mind, what is the empirical evidence about patch parameters and disease extinction and colonization? I searched the Science Citation Index from 1945 to 2004 using the search string "TS = ((patch or fragment) and (area or size or configuration) and (pathogen or disease or parasite) and (intensity or prevalence))". Among a large number of nonrelevant citations, usually referencing human disease, I located the examples listed in table 5.1. From this small number of case studies, a generalization emerges that patch size and isolation can be important factors in the dynamics of parasite metapopulations. However, there is no universal rule that prevalence changes monotonically in a particular direction with patch size or isolation (figure 5.3). For example, in prairie dogs, extinctions induced by plague are lowest in medium-sized colonies (Stapp et al. 2004). This may be a result of stochastic extinction being likely in very small patches, in which most of the force of infection may come from infected immigrants, whereas epizootics may propagate more widely within large colonies. In deer mice,

TABLE 5.1
Studies of patch parameters and disease dynamics

Pathogen	Host	Source	Conclusion
Baylisascaris procyonis (nematode)	Definitive: raccoon, *Procyon lotor* Intermediate: white-footed mouse, *Peromyscus leucopus*	Page et al. (2001)	Both prevalence and intensity in mice are an increasing but convex function of forest patch area. Both are negatively related to the fraction of forested habitat nearby.
Plague, *Yersinia pestis*	Prairie dogs, *Cynomys ludovicianus*	Stapp et al. (2004)	Pathogen-induced extinction is lowest in intermediate-sized colonies. The fate of adjacent colonies is important, but no distance or isolation measures have proved significant.
Sin Nombre (hantavirus)	Deer mice, *Peromyscus maniculatus*	Langlois et al. (2001)	Seroprevalence is lowest in patches with an intermediate proportion of preferred habitat within a 1-km radius. Seroprevalence is increased with fragmentation.
Mermithids (nematodes)	Hawaiian spiders, *Tetragnatha* spp.	Vandergast and Roderick (2003)	Prevalence is increased with size of patch area.
Lyme disease bacterium, *Borrelia burgdorferi*	Vector: ticks, *scapularis; Ixodes* main reservoir, white-footed mouse, *P. leucopus*	Allan et al. (2003)	Prevalence in ticks declines linearly with size of forest patch area.

hantavirus prevalence is least when an intermediate amount of suitable habitat is available within a 1-km radius (Langlois et al. 2001). This pattern may be related to increased rates of movement of mice between patches, facilitating transmission, when there is little suitable habitat, and higher population densities increasing transmission when there is a large amount of suitable habitat.

These differing conclusions are not surprising, given the complexity of transmission dynamics. All things being equal, large patches should

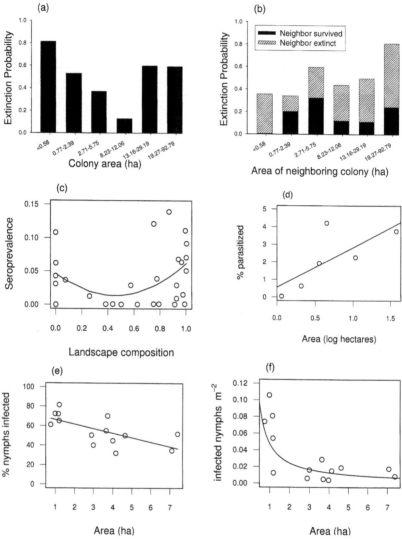

Figure 5.3. Infectious disease and patch parameters. These examples show the broad range of possible relationships between pathogen impact or prevalence and patch parameters. (a, b) probability of extinction of prairie dog colonies during plague epizootic years as a function of colony area (a) and neighboring colony area (b). (After Stapp et al. 2004.) The Probability of colony extinction is minimized at an intermediate size of the colony itself but increases as the size of the neighboring colony increases. (c) Seroprevalence to Sin Nombre virus in deer mice as a quadratic function of landscape composition (preferred habitat in 1km radius). Here, seroprevalence is minimized at intermediate amounts of preferred habitat in a 1-km radius. (After Langlois et al. 2001.) (d) Percentage of spiders parasitized by a mermithid as an increasing linear function of fragment area. (After Vandergast and Roderick [2003].) (e) Prevalence of the Lyme bacterium in nymphal ticks as a decreasing linear function of forest patch area. (f) Density of infected nymphal ticks as an inverse power function of forest area. (After Allan et al. [2003].)

support larger host populations, but these will not necessarily be at a higher density per unit area. There has been a continued controversy about whether transmission is more likely to depend on host density or on host numbers in a population (McCallum et al. 2001). In theory, there should be a positive relationship between microparasite prevalence or the number of macroparasites per host and host density, and a comparative study has shown that this relationship exists empirically if confounding factors are corrected for (Arneberg et al. 1998).

ANTHROPOGENIC CHANGES IN LANDSCAPE AND DISEASE

Several reviews have investigated the effect of landscape changes, in particular habitat destruction and fragmentation, on human disease (Daszak et al. 2000; Patz et al. 2004). One of the major ways in which diseases can emerge in the human population is through habitat fragmentation, bringing animal reservoirs of pathogens into increased contact with human populations. Similarly, pathogens may emerge in wildlife through habitat destruction or fragmentation, bringing wildlife species into closer contact with domestic or feral animals that carry disease (Power and Mitchell 2004). For example, rabies outbreaks have led to the disappearance of African wild dog packs in both Kenya and Tanzania (Alexander and Appel 1994; Kat et al. 1995). These outbreaks are thought to have occurred because of increasing contact between wild dogs and domestic or feral dogs.

Another major way in which landscape changes can lead to disease emergence both in humans and in other animals is through habitat changes leading to increases in the abundance of disease vectors or to spread in their distributions. In particular, forestry and land clearing can produce large numbers of temporary pools, which are ideal breeding habitat for *Anopheles* mosquitoes, the vectors of malaria (Molyneux 2002).

Fragmentation and habitat destruction often lead to changes in foraging and ranging behavior that may have major effects on parasite infection rates. For example, Gillespie et al. (2005) found that redtail guenons (*Cercopithecus ascanius*) in Uganda had both a higher intensity of infection and more macroparasite species per individual in logged forest than in unlogged forest. Interestingly, they found no such effect in two other primates (red colobus, *Piliocolobus tephrosceles*, and black-and-white colobus, *Colobus guerezain*) living in the same area, although the parasite fauna was largely shared between the three species. The logging that had taken place in the late 1960s resulted in the destruction of about 50% of all trees in the logged area, including primate food trees (Gillespie et al. 2005). The reason for the different responses of the parasite communities in the primates to logging may be that the red colobus had larger home ranges in the logged forests than in the unlogged forests, whereas the other primate species did not. The red colobus may therefore encounter patches with high densities of infective stages more frequently in logged than in unlogged, forest.

LANDSCAPE STRUCTURE AND DISEASE CONTROL

Options for disease control in free-ranging populations are limited (Wobeser 2002). Oral vaccination of wild carnivores has been remarkably successful in controlling rabies epizootics in both Europe and North America (Rupprecht et al. 2004). However, other attempts to treat or vaccinate wild animals have been less successful (Burrows et al. 1995; Suppo et al. 2000). Broad-scale culling, whether targeted at diseased animals or not, introduces a variety of ethical and technical problems. For example, a landscape-scale experiment in the United Kingdom investigating the potential of badger culling to reduce bovine tuberculosis found that prevalence in cattle actually increased in areas with culled badger populations because removals disrupted the social structure of this highly territorial animal (Donnelly et al. 2003).

One possibility for controlling disease in wildlife is thus to modify the landscape so that transmission is reduced. The most suitable strategy will depend on the biology of the host and pathogen, but the general principle is that the majority of transmission often occurs at relatively small, discrete parts of the landscape, and if these can be appropriately managed or isolated, the disease can be controlled. Manipulations that have been successful include burning of habitat patches in which transmission occurs and draining or cleaning of waterholes (Wobeser 2002).

Incorporating landscape structure into the design of programs to control disease by vaccination or culling is also of critical importance. The strategy to control rabies in wild carnivores by oral vaccination entails creating a vaccination *cordon sanitaire* along the leading edges of the epizootic. Using vaccination to reinforce natural barriers to the spread of infected animals, such as rivers or mountain ranges, has been essential to the success of such programs (Russell et al. 2005; Sidwa et al. 2005).

PARASITES, PATHOGENS, AND THE DESIGN OF NATURE RESERVE SYSTEMS

Establishing systems of nature reserves is a major landscape-level manipulation for conservation purposes. There is a massive literature on the optimal design of reserve systems, both from the perspective of maintaining as many species as possible from the total pool in a region and from the point of view of conservation of a single focal species (Williams et al. 2004). However, discussion of the implications of parasites and pathogens for the design of reserve systems has received much less attention. Despite the best efforts of some parasitologists to promote the importance of parasites as components of biodiversity (Marcogliese 2005; Poulin and Mouillot 2004), the small amount of literature that does exist is from the perspective of protecting animals in reserves from parasite threats rather than from the perspective of maintaining parasite diversity.

Hess (1994, 1996) pointed out that the generally accepted strategy of increasing connectivity between habitat patches by establishing corridors may not be sensible in the face of disease threats to endangered species, as the corridors also enhance transmission of infectious disease. Using a simple deterministic metapopulation model, he showed that the number of patches occupied by a species affected by a pathogen that increased mortality initially increased with connectivity, but declined at high levels of connectivity. However, his model considered only a single host and a single pathogen, whereas almost all pathogens that have caused conservation problems infect multiple hosts. In almost all cases, a reservoir host that is relatively unaffected by the pathogen maintains a high force of infection, so that the endangered species can be driven to very low levels (Gog et al. 2002; McCallum and Dobson 1995). In this situation, the distribution of the reservoir hosts and the endangered hosts in the landscape is of critical importance. One possibility is that the reservoirs occupy the matrix between the patches occupied by the endangered species that is highly susceptible to disease. In this case, the basic management problem is to control the transfer of the pathogen across the patch boundaries from reservoirs in the external matrix to the highly susceptible hosts within the patches. This is the situation with canine distemper being transferred to lions within African nature reserves from domestic dog populations outside the reserves (Roelke-Parker et al. 1996). Connections between the patches are therefore relatively unimportant for transmission of the disease but will be necessary for maintenance of the endangered hosts as a metapopulation. This was the main conclusion reached by Gog et al. (2002): concern about disease should not, in this case, preclude the establishment of corridors.

The second possibility is that both reservoirs and the highly susceptible species occupy the same habitat patches. The management problem is then whether maintenance of connections between the patches leads to transmission of the pathogen and hence to extinction of the susceptible host metapopulation. This was the situation considered by McCallum and Dobson (2002). We considered the worst-case scenario in which the effect of a pathogen on a highly susceptible host species is so severe that the host species cannot coexist on a patch together with infected reservoir hosts. This appears to be the situation with some frog species and the chytrid fungus *Batracochytrium dendrobatidis*.

In the absence of any spatial structure, the "metapopulation" is a single patch, and our starting assumption means that the pathogen will drive the highly susceptible species to extinction. Whether or not global extinction also occurs at high levels of connectivity in a subdivided habitat depends on the relative colonization and extinction rates of the reservoir host and the highly susceptible host species. The highly susceptible species may be able to function as a "fugitive" species, occupying patches before the reservoir and the consequent infection arrives, provided the ratio of the endangered species' colonization rate to its extinction rate exceeds that of the reservoir. Perhaps less obviously, the highly susceptible species may be able to use reservoir patches from which the disease has faded out as refuges. If the reservoirs are long-lived and have lasting immunity, then a patch in which there has been an epidemic among the reservoirs will not be able to support another epidemic for some time. In this case, the proportion of patches occupied by the highly susceptible species will increase at high connectivity. There are certainly some parameter combinations in which the susceptible species persists at intermediate levels of connectivity and others for which persistence is impossible in the face of disease for all levels of connectivity. Nevertheless, the key point is that if a population functions as a metapopulation, too little connectivity will always lead to extinction, whereas high levels of connectivity, even in this worst-case scenario, are not necessarily a bad thing. Pathogen threats need to be considered when planning reserve systems, but they should not necessarily preclude establishing corridors.

CONCLUSIONS

Landscape has profound effects on both qualitative and quantitative aspects of wildlife disease. However, there are no simple rules, such as "fragmentation increases disease threats" or "corridors increase disease transmission." The examples discussed in this chapter show clearly that the details of the host-parasite interaction are essential to understanding the effects of fragmentation. For example, figure 5.3 shows that parasite prevalence increased with forest patch size in Hawaiian spiders but decreased with forest patch size in ticks in the northeastern United States. Frequently, the effects of patch size or connectivity are non-monotonic. Theory suggests pathogens persist longest in systems with intermediate connectivity, and empirical data show plague-related extinctions are lowest in intermediate-size prairie dog colonies.

It is clear that understanding disease dynamics requires linking spatially explicit models with multiple-host, multiple-pathogen systems. Only a start has been made on this (Carlsson-Graner and Thrall 2002; Dobson 2004; Kamo and Boots 2004; but see Keeling et al. 2001). There is a clear need for a better understanding of the way in which pathogens spread across realistic host network structures. In particular, it is vital that any theoretical work on networks build on a strong empirical base.

ACKNOWLEDGMENTS

Work was funded by the Australian Research Council and partially completed while the author was an Academic Visitor at Imperial College, Silwood Park.

LITERATURE CITED

Alexander, K. A., and M. J. G. Appel. 1994. African wild dogs (*Lycaon pictus*) endangered by a canine distemper epizootic among domestic dogs near the Masai Mara National Reserve, Kenya. Journal of Wildlife Diseases 30:481–85.

Allan, B. F., F. Keesing, and R. S. Ostfeld. 2003. Effect of forest fragmentation on Lyme disease risk. Conservation Biology 17:267–72.

Anderson, R. M., and R. M. May. 1978. Regulation and stability of host-parasite interactions. I. Regulatory processes. Journal of Animal Ecology 47:219–47.

Anderson, R. M., and R. M. May. 1979. Population biology of infectious-diseases. I. Nature 280:361–67.

Anderson, R. M., and R. M. May. 1986. The invasion, persistence and spread of infectious diseases within animal and plant communities. Philosophical Transactions of the Royal Society of London. Series B, Biological Sciences 314:533–70.

Anderson, R. M., and R. M. May. 1991. Infectious Diseases of Humans. Oxford, UK: Oxford University Press.

Arneberg, P., A. Skorping, B. Grenfell, and A. F. Read. 1998. Host densities as determinants of abundance in parasite communities. Proceedings of the Royal Society of London. Series B, Biological Sciences 265:1283–89.

Bartlett, M. S. 1957. Measles periodicity and community size. Journal of the Royal Statistical Society. Series A 120:48–70.

Bolker, B., and B. Grenfell. 1995. Space, persistence and dynamics of measles epidemics. Philosophical Transactions of the Royal Society of London Series. Series B, Biological Sciences 348:309–20.

Bolker, B. M., and B. T. Grenfell. 1996. Impact of vaccination on the spatial correlation and persistence of measles dynamics. Proceedings of the National Academy of Sciences of the United States of America 93: 12648–653.

Boots, M., and A. Sasaki. 2002. Parasite-driven extinction in spatially explicit host-parasite systems. American Naturalist 159:706–13.

Burrows, R., H. Hofer, and M. L. East. 1995. Population dynamics, intervention and survival in African wild dogs (lycaon pictus). Proceedings of the Royal Society of London. Series B, Biological Sciences 262:235–45.

Carlsson-Graner, U., and P. H. Thrall. 2002. The spatial distribution of plant populations, disease dynamics and evolution of resistance. Oikos 97:97–110.

Cooke, B. D. 2002. Rabbit haemorrhagic disease: Field epidemiology and the management of wild rabbit populations. Revue Scientifique et Technique de l'Office International des Epizooties 21:347–58.

Cooke, B. D., and F. Fenner. 2002. Rabbit haemorrhagic disease and the biological control of wild rabbits, Oryctolagus cuniculus, in Australia and New Zealand. Wildlife Research 29:689–706.

Daszak, P., A. A. Cunningham, and A. D. Hyatt. 2000. Emerging infectious diseases of wildlife: Threats to biodiversity and human health. Science 287:443–49.

de Castro, F., and B. Bolker. 2005. Mechanisms of disease-induced extinction. Ecology Letters 8:117–26.

De La Rocque, S., V. Michel, D. Plazanet, and R. Pin. 2004. Remote sensing and epidemiology: Examples of applications for two vector-borne diseases. Comparative Immunology, Microbiology and Infectious Diseases 27:331–41.

Dietz, K. 1993. The estimation of the basic reproductive number for infectious diseases. Statistical Methods in Medical Research 2:23–41.

Dobson, A. P. 2004. Population dynamics of pathogens with multiple hosts. American Naturalist 164:S64–78.

Donnelly, C. A., R. Woodroffe, D. R. Cox, J. Bourne, G. Gettinby, A. M. Le Fevre, J. P. McInerney, and W. I. Morrison. 2003. Impact of localized badger culling on tuberculosis incidence in British cattle. Nature 426:834–37.

Gillespie, T. R., C. A. Chapman, and E. C. Greiner. 2005. Effects of logging on gastrointestinal parasite infections and infection risk in African primates. Journal of Applied Ecology 42:699–707.

Gog, J., R. Woodroffe, and J. Swinton. 2002. Disease in endangered metapopulations: The importance of alternative hosts. Proceedings of the Royal Society of London. Series B, Biological Sciences 269:671–76.

Grenfell, B. T., and B. M. Bolker. 1998. Cities and villages: Infection hierarchies in a measles metapopulation. Ecology Letters 1:63–70.

Grenfell, B., and J. Harwood. 1997. (Meta)population dynamics of infectious diseases. Trends in Ecology & Evolution 12:395–99.

Grenfell, B. T., O. N. Bjørnstad, and J. Kappey. 2001. Travelling waves and spatial hierarchies in measles epidemics. Nature 414:716–23.

Hagenaars, T. J., C. A. Donnelly, and N. M. Ferguson. 2004. Spatial heterogeneity and the persistence of infectious diseases. Journal of Theoretical Biology 229:349–59.

Hanski, I. 1994. A practical model of metapopulation dynamics. Journal of Animal Ecology 63:151–62.

Hanski, I. 1997. Metapopulation dynamics: from concepts and observations to predictive models. In Metapopulation Biology: Ecology, Genetics and Evolution, ed. I. Hanski and M. E. Gilpin, 69–91 London: Academic Press.

Harding, K. C., and J. M. McNamara. 2002. A unifying framework for meta-population dynamics. American Naturalist 160:173–85.

Haydon, D. T., S. Cleaveland, L. H. Taylor, and M. K. Laurenson. 2002. Identifying reservoirs of infection: A conceptual and practical challenge. Emerging Infectious Diseases 8:1468–73.

Hess, G. 1996. Disease in metapopulation models: Implications for conservation. Ecology 77:1617–32.

Hess, G. R. 1994. Conservation corridors and contagious disease: A cautionary note. Conservation Biology 8:256–62.

Hochachka, W. M., and A. A. Dhondt. 2000. Density-dependent decline of host abundance resulting from a new infectious disease. Proceedings of the National Academy of Sciences of the United States of America 97: 5303–06.

Holdenrieder, O., M. Pautasso, P. J. Weisberg, and D. Lonsdale. 2004. Tree diseases and landscape processes: The challenge of landscape pathology. Trends in Ecology & Evolution 19:446–52.

Jeger, M. J. 2000. Theory and plant epidemiology. Plant Pathology Oxford 49:651–58.

Kamo, M., and M. Boots. 2004. The curse of the pharaoh in space: Free-living infectious stages and the evolution of virulence in spatially explicit populations. Journal of Theoretical Biology 231:435–41.

Kat, P. W., K. A. Alexander, J. S. Smith, and L. Munson. 1995. Rabies and African wild dogs in Kenya. Proceedings of the Royal Society of London. Series B, Biological Sciences 262:229–33.

Keeling, M. J. 1999. The effects of local spatial structure on epidemiological invasions. Proceedings of the Royal Society of London. Series B, Biological Sciences 266:859–67.

Keeling, M. J., and B. T. Grenfell. 1997. Disease extinction and community size: Modeling the persistence of measles. Science 275:65–67.

Keeling, M. J., M. E. J. Woolhouse, D. J. Shaw, L. Matthews, M. Chase-Topping, D. T. Haydon, S. J. Cornell, J. Kappey, J. Wilesmith, and B. T. Grenfell. 2001. Dynamics of the 2001 UK foot and mouth epidemic: Stochastic dispersal in a heterogeneous landscape. Science 294:813–17.

Kitron, U. 1998. Landscape ecology and epidemiology of vector-borne diseases: Tools for spatial analysis. Journal of Medical Entomology 35:435–45.

Kovaliski, J. 1998. Monitoring the spread of rabbit hemorrhagic disease virus as a new biological agent for control of wild European rabbits in Australia. Journal of Wildlife Diseases 34:421–28.

Kruse, H., A.-M. Kirkemo, and K. Handeland. 2004. Wildlife as source of zoonotic infections. Emerging Infectious Diseases 10:2067–72.

Langlois, J. P., L. Fahrig, G. Merriam, and H. Artsob. 2001. Landscape structure influences continental distribution of hantavirus in deer mice. Landscape Ecology 16:255–66.

Marcogliese, D. J. 2005. Parasites of the superorganism: Are they indicators of ecosystem health? International Journal for Parasitology 35:705–16.

May, R. M., and R. M. Anderson. 1978. Regulation and stability of host-parasite interactions. II. Destabilizing processes. Journal of Animal Ecology 47:249–67.

May, R. M., and R. M. Anderson. 1979. Population biology of infectious diseases. Part II. Nature 280:455–61.

McCallum, H., N. D. Barlow, and J. Hone. 2001. How should transmission be modelled? Trends in Ecology and Evolution 16:295–300.

McCallum, H. I., and A. P. Dobson. 1995. Detecting disease and parasite threats to endangered species and ecosystems. Trends in Ecology & Evolution 10:190–94.

McCallum, H., and A. Dobson. 2002. Disease, habitat fragmentation and conservation. Proceedings of the Royal Society of London. Series B, Biological Sciences 269:2041–49.

Meyers, L. A., B. Pourbohloul, M. E. J. Newman, D. M. Skowronski, and R. C. Brunham. 2005. Network theory and SARS: Predicting outbreak diversity. Journal of Theoretical Biology 232:71–81.

Molyneux, D. H. 2002. Vector-borne infections and health related to landscape changes. In Conservation Medicine: Ecological Health in Practice, ed. A. A. Aguirre, R. S. Ostfeld, G. M. Tabor, C. House, and M. C. Pearl, 194–206. Oxford: Oxford University Press.

Morgan, E. R., E. J. Milner-Gulland, P. R. Torgerson, and G. F. Medley. 2004. Ruminating on complexity: Macroparasites of wildlife and livestock. Trends in Ecology & Evolution 19:181–88.

Mutze, G., P. Bird, J. Kovaliski, D. Peacock, S. Jennings, and B. Cooke. 2002. Emerging epidemiological patterns in rabbit haemorrhagic disease, its interaction with myxomatosis, and their effects on rabbit populations in South Australia. Wildlife Research 29:577–90.

Newman, M. E. J. 2003. The structure and function of complex networks. Siam Review 45:167–256.

Olinky, R., and L. Stone. 2004. Unexpected epidemic thresholds in heterogeneous networks: The role of disease transmission. Physical Review E 70: 030902(R).

Ostfeld, R. S., G. E. Glass, and F. Keesing. 2005. Spatial epidemiology: An emerging (or re-emerging) discipline. Trends in Ecology & Evolution 20:328–36.

Page, L. K., R. K. Swihart, and K. R. Kazacos. 2001. Changes in transmission of Baylisascaris procyonis to intermediate hosts as a function of spatial scale. Oikos 93:213–20.

Park, A. W., S. Gubbins, and C. A. Gilligan. 2001. Invasion and persistence of plant parasites in a spatially structured host population. Oikos 94:162–74.

Park, A. W., S. Gubbins, and C. A. Gilligan. 2002. Extinction times for closed epidemics: The effects of host spatial structure. Ecology Letters 5:747–55.

Patz, J. A., P. Daszak, G. M. Tabor, A. A. Aguirre, M. Pearl, J. Epstein, N. D. Wolfe, A. M. Kilpatrick, J. Foufopoulos, D. Molyneux, and D. J. Bradley. 2004. Unhealthy landscapes: Policy recommendations on land use change and infectious disease emergence. Environmental Health Perspectives 112: 1092–98.

Pfeiffer, D. U., and M. Hugh-Jones. 2002. Geographic information systems as a tool in epidemiological assessment and wildlife disease management. Revue Scientifique et Technique de l'Office International des Epizooties 21:91–102.

Poulin, R., and D. Mouillot. 2004. The evolution of taxonomic diversity in helminth assemblages of mammalian hosts. Evolutionary Ecology 18:231–47.

Power, A. G., and C. E. Mitchell. 2004. Pathogen spillover in disease epidemics. American Naturalist 164:S79–S89.

Ratcliffe, F. N., K. Myers, B. V. Fennessy, and J. H. Calaby. 1952. Myxomatosis in Australia: A step towards biological control of the rabbit. Nature 170:7–13.

Rhodes, C. J., and R. M. Anderson. 1997. Epidemic thresholds and vaccination in a lattice model of disease spread. Journal of Theoretical Biology 52:101–18.

Rodriguez, D. J., and L. Torres-Sorando. 2001. Models of infectious diseases in spatially heterogeneous environments. Bulletin of Mathematical Biology 63:547–71.

Roelke-Parker, M. E., L. Munson, C. Packer, R. Kock, S. Cleaveland, M. Carpenter, S. J. O'Brien, A. Pospischil, R. Hofmann-Lehmann, et al. 1996. A canine distemper virus epidemic in Serengeti lions (*Panthera leo*). Nature 379:441–45.

Rupprecht, C. E., C. A. Hanlon, and D. Slate. 2004. Oral vaccination of wildlife against rabies: opportunities and challenges in prevention and control. *In* Control of Infectious Animal Diseases by Vaccination, ed. A. Schudel and M. Lombard, Basel: S. Karger. 173–84.

Russell, C. A., D. L. Smith, J. E. Childs, and L. A. Real. 2005. Predictive spatial dynamics and strategic planning for raccoon rabies emergence in Ohio. PLoS Biology 3:382–88.

Sato, K., H. Matsuda, and A. Sasaki. 1994. Pathogen invasion and host extinction in lattice structured populations. Journal of Mathematical Biology 32:251–68.

Sidwa, T. J., P. J. Wilson, G. M. Moore, E. H. Oertli, B. N. Hicks, R. E. Rohde, and D. H. Johnston. 2005. Evaluation of oral rabies vaccination programs for control of rabies epizootics in coyotes and gray foxes: 1995–2003. Journal of the American Veterinary Medical Association 227:785–92.

Smith, D. L., B. Lucey, L. A. Waller, J. E. Childs, and L. A. Real. 2002. Predicting the spatial dynamics of rabies epidemics on heterogeneous landscapes. Proceedings of the National Academy of Sciences of the United States of America 99:3668–72.

Stapp, P., M. F. Antolin, and M. Ball. 2004. Patterns of extinction in prairie dog metapopulations: Plague outbreaks follow El Niño events. Frontiers in Ecology and the Environment 2:235–40.

Story, G., D. Berman, R. Palmer, and J. Scanlan. 2004. The impact of rabbit haemorrhagic disease on wild rabbit (*Oryctolagus cuniculus*) populations in Queensland. Wildlife Research 31:183–93.

Suppo, C., J. M. Naulin, M. Langlais, and M. Artois. 2000. A modelling approach to vaccination and contraception programmes for rabies control in fox populations. Proceedings of the Royal Society of London. Series B, Biological Sciences 267:1575–82.

Swinton, J., J. Harwood, B. T. Grenfell, and C. A. Gilligan. 1998. Persistence thresholds for phocine distemper virus infection in harbour seal *Phoca vitulina* metapopulations. Journal of Animal Ecology 67:54–68.

Swinton, J., M. E. J. Woolhouse, M. E. Begon, A. P. Dobson, E. Ferroglio, B. T. Grenfell, V. Guberti, R. S. Hails, J. A. P. Heesterbeck, A. Lavazza, M. G. Roberts, P. J. White, and K. Wilson. 2002. Microparasite transmission and persistence. *In* The Ecology of Wildlife Diseases, ed. P. J. Hudson, A. Rizzoli, B. T. Grenfell, H. Heesterbeck, and A. P. Dobson, 83–101. Oxford: Oxford University Press.

United States Geological Survey. 2005 West Nile Virus Maps. Available: http://westnilemaps.usgs.gov/historical.html (accessed June 6, 2005).

van den Bosch, F., F. R. Hengeveld, and J. A. J. Metz. 1992. Analysing the velocity of animal range expansion. Journal of Biogeography 19:135–50.

Vandergast, A. G., and G. K. Roderick. 2003. Mermithid parasitism of Hawaiian Tetragnatha spiders in a fragmented landscape. Journal of Invertebrate Pathology 84:128–36.

Watts, D. J., R. Muhamad, D. C. Medina, and P. S. Dodds. 2005. Multiscale, resurgent epidemics in a hierarchical metapopulation model. Proceedings of the National Academy of Sciences of the United States of America 102: 11157–62.

White, A., M. Begon, and R. G. Bowers. 1996. Host-pathogen systems in a spatially patchy environment. Proceedings of the Royal Society of London. Series B, Biological Sciences 263:325–32.

Williams, J. C., C. S. ReVelle, and S. A. Levin. 2004. Using mathematical optimization models to design nature reserves. Frontiers in Ecology and the Environment 2:98–105.

Wobeser, G. 2002. Disease management strategies for wildlife. Revue Scientifique et Technique de l'Office International des Epizooties 21:159–78.

Woolhouse, M. E., C. Dye, J. F. Etard, T. Smith, J. D. Charlwood, G. P. Garnett, P. Hagan, J. L. Hii, P. D. Ndhlovu, et al. 1997. Heterogeneities in the transmission of infectious agents: implications for the design of control programs. Proceedings of the National Academy of Sciences of the United States of America 94:338–42.

Xia, Y. C., O. N. Bjornstad, and B. T. Grenfell. 2004. Measles metapopulation dynamics: A gravity model for epidemiological coupling and dynamics. American Naturalist 164:267–81.

PART II

Effects of Disease on Ecosystems

Introduction

Valerie T. Eviner

WHILE MANY STUDIES DEMONSTRATE that pathogens can have dramatic impacts on their hosts, substantially fewer studies have explored the ecological consequences of these pathogen-induced changes in hosts. Pathogen effects on host behavior, reproduction, and mortality can influence community interactions such as competition, facilitation, predation, and invasion. These pathogen-induced changes in host individuals and communities can have strong impacts on ecosystem processes (e.g., productivity, nutrient cycling) and landscape structure and function (e.g., disturbance regimes, land use, land-atmosphere interactions). While we know that pathogens have the *potential* to shape communities, ecosystems, and landscapes, we still have poor ability to predict the magnitude and duration of the impacts of any specific pathogen, or how the impacts of that pathogen will vary depending on biotic and abiotic factors. Our goal in this section was to incorporate the role of disease into our conceptual models for community, ecosystem, and landscape ecology in order to enhance our ability to predict and manage ecosystems and pathogens. Such an endeavor requires going beyond the expertise of any one research group, integrating many different perspectives. A rough roadmap of how these different areas of ecology intersect (figure II.1) was provided at the outset, and each author was charged with synthesizing knowledge on how pathogens impact a subset of the boxes and arrows in figure II.1.

In chapter 6, the first chapter of this section, Sharon Collinge and colleagues discuss the impacts of pathogens on keystone species, dominant species, and feedbacks between hosts, communities, and pathogens. This chapter highlights how spatial variation in the ecological role of keystone species can cause spatial variation in disease dynamics. Keith Clay and colleagues in chapter 7 explore how the role of pathogens in fostering host diversity depends on host frequency, density, and distance, and temporal oscillations due to factors such as time lags, pathogen cycles, and selection. Both chapter 7 and chapter 8 point out that pathogens can strongly influence the fate of a new species arriving in an ecosystem, and how a new species interacts with the resident community. This is particularly relevant to the topic of invasive species, addressed in chapter 8 by Sarah Perkins and colleagues. Although

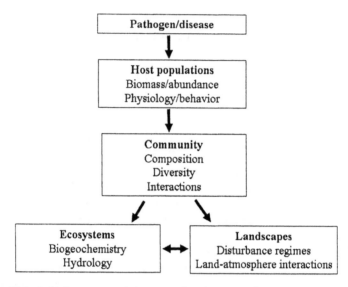

Figure II.1. Initial conceptual framework relating pathogens to communities, ecosystems, and landscapes.

considerable recent attention has focused on how invasions may be facilitated by escape from pathogens, this chapter addresses the tough questions: which pathogens are likely to be lost and gained during invasion, and how that influences the magnitude and duration of invasive species' release from pathogens. These authors also discuss how invasive species may alter pathogen dynamics in native host communities. Kevin Lafferty in chapter 9 shows that parasites play an integral role in determining a large suite of community interactions, including competition, predation, parasitism, and mutualism. The explicit inclusion of parasites in our conceptualization of communities thus greatly increases the connectivity and complexity of these communities. Spencer Hall and colleagues in chapter 10 explore the extent to which pathogen-host interactions and predator-prey interactions mirror one another or are unique, carefully detailing key life history characteristics that can be used to group consumer-resource interactions into qualitatively distinct processes.

The ecosystem consequences of pathogen impacts on populations and communities are explored in the final four chapters of this section. Mathias Middelboe in chapter 11 shows that the conceptual model of marine biogeochemical cycling is transformed by considering how pathogens drive bacterial turnover in marine systems, which accounts for the previously unexplained high ratio of bacteria to NPP in oligotrophic waters. Valerie Eviner and Gene Likens in chapter 12 provide examples of pathogens that play integral roles in terrestrial biogeochemistry through mediating processes such as self-thinning, competition, and succession, while also exploring the ecosystem impacts of pathogens that act as disturbance agents. They articulate some general principles that determine the magnitude and duration of pathogen impacts on ecosystems. Terry Chapin and colleagues in chapter 13 integrate ecological dynamics with socioeconomic systems to demonstrate how landscape processes are affected by human effects on and responses to disease dynamics. They focus on how human activity and pathogens influence the connectivity of socioecological landscapes and on landscape vulnerability to fundamental state changes. Finally, Sharon Deem and colleagues in chapter 14 explore whether the impacts of infectious disease are increasing over time, and the extent to which humans are contributing to these impacts. They carefully assess available evidence

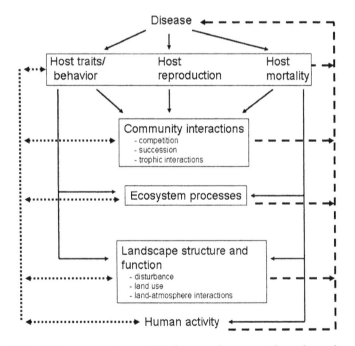

Figure II.2. Full conceptual model linking pathogens with ecological and socioeconomic systems. Pathogen-induced changes in hosts directly influence communities, ecosystems, landscapes, and human activity (solid lines). These direct impacts can then cause large changes in socioecological systems as they feed back to influence other components of the system (dotted lines). Changes in hosts, communities, ecosystems, landscapes, and human activity then feed back to influence pathogens (dashed lines).

and propose a research agenda to tackle these questions by integrating the fields of ecology, veterinary medicine, human medicine, public health, sociology, and economics.

While each of these chapters addresses a different component of ecosystems, many point to a few key factors that regulate the magnitude and duration of pathogen impacts on ecosystems:

- Host specificity of pathogens in relation to disease susceptibility of hosts

- Community interactions within the host community and within the pathogen community

- How pathogen impacts on communities and ecosystems feed back to influence disease dynamics

- Human effects on and responses to disease

Collectively, the chapters in this section underscore the integral role played by pathogens in populations, communities, ecosystems, landscapes, and socioeconomic systems. Our understanding of every level of ecological organization, and the interactions between these levels, is greatly enhanced by the explicit inclusion of pathogens in our conceptual ecological models. This extends our initial conceptual framework in figure II.1 into a much richer, more comprehensive framework that highlights the importance of feedback and interactions in understanding the roles of pathogens in socioecological systems (figure II.2).

CHAPTER SIX

Effects of Disease on Keystone Species, Dominant Species, and Their Communities

Sharon K. Collinge, Chris Ray, and Jack F. Cully, Jr.

SUMMARY

KEYSTONE SPECIES DISPROPORTIONATELY AFFECT THE ABUNDANCE and distribution of other species and, because of their central functional role, may significantly influence the dynamics of infectious diseases in the communities in which they occur. For example, changes in the abundance of keystone species due to disease may cause major shifts in community composition and ecosystem function, thereby providing dynamic feedback to disease transmission processes. We highlight specific examples of the effects of disease on keystone and dominant species, using as examples sea otters in coastal California, chestnut trees in the eastern United States, and prairie dogs in western grasslands. Sea otters are currently experiencing widespread mortality due to parasitic infections. Because sea otters have demonstrable effects on the structure and function of near-shore kelp communities, their population declines due to disease are expected to have pervasive effects on species composition and ecological processes. American chestnut was a dominant species of eastern U.S. forests, and its historic decline due to chestnut blight may have influenced the abundance and distribution of seed predators that affect the present-day dynamics of Lyme disease. Black-tailed prairie dogs, a key species in prairie ecosystems, suffer 95%–99% mortality in colonies affected by plague. We explore the possibilities for ecological feedbacks between prairie dogs, plague, and other rodent species, and use data from several sites throughout the prairie dog's geographic range to consider how spatial variation in the keystone effect of prairie dogs should translate into spatial variation in disease dynamics. We conclude with a general framework for understanding the effects of disease on dominant and keystone species that specifies ecological and epidemiological feedback among host density, community structure, and pathogen prevalence.

INTRODUCTION

Keystone species disproportionately affect the abundance and distribution of other species and, because of their functional role, may significantly influence the dynamics of infectious diseases in the communities in which they occur. For example, changes in the abundance of keystone species due to disease may cause major shifts in community composition and ecosystem function, thereby providing dynamic feedback to disease transmission processes. Although the keystone species concept has been central to community and ecosystem ecology since its formalization in the late 1960s (Paine 1966, 1969), the epidemiological effects of keystone species have only recently been considered.

Parasites and pathogens may themselves be considered keystone species (e.g., Dobson and Crawley 1994; McCallum and Dobson 1995), given their ability to reduce the abundance of particular species and thereby cause shifts in community structure. There are several excellent examples of parasites as keystone species, such as avian malaria in Hawaiian bird communities (Van Riper et al. 1986) and the rinderpest virus in Africa (Dobson and Hudson 1986). Less well understood is how the parasites or pathogens of a keystone *host* species may affect community-level changes that are mediated through the functional roles of the host.

In this chapter, we present a brief history of the keystone species concept, then review examples from three pathogen-host systems to emphasize several points regarding the effects of disease on keystone species and their communities. We highlight specific examples of effects of disease on keystone and dominant host species in North America, including sea otters in waters off the West Coast, the American chestnut tree in eastern forests, and black-tailed prairie dogs in western grasslands. Taken together, these examples suggest three important ways that dominant or keystone hosts may affect disease dynamics within a community. First, because the effects of keystone species vary in both space and time, there are implications for the ultimate effects of disease on ecosystems. Second, the impacts of disease on keystone host species can feed back to affect disease dynamics, depending on how the keystone species affects the abundance and distribution of hosts, vectors, or pathogens. And finally, we suggest a general framework for hypotheses concerning whether keystone species in an ecological sense may also be keystone species in an epidemiological sense.

THE KEYSTONE SPECIES CONCEPT

A Brief History

Paine (1966, 1969) originally conceived of the keystone species concept based on his observations and experiments in intertidal invertebrate communities. Using now classic field experiments, Paine showed that removal of the predatory sea star, *Pisaster ochraceus*, caused major alterations in the composition of invertebrate communities. One year after initiation of Paine's sea star-removal experiment, study plots in which sea stars were present harbored fifteen species of invertebrates, compared with only eight species in sea star-removal plots. Where sea stars were absent, the mussel *Mytilus californianus* dominated. Where sea stars were present, predation reduced the abundance of the competitively dominant mussel, allowing for greater species diversity. Noting the dramatic effect of the predatory sea star on community composition, Paine termed it a keystone species because it preferentially consumed prey that would otherwise dominate the community, thereby enhancing species diversity. Paine's architectural analogy to the keystone of an arch is appropriate, because the keystone is essential for the structural and functional integrity of the ecological community.

Spatial and Temporal Variation in Keystone Effects

Does a keystone species always function as a keystone? Additional research on the impacts of the same sea star, *Pisaster ochraceus*, in similar rocky intertidal ecosystems as those studied by Paine has revealed that these interactions are contingent on a number of ecological factors. Broad-scale studies of marine intertidal communities in Oregon and Washington states have shown that the effects of sea stars on invertebrate communities vary both spatially and temporally (Menge et al. 1994; Navarrete and Menge 1996). For example, the intensity of sea star predation, and thus the keystone effect of sea stars, depended on wave exposure and rates of prey production at particular sites. These studies demonstrated that the impact of a keystone species may be high under some conditions but relatively modest under other conditions.

This spatial and temporal variation in keystone effects has implications for the influence of disease on keystone species. If a keystone species has strong effects on community structure at a particular place and time, then the decline or local extinction of that species due to a pathogen may cause large shifts in community structure. Conversely, if a keystone species has relatively weak impacts on community structure, a pathogen that negatively affects the keystone species may cause relatively small changes in species composition. We discuss the issue of spatial and temporal variation specifically in the context of disease effects on both sea otters and black-tailed prairie dogs.

Refinement of the Keystone Concept

As a result of broad usage of the term "keystone species" in the ecological literature during the years following Paine's publications, some ecologists thought the keystone concept was losing its distinct meaning (e.g., Mills et al. 1993). In response to this concern, and because of the perceived need to incorporate the keystone species concept into biological conservation efforts, a group of ecologists met to consider the concept and arrive at an operational definition of a keystone species. This group proposed an expanded definition that included interactions other than trophic interactions: "a keystone species is a species whose impacts on its community or ecosystem are large, and much larger than would be expected from its abundance" (Power and Mills 1995: 184).

Kotliar (2000), using prairie dogs as an example of a keystone species, questioned the emphasis on low species relative abundance put forth by Power and Mills (1995). She suggested that ecologists abandon the duality of the keystone versus non-keystone species construct and instead focus on the relationship between the abundance of a species and its importance in the community. Kotliar further argued that keystone species should be identified as those that play unique versus redundant roles in ecological communities. We discuss the issue of species uniqueness versus redundancy later in reference to the American chestnut.

These expanded views characterize as keystones those species that are relatively rare but have large and unique effects on communities and ecosystems. This distinguishes them from dominant species, which may have similarly strong interactions but are more abundant members of communities. In the context of disease effects, the critical issue is that keystone and dominant species play central functional roles in ecological communities, and their decline may prompt a cascade of shifts in community structure and function.

RECENT EXAMPLES OF KEYSTONE SPECIES

Since the publication of Paine's classic work and the expanded definition of keystone species put forth by Power and Mills (1995) and Kotliar (2000), many ecologists have described species that strongly interact and have large effects on communities and ecosystems. Studies have addressed diverse species, from tree ferns (Ough and Murphy 2004), for their effects on local microclimate and forest processes, to harvester ter-

TABLE 6.1

Results of ISI Web of Science search for records containing "keystone species AND disease"

Species	Disease	Citation
Sea fan, reef-building corals	Aspergillosis, yellow band disease	Bruno et al. (2003)
Sugar maple	*Armillaria* (root disease)	Horsley et al. (2002)
Acacia koa	Vascular wilt fungus	Anderson et al. (2002)
Whitebark pine	White pine blister rust	Murray and Rasmussen (2003)
Endemic parasites and pathogens	Many	McCallum and Dobson (1995)

Note: The ISI web of Science search covered the period 1971–2005 and yielded seven records, five of which were empirical papers illustrating effects in the host-pathogen systems listed in the table.

mites (Zaady et al. 2003), for their influence on nutrient dynamics, to the Louisiana red swamp crayfish (Smart et al. 2002), for its impacts on aquatic plant community structure.

As an indication that the study of keystone species is an active area of ecological research, a recent ISI Web of Science search for "keystone species" yielded 262 records. A more refined search for "keystone species and disease" yielded only seven records. Of the seven, five were empirical papers that described the effects of disease on keystone species, and subsequent shifts in community structure and function (table 6.1). These keystones included several dominant tree species that are currently declining as a result of pathogens. For example, one tree endemic to Hawaii, *Acacia koa*, is affected by a vascular wilt fungus (similar to *Fusarium oxysporum*), which has caused die-offs of mature stands (Anderson et al. 2002). This pathogen clearly has the potential to create major shifts in community structure through its effects on this dominant tree species.

Impacts of Disease on Keystone and Dominant Species

Southern Sea Otters and Parasites

The southern sea otter (*Enhydra lutris nereis*) was one of the first species to be explicitly described as a keystone, based on research conducted in the early 1970s (Estes and Palmisano 1974). The sea otter was considered a keystone species because of its observed dramatic impacts

on Pacific nearshore kelp communities. The authors used a natural experiment to document the impacts of sea otters on kelp communities. In
Alaska's Aleutian Islands, sea otters had been locally extirpated by fur
traders in the early twentieth century. Along some stretches of coastline,
sea otters had recovered from overexploitation, but along other parts of
the coastline they were still absent. Estes and Palmisano (1974) observed
that where sea otters were present, kelp forests and their associated flora
and fauna flourished, presumably because sea otters fed on sea urchins,
which were grazers of kelp. In areas where sea otters had not recovered,
kelp forests had been decimated by grazers, primarily sea urchins. These
observations clearly fit the concept of a keystone species as described by
Paine (1969).

Sea otters have continued to recover in Alaska and in the coastal waters of California since their harvest was restricted, but population
recovery appears to have recently stalled. Researchers from the USGS
Western Ecological Research Center have noted a lack of population increase since 1995, and actual declines from 1996 to 2002. This has
prompted research into the causes of decline. Several culprits have been
suspected, but recent research has revealed high levels of parasitic infections in California populations of the sea otter. For example, Thomas
and Cole (1996) observed that infectious disease and parasites accounted
for 40% of the mortality of necropsied otters that died between 1992
and 1995.

The protozoan parasite *Toxoplasma gondii* occurred in 52% of dead
sea otters (Conrad et al. 2005; Jessup et al. 2004; Miller et al. 2002,
2004). The definitive hosts of *T. gondii* are felids. It is suspected that domestic cats may contribute to the high prevalence of infection in sea otters, through an accident of hydrodynamics. Sea otters sampled near
areas of high freshwater runoff exhibited a threefold increase in sero-
positivity to *T. gondii* compared with otters sampled in areas more distant from these possible sources of infection (Miller et al. 2002). Cat
feces may be entering marine systems at these maximal runoff areas and
concentrating these sources of infection for the sea otters.

It is not yet clear whether *T. gondii* or other parasites that affect sea otters, such as *Sarcocystis neurona* (whose definitive host is the opossum),
are severe enough to regulate populations of sea otters, but the potential
exists for impacts to reverberate throughout these nearshore marine communities. Because sea otters play a keystone role in these ecosystems, population declines due to infectious disease and parasites are likely to affect
the integrity of kelp forests in these systems. Sea otters may feed preferentially on large sea urchins, which other predatory species are unable to
eat. These large urchins tend to have the greatest impact on kelp mortality
as a result of their distinctive feeding pattern. Large urchins tend to feed

at the base of the kelp, near the holdfast. This feeding pattern tends not only to reduce biomass of the kelp but also to weaken it so that the holdfast becomes detached and the entire frond dies.

There is also some evidence that sea otters exhibit spatial and temporal variation in their effects on nearshore kelp communities. Dean et al. (2000) studied interactions between and among sea otters, urchins, and kelp following the *Exxon Valdez* oil spill in Prince William Sound, Alaska. Sea otter abundance was reduced by the oil spill by approximately 50%, and following the reduction, the affected areas harbored larger sea urchins than unaffected areas, where sea otter abundance was higher. However, both the density and the biomass of sea urchins were similar in areas with reduced and intact sea otter densities. Further, kelp abundance did not appear to be affected in areas of lower sea otter density. This contrasts with previous observations (e.g., Estes and Palmisano 1974) on other parts of the Alaskan coast, where sea otters appeared to exert strong positive effects on kelp communities.

American Chestnut and Chestnut Blight

The American chestnut, *Castanea dentata*, was historically dominant within deciduous forests in the eastern United States. Paillet (2002) used palynological studies and reviews of historical documents to conclude that the "chestnut comprised over 50% of timber by volume on well-drained slopes" (2002: 1519). In the early twentieth century, a fungal pathogen called chestnut blight, *Cryphonectria parasitica,* was inadvertently introduced from Asia. This pathogen effectively rendered American chestnut ecologically extinct. Although chestnut can regenerate asexually via sprouting, these shoots do not usually produce viable seeds. Chestnuts require cross-fertilization for successful reproduction, and most shoots do not survive long enough to flower and produce seed. Early silviculturists had tried to propagate chestnut by seed, but, as Paillet noted, researchers "concluded that chestnut seed predation was so extensive . . . that propagation by seed was just not feasible" (Paillet 2002: 1519).

The observation that seed predation was so extensive suggests a possibly critical role for chestnut in eastern deciduous forest ecosystems. Ecologists often refer to figs in both New and Old World rainforests as keystone species for their role in providing food to dozens of animals (e.g., Terborgh 1986), and it is possible that American chestnut may have similarly provided a crucial and predictable food supply for many forest animals. Because chestnuts are now effectively absent, it may be important to consider whether oaks in current eastern deciduous forests serve a functionally redundant role to that previously played by chestnuts.

The potential for oaks to play this role is reduced because oaks are masting species, producing acorns in abundance only during mast years, which occur unpredictably in time. The American chestnut is not a mast-ing species, so it produced consistent seed crops each year. This difference may have been critical to population, community, and disease dynamics within these ecosystems. For example, recent research suggests that oak mast is followed by a cascade of processes affecting disease prevalence and defoliation in eastern forests (Jones et al. 1998). Oak mast leads to an in-crease in the abundance of the white-footed mouse, *Peromyscus leucopus*, which is both a predator of the gypsy moth and a host to the spirochete that causes Lyme disease. Through its effects on mouse density, oak mast increases the prevalence of Lyme disease and decreases the defoliation caused by larvae of the gypsy moth. We can only speculate on how the de-cline of the chestnut and rise of the oak has altered community and disease dynamics, since we know little about the dynamics of these forests prior to the introduction of chestnut blight.

Prairie Dogs and Plague

Black-tailed prairie dogs have been described as keystone species in the mixed- and shortgrass prairie ecosystems in western North America, because they influence the abundance and distribution of many native plant and animal species, including black-footed ferrets, burrowing owls, and mountain plovers (Kotliar 2000; Kotliar et al. 1999). Through their burrowing activities, prairie dogs also change nutrient availability and water infiltration in grassland soils (Whicker and Detling 1988) and provide refuges for other species, such as tiger salamanders and rattle-snakes (Kretzer and Cully 2001).

Prairie dog populations have declined throughout the West owing to habitat conversion, recreational shooting, and poisoning, but plague has also played a major role (Miller and Cully 2001). Prairie dogs are highly susceptible to plague caused by the bacterial pathogen *Yersinia pestis*. Sylvatic plague was introduced to North America from Asia around 1900 and is now present throughout most western states of the United States (Antolin et al. 2002; Barnes 1982; Gage et al. 1995). Plague spreads through contact between flea vectors and mammalian hosts (Barnes 1982; Perry and Featherston 1997). The occurrence of plague in small mammals that appear to be moderately resistant to the disease, such as deer mice (*Peromyscus maniculatus*), northern grasshopper mice (*Onychomys leucogaster*), and wood rats (*Neotoma* spp.), suggests that these animals may serve as maintenance or reservoir hosts in which plague persists in the enzootic portion of the plague cycle (Barnes 1982; Gage et al. 1995). When plague enters prairie dog colonies, it causes nearly 100% mortality in black-tailed (*C. ludovicianus*) and Gunnison's

(*C. gunnisoni*) prairie dogs (Cully 1997; Cully and Williams 2001), and high but more variable mortality in white-tailed (*C. leucurus*) (Anderson and Williams 1997) and Utah (*C. parvidens*) prairie dogs (Biggins, personal communication).

The role of prairie dogs in plague dynamics is the subject of current study, but it is likely that black-tailed prairie dogs influence the abundance and distribution of several rodent species that act as alternative hosts for plague (Cully et al. 1997, 2000; Gage et al. 1995). If prairie dogs influence the abundance of other plague hosts or the frequency of interactions among hosts and vectors of plague, they may play a keystone role in the epidemiology of plague (Ray and Collinge 2006). Prairie dogs may affect the frequency of host and vector interactions if the resources available in prairie dog colonies (e.g., burrows) attract and concentrate hosts and vectors (Cully and Williams 2001).

Our research goals include an assessment of the extent to which the black-tailed prairie dog as a keystone species may alter disease dynamics in prairie communities. We anticipate that two aspects of prairie dog ecology figure prominently in this keystone role. First, we have observed that black-tailed prairie dogs influence the distribution and abundance of associated rodent species that serve as putative reservoir hosts for disease (Cully et al. 2008; Ray and Collinge 2006; figure 6.1). Second, because prairie dogs suffer high mortality when infected with plague, we are investigating whether they may amplify or dilute plague transmission in native rodent communities (Cully et al. 1997; Gage et al. 1995). For example, massive prairie dog mortality may concentrate scavengers, amplifying plague transmission; conversely, rapid mortality may reduce host or host-vector contact rates, diluting plague transmission.

Further, we have observed both spatial and temporal variation in this keystone effect of prairie dogs on other grassland rodent species. As part of our research on plague ecology, we have surveyed small mammals at more than seventy study sites in four study areas located in Wyoming, Colorado, Kansas, and South Dakota. By sampling rodents at sites of prairie dog colonies as well as on adjacent grasslands that do not have prairie dog colonies, we can infer the effects of prairie dogs on grassland rodent assemblages. In 2003, for example, we observed much higher rodent abundance on prairie dog colonies than at adjacent grassland sites for two of the four study areas (Boulder, Colorado, and the Badlands, South Dakota; see figure 6.1a). At the other two study areas, we observed similar numbers of small mammals on prairie dog colony sites as at off-colony sites. It is clear from this broad-scale survey that this aspect of the keystone effect of prairie dogs varies spatially. The abundance of other rodents within prairie dog colonies may critically affect the likelihood of prairie dogs contracting plague, if the main route of transmission is through contact with other rodent hosts.

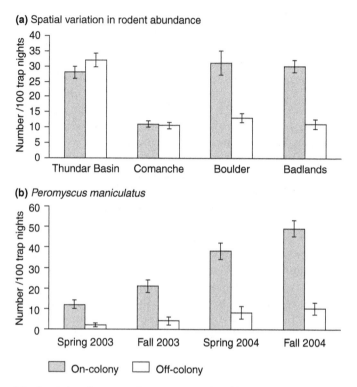

Figure 6.1. Spatial and temporal variation in the effects of prairie dogs on small mammals. (a) Average number of rodents captured per 100 trap-nights on and off prairie dog colonies at four grassland sites. Thunder Basin = Thunder Basin National Grassland, Wyoming; Comanche = Comanche National Grassland, Colorado; Boulder = grassland sites in Boulder County, Colorado; Badlands = Badlands National Park, South Dakota. Means ± SE are presented. (b) Abundance (number of captures/100 trap-nights) of the deer mouse, *Peromyscus maniculatus*, at on- and off-colony sites in Boulder County, Colorado, in 2003 and 2004. Spring trapping sessions in both years occurred from mid-May to mid-June; fall trapping sessions in both years occurred from mid-August to mid-September. Means ± SE are presented. (Data from Cully et al. 2008.)

Similarly, we have observed wide fluctuations among trapping sessions in the extent to which prairie dogs influence rodent abundance. For example, the abundance of the deer mouse in grasslands of Boulder County, Colorado, was significantly higher on prairie dog colonies than on adjacent grassland sites in each of four trapping sessions in 2003 and 2004 (figure 6.1b), but the magnitude of the difference between on- and off-colony sites varied tremendously. Deer mouse abundance has increased significantly across the study area since spring 2003, but it has increased to a much greater extent on prairie dog colonies than in adjacent grasslands. This temporal variation in a keystone effect of prairie dogs is likely to influence plague dynamics. If deer mice are a reservoir host for plague, it is reasonable to expect that as deer mouse abundance increases, the transmission and spread of plague to other species become more likely. But this increase is exacerbated in prairie dog colonies, suggesting that prairie dogs may play a keystone role in plague epidemiology.

In addition to their keystone role in influencing reservoir host abundance, prairie dogs also appear to affect the abundance and distribution of the fleas that serve as vectors for plague transmission. In the course of our studies, we recorded the number and identity of all fleas observed on small mammals captured on and off prairie dog colonies. Both flea abundance (the number of fleas per host) and flea species richness (the number of flea species present per site) were significantly higher on mice captured within prairie dog colonies than at adjacent off-colony sites (figure 6.2). This abundance of fleas within prairie dog colonies is likely due to the suitability of prairie dog burrows for fleas (Gage and Kosoy 2005), and provides further evidence that prairie dogs may play a keystone role in plague epidemiology (Ray and Collinge 2006). Because fleas serve as vectors for pathogen transmission, the greater number of

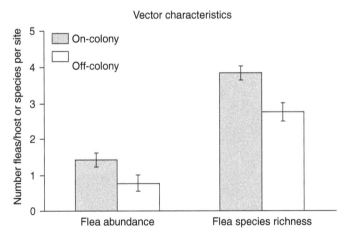

Figure 6.2. Characteristics of flea vectors on mice at prairie dog colony sites and adjacent grasslands in Boulder County, Colorado, in 2003. Flea abundance is the number of fleas per mouse and flea species richness is the number of flea species per study site. Means ± SE are presented. (Unpublished data from S. K. Collinge.)

fleas present on hosts within prairie dog colonies should increase the transmission and spread of the pathogen between animals.

CONCLUSIONS

We present a conceptual framework (figure 6.3) that may be useful in guiding hypotheses concerning the ecological effects of disease in key host species. In systems that feature keystone or dominant species, disease dynamics may be controlled by several feedback pathways that include both ecological and epidemiological effects. First, dynamics may be controlled by feedback between pathogen prevalence and the density of the keystone or dominant host species (arrows a and b in figure 6.3). There may be positive or negative feedback to disease dynamics if pathogens alter spatial or temporal patterns of host density and if pathogen transmission is density dependent. Second, key species have ecological effects on community structure (arrow c in figure 6.3), as discussed in this chapter for both sea otters and prairie dogs. Community structure

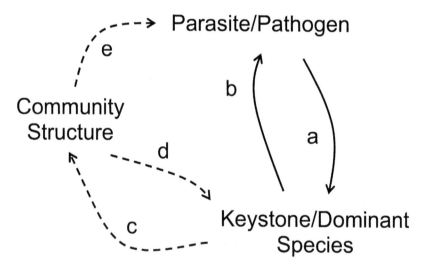

Figure 6.3. Potential feedbacks involving the epidemiological (solid arrows) and ecological (dashed arrows) effects of disease mediated through a keystone or dominant (key) host species. Dynamics may be controlled by feedback between pathogen prevalence and key host density (a and b), by ecological effects of key species on community structure (c), and by effects of community structure on the potential for (re)invasion of the key species (d) and the pathogen (e). Outcomes should depend on the relative strength and frequency of each effect.

likewise affects the potential for (re)invasion of the key species (arrow d in figure 6.3). Finally, community structure, mediated by the keystone species, affects the potential for persistence or (re)invasion of the pathogen (arrow e in figure 6.3).

We expect that when diseases affect key (keystone or dominant) species, there will be large impacts on community structure and ecosystem function. The most intuitive effect relates to maintenance of the keystone species in the ecosystem. Disease in a key host will decrease the density of that host and, subsequently, the potential for disease. An excellent example of an exception to this rule, however, is the chestnut system, in which the disease persists despite a severe reduction in the biomass of the host. Nevertheless, the intuitive hypothesis is that the functional role of the key species should be diminished by disease, and this hypothesis is apparently borne out even in the chestnut case. Following a reduction in key host density, the disease should generally diminish, allowing the key species to increase and resume its functional role. A potential caveat to this scenario, especially for keystone species, is that the state of a community may change in the absence of a keystone species, such that the keystone species cannot regain its former role.

A second, less intuitive effect of disease on a key species involves feedbacks generated through the ecological effects of the key host on the structure of its community. In this scenario, a disease-related decline in the key host causes a shift in community structure that may either increase or decrease the potential for maintenance of the disease within the community. This scenario is most likely for generalist parasites or pathogens that depend on interspecific interactions for maintenance within a community, or parasites or pathogens with complex life cycles that depend on transmission between species on different trophic levels (see, e.g., Collinge and Ray 2006).

We have little doubt that species considered keystone or dominant in an ecological sense have the potential to act as key hosts in an epidemiological sense, that is, to mediate the impact of disease within a community. But the potential for key host species to mediate the effects of parasites or pathogens on the structure of a community should depend on the strength of the interaction between key hosts and other species within the community. These interaction strengths vary in space and time. Therefore, the epidemiological importance of dominant or keystone species is likely variable, which may be one reason why this link has not been recognized more frequently (recall that only seven studies were found that refer both to keystone species and disease; see table 6.1). Given the recent emergence of important diseases in several complex communities, more attention should be given to the potential for this dual effect of keystone species.

Acknowledgments

We thank Pat Conrad, who provided critical insights on sea otter parasitism, and the many fieldworkers who collected the rodent and flea data for the prairie dog study, especially Amelia Markeson, David Conlin, Jory Brinkerhoff, Bala Thiagarajan, and Ben Erie. Research was supported by a grant from the National Center for Environmental Research STAR program of the U.S. EPA (R-82909101-0) and a grant from the NSF/NIH joint program in Ecology of Infectious Diseases (DEB-0224328).

Literature Cited

Anderson, R. C., D. E. Gardner, C. C. Daehler, and F. C. Meinzer. 2002. Dieback of *Acacia koa* in Hawaii: Ecological and pathological characteristics of affected stands. Forest Ecology and Management 162:273–86.
Anderson, S. H., and E. S. Williams. 1997. Plague in a complex of white-tailed prairie dogs and associated small mammals in Wyoming. Journal of Wildlife Diseases 33:720–32.
Antolin M. F., P. Gober, B. Luce, D. E. Biggins, W. E. Van Pelt, D. B. Seery, M. Lockhart, and M. Ball. 2002. The influence of sylvatic plague on North American wildlife at the landscape level, with special emphasis on black-footed ferret and prairie dog conservation. *In* Transactions of the 67th North American Wildlife and Natural Resources Conference, 104–27.
Barnes, A. M. 1982. Surveillance and control of plague in the United States. *In* Animal Disease in Relation to Animal Conservation, ed. M. A. Edwards and U. McDonnell, 237–70. New York: Academic Press.
Bruno, J. F., L. E. Petes, C. D. Harvell, and A. Hettinger. 2003. Nutrient enrichment can increase the severity of coral diseases. Ecology Letters 6:1056–61.
Collinge, S. K., and C. Ray. (Eds.). 2006. Disease Ecology: Community Structure and Pathogen Dynamics. Oxford: Oxford University Press.
Conrad, P. A., M. E. Grigg, C. Kreuder, E. R. James, J. Mazet, H. Dabritz, D. A. Jessup, F. Gulland, and M. A. Miller. 2005. Sea otters serve as sentinels for protozoal pathogens transmitted from the terrestrial hosts to marine mammals (abstract). *In* Cary Conference 2005: Infectious Disease Ecology, 56. Millbrook: Institute for Ecosystem Studies.
Cully, J. F., Jr. 1997. Growth and life-history changes in Gunnison's prairie dogs after a plague epizootic. Journal of Mammalogy 78:146–57.
Cully, J. F., Jr., A. M. Barnes, T. J. Quan, and G. Maupin. 1997. Dynamics of plague in a Gunnison's prairie dog colony complex from New Mexico. Journal of Wildlife Diseases 33:706–719.
Cully, J. F., Jr., L. G. Carter, and K. L. Gage. 2000. New records of sylvatic plague in Kansas. Journal of Wildlife Diseases 36:389–92.
Cully, J. F., Jr., S. K. Collinge, C. Ray, B. Holmes, D. Conlin, W. C. Johnson, and T. Johnson. 2008. Spatial variation in small mammal communities associated with black-tailed prairie dogs. Unpublished manuscript.

Cully, J. F., Jr., and E. S. Williams. 2001. Interspecific comparisons of sylvatic plague in prairie dogs. Journal of Mammalogy 82:894–905.

Dean, T. A., J. L. Bodkin, S. C. Jewett, D. H. Monson, and D. Jung. 2000. Changes in sea urchins and kelp following a reduction in sea otter density as a result of the *Exxon Valdez* oil spill. Marine Ecology Progress Series 199:281–91.

Dobson, A. P., and M. Crawley. 1994. Pathogens and the structure of plant communities. Trends in Ecology & Evolution 9:393–98.

Dobson, A. P., and P. J. Hudson. 1986. Parasites, disease, and the structure of ecological communities. Trends in Ecology & Evolution 1:11–15.

Estes, J. A., and J. F. Palmisano. 1974. Sea otters: Their role in structuring nearshore communities. Science 185:1058–60.

Gage, K. L., and M. Y. Kosoy. 2005. Natural history of plague: Perspectives from more than a century of research. Annual Review of Entomology 50:505–28.

Gage, K. L., R. S. Ostfeld, and J. G. Olson. 1995. Nonviral vector-borne zoonoses associated with mammals in the United States. Journal of Mammalogy 76:695–715.

Horsley, S. B., R. P. Long, S. W. Bailey, R. A. Hallett, and P. M. Wargo. 2002. Health of eastern North American sugar maple forests and factors affecting decline. Northern Journal of Applied Forestry 19:34–44.

Jessup, D. A., M. Miller, J. Ames, M. Harris, C. Kreuder, P. A. Conrad, and J. A. K. Mazet. 2004. Southern sea otter as a sentinel of marine ecosystem health. EcoHealth 1:239–45.

Jones, C. G., R. S. Ostfeld, M. P. Richard, E. M. Schauber, and J. O. Wolff. 1998. Chain reactions linking acorns to gypsy moth outbreaks and Lyme disease risk. Science 279:1023–26.

Kotliar, N. B. 2000. Application of the new keystone species concept to prairie dogs: How well does it work? Conservation Biology 14:1715–21.

Kotliar, N. B., B. W. Baker, A. D. Whicker, and G. Plumb. 1999. A critical review of assumptions about the prairie dog as a keystone species. Environmental Management 24:177–92.

Kretzer, J. E., and J. F. Cully. 2001. Effects of black-tailed prairie dogs on reptiles and amphibians in Kansas shortgrass prairie. Southwestern Naturalist 46:171–77.

McCallum, H., and A. P. Dobson. 1995. Detecting disease and parasite threats to endangered species and ecosystems. Trends in Ecology & Evolution 10: 190–93.

Menge, B. A., E. L. Berlow, C. A. Blanchette, S. A. Navarrete, and S. B. Yamada. 1994. The keystone species concept: Variation in interaction strength in a rocky intertidal habitat. Ecological Monographs 64:249–86.

Miller, M. A., I. A. Gardner, C. Kreuder, D. M. Paradies, K. R. Worcester, D. A. Jessup, E. Dodd, M. D. Harris, J. A. Ames, A. E. Packham, and P. A. Conrad. 2002. Coastal freshwater runoff is a risk factor for *Toxoplasma gondii* infection of southern sea otters (*Enhydra lutris nereis*). International Journal for Parasitology 32:997–1006.

Miller, M. A., M. E. Grigg, C. Kreuder, E. R. James, A. C. Melli, P. R. Crosbie, D. R. Jessup, J. C. Boothroyd, L. D. Brownstein, et al. 2004. An unusual genotype of *Toxoplasma gondii* is common in California sea otters (*Enhydra lutris nereis*) and is a cause of mortality. International Journal for Parasitology 34:275–84.

Miller, S. D., and J. F. Cully, Jr. 2001. Conservation of black-tailed prairie dogs (*Cynomys ludovicianus*). Journal of Mammalogy 82:889–93.

Mills, L. S., M. E. Soule, and D. F. Doak. 1993. The keystone species concept in ecology and conservation. Bioscience 43:219–24.

Murray, M. P., and M. C. Rasmussen. 2003. Non-native blister rust disease on whitebark pine at Crater Lake National Park. Northwest Science 77:87–91.

Navarrete, S. A., and B. A. Menge. 1996. Keystone predation and interaction strength: Interactive effects of predators on their main prey. Ecological Monographs 66:409–29.

Ough, K., and A. Murphy. 2004. Decline in tree-fern abundance after clearfell harvesting. Forest Ecology and Management 199:153–63.

Paillet, F.L. 2002. Chestnut: History and ecology of a transformed species. Journal of Biogeography 29:1517–30.

Paine, R. T. 1966. Food web complexity and species diversity. American Naturalist 100:65–75.

Paine, R. T. 1969. A note on trophic complexity and community stability. American Naturalist 103:91–93.

Perry, R. D., and J. D. Featherston. 1997. *Yersinia pestis*—etiologic agent of plague. Clinical Microbiology Reviews 10:35–66.

Power, M. E., and L. S. Mills. 1995. The Keystone cops meet in Hilo. Trends in Ecology & Evolution 10:182–84.

Ray, C., and S. K. Collinge. 2006. Potential effects of a keystone species on the dynamics of sylvatic plague. *In* Disease Ecology: Community Structure and Pathogen Dynamics, ed. S. K. Collinge and C. Ray, 204–18. Oxford: Oxford University Press.

Smart, A. C., D. M. Harper, F. Malaisse, S. Schmitz, S. Coley, and A. C. G. de Beauregard. 2002. Feeding of the exotic Louisiana red swamp crayfish, *Procambarus clarkii* (Crustacea, Decapoda), in an African tropical lake: Lake Naivasha, Kenya. Hydrobiologia 488:129–42.

Terborgh, J. 1986. Keystone plant resources in the tropical forest. *In* Conservation Biology: The Science of Scarcity and Diversity, ed. M. E. Soulé, 330–44. Sunderland, MA: Sinauer Associates.

Thomas, N. J., and R. A. Cole. 1996. Biology and status of the southern sea otter: The risk of disease and threats to the wild population. Endangered Species Update 13:23–27.

Van Riper, C., S. G. Van Riper, M. L. Goff, and M. Laird. 1986. The epizootiology and ecological significance of malaria in Hawaiian land birds. Ecological Monographs 56:327–44.

Whicker, A. D., and J. K. Detling. 1988. Ecological consequences of prairie dog disturbances. Bioscience 38:778–85.

Zaady, E., P. M. Groffman, M. Shachak, and A. Wilby. 2003. Consumption and release of nitrogen by the harvester termite *Anacanthotermes ubachi navas* in the northern Negev desert, Israel. Soil Biology and Biochemistry 35:1299–303.

CHAPTER SEVEN
Red Queen Communities

*Keith Clay, Kurt Reinhart, Jennifer Rudgers, Tammy Tintjer,
Jennifer Koslow, and S. Luke Flory*

SUMMARY

WE EXTEND THE RED QUEEN HYPOTHESIS, a mechanism proposed to
maintain genetic diversity at the population level, to explain diversity at
the community level. The Red Queen hypothesis assumes that pathogens
become specialized on common host genotypes and reduce their fitness,
thereby favoring genetic mechanisms generating rare genotypes. Here
we develop the idea that pathogens favor diversity in communities by an
analogous mechanism of frequency-dependent selection against common
species. Empirical evidence from a variety of studies suggests that host-
specific pathogens are more likely to attack and reduce population sizes
of common species relative to rare species. Pathogens could therefore
counteract competitive exclusion and maintain species diversity in communities. We also postulate an indirect relationship between pathogens
and ecosystem function because field and laboratory studies suggest a
link between species diversity and improved functioning of ecosystems.

INTRODUCTION

The interaction between pathogens and plants has traditionally been the
purview of agricultural plant pathologists, but because crop species are
typically grown in monocultures, there has been little consideration of
the role of pathogens in multispecies plant communities. However, ecologists have become increasingly appreciative of the potential role of
pathogens in natural communities and ecosystems (Burdon 1982; Holah
and Alexander 1999; Jarosz and Davelos 1995; Mitchell et al. 2002; van
der Kamp 1991; Van der Putten et al. 1993). Over the past several decades a series of related hypotheses and concepts on frequency, density,
and distance dependence of disease and their effects on plant diversity
have been developed (Bever 1994; Burdon and Chilvers 1982; Connell
1971; Gilbert et al. 1994; Hansen and Goheen 2000; Janzen 1970; Van
der Putten et al. 1993). A pathogen that causes the local extinction of its
host species will reduce diversity by reducing species richness, but more

commonly pathogens reduce the density and dominance of host populations, potentially increasing community evenness and diversity. Besides
affecting relative species abundances, pathogens can create environmental heterogeneity by affecting species composition (Mills and Bever
1998; Olff et al. 2000; Packer and Clay 2000) and successional stage in
localized patches (Hansen and Goheen 2000; Holah et al. 1993), and by
altering microenvironmental conditions through changes in canopy
structure (Agrios 1988; Burdon 1987). By affecting the diversity of ecological communities, plant pathogens may indirectly influence ecosystem processes responsive to diversity.

The idea that pests, parasites and pathogens can affect the diversity of
plant communities is not new. For example, Ridley (1930) wrote: "When
plants are too close together, disease can spread from one to another and
become fatal to all. Where plants of one kind are separated by plants of
other kinds, the pest, even if present, cannot spread." Similarly, Hendrix
and Campbell (1973) observed, "These pathogens (*Pythium* spp.), through
the selective elimination of susceptible plant species, may act as potent determinants of forest and plant vegetational types." Moreover, Harper
(1977) stated that "the host specific pathogen may be an important agent
in the diversification of vegetation." If the density or relative abundance
of a given species increases to a high level, such as might be expected
with a competitive dominant, and the probability of pathogen attack increases, then pathogens could maintain diversity in ecological communities through negative frequency-dependent selection. And as a corollary,
species diversity can minimize pathogen damage at the level of the entire
community (Elton 1958; Mitchell et al. 2002; Van der Plank 1963).

Ridley (1930) should be compared with Lively's (1996) comments on
the Red Queen hypothesis for pathogen selection for sexual reproduction
in hosts: "Outcrossing and recombination are favored because they allow
for the production of rare offspring (genotypes) that are expected to have
a greater chance of escaping parasites." Thus, the ecological hypothesis of
pathogens maintaining species diversity within communities and the evolutionary hypothesis of pathogens maintaining genetic diversity within
populations are unified by the concept of frequency dependence. Because
of their abundance, both species and genotypes may become increasingly
disadvantaged by pathogen attack, while less abundant species and genotypes are more likely to escape infection. The parallels between maintenance of diversity at these two levels of biological organization are
intriguing, and there may be important linkages between them.

Species diversity within communities and genetic diversity within pop-
ulations have historically been considered within distinct fields of biology.
However, there have been some attempts at a conceptual reconciliation.
Antonovics (1976) suggested that the forces maintaining species diversity

and genetic diversity should not be treated as separate phenomena (see also Antonovics 2003; Huston 1994; Hubbell 2001). Few empirical studies have examined relationships between these two levels of diversity in nature (but see Vellend 2003, 2004). Here we argue that host-specific pathogens may act to maintain diversity at both levels of organization by similar processes. In particular, we evaluate a series of related concepts applicable both to pathogen-driven maintenance of genetic diversity within host populations and species diversity within ecological communities: (1) host specificity, (2) frequency-dependent selection, (3) the costs of resistance in hosts and virulence in pathogens, and (4) oscillating dynamics. Each section begins with a short overview of one component of the evolutionary Red Queen hypothesis, followed by application of the analogous concept to species diversity. We also extend the relationship between genetic diversity and population stability to species diversity and ecosystem stability, thereby developing a mechanistic link between plant disease and ecosystem function. By extending the well-developed ideas of Red Queen interactions at the population level (e.g., Clay and Kover 1996; Hamilton et al. 1990, Lively 1996), we hope to stimulate thinking and research on similar interactions at the community level.

The Red Queen Hypothesis

The Red Queen hypothesis was originally conceived to explain the maintenance of sexual reproduction despite the intrinsic advantages of asexual reproduction, namely, that offspring are 100% related and no males are required. The Red Queen hypothesis posits that host-specific parasites or pathogens maintain sexual reproduction through time-lagged, negative frequency-dependent selection on hosts. Pathogens cause frequency-dependent selection by infecting common host genotypes. Rare genotypes thereby benefit and increase in frequency until they too become common and suffer increasing pathogen attack (Haldane 1949). Thus, under parasite pressure, genotypes are expected to oscillate in frequency over time (figure 7.1; Hamilton 1980; Jaenike 1978). An important assumption is that resistance and virulence have a genetic basis such that pathogen populations can respond to changes in host genotype frequencies, and vice versa. Sexual reproduction may offer protection from natural enemies by generating diverse progeny that differ from their parents as a result of genetic recombination (Adams et al. 1971; Clarke 1976; Hamilton et al. 1990; Kelley 1994; Lively 1996). Supporting this idea, a higher proportion of outcrossing plant species tend to be found in less disturbed, biologically complex tropical habitats where pathogens and other natural enemies are more prevalent (Levin

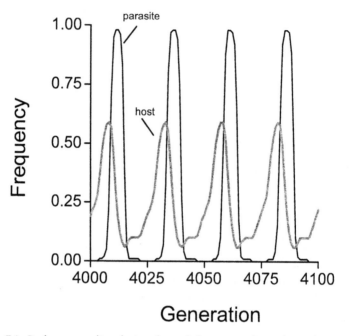

Figure 7.1. Pathogen-mediated, time-lagged, frequency-dependent selection plotted over time. Generations 4000–4100 are shown. Computer simulation by C. Lively, Indiana University (unpublished).

1975; see also Glesner and Tilman 1978). Similarly, outcrossing weeds are harder to control with biological control agents (Burdon and Marshall 1981), and there is a positive correlation between outcrossing rate and the number of fungal pathogen species infecting individual plant species (Busch et al. 2004). Under pathogen pressure, sexual reproduction may be favored because it maintains genetic diversity and the production of rare genotypes.

There is no direct parallel with sexual recombination within populations at the level of the community, but novel taxa may be introduced into communities through the processes of immigration, hybridization, or in situ speciation. Thus, the makeup of the local species pool may be continually changing, just as the genetic makeup of the host population may be dynamic. Although most of these mechanisms for introducing new taxa would be independent of pathogens, pathogen pressure could facilitate the speciation process under some conditions (Burger 1992). Here we focus not on mechanisms that generate new species but rather on the role of pathogens in facilitating the persistence of newly arrived and resident species.

HOST SPECIFICITY: A KEY REQUIREMENT FOR PATHOGENS
TO AFFECT DIVERSITY IS THAT THEY HAVE DIFFERENTIAL EFFECTS
ON INDIVIDUAL GENOTYPES AND SPECIES

RED QUEEN HYPOTHESIS

Although quantitative and nonspecific resistance and tolerance play important roles in host-pathogen dynamics (Alexander 1992; Price et al. 2004), single, dominant resistance alleles of major effect can confer resistance to specific pathogen races in both natural and agricultural systems (Burdon 1987; Flor 1956; Jarosz and Burdon 1991; Parker 1988; Thompson and Burdon 1992), providing strong evidence for specificity of the interaction. The interaction between single plant genes for resistance and single genes for virulence in pathogens has been called a gene-for-gene interaction (Flor 1956). Although critics argue that gene-for-gene interactions are an artifact of artificial selection (Barrett 1985; Thompson 1994), they have been found in a variety of nonagricultural systems (reviewed in Burdon 1987; Thompson and Burdon 1992). Red Queen hypothesis models have historically assumed a genetic mechanism of matching alleles (e.g., Hamilton et al. 1990; Howard and Lively 1994) rather than gene-for-gene interactions. With matching alleles, infection results when a pathogen genotype exactly matches the host genotype. With gene-for-gene interactions, a given host plant can be resistant to multiple (rather than single) pathogen races (Agrios 1988), and a single pathogen genotype can be virulent on multiple host genotypes (Parker 1994). There has not been any attempt to determine whether there is evidence for a matching allele interaction in natural pathosystems.

RED QUEEN COMMUNITIES

If pathogens attack and harm all plant species equally, then there will be no net change in community composition. However, most pathogens exhibit some level of specificity at the species, genus, or familial level (Farr et al. 1989; Parker and Gilbert 2004). Similar to host-pathogen interactions within populations, the host range of the pathogen is determined both by traits of the pathogen that confer virulence and by traits of the host that confer resistance (Osbourn 1996; Staats et al. 2005).

Plant pathogens are variable, but most exhibit some level of specificity, infecting only certain genotypes of a host species (Chaboudez and Burdon 1995; Nsarellah et al. 2003), single species (Martinez et al. 2004), or several closely related taxa (Staats et al. 2005) in the same genus (Daughtrey and Hibben 1994) or family (Skinner and Stuteville 1995). Plant pathogens may have narrower host ranges than originally assumed because single species have been shown to represent complexes of undescribed or nascent species specific to certain (often closely related) host plants (Baker et al. 2003; Gudelj et al. 2004) or host races (Blok and Bollen 1997; Delgado et al. 2001). For example, diversity of rust fungi infecting grasses is strongly concordant with host phylogeny (Hijwegen 1979; Savile 1987). Similarly, endophytic fungi in grasses, which include both pathogens and mutualists, exhibit a high degree of host specificity related to their mode of transmission, and some show co-cladogenesis with their hosts (figure 7.2; Clay and Schardl 2002; Leuchtmann 1992; Schardl et al. 1997). Alternatively, some pathogens are limited to a particular geographic range but can infect many species in that range (e.g., root pathogens of western forests; Hansen and Goheen 2000; Holah et al. 1997; Rizzo and Garbelotto 2003). The ability to infect hosts is often broader than the range of hosts on which pathogenic effects occur (Barton et al. 2003; Blok and Bollen 1997) or the range of hosts from which the pathogen can be transmitted (Leon-Ramirez et al. 2004).

Differences in the degree of host specificity may reflect the way in which the pathogen infects and harms the host plant. Biotrophic plant pathogens (those that feed on live plant tissue) are expected to be more host-specific than necrotrophic or saprophytic fungi, which feed on dead or dying tissue (Mundt 2002; Parker and Gilbert 2004). Biotrophic pathogens (e.g., *Pseudomonas* bacteria) need living cells for nutrition and can be suppressed by plants' hypersensitive response. In contrast, necrotrophic pathogens (e.g., *Botrytis* fungi) require cell death to access plant nutrients and may use the plant cell death (hypersensitive) response to their benefit. Necrotrophs showing little or no host specificity include *Phytophthora cinnamoni*, where there are no patterns of infectivity at the genus or family level (Shearer et al. 2004), *Phytophthora capsici*, which infects a broad range of crop plants (Tian et al. 2003), and *Sclerotinia sclerotiorum*, which infects at least 408 plant species in 278 genera and 75 families (Boland and Hall 1994). However, even within necrotrophs there is often specificity of pathogenic effect (Tian et al. 2003).

Like pathogens, host species, too, may vary in the level or intensity of disease once infected, and may vary in tolerance to a given level of disease (Alexander 1992). Pathogens that infect multiple hosts but have a more detrimental effect on a particular host are common (Barton et al. 2003; Blok and Bollen 1997; Holah and Alexander 1999; Power and Mitchell 2004; Tian et al. 2003). This can lead to changes in the competitive hierarchy between the species. For example, a model with Lotka-Volterra-type dynamics of two hosts interacting via a shared pathogen predicted either competitive exclusion or coexistence of host species, depending on initial densities (Alexander and Holt 1998). For example, attack by a shared generalist pathogen allowed a competitively inferior

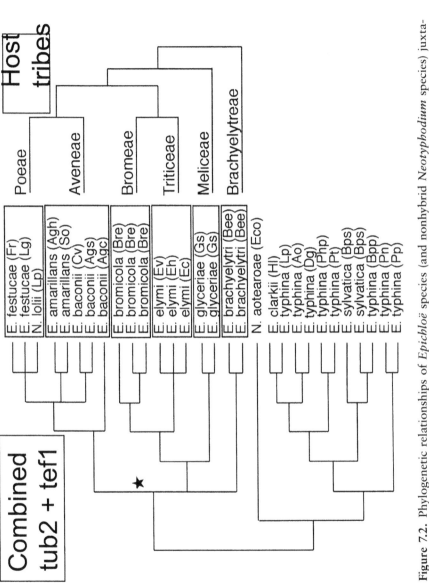

Figure 7.2. Phylogenetic relationships of *Epichloë* species (and nonhybrid *Neotyphodium* species) juxtaposed against the phylogeny of their host grass tribes. (From Clay and Schardl 2002.) Fungal sequences from elongation factor 1-alpha (tef1) and beta-tubulin (tub2) genes.

prairie species to coexist with an otherwise superior competitor (Holah and Alexander 1999). Spillover of barley yellow mosaic virus from a highly susceptible reservoir species decreased the abundance of two other host species through pathogen-mediated apparent competition (Power and Mitchell 2004). Another outcome of such shared infection dynamics might be habitat partitioning or non-overlapping geographic ranges (Holt and Pickering 1985).

In summary, there is strong evidence that many pathogens exhibit host specificity to some degree. Similarly, plant hosts show specificity both in resistance and in tolerance to pathogen attack. These lines of evidence support the underlying assumption of the Red Queen hypothesis that specificity exists in both attack and resistance.

FREQUENCY-DEPENDENT SELECTION: PATHOGENS WILL MAINTAIN DIVERSITY IF ABUNDANT GENOTYPES AND SPECIES ARE MORE FREQUENTLY ATTACKED BY PATHOGENS THAN RARE GENOTYPES AND SPECIES

RED QUEEN HYPOTHESIS

The Red Queen hypothesis predicts that frequency-dependent selection by co-adapted pathogens favors rare host genotypes. However, as rare genotypes increase in frequency because of their lack of disease, their fitness advantage should decline. A few empirical studies have quantified variation in infection among rare and common genotypes within populations. In agricultural systems, monocultures of a single genetic variety are often prone to epidemics (Adams et al. 1971). Novel resistant crop varieties eventually become less productive when new virulent pathogens evolve (McDonald and McDermott 1993; Mundt 2002). Similarly, cross-pollinated crops, which exhibit the greatest genetic diversity, typically suffer less pathogen damage than self-pollinated crops, which in turn sustain less damage than asexual or clonal crops, which exhibit the least genetic diversity (Stevens 1948). There have been few tests in natural plant populations. In one study, clones of the apomictic crucifer *Arabis holboellii* varied in susceptibility to *Puccinia* spp, but there was no relationship between clone frequency and disease incidence within any population examined (Roy 1993).

RED QUEEN COMMUNITIES

Current evidence supports the existence of frequency-dependent attack within communities. Pathogen effects on monocultures or low-diversity communities often result from the higher density or relative abundance of individual species in those environments (Adams et al. 1971; Harper 1977). Epidemics caused by exotic pathogens such as Chestnut blight (Burdon and Shattock 1980; Harlan 1976) and Dutch elm disease (Harlan 1976; reviewed by Burdon and Chilvers 1982) tend to occur on abundant hosts. Conversely, reduced density of the host plant as a result of increased abundance of other plant species can reduce inoculum produced on that host (Burdon and Chilvers 1982; Mundt 2002). Plant density is often, although not always, positively correlated with disease incidence (Burdon and Chilvers 1982; Burdon et al. 1992; Gilbert et al. 1994; Thrall et al. 2001). Density- or distance-dependent mortality from pathogens has been reported (or inferred) in a number of tropical forest systems (Augspurger 1983; Gilbert et al. 1994; Harms et al. 2000; Hood et al. 2004; Wills et al. 1997). However, the relationship between host plant density and disease may depend strongly on when density is measured in the epidemic cycle (Burdon et al. 1992). In one epidemic, several years of low disease incidence and increasing host population size were followed by a crash in which 79% of the local host population died due to attack (Jarosz and Burdon 1991, 1992).

Several mechanisms may promote a proportionally greater success of pathogens in dense compared to sparse host populations. Direct effects can result from a higher number of hosts to intercept inoculum or a shorter distance for inoculum to disperse to a new host plant. Dispersal ability can be an important factor limiting pathogen populations (Garrett and Mundt 1999). Dense host populations may also affect microenvironmental conditions such as humidity, temperature, and air movement that could influence pathogen transmission (Burdon and Chilvers 1982). Indirect effects of density may also alter disease development. Further, increased host density increases intraspecific competition relative to interspecific competition. Increased stress with competition may enhance host susceptibility in dense stands. In addition, a larger, denser host populations may be more likely to maintain a pathogen if host-specific pathogens require a critical host population size to persist (Carlsson and Elmqvist 1992). Further, if the infection goes extinct locally, then large populations may be more likely to become reinfected. Infection is therefore expected to be more transient in smaller populations (Burdon and Thrall 2003).

Distance-dependent mortality adds a spatial dimension to the Red Queen hypothesis. Specifically, the Janzen-Connell hypothesis proposes that host-specific natural enemies maintain high diversity of tropical forests by limiting local recruitment around adult conspecifics (Connell 1971; Janzen 1970). The "dead zone" around individual trees can be colonized by other species. Although some studies have found little or limited evidence for this mechanism (Clark and Clark 1984; Condit et al. 1992; Hubbell 1979), the results of other studies are strongly supportive. For example, sixty-seven of eighty-four common tree species on Barro Colorado Island, Panama, exhibited strong density-dependent effects on seedling recruitment (Wills et al. 1997). It was suggested that these results were consistent with pathogen activity, similar to the results of Augspurger (1983), although Wills et al. (1997) did not actually measure disease mortality. The mortality of *Milicia regia*, a rainforest tree of western African, was significantly higher in soil collected below adult *Milicia* than in soil collected 100 m away (Hood et al. 2004). Seedling survival of *Prunus serotina* increased with distance from adult trees owing to oomycete *Pythium* pathogens (Packer and Clay 2000). Seedling density also decreased with distance, but multiple regression models indicate that distance from adult trees was a better predictor of seedling survival than seedling density (figure 7.3). These studies on individual species support the idea that pathogens maintain diversity by limiting the density or aggregation of individual species.

Density, distance, and frequency dependence are not necessarily mutually exclusive, and under some circumstances they are the same. If plant

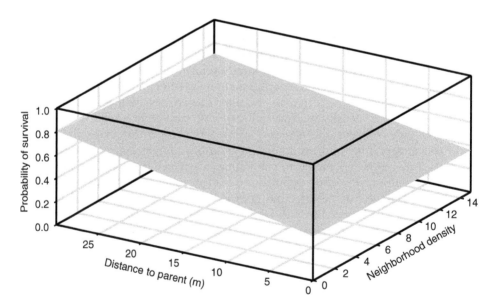

Figure 7.3. Logistic regression models of the probability of black cherry (*Prunus serotina*) seedling survival in relation to distance to parent and neighborhood density. (From Packer and Clay 2000.)

communities are homogeneous and if individuals of particular species are randomly distributed, then plant density will be highly correlated with its frequency (or relative abundance) and with distance to conspecifics. However, if plant density varies and if most seed dispersal occurs near adult trees, then the correlation of these measures will be lower. Further, insect vectors may transmit plant pathogens in a frequency-dependent manner independent of plant density (Alexander and Antonovics 1988), or even in a negative density-dependent manner (Burdon and Chilvers 1982; Power 1987), in contrast to our argument of a Red Queen process at the community level.

Thus far we have focused on individual plant responses, but some experimental studies have been conducted at the level of whole communities. In the experimental biodiversity plots at Cedar Creek, the level of foliar fungal disease was negatively correlated with plant species richness (Knops et al. 1999). This effect was a result of reduced species density in low-diversity plots and not a result of diversity per se. Similarly, Mitchell et al. (2002) found that mean pathogen load was three times higher in monoculture plots relative to plots planted with twenty-four species, which approximated natural diversity levels (figure 7.4). Ten of eleven host-specific pathogens exhibited significantly higher infection levels (measured as percent leaf area infected) with decreasing diversity. When host abundance was included as a covariate, the effect

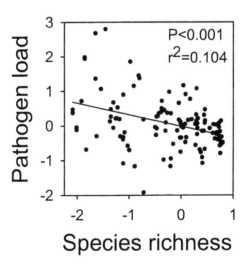

Figure 7.4. Added-variable plot of the effect of ln(species richness treatment) on the square root of pathogen load, after controlling for the effects of disease proneness and dominant species presence. (From Mitchell et al. 2002.)

of species diversity disappeared, indicating the importance of individual host species abundance. This work indicates that rare species are afforded protection from pathogen attack as a result. Long-term studies would be useful for determining whether species oscillate in relative abundance over time.

In summary, most evidence shows that pathogens often have a greater effect on abundant plant species than on less abundant species. Dependency of pathogen impact on relative species abundance can counteract competitive exclusion and prevent domination by one or more species.

Costs of Resistance and Virulence: Resistance to One Pathogen (or Virulence for One Host) Entails Fitness Costs, Preventing Dominance by a Single Resistant (or Virulent) Type

RED QUEEN HYPOTHESIS

Gene-for-gene interactions can maintain genotype cycling of both host and pathogen only if costs of resistance and virulence exist (e.g., Sasaki 2000). The prevalence of polymorphism for resistance in natural plant populations implies that resistance must be costly or else all plants should be resistant (Parker 1992; Simms 1992). Similarly, Brown et al. (2002) argued that disease resistance in crops entails a yield cost. An extensive review of plants resistant to pathogens found a cost of resistance in 50% of the studies surveyed (Bergelson and Purrington 1996), although costs are difficult to measure and may occur only in certain circumstances. An illustrative example of the cost of resistance was a 9% cost of resistance in the absence of disease for transgenic *Arabidopsis thaliana* that differed from control plants by only a single resistance gene (Tian et al. 2003). Considering pathogens, virulence is often costly compared to avirulence when pathogens are infecting nonresistant hosts (Thompson and Burdon 1992; Van der Plank 1968). For example, Van der Plank suggested that pathogen populations soon lose virulence genes that have no matching resistance genes in newly planted crop cultivars. Parker and Gilbert (2004) provide a review and additional examples of costs of virulence.

RED QUEEN COMMUNTIES

Within populations, trade-offs may result from negative genetic correlations among traits or pleiotropy (direct costs) or from trade-offs between fitness and resistance that are mediated by interactions with other organisms (indirect, or ecological, costs) (Strauss et al. 2002). At the community level, trade-offs are evidenced by life history or physiological differences among species that are correlated either directly or indirectly with pathogen susceptibility. For example, shade-tolerant species gain a competitive advantage in low light, but may become more susceptible to damping-off pathogens that prefer moist organic soils in shaded habitats. Species that exhibit high levels of resistance to pathogens and dominate when pathogen attack is high may be competitively inferior when attack is low. Similarly, species with high levels of pathogen resistance may be more susceptible to herbivore attack (Felton and Korth 2000). Holah and Alexander (1999) reported that big bluestem (*Andropogon gerardii*), a prairie dominant, was negatively affected by soil microbes associated with an annual legume, perhaps allowing the annual to persist in mixtures with an otherwise stronger competitor. More generally, pathogens may mitigate or reverse the outcome of interspecific competition (Alexander and Holt 1998; Chilvers and Brittain 1972; Clay 1990), consistent with the idea that competitive ability and disease resistance are negatively correlated. However, very few studies have specifically looked for trade-offs of pathogen resistance with other traits beyond quantifying the cost of resistance genes. There is more evidence for trade-offs for susceptibility to insect herbivory (Coley et al. 1985; Strauss et al. 2002). A broader look for direct and ecological costs of pathogen resistance across (rather than within) species may reveal similar community-level effects. It would be important to control for phylogenetic relatedness when making such cross-species comparisons.

In summary, ecological trade-offs among resistance and other life history or physiological traits may restrict the dominance of species in communities, just as the costs of resistance can reduce the dominance of resistant genotypes and promote polymorphism. These trade-offs across species at the community level prevent one (or a few) species from dominating communities.

Oscillating Dynamics: Diversity Is Maintained When Rare Genotypes or Species Tend to Become More Abundant and Common Genotypes or Species Tend to Become Rarer

RED QUEEN HYPOTHESIS

The Red Queen hypothesis predicts dynamic changes in host resistance and pathogen virulence over time within a single population. Investigations of the spatial and temporal dynamics of host resistance and pathogen virulence within populations have largely focused on resistance in hosts (Alexander 1992; Burdon et al. 1995; Parker 1985) rather than on virulence in pathogens (but see Watson 1970). Longitudinal studies are necessary to reveal correlated changes in the dynamics of both host and pathogen populations. Time-lagged host and pathogen cycling results in periods when a rare host genotype becomes common yet still has a low frequency of pathogen infection (see figure 7.1), while at other times in the cycle, common genotypes have been driven to low frequencies by heavy pathogen attack but still support high levels of infection (Dybdahl and Lively 1995; Kaltz and Shykoff 1998). A long-term study of the *Linum marginale/Melampsora lini* system in Australia revealed that over a six-year period there was a major shift in relative abundances of resistance genotypes in the population, but the changes were not related to pathogen populations or frequency-dependent selection (Burdon and Thompson 1995). Instead, other selective factors, such as linkage to other traits under selection and dispersal of pathogens from outside the host populations, may be responsible for observed changes in pathogen virulence (Burdon and Thompson 1995). In agricultural systems, coupled changes in host and pathogen populations have been found (Johnson 1987; Leonard 1987), but are often confounded by human influences. Genetically heterogeneous closed populations of barley (*Hordeum vulgare*) have been grown for more than fifty years without any human interference (McDonald et al. 1989). The frequency of families resistant to *Rhynchosporium secalis*, the agent of scald disease, increased over time, including families resistant to multiple scald races (Maroof et al. 1983). Simultaneous comparisons of the pathogen populations from barley generation 56 and generation 44 demonstrated that there was genetic divergence of both populations over time, with higher virulence in the more resistant host population (McDonald et al. 1989). The results of these studies indicate that plants and pathogens have oscillating dynamics in resistance and virulence consistent with the Red Queen hypothesis.

RED QUEEN COMMUNITIES

Time-lagged oscillations have been demonstrated both theoretically and empirically in a variety of predator-prey, host-parasitoid, and host-pathogen systems (Hassell 2000; Hudson et al. 1998; Turchin et al. 1999). Theoretical investigations of negative frequency dependence and feedback in plant-soil interactions also suggest that these processes can generate oscillations in the abundances of plant species (Bever 2003; Molofsky et al. 2002). Observations of oscillatory cycles of host increase, decrease, and increase again correlated with pathogen activity are, not surprisingly, few, given the extended time frame necessary for documentation. Empirical evidence for temporal oscillations in relative species

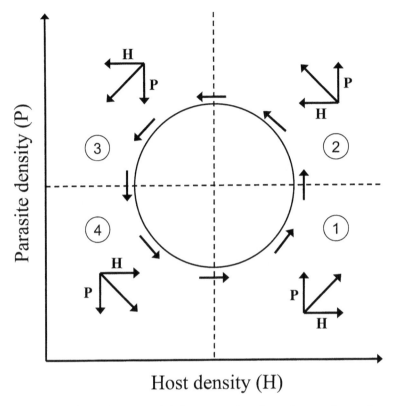

Figure 7.5. Lotka-Volterra model demonstrating four phases of time-lagged oscillations.

abundance was found by Olff et al. (2000) in grazed grasslands in the Netherlands. The two dominant species, *Festuca rubra* and *Carex arenaria*, alternated in abundance with neighboring sites cycling out of phase, creating a shifting mosaic of species abundance. Experimental sterilization of soil demonstrated that soil-borne pathogens had a stronger negative effect where each species had been locally decreasing in abundance and a weaker effect where they were increasing. Van der Putten et al. (1993) found a similar effect of soil-borne pathogens but in a linear temporal sequence associated with sand dune succession. In forest systems of the Pacific Northwest, laminated root rot caused by the soil-borne fungus *Phellinus weirii* causes extensive mortality of native conifers in patchy infection centers scattered through the forest (Hansen and Goheen 2000; Holah et al. 1993). Earlier successional species such as pines and hardwoods, which are more resistant to *P. weirii*, colonize these infection centers and increase forest and landscape diversity. Later these infection centers are often recolonized by susceptible, later successional species, which may again succumb to the pathogen.

For plant-pathogen interactions in natural systems, this limited number of studies demonstrates ebb-and-flow dynamics of species, but a wider range of data supports one or more components of oscillatory dynamics. Time-lagged oscillations of plant and pathogen populations exhibit four phases: (1) host species increasing, pathogen species decreasing, (2) host species increasing, pathogen increasing, (3) host species decreasing, pathogen increasing, and (4) host species decreasing, pathogen decreasing (figure 7.5). Evidence is presented for each of these phases with examples from natural plant communities, with a particular focus on novel interactions such as occur when exotic plants or pathogens establish in a new communities (Parker and Gilbert 2004). Invasions by exotic plants and pathogens represent well-documented natural experiments. Together, these examples of different phases suggest that interactions with pathogens contribute to oscillating dynamics that may maintain diversity in plant communities.

(**1**) *Host species increasing, pathogen species decreasing or absent.* Darwin observed that plant and animal species brought to new regions of the world often experienced dramatic population growth, and surmised that these species escaped regulation by "natural enemies" (Darwin 1872). This original observation continues to be one of the leading hypotheses (i.e., enemy release hypothesis) to explain the success of exotic species (Keane and Crawley 2002; Maron and Vilá 2001; Mitchell and Power 2003). Much empirical data support the hypothesis of enemy escape. Mitchell and Power (2003) reported on infection by foliar viruses and fungi infecting 473 plant species in their native and non-native ranges in the United States and Europe; on average, plant species had 84% fewer fungi and 24% fewer virus species in their non-native than in their native range. Plant species categorized as harmful invaders experienced a stronger decline in pathogen infection than weak invaders. These results suggest that invasive non-native plants have escaped many species of pathogens, but they do not provide any information about the density or virulence of the pathogens. Several experimental studies do suggest that non-native plants have escaped the high pathogen abundance and virulence found in native ranges (figure 7.6; DeWalt et al. 2004; Knevel et al. 2004; Reinhart et al. 2003; but see Beckstead and Parker 2003 for an exception), and that escape has been one factor contributing to their success.

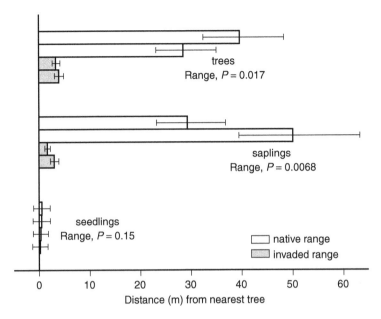

Figure 7.6. Distances from conspecifics of black cherry (*Prunus serotina*) in their native range in Indiana (white bars) and the invaded range in the Netherlands (gray bars). (From Reinhart et al. 2003.)

(2) *Host species increasing, pathogen increasing.* The second phase is evidenced by the accumulation of pathogens and other enemies as invasive species become naturalized in a new area. For example, pathogens native to the new geographic ranges of the host plant may limit the invasion potential of introduced species (biotic resistance hypothesis) (Knevel et al. 2004; Mack 1996; Parker and Gilbert 2004; Sceffer 2003). Biotic resistance may be more likely where the invasive species is closely related to natives (Parker and Gilbert 2004). For example, Blaney and Kotanen (2001) tested the effects of soil fungi (using fungicides) on buried seeds of fifteen pairs of congeneric native and non-native herbaceous plant species in Ontario, Canada, and found no difference between native and non-native species. Although their study did not specifically compare pathogen effects on related versus unrelated non-native species, the results demonstrate that non-native species accumulate pathogens. Similarly, Parker and Gilbert found no difference in disease levels in native versus introduced clovers at Bodega Bay (cited in Parker and Gilbert 2004). Mitchell and Power (2003) also found that invasive plant

species accumulated some fungal pathogens in their new habitat. Similarly, negative feedback between plants and their soil communities, manifested as reduced growth of plants in soil that has been "cultured" by the same species, may reflect buildup of soil-borne pathogens as the host species increases (Bever 1994; Bever et al. 1997; Mills and Bever 1998; Packer and Clay 2000; Van der Putten et al. 1993).

(3) *Host species decreasing, pathogen increasing.* The third phase is most apparent during disease epidemics caused by introduced pathogens. Microbial pathogens introduced into novel habitats have devastated populations of resident species in their new range (e.g., chestnut blight fungi, *Cryphonectria parasitica*, and Dutch elm pathogens, *Ophiostoma ulmi* and *O. nova-ulmi*). Other exotic pathogens with potentially devastating effects include those causing sudden oak disease (*Phytophthora ramorum*) (Rizzo and Garbelotto 2003), white pine blister rust (*Cronartium ribicola*) (Campbell and Antos 2000), and Port-Orford-cedar root disease (*Phytophthora lateralis*) (Jules et al. 2002). At the population level, these pathogens may arise through mutation or selection; however, at the community level, virulent pathogen species may be most likely to appear by immigration rather than through speciation or evolution.

A second line of evidence for pathogens reducing host abundance comes from the biological control of weedy and invasive plant species. Classic biocontrol identifies suitable host-specific enemies (pathogens or herbivores) from the invader's native range and introduces these enemies to control the plant in its new range (Caesar 2000; Newman et al. 1998; Randall and Tu 2001). The best case study of biocontrol by a pathogen is that of the foliar rust, *Puccinia chondrillina*, which significantly reduced populations of *Chondrilla juncea*, a perennial weed of cereal crops native to Europe and introduced to Australia and North America (Burdon et al. 1981; Panetta and Dodd 1995).

(4) *Host species decreasing, pathogen decreasing.* The fourth phase of a time-lagged cycle occurs when host species are driven to a low level and the pathogen population also declines, setting the stage for resurgence of the host. Examples of density- and frequency-dependent pathogen attack, discussed earlier, are consistent with the fourth phase (e.g., Burdon and Chilvers 1982; Mitchell et al. 2002; see figure 7.4). The agricultural practice of crop rotation (growing another crop species results in a decline in pathogen populations of the first) also supports the existence of pathogen decline following host decline. Paleoecological studies also suggest past pathogen epidemics followed by host recovery (Davis 1981), although other unknown environmental changes may have also occurred.

In summary, many lines of evidence suggest oscillating dynamics at the community level, consistent with extension of the Red Queen hypothesis to communities. Fewer studies have documented long-term repeated oscillations in species and pathogen abundance. Additional studies in a broader range of systems would greatly augment our knowledge.

PATHOGENS, SPECIES DIVERSITY AND ECOSYSTEM: BY MAINTAINING DIVERSE ECOLOGICAL COMMUNITIES, PLANT PATHOGENS MAY AFFECT ECOSYSTEM PROCESSES RESPONSIVE TO DIVERSITY

RED QUEEN HYPOTHESIS

Genetic diversity within plant populations can promote greater productivity and greater resistance (or resilience) to disease outbreaks. In agriculture, mixed plantings of multiple plant cultivars can increase disease control by reducing host density to limit host-specific plant pathogens (Browning and Frey 1969; Jensen 1952; Mundt 2002; Wolfe and Barrett 1985; Zhu et al. 2000). Advantages of genetic diversity include inoculum reduction (fewer susceptible plants result in less pathogen spread to other susceptible hosts) and compensation by resistant hosts, which co-occur with but outcompete susceptible genotypes to maintain high yields remain despite some disease. In one case study of powdery mildew control using mixtures, only 10% of fields previously planted in monocultures became severely infected, whereas 50% of monoculture fields developed severe infections (Mundt 2002). However, mixtures work less well when there is a large amount of long-distance inoculum dispersal such that quantity of pathogen inoculum is independent of local conditions (e.g., Garrett et al. 2001).

RED QUEEN COMMUNITIES

Pathogens may have many direct effects on ecosystem processes (see Eviner and Likens, chapter 12, this volume) as well as indirect effects through their effects on species diversity. The effects of disease in maintaining species diversity may have important consequences at the ecosystem scale, when ecosystem processes and services are enhanced by increased species diversity. Theoretical models and experimental tests exploring relationships between diversity and ecosystem properties have proliferated during the past decade (Naeem 2002; Schulze and Mooney 1993). Recent models predict positive correlations between diversity and both the productivity (Mouquet et al. 2002; Tilman, Knops, et al. 1997) and the stability of communities (Doak et al. 1998; Ives and Hughes 2002; Lehman and Tilman 2000; Yachi and Loreau 1999). Experiments have revealed important functions for diversity in some cases, including the enhancement of primary productivity (Hector et al. 1999; Mikola et al. 2002;

Tilman, Lehman, et al. 1997; Wilsey and Potvin 2000), nutrient retention
(Tilman, Knops, et al. 1997; van Ruijven and Berendse 2005), nutrient
flow (Cardinale et al. 2002), water availability (Caldeira et al. 2001), and
resistance to invasion (Levine 2000; Zavaleta and Hulvey 2004). Mechanisms proposed to underlie diversity-ecosystem functioning relationships
include the sampling effect, whereby communities with higher diversity
have a greater probability of including a species with a strong effect on the
community, and complementarity effects, in which the addition of greater
phenotypic diversity to species assemblages enhances resource utilization
and fills ecological niches (Dimitrakopoulos and Schmid 2004; Lambers
et al. 2004; Loreau and Hector 2001; Wardle 1999). Functional redundancy of species may also be important for maintaining some ecosystem
processes (Wohl et al. 2004).

Integrating the results from models, experiments, and observations
into a general theory has been challenging. Although across studies, the
pattern is that diversity increases ecosystem functioning, discrepancies
remain concerning both the nature of diversity-ecosystem functioning
relationships and the mechanisms that underlie them. Not all studies
have uncovered a clear relationship between diversity and ecosystem
properties (Grime 1998; Hooper and Vitousek 1997; Wardle 1999), and
correlations often saturate at relatively low levels of diversity (Schwartz
et al. 2000). Furthermore, observational and experimental studies frequently give conflicting results.

Recent work has begun to explore whether microbial symbionts of
plants influence relationships between diversity and ecosystem functioning (van der Heijden et al. 1998; Klironomos et al. 2000; Jonsson
et al. 2001). These symbionts, which often go unnoticed, may explain
some of the inconsistencies among studies on diversity. For example,
Klironomos et al. (2000) showed that mycorrhizae caused primary productivity to level off more quickly with increasing diversity than when
mycorrhizae were absent. Similarly, other symbionts, such as nitrogen-
fixing bacteria, can alter ecosystem functioning as well as the composition of plant communities (Maron and Connors 1996; Vitousek et al.
1987). Similarly, the presence of fungal endophytes in grasses alters the
relationship between plant species diversity and both primary productivity and resistance to invasion (figure 7.7; Rudgers et al. 2004; Rudgers et al. 2005). Thus, mutualistic symbionts may have major effects on
the relationship between diversity and ecosystem functioning.

Similar effects may also be found for pathogens of plants. A diverse
assemblage of host-specific pathogens in an ecosystem may be essential
to maintaining plant species diversity and resisting invasion by exotic
plants (and pathogens). Total pathogen attack (ca. 9% of leaf area was
infected) was reduced at the plot level with foliar fungicide and resulted

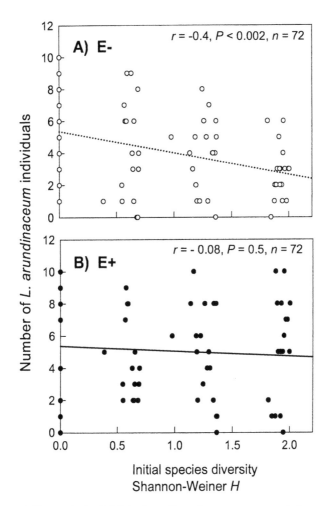

Figure 7.7. Effects of the initial plant diversity treatment and the endophyte treatment (E+ = infected, E− = experimentally disinfected) on the number of surviving *L. arundinaceum*. (a) Without the endophyte, initial diversity significantly reduced the invader. (b) With the endophyte, establishment was independent of diversity. (From Rudgers et al. 2004.)

in increased root biomass and leaf longevity across plant species (Mitchell 2003). Similarly, De Deyn et al. (2004) found that soil biota, including pathogens, nematodes, mycorrhizal fungi, and so forth, enhanced species diversity relative to sterilized soils. This is suggestive evidence of the role of soil pathogens (although other interactions may explain the correlation) in regulating plant diversity by affecting evenness. However, little is known about the role of pathogen diversity and the importance of pathogen host specificity at the community or ecosystem level. For example, a recent search of Web of Science detected no articles containing the key words "diversity and pathogen and ecosystem," which highlights the lack of attention to this topic. Manipulative experiments altering the diversity of plant pathogens (particularly host-specific pathogens) and monitoring both responses in the plant community and ecosystem variables (e.g., productivity, nutrient retention) would aid in forging connections between the community and ecosystem scales in disease ecology. In addition, this work would inform efforts to conserve pathogen biodiversity (Ingram 1999).

In summary, a variety of field and laboratory-based studies suggest a link between species diversity and improved functioning of ecosystems. Given the potentially strong role that pathogens have in maintaining species diversity, we posit an important indirect link between pathogens and ecosystem function. Further, by directly affecting plant growth, pathogens can directly affect ecosystem processes as well.

CONCLUSIONS

The Red Queen hypothesis proposes that pathogens specialized on common host genotypes favor sexual reproduction in hosts because it generates rare genotypes. We argue that pathogens favor species diversity in communities by an analogous mechanism of frequency-dependent selection against common species. More generally, forces maintaining species diversity and genetic diversity may be similar (Velland and Geber 2005). A wider range of theory for population genetic diversity may be applicable to community ecology. By way of their effect on diversity, plant pathogens may have an indirect but important role in the functioning of terrestrial ecosystems.

Our ideas are supported by some results and inferences from other studies. However, there is a need for more empirical research under natural field conditions to explore more directly the effects of pathogens on ecosystem processes. The most obvious approach is to eliminate or reduce pathogen populations by pesticide applications, although this introduces other potential confounding effects. Advances in molecular biology and plant breeding may allow more field studies with plants genetically modified for disease resistance. The dramatic effects of pathogen epidemics on ecosystems are obvious; a more difficult challenge will be to elucidate the effects of native pathogens in undisturbed ecosystems. We hope that this contribution may serve as a stimulus to address this challenge.

LITERATURE CITED

Adams, M. W., A. H. Ellingboe, and E. C. Rossman. 1971. Biological uniformity and disease epidemics. Bioscience 21:1067–70.

Agrios, G. N. 1988. Plant Pathology, 3rd ed. San Diego: Academic Press.

Alexander, H. M. 1992. Evolution of disease resistance in natural plant populations. *In* Plant Resistance to Herbivores and Pathogens: Ecology, Evolution, and Genetics, ed. R. S. Fritz and E. L. Simms, 326–44. Chicago: University of Chicago Press.

Alexander, H. M., and J. Antonovics. 1988. Disease spread and population dynamics of anther-smut infestation of *Silene alba* caused by the fungus *Ustilago violacea*. Journal of Ecology 76:91–104.

Alexander, H. M., and R. D. Holt. 1998. The interaction between plant competition and disease. Perspectives in Plant Ecology, Evolution and Systematics 1:206–20.

Antonovics, J. 1976. The nature of limits to natural selection. Annals of the Missouri Botanical Garden 63:224–47.

Antonovics, J. 2003. Toward community genomics? Ecology 84:224–47.

Augspurger, C. K. 1983. Seed dispersal of the tropical tree, *Platypodium elegans*, and the escape of its seedlings from fungal pathogens. Journal of Ecology 71:759–71.

Baker, C. J., T. C. Harrington, U. Krauss, and A. C. Alfenas. 2003. Genetic variability and host specialization in the Latin American clade of *Ceratocystis fimbriata*. Phytopathology 93:1274–84.

Barrett, J. 1985. The gene-for-gene hypothesis: Parable or paradigm. *In* Ecology and Genetics of Host-Parasite Interactions, ed. D. Rollinson and R. M. Anderson, 215–25. London: Academic Press.

Barton, J., A. F. Gianotti, L. Morin, and R. A. Webster. 2003. Exploring the host range of *Fusarium tumidum*, a candidate bioherbicide for gorse and broom. Australasian Plant Pathology 32:203–11.

Beckstead, J., and I. M. Parker. 2003. Invasiveness of *Ammophila arenaria*: Release from soil-borne pathogens? Ecology 84:2824–31.

Bergelson, J., and C. B. Purrington. 1996. Surveying patterns in the cost of resistance in plants. American Naturalist 148:536–58.

Bever, J., K. Westover, and J. Antonovics. 1997. Incorporating the soil community into plant population dynamics: The utility of the feedback approach. Journal of Ecology 85:561–74.

Bever, J. D. 1994. Feedback between plants and their soil communities in an old field community. Ecology 75:1965–77.

Bever, J. D. 2003. Soil community feedback and the coexistence of competitors: Conceptual frameworks and empirical tests. New Phytologist 157: 465–73.

Blaney, C. S., and P. M. Kotanen. 2001. Effects of fungal pathogens on seeds of native and exotic plants: A test using congeneric pairs. Journal of Applied Ecology 38:1104–13.

Blok, W. J., and G. J. Bollen. 1997. Host specificity and vegetative compatibility of Dutch isolates of *Fusarium oxysporum* f. sp. *asparagi*. Canadian Journal of Botany 75:383–93.

Boland, G. J., and R. Hall. 1994. Index of plant hosts of *Sclerotinia sclerotiorum*. Canadian Journal of Plant Pathology 16:93–108.

Brown, S. P., M. E. Hochberg, and B. T. Grenfell. 2002. Does multiple infection select for raised virulence? Trends in Microbiology 10:401–5.

Browning, J. A., and K. J. Frey. 1969. Multiline cultivars as a means of disease control. Annual Review of Phytopathology 7:355–82.

Burdon, J. J. 1982. The effect of fungal pathogens on plant communities. *In* The Plant Community as a Working Mechanism, ed. E.I. Newman, 99–112. London: Blackwell.

Burdon, J. J. 1987. Diseases and Plant Population Biology. New York: Cambridge University Press.

Burdon, J. J., and G. A. Chilvers. 1982. Host density as a factor in plant disease ecology. Annual Review of Phytopathology 20:143–66.

Burdon, J. J., L. Ericson, and W. J. Muller. 1995. Temporal and spatial changes in a metapopulation of the rust pathogen *Triphragmium ulmariae* and its host, *Filipendula ulmaria*. Journal of Ecology 83:979–90.

Burdon, J. J., R. H. Groves, and J. M. Cullen. 1981. The impact of biological control on the distribution and abundance of *Chondrilla juncea* in southeastern Australia. Journal of Applied Ecology 18:957–66.

Burdon, J. J., and A. M. Jarosz. 1991. Host-pathogen interactions in natural populations of *Linum-marginale* and *Melampsora lini*. 1. Patterns of resistance and racial variation in a large host population. Evolution 45:205–17.

Burdon, J. J., and D. R. Marshall. 1981. Biological control and the reproductive mode of weeds. Journal of Applied Ecology 18:649–58.

Burdon, J. J., and R. C. Shattock. 1980. Disease in plant communities. Annals of Applied Biology 5:145–219.

Burdon, J. J., and J. N. Thompson. 1995. Changed patterns of resistance in a population of *Linum marginale* attacked by the rust pathogen *Melampsora lini*. Journal of Ecology 83:199–206.

Burdon, J. J., and P. H. Thrall. 2003. The fitness costs to plants of resistance to pathogens. Genome Biology 4, art. no. 227.

Burdon, J. J., A. Wennstrom, L. Ericson, W. J. Moller, and R. Morton. 1992. Density-dependent mortality in *Pinus sylvestris* caused by the snow blight pathogen *Phacidium infestans*. Oecologia 90:74–79.

Burger, W. 1992. Parapatric close-congeners in Costa Rica: Hypotheses for pathogen-mediated plant distribution and speciation. Biotropica 24:567–70.

Busch, J. W., M. Neiman, and J. M. Koslow. 2004. Evidence for maintenance of sex by pathogens in plants. Evolution 58:2584–90.

Caesar, A. J. 2000. Insect-pathogen interactions are the foundation of weed biocontrol. *In* Proceedings of the Tenth International Symposium on Biological

Control of Weeds, July 4–14, 1999, Montana State University, Bozeman, MT, 793–98.

Caldeira, M. C., R. J. Ryel, J. H. Lawton, and J. S. Pereira. 2001. Mechanisms of positive biodiversity-production relationships: Insights provided by delta C-13 analysis in experimental Mediterranean grassland plots. Ecology Letters 4:439–43.

Campbell, E. M., and J. A. Antos. 2000. Distribution and severity of white pine blister rust and mountain pine beetle on whitebark pine in British Columbia. Canadian Journal of Forest Research 30:1051–59.

Cardinale, B. J., M. A. Palmer, and S. L. Collins. 2002. Species diversity enhances ecosystem functioning through interspecific facilitation. Nature 415: 426–29.

Carlsson, U., and T. Elmqvist. 1992. Epidemiology of anther-smut disease (*Microbotryum violaceum*) and numeric regulation of populations of *Silene dioica*. Oecologia 90:509–17.

Chaboudez, P., and J. J. Burdon. 1995. Frequency-dependent selection in a wild plant-pathogen system. Oecologia 102:490–93.

Charudattan, R., and A. Dinoor. 2000. Biological control of weeds using plant pathogens: accomplishments and limitations. Crop Protection 19:691–95.

Chilvers, G. A., and E. G. Brittain. 1972. Plant competition mediated by host-specific parasites: A simple model. Australian Journal of Biological Science 25:749–56.

Clark, D. A., and D. B. Clark. 1984. Spacing dynamics of a tropical rain forest tree: Evaluation of the Janzen-Connell model. American Naturalist 124: 769–88.

Clarke, B. 1976. The evolution of genetic diversity. Proceedings of the Royal Society of London. Series B, Biological Sciences 205:453–74.

Clay, K. 1990. The impact of parasitic and mutalistic fungi on competitive interactions among plants. *In* Perspectives on Plant Competition, ed. J. B. Grace and D. Tilman, 391–412. San Diego: Academic Press.

Clay, K., and P. Kover. 1996. The Red Queen hypothesis and plant/pathogen interactions. Annual Review of Phytopathology 34:29–50.

Clay, K., and C. Schardl. 2002. Evolutionary origins and ecological consequences of endophyte symbiosis with grasses. American Naturalist 160:S99–127.

Coley, P. D., J. P. Bryant, and F. S. Chapin. 1985. Resource availability and plant antiherbivore defense. Science 230:895–99.

Condit, R., S. P. Hubbell, and R. B. Foster. 1992. Recruitment near conspecific adults and the maintenance of tree and shrub diversity in a neotropical forest. American Naturalist 140:261–86.

Connell, J. H. 1971. On the role of natural enemies in preventing competitive exclusion in some marine animals and in rain forest trees. *In* Dynamics of Numbers in Populations, ed. P. J. de Boer and G. R. Gradwell, 298–310. Wagenigen, Netherlands: Center for Agricultural Publishing and Documentation, PUDOC.

Darwin, C. 1872. On the Origin of Species, 6th ed. Chicago: Thompson and Thomas.

Daughtrey, M. L., and C. R. Hibben. 1994. Dogwood anthracnose: A new disease threatens two native *Cornus* species. Annual Review of Phytopathology 32:61–73.

Davis, M. B. 1981. Outbreaks of forest pathogens in Quaternary history. *In* Proceedings of the Fourth International Palynological Conference, Vol. 3, ed. D. Bharadwaj and H. Maheshwari, 216–27. Lucknow, India: Birbal Sahni Institute of Paleobotany.

De Deyn, G. B., C. E. Raaijmakers, and W. H. Van der Putten. 2004. Plant community development is affected by nutrients and soil biota. Journal of Ecology 92:824–34.

Delgado, N. J., C. R. Grau, and M. D. Casler. 2001. Host range and alternate host of a *Puccinia coronata* population from smooth brome grass. Plant Disease 85:513–16.

DeWalt, S. J., J. S. Denslow, and K. Ickes. 2004. Natural-enemy release facilitates habitat expansion of the invasive tropical shrub *Clidemia hirta*. Ecology 85:471–83.

Dimitrakopoulos, P. G., and B. Schmid. 2004. Biodiversity effects increase linearly with biotope space. Ecology Letters 7:574–83.

Doak, D. F., D. Bigger, E. K. Harding, M. A. Marvier, R. E. O'Malley, and D. Thomson. 1998. The statistical inevitability of stability-diversity relationships in community ecology. American Naturalist 151:264–76.

Dybdahl, M. F., and C. M. Lively. 1995. Host-parasite interactions: Infection of common clones in natural populations of a freshwater snail (*Potamopyrgus antipodarum*). Proceedings of the Royal Society of London. Series B, Biological Sciences 260:99–103.

Elton, C. S. 1958. The Ecology of Invasions by Animals and Plants. London: Methuen.

Farr, D. F., G. F. Bills, G. P. Chamuris, and A. Y. Rossman. 1989. Fungi on Plants and Plant Products in the United States. St. Paul, MN: APS Press.

Felton, G. W., and K. L. Korth. 2000. Trade-offs between pathogen and herbivore resistance. Current Opinion in Plant Biology 3:309–14.

Flor, H. H. 1956. The complimentary genetic systems in flax and flax rust. Advances in Genetics 8:29–54.

Garrett, K. A., and C. C. Mundt. 1999. Epidemiology in mixed host populations. Phytopathology 89:984–90.

Garrett, K. A., R. J. Nelson, C. C. Mundt, G. Chacon, R. E. Jaramillo, and G. A. Forbes. 2001. The effects of host diversity and other management components on epidemics of potato late blight in the humid highland tropics. Phytopathology 91:993–1000.

Gilbert, G. S. 2002. Evolutionary ecology of plant diseases in natural ecosystems. Annual Review of Phytopathology 40:13–43.

Gilbert, G. S., S. P. Hubbell, and R. B. Foster. 1994. Density and distance-to-adult effects of a canker disease of trees in a moist tropical forest. Oecologia 98:100–108.

Glesner, R. R., and D. Tilman. 1978. Sexuality and the components of environmental uncertainty: Clues from geographical parthenogenesis in terrestrial animals. American Naturalist 112:659–73.

Grime, J. P. 1998. Benefits of plant diversity to ecosystems: immediate, filter and founder effects. Journal of Ecology 86:902–10.

Gudelj, I., F. van den Bosch, and C. A. Gilligan. 2004. Transmission rates and adaptive evolution of pathogens in sympatric heterogeneous plant populations. Proceedings of the Royal Society of London. Series B, Biological Sciences 271:2187–94.

Haldane, J. B. S. 1949. Disease and evolution. La Ricerca Scientifica Supplementa 19:68–76.

Hamilton, W. D. 1980. Sex versus non-sex parasite. Oikos 35:282–90.

Hamilton, W. D., R. Axelrod, and R. Tanese. 1990. Sexual reproduction as an adaptation to resist parasites (a review). Proceedings of the National Academy of Science of the United States of America 87:3566–73.

Hansen, E. M., and E. M. Goheen. 2000. *Phellinus weirii* and other native root pathogens as determinants of forest structure and process in western North America. Annual Review of Phytopathology 38:515–39.

Harlan, J. R. 1976. Diseases as a factor in plant evolution. Annual Review of Phytopathology 14:31–51.

Harms, K. E., S. J. Wright, O. Calderon, A. Hernandez, and E. A. Herre. 2000. Pervasive density-dependent recruitment enhances seedling diversity in a tropical forest. Nature 404:493–95.

Harper, J. L. 1977. Population Biology of Plants. New York: Academic Press.

Hassell, M. P. 2000. The Spatial and Temporal Dynamics of Host-Parasitoid Interactions. Oxford: Oxford University Press.

Hector, A., B. Schmid, C. Beierkuhnlein, M. C. Caldeira, M. Diemer, P. G. Dimitrakopoulos, J. A. Finn, H. Freitas, P. S. Giller, et al. 1999. Plant diversity and productivity experiments in European grasslands. Science 286:1123–27.

Hendrix, F. F., and W. A. Campbell. 1973. Pythiums as plant pathogens. Annual Review of Phytopathology 11:77–98.

Hijwegen, T. 1979. Fungi as plant taxonomists. Symbolae Botanicae Upsalienses 22:146–65.

Holah, J. C., and H. M. Alexander. 1999. Soil pathogenic fungi have the potential to affect the coexistence of two tallgrass prairie species. Journal of Ecology 87:598–608.

Holah, J. C., M. V. Wilson, and E. M. Hansen. 1993. Effects of a native forest pathogen, *Phellinus weirii*, on Douglas fir forest composition in western Oregon. Canadian Journal of Forestry Research 23:2473–80.

Holah, J. C., M. V. Wilson, and E. M. Hansen. 1997. Impacts of a native root-rotting pathogen on successional development of old-growth Douglas fir forests. Oecologia 111:429–33.

Holt, R. D., and J. Pickering. 1985. Infectious disease and species coexistence: A model of Lotka-Volterra form. American Naturalist 126:196–211.

Hood, L. A., M. D. Swaine, and P. A. Mason. 2004. The influence of spatial patterns of damping-off disease and arbuscular mycorrhizal colonization on tree seedling establishment in Ghanaian tropical forest soil. Journal of Ecology 92:816–23.

Hooper, D. U., and P. M. Vitousek. 1997. The effects of plant composition and diversity on ecosystem processes. Science 277:1302–5.

Howard, R. S., and C. M. Lively. 1994. Parasitism, mutation accumulation and the maintenance of sex. Nature 367:554–57.

Hubbell, S. P. 1979. Tree dispersion, abundance, and diversity in a tropical dry forest. Science 203:1299–309.

Hubbell, S. P. 2001. The Unified Neutral Theory of Biodiversity and Biogeography. Princeton, NJ: Princeton University Press.

Hudson, P. J., A. P. Dobson, and D. Newborn. 1998. Prevention of population cycles by parasite removal. Science 282:2256–58.

Huston, M. A. 1994. Biodiversity. Cambridge: Cambridge University Press.

Ingram, D. S. 1999. Biodiversity, plant pathogens and conservation. Plant Pathology 48:433–42.

Ives, A. R., and J. B. Hughes. 2002. General relationships between species diversity and stability in competitive systems. American Naturalist 159:388–95.

Jaenike, J. 1978. A hypothesis to account for the maintenance of sex within populations. Evolutionary Theory 3:191–94.

Janzen, D. H. 1970. Herbivores and the number of tree species in tropical forests. American Naturalist 104:501–28.

Jarosz, A. M., and J. J. Burdon. 1991. Host-pathogen interactions in natural populations of *Linum marginale* and *Melampspora lini*. II. Local and regional variation in patterns of resistance and racial structure. Evolution 45:1618–27.

Jarosz, A. M., and J. J. Burdon. 1992. Host pathogen interactions in natural populations of *Linum marginale* and *Melampsora lini*. 3. Influence of pathogen epidemics on host survivorship and flower production. Oecologia 89: 53–61.

Jarosz, A. M., and A. L. Davelos. 1995. Effects of disease in wild plant populations and evolution of pathogen aggressiveness. New Phytologist 129:371–87.

Jasieniuk, M., A. L. Brule-babel, and I. N. Morrison. 1996. The evolution and genetics of herbicide resistance in weeds. Weed Science 44:176–93.

Johnson, R. 1987. Selected examples of relationships between pathogenicity in cereal rusts and resistance in their hosts. *In* Populations of Plant Pathogens: Their Dynamics and Genetics, ed. M. S. Wolfe and C. E. Caten, 181–92. Oxford: Blackwell Scientific Publications.

Jonsson, L. M., Nilsson, M. C., Wardle, D. A., Zackrisson, O. 2001. Context dependent effects of ectomycorrhizal species richness on tree seedling productivity. Oikos 93:353–64.

Jules, E. S., M. J. Kauffman, W. D. Ritts, and A. L. Carroll. 2002. Spread of an invasive pathogen over a variable landscape: A nonnative root rot on Port Orford cedar. Ecology 83:3167–81.

Kaltz, O., and J. A. Shykoff. 1998. Local adaptation in host-parasite systems. Heredity 81:361–70.

Keane, R. M., and M. J. Crawley. 2002. Exotic plant invasions and the enemy release hypothesis. Trends in Ecology & Evolution 17:164–70.

Kelley, S. 1994. Viral pathogens and the advantage of sex in the perennial grass *Anthoxanthum odoratum*. Philosophical Transactions of the Royal Society of London. Series B, Biological Sciences 346:295–302.

Klironomos, J. N., J. McCune, M. Hart, and J. Neville. 2000. The influence of arbuscular mycorrhizae on the relationship between plant diversity and productivity. Ecology Letters 3:137–41.

Knevel, I. C., T. Lans, F. B. J. Menting, U. M. Hertling, and W. H. van der Putten. 2004. Release from native root herbivores and biotic resistance by soil pathogens in a new habitat both affect the alien *Ammophila arenaria* in South Africa. Oecologia 141:502–10.

Knops, J. M. H., D. Tilman, N. M. Haddad, S. Naeem, C. E. Mitchell, J. Haarstad, M. E. Ritchie, K. M. Howe, P. B. Reich, et al. 1999. Effects of plant species richness on invasion dynamics, disease outbreaks, insect abundances and diversity. Ecology Letters 2:286–93.

Lambers, J. H. R., W. S. Harpole, D. Tilman, J. Knops, and P. B. Reich. 2004. Mechanisms responsible for the positive diversity-productivity relationship in Minnesota grasslands. Ecology Letters 7:661–68.

Lehman, C. L., and D. Tilman. 2000. Biodiversity, stability, and productivity in competitive communities. American Naturalist 156:534–52.

Leonard, K. J. 1987. The host population as a selective factor (including stabilizing selection). *In* Populations of Plant Pathogens: Their Dynamics and Genetics, ed. M. S. Wolfe and C. Caten, 163–78. Oxford: Blackwell Scientific Publications.

Leon-Ramirez, C. G., J. L. Cabrera-Ponce, A. D. Martinez-Espinoza, L. Herrera-Estrella, L. Mendez, C. G. Reynaga-Pena, and J. Ruiz-Herrera. 2004. Infection of alternative host plant species by *Ustilago maydis*. New Phytologist 164:337–46.

Leuchtmann, A. 1992. Systematics, distribution and host specificity of grass endophytes. Natural Toxins 1:150–62.

Leuchtmann, A., and K. Clay. 1996. Isozyme evidence for host races of the fungus *Atkinsonella hypoxylon* (Clavicipitaceae) infecting the *Danthonia* complex in the southern Appalachians. American Journal of Botany 83:1144–52.

Levin, D. A. 1975. Pest pressure and recombination systems in plants. American Naturalist 109:437–51.

Levine J. M. 2000. Species diversity and biological invasions: Relating local process to community pattern. Science 288:852–54.

Lively, C. M. 1996. Host-parasite coevolution and sex. Bioscience 46:107–14.

Loreau, M., and A. Hector. 2001. Partitioning selection and complementarity in biodiversity experiments. Nature 412:72–76.

Mack, R. N. 1996. Predicting the identity and fate of plant invaders: Emergent and emerging approaches. Biological Conservation 78:107–21.

Maron, J. L., and P. G. Connors. 1996. A native nitrogen-fixing shrub facilitates weed invasion. Oecologia 105:302–12.

Maron J. L., and M. Vilá. 2001. When do herbivores affect plant invasion? Evidence for the natural enemies and biotic resistance hypotheses. Oikos 95:361–73.

Maroof, M. A. S., R. K. Webster, and R. W. Allard. 1983. Evolution of resistance to scald, powdery mildew, and net blotch in barley composite cross II populations. Theoretical and Applied Genetics 66:279–83.

Martinez, S. P., R. Snowdon, and J. Pons-Kuhnemann. 2004. Variability of Cuban and international populations of *Alternaria solani* from different hosts and localities: AFLP genetic analysis. European Journal of Plant Pathology 110:399–409.

McDonald, B. A., and J. M. McDermott. 1993. Population genetics of plant pathogenic fungi. Bioscience 43:311–19.

McDonald, B. A., J. M. McDermott, R. W. Allard, and R. K. Webster. 1989. Coevolution of host and pathogen populations in the *Hordeum vulgare-Rhynchosporium secalis* pathosystem. Proceedings of the National Academy of Sciences of the United States of America 86:3924–27.

Mikola, J., V. Salonen, H. Setala. 2002. Studying the effects of plant species richness on ecosystem functioning: Does the choice of experimental design matter? Oecologia 133:594–98.

Mills, K. E., and J. D. Bever. 1998. Maintenance of diversity within plant communities: Soil pathogens as agents of negative feedback. Ecology 79: 1595–601.

Mitchell, C. E. 2003. Trophic control of grassland production and biomass by pathogens. Ecology Letters 6:147–55.

Mitchell, C. E., and A. G. Power. 2003. Release of invasive plants from fungal and viral pathogens. Nature 421:625–27.

Mitchell, C. E., D. Tilman, and J. V. Groth. 2002. Effects of grassland plant species diversity, abundance, and composition on foliar fungal disease. Ecology 83:1713–26.

Molofsky, J., J. D. Bever, J. Antonovics, and T. J. Newman. 2002. Negative frequency dependence and the importance of spatial scale. Ecology 83:21–27.

Mouquet, N., J. L. Moore, and M. Loreau. 2002. Plant species richness and community productivity: Why the mechanism that promotes coexistence matters. Ecology Letters 5:56–65.

Mundt, C. C. 2002. Use of multiline cultivars and cultivar mixtures for disease management. Annual Review of Phytopathology 40:381.

Naeem, S. 2002. Ecosystem consequences of biodiversity loss: The evolution of a paradigm. Ecology 83:1537–52.

Newman, R. M., D. C. Thompson, and D. B. Richman. 1998. Conservation strategies for the biological control of weeds. *In* Conservation Biological Control, ed. P. Barbosa. San Diego: Academic Press.

Nsarellah, N., E. M. Elias, and R. G. Cantrell. 2003. Variation in virulence and host specificity of *Pyrenophora tritici-repentis* on common and durum wheat in North Dakota, USA and Morocco. Cereal Research Communications 31:121–28.

Olff, H., B. Hoorens, R. G. M. de Goede, W. H. van der Putten, and J. M. Gleichman. 2000. Small-scale shifting mosaics of two dominant grassland species: The possible role of soil-borne pathogens. Oecologia 125:45–54.

Osbourn, A. 1996. Saponins and plant defense: A soap story. Trends in Plant Science 1:4–9.

Packer, A., and K. Clay. 2000. Soil pathogens and spatial patterns of seedling mortality in a temperate tree. Nature 404:278–81.

Panetta, E. D., and J. Dodd. 1995. *Chondrilla juncea. In* The Biology of Australian Weeds, ed. R. H. Groves, R. C. H. Sheperd, and R. G. Richardson, 67–84. Melbourne, Australia: R. G. and F. J. Richardson.

Parker, I. M., and G. S. Gilbert. 2004. The evolutionary ecology of novel plant-pathogen interactions. Annual Review of Ecology Evolution and Systematics 35:675–700.

Parker, M. A. 1985. Local population differentiation for compatibility in an annual legume and its host-specific fungal pathogen. Evolution 39:713–23.

Parker, M. A. 1988. Polymorphism for disease resistance in the annual legume *Amphicarpaea bracteata*. Heredity 60:27–31.

Parker, M. A. 1992. Disease and plant population genetic structure. *In* Plant Resistance to Herbivores and Pathogens: Ecology, Evolution and Genetics, ed. R. S. Fritz and E. L. Simms, 345–62. Chicago: University of Chicago Press.

Parker, M. A. 1994. Pathogens and sex in plants. Evolutionary Ecology 8:560–84.

Power, A. G. 1987. Plant community diversity, herbivore movement, and an insect transmitted disease of maize. Ecology 68:1658–69.

Power, A. G., and C. E. Mitchell. 2004. Pathogen spillover in disease epidemics. American Naturalist 164:S79–89.

Price, J. S., J. D. Bever, and K. Clay. 2004. Genotype, environment, and genotype by environment interactions determine quantitative resistance to leaf rust (*Coleosporium asterum*) in *Euthamia graminifolia* (Asteraceae). New Phytologist 162:729–43.

Randall, J.M., and M. Tu. 2001. Biological Control. Available: http://tncweeds .ucdavis.edu/products/handbook/06.BiologicalControl.doc.

Reinhart, K. O., A. Packer, W. H. Van der Putten, and K. Clay. 2003. Plant-soil biota interactions and spatial distribution of black cherry in its native and invasive ranges. Ecology Letters 6:1046–50.

Ridley, H. N. 1930. The Dispersal of Plants Throughout the World. Ashford, Kent, UK: L. Reeve.

Rizzo, D. M., and M. Garbelotto. 2003. Sudden oak death: Endangering California and Oregon forest ecosystems. Frontiers in Ecology and the Environment 1:197–204.

Roy, B. A. 1993. Patterns of rust infection as a function of host genetic diversity and host density in natural populations of the apomictic crucifer *Arabis holboellii*. Evolution 47:111–24.

Rudgers, J. A., J. M. Koslow, and K. Clay. 2004. Endophytic fungi alter relationships between diversity and ecosystem properties. Ecology Letters 7:42–51.

Rudgers, J. A., W. B. Mattingly, and J. M. Koslow. 2005. Mutualistic fungus promotes plant invasion into diverse communities. Oecologia 144:463–71.

Sasaki, A. 2000. Host-parasite coevolution in a multilocus gene-for-gene system. Proceedings of the Royal Society of London. Series B, Biological Sciences 267:2183–88.

Savile, D. B. O. 1987. Use of rust fungi (*Uredinales*) in determining ages and relationships in Poaceae. *In* Grass Systematics and Evolution, ed. T. R. Soderstrom, K. W. Hilu, C. Campbell, and M. E. Barkworth, 168–78. Grass Systematics and Evolution. Washington, DC: Smithsonian Institution Press.

Schardl, C. L., A. Leuchtmann, K. R. Chung, D. Penny, and M. R. Siegel. 1997. Coevolution by common descent of fungal symbionts (*Epichloë* spp) and grass hosts. Molecular Biology and Evolution 14:133–43.

Schulze, E. D., and H. A. Mooney, editors. 1993. Biodiversity and Ecosystem Function. Berlin: Springer-Verlag.

Schwartz, M. W., C. A. Brigham, J. D. Hoeksema, K. G. Lyons, M. H. Mills, and P. J. van Mantgem. 2000. Linking biodiversity to ecosystem function: Implications for conservation ecology. Oecologia 122:297–305.

Shearer, B. L., C. E. Crane, and A. Cochrane. 2004. Quantification of the susceptibility of the native flora of the South-West Botanical Province, Western Australia, to *Phytophthora cinnamomi*. Australian Journal of Botany 52: 435–43.

Scheffer, R. P. 2003. The Nature of Disease in Plants. Cambridge: Cambridge University Press.

Siemann, E., and W. E. Rogers. 2001. Genetic differences in growth of an invasive tree species. Ecology Letters 4:514–18.

Simms, E. L. 1992. Costs of plant resistance to herbivory. *In* Plant Resistance to Herbivores and Pathogens: Ecology, Evolution, and Genetics, ed. R. S. Fritz and E. L. Simms, 392–425. Chicago: University of Chicago Press.

Skinner, D. Z., and D. L. Stuteville. 1995. Host-range expansion of the alfalfa rust. Plant Disease 79:456–60.

Staats, M., P. van Baarlen, and J. A. L. van Kan. 2005. Molecular phylogeny of the plant pathogenic genus *Botrytis* and the evolution of host specificity. Molecular Biology and Evolution 22:333–46.

Stevens, N. E. 1948. Disease damage in clonal and self-pollinated crops. Journal of the American Society of Agronomists 40:841–44.

Strauss, S. Y., J. A. Rudgers, J. A. Lau, and R. E. Irwin. 2002. Direct and ecological costs of resistance to herbivory. Trends in Ecology & Evolution 17: 278–85.

Strong, D. R., and D. A. Levin. 1975. Species richness of the parasitic fungi of British trees. Proceedings of the National Academy of Sciences of the United States of America 72:2116–19.

Thompson, J. N. 1994. The Coevolutionary Process. Chicago: University of Chicago Press.

Thompson, J. N., and J. J. Burdon. 1992. Gene-for-gene coevolution between plants and parasites. Nature 360:121–25.

Thrall, P. H., J. J. Burdon, and A. Young. 2001. Variation in resistance and virulence among demes of a plant host-pathogen metapopulation. Journal of Ecology 89:736–48.

Tian, D., M. B. Traw, J. Q. Chen, M. Kreitman, and J. Bergelson. 2003. Fitness costs of R-gene-mediated resistance in *Arabidopsis thaliana*. Nature 423: 74–77.

Tilman, D., J. Knops, D. Wedin, P. Reich, M. Ritchie, and E. Siemann. 1997. The influence of functional diversity and composition on ecosystem processes. Science 277:1300–302.

Tilman, D., C. L. Lehman, and K. T. Thomson. 1997. Plant diversity and ecosystem productivity: Theoretical considerations. Proceedings of the National Academy of Sciences of the United States of America 94:1857–61.

Turchin, P., A. D. Taylor, and J. D. Reeve. 1999. Dynamical role of predators in population cycles of a forest insect: An experimental test. Science 285: 1068–71.

van der Heijden, M. G. A., J. N. Klironomos, M. Ursic, P. Moutoglis, R. Streitwolf-Engel, T. Boller, A. Wiemken, and I. R. Sanders. 1998. Mycorrhizal fungal diversity determines plant biodiversity, ecosystem variability and productivity. Nature 396:69–72.

van der Kamp, B. J. 1991. Pathogens as agents of diversity in forested landscapes. Forestry Chronicle 67:353–54.

Van der Plank, J. E. 1963. Plant Disease: Epidemics and Control. New York: Academic Press.

Van der Plank, J. E. 1968. Disease Resistance in Plants. New York: Academic Press.

Van der Putten, W. H., C. Van Dijk, and B. A. M. Peters. 1993. Plant-specific soil-borne disease contribute to succession in foredune vegetation. Nature 362:53–56.

Van Horn, R., and K. Clay. 1995. mtDNA variation in the fungus *Atkinsonella hypoxylon* infecting sympatric *Danthonia* grasses. Evolution 49:360–71.

van Ruijven, J., and F. Berendse. 2005. Diversity-productivity relationships: Initial effects, long-term patterns, and underlying mechanisms. Proceedings of the National Academy of Sciences of the United States of America 102:695–700.

Vellend, M. 2003. Island biogeography of genes and species. American Naturalist 162:358–65.

Vellend, M. 2004. Parallel effects of land-use history on species diversity and genetic diversity of forest herbs. Ecology 85:3043–55.

Vellend, M., and M. Geber. 2005. Connections between species diversity and genetic diversity. Ecology Letters 8:767–81.

Vitousek, P. M., L. R. Walker, L. D. Whiteaker, D. Muellerdombois, and P. A. Matson. 1987. Biological invasion by *Myrica faya* alters ecosystem development in Hawaii. Science 238:802–4.

Wardle, D. A. 1999. Is "sampling effect" a problem for experiments investigating biodiversity-ecosystem function relationships? Oikos 87:403–7.

Watson, I. A. 1970. Changes in virulence and population shifts in plant pathogens. Annual Review of Phytopathology 8:209–30.

Wills, C., R. Condit, R. B. Foster, and S. P. Hubbell. 1997. Strong density- and diversity related effects help maintain tree species diversity in a neotropical forest. Proceedings of the National Academy of Sciences of the United States of America 94:1252–57.

Wilsey, B. J., and C. Potvin. 2000. Biodiversity and ecosystem functioning: Importance of species evenness in an old field. Ecology 81:887–92.

Wohl, D. L., S. Arora, and J. R. Gladstone. 2004. Functional redundancy supports biodiversity and ecosystem function in a closed and constant environment. Ecology 85:1534–40.

Wolfe, M. S., and J. A. Barrett. 1985. The current status and prospects in multiline cultivars and variety mixtures for disease resistance. Annual Review of Phytopathology 23:251–73.

Yachi, S., and M. Loreau. 1999. Biodiversity and ecosystem productivity in a fluctuating environment: The insurance hypothesis. Proceedings of the National Academy of Sciences of the United States of America 96:1463–68.

Zavaleta, E. S., and K. B. Hulvey. 2004. Realistic species losses disproportionately reduce grassland resistance to biological invaders. Science 306:1175–77.

Zhu, Y. Y., H. R. Chen, J. H. Fan, Y. Y. Wang, Y. Li, J. B. Chen, J. X. Fan, S. S. Yang, L. P. Hu, et al. 2000. Genetic diversity and disease control in rice. Nature 406:718–22.

CHAPTER EIGHT
Invasion Biology and Parasitic Infections

*Sarah E. Perkins, Sonia Altizer, Ottar Bjornstad, Jeremy J. Burdon,
Keith Clay, Lorena Gómez-Aparicio, Jonathan M. Jeschke,
Pieter T. J. Johnson, Kevin D. Lafferty, Carolyn M. Malmstrom,
Patrick Martin, Alison Power, David L. Strayer, Peter H. Thrall,
and Maria Uriarte*

SUMMARY

PARASITIC INFECTIONS CAN STRONGLY affect invasion success and the impact of invasive species on native biota. A key mechanism facilitating invasion is escape from regulation by natural enemies—the enemy release hypothesis. The level and duration of release depend on the types of parasites lost and gained, with highly regulating acute infections most likely to be lost and, over time, pathogenic RNA viruses likely to be gained. The rate at which hosts accumulate parasites depends on multiple factors, including the biotic resistance of the community and the ecosystem changes induced by the invasive species themselves. We discuss several examples of how invasive species may increase parasite susceptibility of the community by increasing parasite reservoir densities or by altering parasite flow via apparent competition. We then consider the evolutionary implications on a longer time scale if the susceptibility of invasive species is enhanced by loss of parasite resistance. Finally, we discuss whether parasites should be considered a special class of invader and conclude by identifying approaches, challenges, and priorities for future research in parasite dynamics of introduced species.

INTRODUCTION

With an estimated cost of U.S. $120 billion in environmental damage and economic loss in the United States alone (Pimentel et al. 2005), invasive species and their ecosystem effects are a major focus of research in ecology (Elton 1958; Kolar and Lodge 2001; Shea and Chesson 2002). Invasive species are those that have been moved beyond natural dispersal barriers and introduced into habitat outside their native range in which they become established and prolific, often with substantial consequences for native biota, human health, and ecosystem functioning (Kolar and

Lodge 2001; Vitousek et al. 1997). Not all introduced species become invasive, and as such, a focus in invasion ecology has been to determine the mechanisms facilitating species invasions. One of the major hypotheses is that of enemy release, which posits that in their native range, populations are regulated by enemies, but these enemies are reduced in number or absent from populations outside their natural dispersal range, thereby allowing introduced species to escape regulatory forces and become invasive (Darwin 1859; Elton 1958; Hierro et al. 2005). The enemy release hypothesis tends to focus on predators as enemies, but recent research recognizes the role of parasites as enemies and has documented high loss of parasites from host species in invaded compared with native ranges (Mitchell and Power 2003; Torchin et al. 2003, Torchin and Mitchell 2004).

We start this chapter by reviewing the empirical evidence for enemy release with respect to parasites and posit which parasites are likely to be lost and gained and the implications this has for host invasion success. We further explore the intersection between the ecology of species invasions and parasitic infections by examining what effect the parasites of introduced species may have on native biota and the knock-on ecosystem effects. In addition, we address evolutionary aspects of introduced species and their parasite fauna. Finally, we consider parasites as invaders themselves. Our aim is a review of invasive species and their interactions with parasites and the impact this has for native biota, ecosystem functioning, evolution, and human health. Throughout this chapter, we refer to parasites and pathogens as simply "parasites" (see Hall et al., chapter 10, this volume, for definitions) unless explicitly stated otherwise, and "enemies" in this context refers specifically to parasites, rather than also including predators.

The Enemy Release Hypothesis

To cite the enemy release hypothesis as a mechanism for facilitating successful invasion of introduced species, a combination of both escape and release from parasites must be demonstrated. *Enemy escape* quantifies the extent to which the parasite diversity and prevalence of introduced species are reduced relative to those in the native range. European plants that become established in the United States are infected by, on average, 77% fewer pathogen species than in their native range (Mitchell and Power 2003), whereas introduced animals are, on average, infected by 53% fewer helminth species than in their native range (Torchin et al. 2003). For successful invasion, we must determine whether this documented escape translates into *release*, which would require that the parasites have measurable negative impacts on species abundance or fitness so that escape from them results in increased success of the invaders. From disease ecology studies, we know that parasites cause harm to and regulate their host populations in terms of altering host density, fecundity, or growth (Hudson et al. 2002; Tompkins and Begon 1999; Torchin et al. 2001). Bringing together both invasion and disease ecology, we find evidence for an increase in demographic parameters in introduced species that have been released from parasites (Lafferty et al. 2005). A well-studied example is that of the European green crab, which performed better in an invaded range than the average parasitized European population but similar to an uninfected European population (Torchin et al. 2001). More empirical examples are required, but challenges exist in documenting enemy release empirically with respect to the role of parasites, as opposed to other enemies.

One challenge in documenting enemy release is that unsuccessful invasions often go unnoticed and assessment must rely on comparisons of established populations or species in their native versus introduced ranges (but see Jeschke and Strayer 2005; Suarez et al. 2005). This requires comprehensive and rigorous sampling of parasites in both the native and introduced range. Furthermore, given the complex number of ecological processes that influence the success of invaders, how do we measure the importance of enemy release by parasites relative to predators and other variables that influence invasion success? The approach thus far has been to use common garden or reciprocal transplant experiments and biogeographical studies, which have yielded insights into the success of introduced species with specific reference to parasites. However, both approaches suffer from a number of experimental difficulties.

Biogeographical studies (e.g., Mitchell and Power 2003; Torchin et al. 2001, 2003) suffer from confounding variables that are difficult to control for and usually involve a small number of randomly chosen populations inhabiting a restricted area in the native range of the species (e.g., Edwards et al. 2003; Grigulis et al. 2001; Reinhart et al. 2003; but see Buckley et al. 2003). As a result, it is difficult to discern to what extent the results are highly specific to the study sites. Often when several populations are studied in the same area in the native range, important differences in the role of enemies appear (Callaway et al. 2004). However, in contrast, Reinhart et al. (2005) sampled black cherry populations widely from throughout their range in eastern North America and found a consistent negative effect of soil-borne pathogens on seedling survival. Common garden experiments have been less frequently applied with respect to elucidating the effect of parasites as enemies and have the issue of being potentially confounded by variation among species in traits such as resistance and tolerance.

With these issues in mind, we suggest that conclusive evidence of enemy release would require replicated experimental manipulations and descriptive studies of two types. The first would be removal experiments, where one could show for a specific parasite that removal from the host in its native range resulted in demographic release, that is, greater host fitness, an increase in population size, or a range expansion. Some evidence for this exists with macroparasites (Tompkins and Begon 1999), and although these experiments have not focused specifically on hosts that are invasive species, they do provide evidence of enemy release with respect to parasites. The second type of conclusive evidence would be addition experiments in which parasites, when introduced into invasive populations, resulted in lower fitness, decreased population growth, or range contraction of the invader. Such "experiments" are commonly done during biological control (Lafferty et al. 2005). However, the parasites used in biocontrol often are not native to the invasive species in their home range. For example, myxomavirus, endemic to South America, is used to control European rabbits invasive in Australia.

In summary, the challenge is to disentangle the relative importance of an array of interactions with parasites, other enemies, and environmental factors to which an invasive species is exposed in its introduced range as compared with those in its native range. Given the broad range of species that occur in natural communities and the diverse ways these interact with other biotic and abiotic elements of the environment, we might not expect simple pairwise host-parasite interactions, and the effect of a single enemy, in our case a parasite, may be diluted (Keane and Crawley 2002). More demographic data demonstrating strong population-level suppression by parasites is necessary to evaluate the ecological significance of parasites in the enemy release hypothesis. Given the potential limitations of experimental tests, we advocate a pluralistic approach to the study of invaders using a combination of molecular techniques, field surveys, experiments, and historical records to understand the role of parasites in facilitating species invasions.

PARASITES LOST AND GAINED

Invasion success can be a function of escaping parasitic regulation during the different transition phases, which include transport, establishment, and spread (Jeschke and Strayer 2005; Kolar and Lodge 2001). Considering the variation in host regulation associated with different parasites an important question to ask is, which types of parasites might be lost and gained during invasion? For example, if highly regulatory parasites are lost during host transport, then we may expect immediate demographic release and increases in fecundity and survival, thus facilitating invasion. Conversely, if regulatory parasites are gained, then we may expect reduced invasion success.

Parasites Lost

Based on current knowledge of the ecology of parasites and invasive species, we can make predictions concerning the ecological characteristics of parasites that will be lost (table 8.1). The first phase of invasion, transport, typically occurs with a subset of the host population, and this founder population is expected to be devoid of parasites that are rare, or found at low prevalence, in the host population in its native range (Torchin et al. 2003). If these rare parasites do not have great regulatory control over the host population, then demographic release will be minimal (Colautti et al. 2004).

TABLE 8.1
Parasites predicted to be lost and gained during the invasion process for plants and animals

	Animals	*Plants*
Lost	Ectoparasites and the vector- borne diseases that they transmit	Aphids/nematode vectors and the pathogens they transmit
Lost	Specialists	Specialists
Lost	Rare parasites, low prevalence	Rare parasites, low prevalence
Lost	Acute immunizing pathogens (large threshold and short infectious period)	Pathogens that cause high mortality
Lost	Complex life cycles (e.g., trophically transmitted macroparasites)	Parasites with obligate alternative hosts
Retained	Vertically transmitted parasites (rare in animals)	Seed-borne viruses and some fungal pathogens
Retained	Retroviruses	Retroviruses
Retained	Generalists	Generalists
Gained	RNA viruses	RNA viruses
Gained	Generalists	Generalists
Gained	Parasites and pathogens of phylogenetically similar host species	Parasites and pathogens of phylogenetically similar host species

Indeed, it is the common or prevalent macroparasites that tend to be regulatory (Hudson et al. 2002), but these too may be lost or reduced in intensity, because subsampling hosts from aggregated host-macroparasite distributions tends to sample individuals with low parasite intensity (Shaw et al. 1998). We may expect parasite loss to be accentuated further if the process of selecting invaders favors healthy individuals that are free of parasitic infection, for example functional groups within the population that have low exposure or have acquired immunity to infection.

Immature stages have different (often fewer) parasites than do adults. For example, sexually mature male rodents have been identified as key hosts in the persistence of tick-borne disease (Perkins et al. 2003). Generally, for vertebrate hosts the intensity of macroparasites is male biased (Moore and Wilson 2002); thus if a deliberate introduction favors females, then parasite diversity will be further reduced within the founder population. For plants, a key issue is whether transport occurs as a photosynthetic plant or as a dormant seed or spore, as the latter generally have a smaller subset of parasites than do adults (e.g., Molloy et al. 1997). Adult plants are likely to be transported in soil with a much greater likelihood that associated parasites of leaves, flowers, or roots are simultaneously transported (Keane and Crawley 2002). Invasive plants are more likely to escape from fungal than from viral pathogens (Mitchell and Power 2003), in part because fungal parasites tend to be more host specific, but also because viruses can often persist in seeds and be vertically transmitted (Torchin and Mitchell 2004). Therefore, the mode of parasite transmission has an impact on the likelihood of loss, and retroviruses and other endogenous or vertically transmitted parasites will be extremely difficult to escape from. However, in plants, many vertically transmitted microbes tend to be mutualistic and not parasitic, and so movement of seeds compared with whole plants is much more likely to result in enemy escape.

Intuitively, we expect the high mortality associated with host transport, coupled with parasitic infection, could cause host mortality to be amplified, particularly if the infection is acute (Moller 2005). Therefore, acute infections of high virulence and short infectious periods are likely to be lost, along with their hosts, in the transport stage. Acute infections, which include many microparasitic diseases, are often highly regulating, and release from them may partly explain the increased demographic capability of invaders (figure 8.1).

After transport, some parasites could be lost during establishment, in part due to thresholds for invasion. Parasites require a threshold population size to establish and persist (Lloyd-Smith et al. 2005). However, founder populations of invading species typically fluctuate around

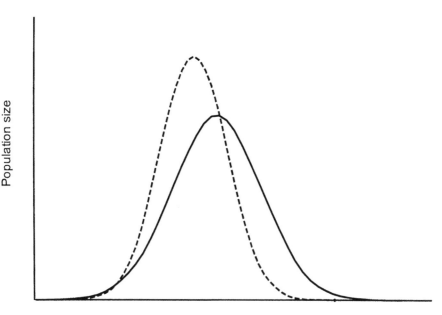

Figure 8.1. A schematic of the change in size over time of an introduced population that experiences parasite release and so increased growth during invasion, followed by parasite regulation coupled with a reduction in population size. Two comparable potential outcomes are shown. The solid line represents moderate parasite escape and release, followed by moderate biotic resistance of the invaded community; the dashed line represents high parasite escape and release, followed by high biotic resistance, or rapid accumulation of regulatory parasites.

relatively small numbers, and therefore parasites cannot establish and persist before the invader reaches population densities that would support endemicity (e.g., de Castro and Bolker 2005).

For hosts with complex life cycles, appropriate alternative hosts may not be present in the invaded range, precluding parasite establishment. Ectoparasites and other vectors also may be lost, since those with short feeding periods and seasonality in host biting behavior (e.g., ticks in temperate zones) are unlikely to be transported with the introduced hosts. Consequently, vector-borne pathogens will be especially vulnerable to loss. Additionally, if the abiotic conditions in the new habitat are not suitable for free-living parasitic stages, then they will be lost during establishment. Indeed, any form of complex life cycle parasite or specialist requirements suggests that specialist parasites will be lost over generalists (Cornell and Hawkins 1993; Kennedy and Bush 1994). The majority of parasites are lost during introduction and establishment.

Once established and spreading or increasing in population size, then the invasive species are no longer likely to lose parasites but instead start to gain them.

Parasites Gained

The accumulation of parasites during the invasion process determines the extent of net release, which in turn depends on the diversity of the parasite community in the invaded range and the susceptibility of the invader to these parasites, which may differ from its susceptibility in its native range. On average, animal invaders accumulate four new parasites from the invaded range, less than a third as many as they escape in their native range (Torchin et al. 2003). Plants accumulate about 13% as many new fungal and viral pathogens as they escape (Mitchell and Power 2003). This accumulation process is key to the biotic resistance hypothesis (Elton 1958; Maron and Vila 2001), which suggests that accumulation of enemies (both parasites and predators) by an introduced species may often prevent them from becoming damaging or invasive. The rate of accumulation and the type of parasites that are gained are likely to have profound effects on whether the species can become established or invasive and the time period over which an invasive species may remain so.

During the establishment phase, parasites from local communities can start to colonize the introduced hosts. Many introduced species might disappear before they are even noticed due to high parasite pressure, particularly if the local community exhibits high biotic resistance (Elton 1958). The phylogenetic similarity between introduced species and resident native species may determine the likelihood of colonization by preadapted parasites. An introduced species from an unrepresented genus or family should have a lower probability of parasite accumulation than an introduced species with many close relatives. However, close relatives might also indicate greater suitability of local environments for the introduced species (Mack 1996).

Encounter with potential natural enemies will increase as the invader expands its range. Parasites with broad host ranges and little host or vector specificity should be more likely to colonize than highly specific parasites (see table 8.1). Examples include RNA viruses, parasites with generalist vectors, or other generalist parasites. If there are no preadapted or generalist pathogens, the probability of de novo pathogen evolution by random mutation will increase as the local abundance of the introduced species increases (e.g., Antia and Koella 2004). Thus, the more successful an invasive species is, the greater the chance that a novel virulent pathogen will arise. Once a pathogen establishes, it may rapidly increase and lead to epidemics and major die-offs in the introduced species, in part because of the high density of susceptible hosts. This may be especially true when the introduced species is genetically uniform. Although natural examples of this are few (but see Hochachka and Dhondt 2000), considerable evidence is available from the biological control of invasive plant and animal species (e.g., Fenner and Fantini 1999).

Evidence for the accumulation of parasites by introduced species comes from many sources: well-documented cases of infection of non-native crop species and domesticated animals (Scheffer 2003), published databases and disease indices (Farr et al. 1989), and comparative studies (Mitchell and Power 2003; Torchin et al. 2003). For example, Pierce's disease of the grape, which caused several devastating epidemics in California vineyards, is caused by a bacterial pathogen common in native grapes from the southeastern United States. Fire blight of pome orchards was also thought to have arisen from native pathogens adapted to native rosaceous trees (Scheffer 2003). Similar examples exist in animals: cattle in East Africa succumb to sleeping sickness caused by trypanosome parasites in native ungulates, and introduced house finches in the United States are attacked by a native pathogen causing mycoplasma conjunctivitis (Dhondt et al. 2005).

Intuitively, release from natural enemies may be temporary, as introduced species gradually accumulate resident parasites (see table 8.1). Parasites from the invader's native range may eventually colonize introduced host populations through repeated introductions or long-distance or human-assisted dispersal. Colonization leads to accumulated gains over time of parasites and corresponding declines in invasive species abundance, and the extent and timing of the decline are a function of the likelihood of acquiring regulatory parasites (see figure 8.1). Therefore, a key question is, will parasite communities of introduced populations ever "catch up" to those of the native populations, and if so, how long does this take, and will the acquired community of parasites limit the fitness of the host species as effectively as in their native range?

For some species, accumulation of parasites can be a relatively rapid process leading to no differences in parasite diversity or infection levels between native and introduced hosts, and so host regulation is also expected to be rapid. For example, Clay (1995) found that introduced grasses in the United States actually supported more fungal pathogens than native grasses, although introduced species also had larger geographical ranges, which was the primary predictor of pathogen load. Strong and Levin (1975) examined species richness of parasitic fungi of British trees and found that there was no difference in fungal species richness between introduced and native trees when corrected for range.

They suggested that fungal species richness rapidly saturates over eco-logical time (i.e., over several hundred years). However, the agents with the greatest potential to have long-term control may not necessarily be the ones with the most dramatic immediate effects. Thus, for example, floral smuts that reduce fecundity are predicted to have an increasingly detrimental effect as host longevity increases, while for pathogens that affect mortality, causing intermediate levels of mortality will reduce population sizes significantly more in the long term, than pathogens that are more destructive (Thrall and Burdon 2004). Data from agricultural experiment stations or forestry plantations might prove useful, given the intensive records of plant introductions. Longer-term historical records may provide some solace in light of the disruption caused by recently in-troduced species, many of which have become problematic only in the past 20–30 years. Introduced species may cause much ecological dam-age in the meantime but nevertheless offer many opportunities for ex-amining the longer-term dynamics of disease.

Effects of Introduced Species on Parasite Dynamics in Native Hosts

Introduced species, regardless of the pathogens they lose or gain, can have large indirect impacts on native hosts by changing the characteris-tics of the ecosystem, by amplifying disease or vector populations, or by changing the spatial distribution of hosts.

Ecosystem-Level Changes

Introduced species can affect ecosystem properties, such as productivity, nutrient status, water balance, physical structure, and disturbance re-gime (e.g., Cox 1999; Dukes and Mooney 2004; Mack et al. 2000; Eviner and Likens, chapter 12, this volume). Modification of ecosystem functions change the rules of the game for every other species in the community (National Research Council 2002), which ultimately may alter host-parasite dynamics. The consequences of ecosystem alterations by introduced species on the dynamics of native parasites seem little studied, but we can suggest some plausible possibilities based on well-studied introduced organisms. Introduced earthworms in eastern North America destroy the forest litter layer, changing the water and nutrient content of the upper soil layers (Hendrix and Bohlen 2002), thereby po-tentially altering conditions for transmission and survival of soil-borne plant pathogens. Another example is that of zebra mussels, widely intro-duced in Europe and North America, which produce nutrient-rich

(Roditi et al. 1997) and presumably microanaerobic conditions in their beds. This may favor anaerobic bacteria such as *Clostridium* and contribute to the recent rise in Type E botulism in Lake Erie (New York Sea Grant 2003). Finally, introduced grasses change the frequency of fires (D'Antonio and Vitousek 1992), which ought to affect the timing and severity of disease caused by plant pathogens. Similarly, parasite outbreaks in native coniferous forests make them more prone to devastating wildfires.

Amplification of Reservoir Host Populations

Invasive species may affect parasite dynamics by amplifying disease or vector populations. For example, in the highly invaded California grasslands, introduced annual grasses attract and amplify the fecundity of cereal aphids that vector barley and cereal yellow dwarf viruses (B/CYDVs). The presence of these introduced grasses has been found to more than double the incidence of barley and cereal yellow dwarf infection in nearby native bunchgrasses (Malmstrom, McCullough, et al. 2005). Because B/CYDV infection can stunt bunchgrasses and increase their mortality (Malmstrom, Hughes, et al. 2005), these findings indicate that virus-mediated apparent competition has the capacity to influence interactions between native and introduced species and contribute to the decline of the natives. In contrast, introduced tall fescue grass is widely distributed and commonly endophyte infected, and is more resistant to aphids as a result (Siegel 1990). Resistance to aphids has the knock-on effect of reduced aphid-borne virus pressure on neighboring native grasses. A good example of an introduced domestic species that has enlarged the pathogen reservoir population comes from the Serengeti, where domestic cattle increased rinderpest prevalence in wild ungulates, a result discovered when cattle vaccination indirectly lowered rinderpest incidence in wild hosts (Sinclair 1979).

Another mechanism by which introduced species can enlarge the reservoir population is by serving as alternative hosts for parasites with a multihost life history strategy. For example, the fungus *Cronartium ribicola*, which causes the devastating white pine blister rust, requires two hosts—one *Pinus*, one *Ribes*—to complete its life cycle. In the United States, the presence of cultivated and escaped *Ribes* (currants and gooseberries) has such a strong influence on white pine blister rust incidence in nearby pines that horticultural use of *Ribes* is severely restricted (Maloy 1997). Similarly, pheasants introduced into the United Kingdom have been shown to successfully feed ticks and support Lyme bacterium transmission, thus contributing to the reservoir host population for Lyme disease (Kurtenbach et al. 1998).

Changes to Host Spatial Distribution or Density

Native host spatial distribution, density, and host contact structure within and among populations and communities can be altered by species introductions. For example, impacts of invasive weeds on soil communities could alter ecosystem functions, with secondary impacts on native host abundances (Kourtev et al. 2002). Thus, changes in parasite dynamics are expected partly through indirect impacts on host population structure, including total host abundance or density, and the movements of hosts or vectors in the environment. One obvious way that introduced host species can have an impact on native community structure is through population explosions. Such invasions may lead to net increases in species diversity, although a decrease is usually more common. A decrease in species diversity can lead to an upsurge in parasite levels primarily by concentrating infection within the most competent reservoir hosts, an amplification effect (Mitchell et al. 2002, 2003; see also Clay et al., chapter 7, and Begon, chapter 7, this volume). Interestingly, as shown by Mitchell and colleagues, it is not just parasite abundance and disease prevalence that may change but also the relative prevalence of particular types of pathogens—that is, the parasite community. The converse of this observation is the basis for the use of varietal mixtures in cereal crops, where parasite levels may be significantly reduced by the construction of random three-component mixtures of susceptible and resistant varieties (Wolfe 1985). Other possible effects of changes in community composition following invasions include shifts in patterns of herbivory or predation, which if they increase or reduce a reservoir host population might further alter disease dynamics.

An intriguing but little considered impact of an invasive species is the potential collateral impact the invading species's parasites may have on the native species's own suite of host-specific parasites (e.g., Torchin et al. 2005). Although this may be minimal when the native species is still reasonably abundant, in the case of introduced pathogens like chestnut blight, *Cryphonectria parasitica*, or root rot, *Phytophthora cinnamomi*, that have devastated the entire population of their host species (chestnut, *Castanea dentata*, and Brown's banksia, *Banksia brownii*), any host-specific native parasites must have undergone a devastating (and undocumented!) decline if not extinction.

Evolutionary Considerations

A range of evolutionary processes should affect the vulnerability of introduced species to parasite accumulation. First, most invasive populations are likely to originate from a few founders, thus limiting their genetic diversity and increasing their susceptibility to parasites (Sakai et al. 2001). Second, introduced species that are released from parasites might reallocate resources away from parasite defense and into growth and reproduction (Colautti et al. 2004; Wolfe et al. 2004), thereby potentially allowing the species to become invasive (Siemann and Rogers 2001; Tilman 1999). Together, these ideas suggest that low overall genetic diversity combined with evolutionary reductions in parasite defense should make introduced species vulnerable targets for future epidemics (DeWalt et al. 2004, Knevel et al. 2004; Reinhart et al. 2003).

Population Bottlenecks and Parasite Susceptibility

Genetic variation tends to be reduced in introduced populations, particularly if colonists come from a single source population or undergo an establishment phase during which population sizes remain small (Sakai et al. 2001). Founder events resulting in extreme genetic drift and inbreeding could lower the fitness of introduced populations and limit their ability to adapt to future challenges (Lee 2002). Host populations characterized by loss of allelic diversity or reduced heterozygosity may be unable to respond evolutionarily to new threats imposed by parasites (Lande 1988; Lyles and Dobson 1993). This issue is more commonly raised in the context of agricultural systems (e.g., Elton 1958) or species of conservation concern (e.g., Acevedo-Whitehouse et al. 2003; Thorne and Williams 1988). However, similar problems could apply to populations of introduced species with low genetic diversity, as has been suggested for the unusually high susceptibility of introduced house finches in eastern North America to ongoing outbreaks of mycoplasmal conjunctivitis (Dhondt et al. 2005).

Yet, despite the fact that many introduced species probably show reduced genetic diversity in their invaded versus native ranges, this has yet to be widely established. A recent comparison of the loss of allelic diversity and heterozygosity in twenty-nine introduced animal species found that, on average, there is little reduction in genetic diversity (Wares et al. 2005). In plants, there may even be greater genetic variation in introduced populations than in populations where they are native (Novack and Mack 2005). In some cases, repeated introductions from multiple native sites could actually cause blending of alleles from different geographic locations in the new habitat, leading to greater genetic variation, rather than less, in the introduced range (as has been demonstrated with brown anole lizards; Kolbe et al. 2004). Hybridization in the new range could also lead to hosts with novel gene combinations that are highly resistant to parasite infections, and such genetically variable populations could serve as problematic sources of introduction for other vulnerable locations (Sakai et al. 2001).

Evolution of Increased Competitive Ability

Invasive species often are larger, more abundant, and more vigorous in their introduced range relative to their native range (Crawley 1987; Grosholz and Ruiz 2003). One explanation for this observation is that following release from their natural enemies, introduced species experience increased growth and reproduction. A related idea, known as the EICA (evolution of increased competitive ability) hypothesis, states that because defenses are often costly and organisms have limited resources, introduced species should adapt to the loss of natural enemies by allocating more energy to growth and reproduction and investing less in pathogen resistance or immune defense (Blossey and Nötzold 1995). This hypothesis predicts that in the native range, growth and reproduction should be lower, natural enemies should be common, and investment in defenses high, whereas in the new range, natural enemies should be less common or absent, defenses should be low, and growth and reproduction should be greater (Wolfe et al. 2004). Furthermore, these phenotypic differences should be genetically based, and parasites and other natural enemies should preferentially attack the invasive phenotypes, two predictions that can be tested using comprehensive common garden and reciprocal transplant experiments. Some recent studies provide support for genetic divergence in enemy defense and reproductive strategies between native and introduced populations of weeds and trees (Siemann and Rogers 2001; Wolfe et al. 2004). Under this scenario, if pathogens are lost from introduced populations, the frequency of resistance should decline over evolutionary time scales, potentially setting the stage for future disease outbreaks.

The EICA hypothesis assumes that resistance is costly—and further depends on whether resistance traits are targeted against generalist or specialist enemies (Joshi and Vrieling 2005). Many studies have demonstrated that resistance-conferring host traits are in fact costly in terms of reductions in growth rates, fecundity, competitive ability, or body size (Simms and Rausher 1987). Invasive species might lose protection against specialist parasites, since these are most likely to be lost, and instead shift resources into defenses against generalist parasites (Joshi and Vrieling 2005), which are likely to be gained in the invaded range.

Importantly, species invasions offer new opportunities to understand the strength of parasites as agents of selection, particularly with respect to the evolution and maintenance of host defenses in the wild (Altizer et al. 2003). Field monitoring studies, reciprocal transplant experiments, and common garden studies of introduced species from locations in the invaded and native range provide prime opportunities for researchers to measure host investment in parasite defenses relative to growth and reproduction, and to compare populations exposed to different levels of attack by a range of specialist and generalist parasites. At the present time, comprehensive studies of the biology of invasive species in both their native and introduced ranges are surprisingly rare (but see Reinhart et al. 2003, 2005 for a counterexample of detailed studies of parasite regulation of black cherry in its native and invasive ranges), despite the potential insights that can be gained from such comparisons. Furthermore, as has been demonstrated by a growing number of "virgin ground" epidemics, emerging pathogens often cause high case fatality rates and stunning reductions in host abundance (reviewed in Daszak et al. 2000). If these epidemics can be buffered by the genetic composition of host populations, then studies of disease outbreaks in populations of introduced species will provide new perspectives on the role of host genetic diversity and investment in immune defense in the outcomes of host-parasite interactions.

Parasite Evolution in Introduced Species

We have reviewed how invasion may be facilitated because parasites, particularly those causing acute infectious diseases, will be lost in transit. An important area for the future ought to focus on the evolutionary adaptive processes of parasites in both invading and native species in the community in the invaded range. Disease-causing pathogens generally have short generation times and high replication rates, and hence a great ability for fast evolution (Frank 2002). The literature on virulence-transmission trade-offs and the rapid evolution of the myxomatosis-causing DNA virus of rabbits provides an excellent testimony (Fenner 1983). Another interesting example is the extreme rate of evolution in the form of both gene loss and genomic organization of *Bordetella pertussis*—the whooping cough–causing bacterium—following its emergence in humans from its ancestral commensal of various mammals (Bjornstad and Harvill 2005; Parkhill et al. 2003). Over and above that, disease-causing RNA viruses may have even higher evolutionary rates because of the high mutation rates during RNA transcription (Grenfell et al. 2004).

There are at least four reasons why studies focused on the postinvasion evolution of parasites might be particularly interesting. The first is the great evolutionary potential of many parasites. The second reason is that the effects of a parasite on the host population depend critically on its virulence (Anderson and May 1978) and infectious period (Grenfell

2001), so that evolutionary changes in these two parameters can drive ecological change at the population level. The third and fourth reasons are that selective gradients for changes in virulence and infectious periods appear to be particularly steep when (3) genetic diversity within the host population is low and (4) contact networks are altered. In terms of contact networks, transmission rates will likely change whenever host densities are higher *or* lower in the invaded range than in the native habitats range, at least for directly transmitted pathogens. Changes in contact rates will generally alter the optimal pathogen strategy because of virulence-transmission trade-offs (Ebert 1998; Frank 2002). The introduction of a species may be enough to alter the contact network of a species. Indeed, several recent theoretical studies have shown that subtle changes in social networks can greatly alter evolution toward enhanced or diminished virulence (Boots et al. 2004; Read and Keeling 2003). The impact of any parasite that *does* survive the invasion process on the introduced host populations' growth and regulation may therefore rapidly diverge from the effects of the same pathogen in the host's native range.

Parasites as Introduced Species

Finally, it is important to keep in mind that parasites themselves can be a special class of introduced species (see also Hudson et al., chapter 16, this volume). Research questions in this area focus on the intersection of two important predictive frameworks: factors that affect the success of introduced species and factors that govern emerging diseases. Introduced parasites will represent a subset of both of these groups.

The introduction of parasites with invading hosts is the most important and widespread driver of disease emergence worldwide (Anderson et al. 2004; Bauer 1991; Daszak et al. 2000; Dobson and May 1986). Such introduced diseases can have a devastating impact on immunologically naïve host populations, often with enormous consequences for human health, the economy, and wildlife conservation. Diseases of humans and wildlife have been traded across the globe with increasing frequency for centuries; common examples of relevance to human health include smallpox, typhus, yellow fever, cholera, schistosomiasis, SARS, West Nile virus, HIV, and influenza. These same patterns are evident in animals and plants. Select examples for animals include salmonid whirling disease, chronic wasting disease, rinderpest, shrimp whitespot disease, crayfish plague, avian malaria, avian cholera, and duck plague. Common examples for plants include wooly hemlock adelgid, dogwood anthracnose, beech bark disease, white pine blister rust, oak wilt, and numerous others.

Unfortunately, even basic epidemiological information is lacking for most pathogens, native or introduced. For example, Taylor et al. (2001) reported that the basic transmission mode was unknown for more than 200 human pathogens, precluding any comparative analyses of the basic reproduction number, R_0. Information on host use, native geographic range, propagule pressure, and failed invasions is correspondingly lacking for many plant and animal pathogens. As such, attempts to analyze introduced pathogens through stages of the invasion process, while very promising, and to develop more quantitative predictions have lagged behind recent efforts with free-living groups, including fish, plants, and birds (see Kolar and Lodge 2001, 2002).

Recent reviews of disease emergence in humans (Taylor et al. 2001), vertebrate wildlife (Dobson and Foufopoulos 2001), and plants (Anderson et al. 2004) offer important first steps to understanding how disease emergence intersects with stages of the invasion process. Do these different stages represent different sorts of barriers for parasites than for other invasive species? Clearly, arrival doesn't lead to establishment for many introduced species (e.g., failed introductions for biocontrol). Is the probability of getting from one stage of the invasion process different for parasites than for plants and animals? What features promote success at each stage? Do plant parasites possess life-history features that make them more or less likely to invade than animal parasites? For example, wind-blown spores may be capable of dispersal over thousands of kilometers (Brown and Hovmoller 2002). Comparative studies could provide a useful approach to answering some of these questions, and are likely to provide further insights into factors influencing the invasion process.

The arrival of parasites into a novel environment is dependent on the invading host population, and we have posited which parasites are most likely to be present in that population. However, important questions remain, including the following: Are parasites a special class of invaders, or do they share features common to other invasive species, such as a generalist host range and simple life cycle? And are the routes by which parasites arrive different in some qualitative way from the ways in which other invaders arrive, so that the probability of successful invasion might also different?

Different types of investigative or applied approaches are more likely to be useful at different stages of an invasion process. For example, the earliest stages of invasion following arrival may be most amenable to eradication, especially if there are sufficiently effective quarantine and monitoring programs in place. Although it is difficult to envisage using experimental manipulations to study early processes associated with parasite arrival, the use of high-resolution molecular markers to study within-host variation or to trace pathways of infection back to sources can be effective. As parasites become established and spread into multiple host populations, and eradication becomes less likely, the applied emphasis may shift from eradication to long-term control. Population and metapopulation modeling approaches become more valuable: one can begin to study general dynamical properties of the system and to test predictions empirically. Developing an understanding of the underlying patterns of host resistance and pathogen infectivity and aggressiveness as part of evolutionary studies (which may include modeling, population genetic, and phylogenetic approaches, as well as experimental studies) will be critical for explaining underlying patterns of disease incidence and prevalence. The lack of empirical evidence is particularly surprising in light of the potential for such variation to affect not only disease dynamics and prevalence, but also when or where new diseases emerge (e.g., canine parvovirus; Parrish 1999). Increasingly, this lack of knowledge has led to calls for an integrated approach to disease management that would incorporate both ecological and evolutionary processes. Addressing such questions will be essential if we are to develop a predictive understanding of diseases as invaders.

Conclusions

Questions concerning the role of parasites in the success of species invasions bring together two rapidly developing research fields, the ecology of species invasions and the population biology of infectious diseases. Recent years have seen an increase in the number of quantitative studies concerning the ecology of invasive species (reviewed in Colautti et al. 2004; Kolar and Lodge 2001), as well as factors determining the spread and impacts of parasites and the infectious diseases they cause in natural systems (reviewed in Hudson et al. 2002). A growing number of studies point to the role of parasites in regulating hosts and driving population dynamics, such that their removal could in part explain increased growth, abundance, and size of exotic organisms in novel habitats. As such, there exists an experimental niche in terms of bringing these disciplines together with regard to specifically assessing the role of parasites in the enemy release hypothesis. However, confounding variables create difficulties for examining empirical evidence of the role that parasites alone, compared with other enemies, play in facilitating or inhibiting the invasion process. Most studies of assessing enemy release thus far have taken comparative approaches, with release of biological control agents providing an alternative set of opportunities to assess the impacts of different parasites on host regulation.

Given the potential limitations of experimental tests of enemy release in introduced populations, we advocate a synergistic approach to study the role of parasites in invasion biology using a combination of molecular techniques, field surveys, experiments, historical records, and multivariate models. Molecular techniques can be used to test for evidence of multiple introductions in the introduced range as well as for genetic variation in both the native and introduced ranges (e.g., founder effects in the introduced range). Field surveys help quantify both enemy pressure and geographic variation in size and fitness of the invader across the native and introduced ranges. Historical records can provide further information on both the scope and timing of the introduction and the number of potential introduction events. Multivariate models that predict the success of invaders (at species or population level) based on overlap between native and introduced range, time since introduction, parasite burden, population density, climate, and existing genetic variation will provide a more comprehensive review of the importance of parasite release in determining invader success. Not only is there a need for further experimental approaches, such as manipulation of hosts and parasites in both natural and invaded range, but there is also a need to incorporate developing areas of disease ecology, such as interactions of parasite species within hosts.

Complementary to determining the role of parasites in enemy release is determining which types of parasites would be lost and gained during invasion. In this chapter we have speculated that highly regulatory parasites are most likely to be lost, but they could also be rapidly gained, especially RNA viruses. Therefore the rate at which parasite accumulations occur will determine the time period under which the invasive host population becomes regulated by parasites (see figure 8.1). We speculate that this time frame may be hundreds of years, but it may be accelerated by high propagule pressure, which in turn will depend on the level of biotic resistance of the invaded community. This could be high for invaded communities that contain phylogenetically close species, or low for species that invade phylogenetically distant communities, for example, deliberate introductions to distant and remote islands.

Owing to the potentially long time period over which accumulation of parasites and thus regulation of invasive species occur, it is essential to take into account evolutionary theory. This suggests that, at first, resource allocation away from costly parasite defense should enhance invasive species growth rates, but that founder effects and loss of resistance should increase host susceptibility to parasitic infection in the long term.

Careful consideration should be given to the changes in ecosystem functioning that invasive species may cause. Ultimately these indirect changes can affect host-parasite dynamics not just of the invading species but also of the native host community and its parasite fauna. For instance, invasive species can increase the reservoir host density, facilitating the persistence of some parasites or increasing transmission through parasite-mediated competition. Indirect effects brought about by invasions could form the focus of future research, for example changes in native reservoir host density though trophic interactions with invasive species. An overall focus for future work is to move beyond case studies, speculation, and inference-based retrospective studies to develop a general understanding of key processes and patterns that reflect the interface between infectious diseases and biology of species invasions. As global trade and travel increase, the number of accidental and deliberate introductions is expected to rise, providing further impetus to elucidate the role of parasites and pathogens in host regulation and the importance this has for invasive species and ecosystem functioning.

LITERATURE CITED

Acevedo-Whitehouse, K., F. Gulland, D. Grieg, and W. Amos. 2003. Inbreeding: Disease susceptibility in California sea lions. Nature 422:35.

Altizer, S., E. Friedle, and C. D. Harvell. 2003. Rapid evolutionary dynamics and disease threats to biodiversity. Trends in Ecology & Evolution 18:589–96.

Anderson, P. K., A. A. Cunningham, N. G. Patel, F. J. Morales, P. R. Epstein, and P. Daszak. 2004. Emerging infectious diseases of plants: Pathogen pollution, climate change and agrotechnology drivers. Trends in Ecology & Evolution 19:535–44.

Anderson, R. M., and R. M. May. 1978. Regulation and stability of host-parasite population interactions. I. Regulatory processes. Journal of Animal Ecology 47:219–47.

Antia, R., and J. Koella. 2004. Theoretical immunology: Parasitic turncoat. Nature 429:511–13.

Bauer, O. N. 1991. Spread of parasites and diseases of aquatic organisms by acclimatization: A short review. Journal of Fish Biology 39:679–86.

Bjornstad, O. N., and E. T. Harvill. 2005. Evolution and emergence of *Bordetella* in humans. Trends in Microbiology 13:355–59.

Blossey, B., and R. Nötzold. 1995. Evolution of increased competitive ability in invasive nonindigenous plants: A hypothesis. Journal of Ecology 83:887–89.

Boots, M., P. J. Hudson, and A. Sasaki. 2004. Large shifts in pathogen virulence relate to host population structure. Science 303:842–44.

Brown, J. K. M., and M. S. Hovmoller. 2002. Epidemiology: Aerial dispersal of pathogens on the global and continental scales and its impact on plant disease. Science 297:537–41.

Buckley, Y. M., P. S. Downey, V. Fowler, R. Hill, J. Memmot, H. Norambuena, M. Pitcairn, R. Shaw, A.W. Sheppard, C. Winks, R. Wittenberg, and M. Rees. 2003. Are invasives bigger? A global study of seed size variation in two invasive shrubs. Ecology 84:1434–40.

Callaway, R. M., G. C. Thelen, A. Rodríguez, and W. E. Holben. 2004. Soil biota and exotic plant invasion. Nature 427:731–33.

Clay, K. 1995. Correlates of pathogen species richness in the grass family. Canadian Journal of Botany 73:542–49.

Colautti, R. I., A. Ricciardi, I. A. Grigorvich, and H. J. MacIsaac. 2004. Is invasion success explained by the enemy release hypothesis? Ecology Letters 7:721–33.

Cornell, H. V., and B. A. Hawkins. 1993. Accumulation of native parasitoid species on introduced herbivores: a comparison of hosts as natives and hosts as invaders. American Naturalist 141:847–65.

Cox, G. W. 1999. Alien Species in North America and Hawaii. Washington, DC: Island Press.

Crawley, M. J. 1987. What makes a community invasible? In Colonization, Succession and Stability, ed. A. J. Gray, M. J. Crawley, and P. J. Edwards. Oxford: Blackwell Scientific Publications.

D'Antonio, C. M. D., and P. M. Vitousek. 1992. Biological invasions by exotic grasses, the grass/fire cycle, and global change. Annual Review of Ecology and Systematics 23:63–87.

Darwin, C. 1859. On the Origin of Species by Means of Natural Selection, or the Preservation of Favoured Races in the Struggle for Life. London: John Murray.

Daszak, P., A. A. Cunningham, and A. D. Hyatt. 2000. Emerging infectious diseases of wildlife: Threats to biodiversity and human health. Science 287:443–49.

de Castro, F., and B. Bolker. 2005. Mechanisms of disease-induced extinction. Ecology Letters 8:117–26.

DeWalt, S. J., J. S. Denslow, and K. Ickes. 2004. Natural-enemy release facilitates habitat expansion of the invasive tropical shrub Clidemia hirta. Ecology 85:471–83.

Dhondt, A. A., S. Altizer, E. G. Cooch, A. K. Davis, A. P. Dobson, M. J. L. Driscoll, B. K. Hartup, D. M. Hawley, W. M. Hochachka, et al. 2005. Dynamics of a novel pathogen in an avian host: Mycoplasmal conjunctivitis in house finches. Acta Tropica 94:77–93.

Dobson, A., and J. Foufopoulos. 2001. Emerging infectious pathogens of wildlife. Philosophical Transactions of the Royal Society of London. Series B, Biological Sciences 356:1001–12.

Dobson, A. P., and R. M. May. 1986. Patterns of invasions by pathogens and parasites. In Ecology of Biological Invasions of North America and Hawaii, ed. H. A. Mooney, and J. A. Drake, 58–76. New York: Springer-Verlag.

Dukes, J. S., and H. A. Mooney. 2004. Disruption of ecosystem processes in western North America by invasive species. Revista Chilena de Historia Natural 77:411–37.

Ebert, D. 1998. Evolution: Experimental evolution of parasites. Science 282:1432–35.

Edwards, K. R., M. S. Adams, and J. Kvit. 2003. Differences between European and American invasive populations of Lythrum salicaria. Journal of Vegetation Science 9:267–80.

Elton, C. S. 1958. The Ecology of Invasions by Animals and Plants. London: Methuen.

Farr, D. F., G. F. Bills, G. P. Chamuris, and A. Y. Rossman. 1989. Fungi on Plants and Plant Products in the United States. St. Paul: MN: APS Press.

Fenner, F. 1983. The Florey Lecture 1983: Biological control, as exemplified by smallpox eradication and myxomatosis. Proceedings of the Royal Society of London. Series B, Biological Sciences 218:259–85.

Fenner, F., and B. Fantini. 1999. Biological Control of Vertebrate Pests. Wallingford, UK: CABI Publishing.

Frank, S. A. 2002. Immunology and Evolution of Infectious Disease. Princeton, NJ: Princeton University Press.

Grenfell, B. T. 2001. Dynamics and epidemiological impact of microparasites. In New Challenges to Health: The Threat of Virus Infection, ed. W. L. I. G. l. Smith, J. W. McCauley, and D. J. Rowlands, 33–52. New York: Cambridge University Press.

Grenfell, B. T., O. G. Pybus, J. R. Gog, J. L. N. Wood, J. M. Daly, J. A. Mumford, and E. C. Holmes. 2004. Unifying the epidemiological and evolutionary dynamics of pathogens. Science 303:327–32.

Grigulis, K., A. W. Sheppard, J. E. Ash, and R. H. Groves. 2001. The comparative demography of the pasture weed Echium plantagineum between its native and invaded ranges. Journal of Applied Ecology 38:281–90.

Grosholz, E. D., and G. M. Ruiz. 2003. Biological invasions drive size increases in marine and estuarine invertebrates. Ecology Letters 6:700–05.

Hendrix, P. F., and P. J. Bohlen. 2002. Exotic earthworm invasions in North America: Ecological and policy implications. BioScience 52:801–11.

Hierro, J. L., J. L. Maron, and R. M. Callaway. 2005. A biogeographical approach to plant invasions: The importance of studying exotics in their introduced and native range. Journal of Ecology 93:5–15.

Hochachka, W. M., and A. A. Dhondt. 2000. Density-dependent decline of host abundance resulting from a new infectious disease. Proceedings of the National Academy of Sciences of the United States of America 97:5303–6.

Hudson, P. J., A. P. Rizzoli, B. T. Grenfell, H. Heesterbeek, and A. P. Dobson. 2002. The Ecology of Wildlife Diseases. Oxford: Oxford University Press.

Jeschke, J. M., and D. L. Strayer. 2005. Invasion success of vertebrates in Europe and North America. Proceedings of the National Academy of Sciences of the United States of America 102:7198–202.

Joshi, J., and K. Vrieling. 2005. The enemy release and EICA hypothesis revisited: Incorporating the fundamental difference between specialist and generalist herbivores. Ecology Letters 8:704–14.

Keane, R. M., and M. J. Crawley. 2002. Exotic plant invasions and the enemy release hypothesis. Trends in Ecology & Evolution 17:164–70.

Kennedy, C. R., and A. O. Bush. 1994. The relationship between pattern and scale in parasite communities: A stranger in a strange land. Parasitology 109:187–96.

Knevel, I. C., T. Lans, F. B. J. Menting, U. M. Hertling, and W. H. van der Putten. 2004. Release from native root herbivores and biotic resistance by soil pathogens in a new habitat both affect the alien *Ammophila arenaria* in South Africa. Oecologia 141:502–10.

Kolar, C. S., and D. M. Lodge. 2001. Progress in invasion biology: Predicting invaders. Trends in Ecology & Evolution 16:199–204.

Kolar, C. S., and D. M. Lodge. 2002. Ecological predictions and risk assessment for alien fishes in North America. Science 298:1233–36.

Kolbe, J. J., R. E. Glor, L. Rodriguez Schettino, A. Chamizo Lara, A. Larson, and J. B. Losos. 2004. Genetic variation increases during biological invasion by a Cuban lizard. Nature 431:177–81.

Kourtev, P. S., J. G. Ehrenfeld, and M. Haggblom. 2002. Exotic plant species alter the microbial community structure and function in the soil. Ecology 83:3152–66.

Kurtenbach, K., D. Carey, A. N. Hoodless, P. A. Nuttall, and S. E. Randolph. 1998. Competence of pheasants as reservoirs for Lyme disease spirochetes. Journal of Medical Entomology 35:77–81.

Lafferty, K. D., K. F. Smith, M. E. Torchin, A. P. Dobson, and A. M. Kuris. 2005. The role of infectious disease in natural communities: What do introduced species tell us? *In* Species Invasions: Insights into Ecology, Evolution, and Biogeography, ed. D. F. Sax, J. J. Stachowicz, and S. D. Gaines. Sunderland, MA: Sinauer.

Lande, R. 1988. Genetics and demography in biological conservation. Science 241:1455–60.

Lee, C. E. 2002. Evolutionary genetics of invasive species. Trends in Ecology & Evolution 17:386–391.

Lyles, A. M., and A. P. Dobson. 1993. Infectious disease and intensive management: Population dynamics, threatened hosts, and their parasites. Journal of Zoo and Wildlife Medicine 24:315–26.

Lloyd-Smith, J. O., P. C. Cross, C. J. Briggs, M. Daugherty, W. M. Getz, J. Latto, M. S. Sanchez, A. B. Smith, and A. Swei. 2005. Should we expect population thresholds for wildlife disease? Trends in Ecology & Evolution 20:511–19.

Mack, R. N. 1996. Predicting the identity and fate of plant invaders: Emergent and emerging approaches. Biological Conservation 78:107–21.

Mack, R. N., D. Simberloff, W. M. Lonsdale, H. Evans, M. Clout, and F.A. Bazzaz. 2000. Biotic invasions: Causes, epidemiology, global consequences, and control. Ecological Applications 10:689–710.

Malmstrom, C. M., C. C. Hughes, L. A. Newton, and C. J. Stoner 2005. Virus infection in remnant native bunchgrasses from invaded California grasslands. New Phytologist 168:217–30.

Malmstrom, C. M., A. J. McCullough, L. A. Newton, H. A. Johnson, and E. T. Borer. 2005. Invasive annual grasses indirectly increase virus incidence in California native perennial bunchgrasses. Oecologia 145:153–64.

Maloy, O. C. 1997. White pine blister control in North America: A case history. Annual Review of Phytopathology 35:87–109.

Maron, J. L., and M. Vila. 2001. When do herbivores affect plant invasion? Evidence for the natural enemies and biotic resistance hypothesis. Oikos 95:361–73.

Mitchell, C. E., and A. G. Power. 2003. Release of invasive plants from fungal and viral pathogens. Nature 421:625–27.

Mitchell, C. E., P. B. Reich, D. Tilman, and J. V. Groth. 2003. Effects of elevated CO_2, nitrogen deposition, and decreased species diversity on foliar fungal plant disease. Global Change Biology 9:438–51.

Mitchell, C. E., D. Tilman, and J. V. Groth. 2002. Effects of grassland plant species diversity, abundance, and composition on foliar fungal disease. Ecology 83:1713–26.

Moller, A. P. 2005. Parasitism and the regulation of host populations. In Parasitism and Ecosystems, ed. F. Thomas, J. F. Guegan and F. Renaud. Oxford: Oxford University Press.

Molloy, D. P., A. Y. Karatayev, L. E. Burlakova, D. P. Kurandina, and F. Laruelle. 1997. Natural enemies of zebra mussels: Predators, parasites and ecological competitors. Reviews in Fisheries Science 5:27–97.

Moore, S. L., and K. Wilson. 2002. Parasites as a viability cost of sexual selection in natural populations of mammals. Science 297:2015–18.

National Research Council. 2002. Predicting Invasions of Nonindigenous Plants and Plant Pests. Washington, DC: National Academy Press.

New York Sea Grant. 2003. Botulism in Lake Erie Workshop Proceedings. Buffalo, NY: New York Sea Grant.

Novack, S. J., and R. N. Mack. 2005. Genetic bottlenecks in alien plant species: influence of mating systems and introduction dynamics. In Species Invasions: Insights into Ecology, Evolution, and Biogeography, ed. D. F. Sax, J. J. Stachowicz, and S.D. Gaines, 201–28. Sunderland, MA: Sinauer.

Parkhill, J., M. Sebaihia, A. Preston, L. D. Murphy, N. Thomson, D. E. Harris, M. T. Holden, C. M. Churcher, S. D. Bentley, et al. 2003. Comparative analysis of the genome sequences of Bordetella pertussis, Bordetella parapertussis and Bordetella bronchiseptica. Nature Genetics 35:32–40.

Parrish, C. R. 1999. Host range relationships and the evolution of canine parvovirus. Veterinary Microbiology 69:29–40.

Perkins, S. E., I. M. Cattadori., V. Tagliapietra, A. P. Rizzoli., and P. J. Hudson. 2003. Empirical evidence for key hosts in persistence of a tick-borne disease. International Journal for Parasitology 33:909–17.

Pimentel, D., R. Zuniga, and D. Morrison. 2005. Update on the environmental and economic costs associated with alien-invasive species in the United States. Ecological Economics 52:273–88.

Read, J. M., and M. J. Keeling. 2003. Disease evolution on networks: The role of contact structure. Proceedings of the Royal Society of London. Series B, Biological Sciences 1516:699–708.

Reinhart, K. O., A. Packer, W. H. van der Putten, and K. Clay. 2003. Plant-soil biota interactions and spatial distribution of black cherry in its native and invasive ranges. Ecology Letters 6:1046–50.

Reinhart, K. O., A. A. Royo, W. H. Van der Putten, and K. Clay. 2005. Soil feedback and pathogen activity in Prunus serotina throughout its native range. Journal of Ecology (in press).

Roditi, H. A., D. L. Strayer, and S. E. G. Findlay. 1997. Characteristics of zebra mussel (Dreissena polymorpha) biodeposits in a tidal freshwater estuary. Archiv für Hydrobiologie 140:207–19.

Sakai, A. K., W. Fred, J. S. Allendorf, D. M. Holt, J. M. Lodge, A. W. Kimberly, S. Baughman, R. J. Cabin, J. E. Cohen, et al. 2001. The population biology of invasive species. Annual Review of Ecology and Systematics 32:305–32.

Scheffer, R. P. 2003. The Nature of Disease in Plants. New York: Cambridge University Press.

Shaw, D. J., B. T., Grenfell, and A. P. Dobson. 1998. Patterns of macroparasite aggregation in wildlife host populations. Parasitology 117:597–610.

Shea, K., and P. Chesson. 2002. Community ecology theory as a framework for biological invasions. Trends in Ecology & Evolution 17:170–76.

Siegel, M. R. 1990. Fungal endophyte-infected grasses: Alkaloid accumulation and aphid response. Journal of Chemical Ecology 16:3301–15.

Siemann, E., and W. E. Rogers. 2001. Genetic differences in growth of an invasive tree species. Ecology Letters 4:514–18.

Simms, E. L., and M. D. Rausher. 1987. Costs and benefits of the evolution of plant defense against herbivory. American Naturalist 130:570–81.

Sinclair, A. R. E. 1979. The eruption of the ruminants. In Serengeti: Dynamics of an Ecosystem, ed. R. Sinclair, and M. Norton-Griffiths, 82–103. Chicago: University of Chicago Press.

Strong, D. R., and D. A. Levin. 1975. Species richness of the parasitic fungi of British trees. Proceedings of the National Academy of Sciences United States of America 72:2116–19.

Suarez, A. S., D. A. Holway, and P. S. Ward. 2005. The role of opportunity in the unintentional introduction of non-native ants. Proceedings of the National Academy of Sciences of the United States of America 102:17032–35.

Taylor, L. H., S. M. Latham, and M. E. J. Woolhouse. 2001. Risk factors in human disease emergence. Philosophical Transactions of the Royal Society of London. Series B, Biological Sciences 356:983–989.

Thorne, E. T., and E. S. Williams. 1988. Disease and endangered species: The black-footed ferret as a recent example. Conservation Biology 2:66–74.

Thrall P. H., and J. J. Burdon. 2004. Host-pathogen life-history interactions affect the success of biological control. Weed Technology 18:S1269–74.

Tilman, D. 1999. The ecological consequences of changes in biodiversity: A search for consequences. Ecology 80:1455–74.

Tompkins, D. M., and M. Begon. 1999. Parasites can regulate wildlife populations. Parasitology Today 15:311–13.

Torchin, M. E., J. E. Byers, and T. C. Huspeni. 2005. Differential parasitism of native and introduced snails: Replacement of a parasite fauna. Biological Invasions 7:885–94.

Torchin, M. E., K. D. Lafferty, A. P. Dobson, V. J. McKenzie, and A. M. Kuris. 2003. Introduced species and their missing parasites. Nature 421:628–30.

Torchin, M. E., K. D. Lafferty, and A. M. Kuris. 2001. Release from parasites as natural enemies: Increased performance of a globally introduced marine crab. Biological Invasions 3:333–45.

Torchin, M. E., and C. E. Mitchell. 2004. Parasites, pathogens and invasions by plants and animals. Frontiers in Ecology and the Environment 4:183–90.

Vitousek, P. M., C. M. DAntonio, L. L. Loope, M. Rejmanek, and R. Westbrooks. 1997. Introduced species: A significant component of human-caused global change. New Zealand Journal of Ecology 21:1–16.

Wares, J. P., A. R. Hughes, and R. K. Grosberg. 2005. Mechanisms that drive evolutionary change: insights from species introductions and invasions. *In* Species Invasions: Insights into Ecology, Evolution, and Biogeography, ed. D. F. Sax, J. J. Stachowicz, and S. D. Gaines, 229–58. Sunderland, MA: Sinauer.

Wolfe, L. M., J. A. Elzinga, and A. Biere. 2004. Increased susceptibility to enemies following introduction in the invasive plant *Silene latifolia*. Ecology Letters 7:813–20.

Wolfe, M. S. 1985. The current status and prospects of multiline cultivars and variety mixtures for disease resistance. Annual Review of Phytopathology 23:251–73.

CHAPTER NINE
Effects of Disease on Community Interactions and Food Web Structure

Kevin D. Lafferty

SUMMARY

INFECTIOUS DISEASES CAN BE POWERFUL forces in natural populations. When diseases affect influential species, the consequences of disease can ramify through communities. For instance, parasites can reverse the outcome of competition and, therefore, alter biodiversity. They may aid or buffer against biological invasions. Parasites permeate food webs and may change communities by altering predator-prey interactions. In particular, they may alter trophic cascades. Infectious human diseases probably limited our influence on the environment in the past. Humans' continued escape from disease through history has contributed to our having the largest influence over natural communities of any species.

INTRODUCTION

This chapter considers how parasitism might affect interacting species. The possibilities are many and involve entire food webs. Before branching out into the complexities of communities, it is worth considering the wide range of effects that parasites have on individuals (In this chapter, I usually use the term parasite to refer to agents of infectious disease). Many parasites do so little that they rate only as an inconvenience. Others have subtle but noticeable effects on host growth and competitive ability, perhaps through host investment in immune function, reduced attractiveness, slowed growth, or increased susceptibility to predators or stress. Sometimes the effect on the host depends on the parasite's life history, of which there are several discrete types (Lafferty and Kuris 2002). Most parasites kill the host at their own peril. However, some, like parasitoids, can be deadly, and this helps them transfer to new hosts. Others, like trematodes in snails, may have no discernible effect on longevity but block reproduction. Because this chapter focuses on the effects of parasites on communities, I consider only those parasites that greatly affect host populations.

For host-specific parasites (and many parasites evolve host specificity), effects at the host population level are a function of the effects of parasitism on infected individuals and the pattern of spread between infected and uninfected individuals. Transmission requires that an infected individual or infective stage contact an uninfected host. The more uninfected hosts that are available, the more likely it is that a parasite will pass its offspring to a new host before the parasite dies, where death of a parasite is usually a result of host defense, host death, or, for free-living stages, too much time waiting in a hostile environment. If a parasite does invade a host population, it will spread until the contact rate between infected and susceptible hosts drops because the epidemic runs short of susceptible hosts and infected hosts die or become immune. This means abundant species are more susceptible to infectious disease. Density-dependent transmission is a recurring theme of this chapter. Parasites that affect host populations are the ones most likely to have impacts at the ecosystem level, so long as the hosts they affect play important roles in an ecosystem (see Collinge et al., chapter 6, this volume). For example, parasites can interact with food webs when they affect species involved in trophic cascades. The next several sections of this chapter take a food web approach to understanding how disease can indirectly affect communities by altering species interactions.

COMPETITION

There are many nonexclusive explanations for how some communities can support many species, including the tendency for overlapping species to diverge in their resource use. It is more difficult to explain the co-existence of similar species because most simple models predict that competitive dominants will always exclude subordinate species. Three factors can help prevent competitive exclusion: indiscriminant disturbance at intermediate levels (Connell 1978), rare species advantage (Roughgarden and Feldman 1975), and impacts to competitive dominants (Paine 1966). Parasites help promote diversity if they differentially affect dominant or common species. Parasites may also reduce biological diversity (particularly as measured by heterogeneity) if they differentially affect subordinate species or lead to apparent competition. Given the increased homogenization of biotic communities through species introductions, it is also worth considering how parasites might help or hinder the invasion of competitive dominants (or generalist predators) that could reduce biodiversity.

Density-dependent transmission allows parasites to disproportionately affect common species (figure 9.1a). This helps maintain rarer

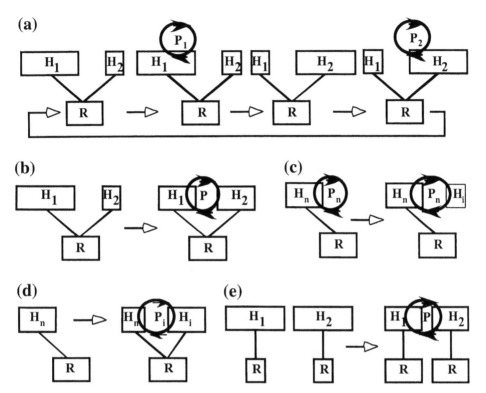

Figure 9.1. Interactions between infectious disease and competition. In this and subsequent figures, rectangles represent free-living species, with the volume of the rectangle proportional to the abundance of the species. Lines from one species to another represent a trophic link of a consumer (above) to a resource (below). Basal resources are indicated as R. Some free-living species are hosts (H) for parasites (P). Open horizontal arrows represent a comparison between two states (often uninfected vs. infected). (a) The parasite attacks an abundant species, releasing a rare species from competition, but another parasite attacks the released species when it becomes abundant. This prevents either species from becoming overwhelmingly common. (b) A shared parasite interferes with competitive dominance, allowing a subordinate species to succeed in areas where the parasite occurs. (c) A native species's (Hn) parasites (Pn) impede an invader (Hi). (d) A parasite (Pi) of an invader (Hi) disproportionately affects a native competitor (Hn), aiding the establishment of the invader. (e) Apparent competition through a shared parasite between two normally noncompeting species (H1 and H2).

competitors, thereby promoting coexistence and stability (Dobson et al. 2005). For example, natural enemies may maintain forest tree diversity (Wright 2002). This can occur if fungal pathogens disproportionately affect seedlings near conspecifics (Augspurger 1984; Packer and Clay 2000). (Clay et al., in chapter 7, this volume, discuss the idea that parasites beget host diversity in more detail.) The above scenario largely assumes parasites are host specific. When hosts share parasites, parasites can be competitive weapons. If subordinate species are tolerant or resistant to infection, parasites could help maintain them in a community (figure 9.1b). In a classic experiment, Park (1948) found that one flour beetle species (*Tribolium castaneum*) could competitively exclude another (*T. confusum*). A sporozoan parasite infects both species but has a bigger impact on the dominant species, reversing the outcome of competitive exclusion. In a more recent study, the fruit fly *Drosophila melanogaster* consistently outcompeted *D. simulans* in vials. Adding a parasitoid to the vial that slightly preferred the dominant *D. melanogaster* allowed the two species to coexist (Bouletreau et al. 1991). Similarly, in two separate studies of different species assemblages, competing amphipods may coexist in nature because a trematode reverses their relative population growth rates (Jensen et al. 1998; Thomas et al. 1995). A natural enemy might evolve to prefer competitive dominants if these dominants are more common and, therefore, the parasite encounters them more frequently. Still, parasite-mediated competition can differentially reduce subordinate species; a larval tapeworm shared by Park's flour beetles increased the rate at which the dominant beetle excluded the subordinate beetle (Yan et al. 1998).

Parasites could tip the balance in competitive interactions between native and introduced species (see also Perkins et al., chapter 8, this volume). On average, an invasive animal species has sixteen recorded parasite species in its home range but brings only three of these to invaded regions, where it picks up an additional four parasite species (Torchin et al. 2003). Leaving parasites behind could give invaders an advantage over natives saddled with a full parasite burden (Torchin and Mitchell 2004). Alternatively, if the invader has no coevolved history with the few new parasites it acquires, it might lack specific defenses, and infection could limit the invasion (figure 9.1c). For example, domestic cattle are very sensitive to the tsetse fly–transmitted trypanosome that causes sleeping sickness. This prevents their introduction to large parts of Africa, where cattle herding would likely exclude native grazers. Similarly, a meningeal nematode of white-tailed deer is highly pathogenic to other cervids and prevents their establishment in whitetail areas (Anderson 1972). We know that most invasions fail, and parasite defense could be one reason.

Of those three parasites that an average invader brings, some likely serve as handicaps, but others could serve as weapons (figure 9.1d). In the United Kingdom, the introduced grey squirrel competes with the native red squirrel. The grey squirrel is a good competitor for food on its own (Bryce 1997; MacKinnon 1978; O'Teangane et al. 2000), but it is aided by the parapoxvirus (Tompkins et al. 2003). This parapoxvirus is a relatively benign disease of grey squirrels. As grey squirrels expand into new habitat, they bring their pathogen along. Red squirrels are naïve hosts and suffer higher pathology. The parapoxvirus can persist even as red squirrels become rare because the more tolerant grey squirrels serve as a reservoir for the virus. A parallel situation occurs when caecal nematodes (*Heterakis gallinarum*) aid introduced pheasants (*Phasianus colchicus*) in their competition with grey partridges (*Perdix perdix*). Here, the effects of competitive exclusion center even more on the effect of the parasite on the native species (Tompkins et al. 2000).

Parasites can cause two species to interact indirectly even if these species do not compete for resources. This is known as apparent competition (figure 9.1e). Holt (1977) formalized this concept, and many derivations are possible. Apparent competition occurs because one host (the more tolerant or resistant) helps maintain the abundance of a natural enemy, which then differentially affects the second species.

Multiple hosts can also affect parasitism (see Power and Flecker, chapter 4, this volume). The predominance of host specificity suggests that host diversity will beget parasite diversity (Hechinger and Lafferty 2005). In other words, communities rich in hosts should also be rich in parasites. However, host diversity can also dilute transmission of a particular infectious agent if some hosts are infected but not suitable (see Begon, chapter 1, this volume).

PREDATION

Between 1846 and 1850, a pathogenic fungus transformed the human ecosystem of Ireland (Donnelly 2001). The resulting Irish potato famine serves as an allegory for how parasites can compete with predators for common food resources and alter entire communities. The adoption of agricultural monocultures increases plant density while reducing species and genetic diversity (Wolfe 2000). This provides a disease with a dense and uniform population through which to spread, conditions that are also conducive to the evolution of high virulence. In the 1800s, Irish peasant farmers produced cash crops (meat, dairy, grain) to pay rent to British landowners. They fed themselves on potatoes. This strategy allowed the population to nearly double in forty years, with two million acres planted

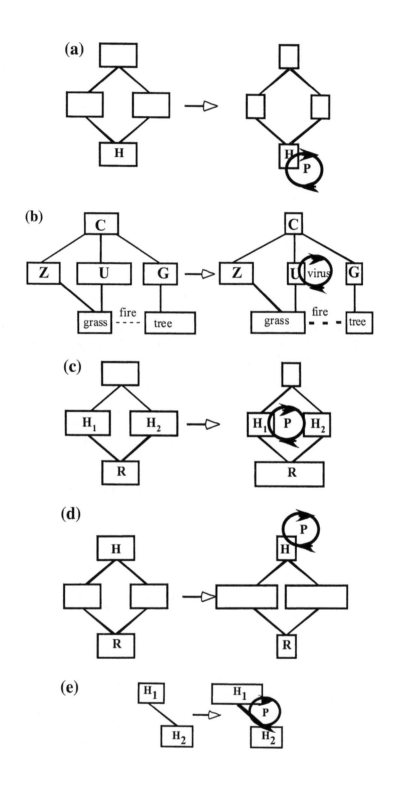

in potatoes. In 1845, a warm, wet winter favored fungal spread, and blight left potatoes rotting in the fields. Many peasants starved, succumbed to disease, or immigrated to North America. This illustrated how when disease affects lower trophic levels, bottom-up effects may cascade through a web (figure 9.2a).

A similar example is the loss of the American chestnut tree to blight. This tree was a dominant species, and chestnuts were an important source of food for wildlife in eastern deciduous forests. This was particularly true for the massive flocks of passenger pigeons (Schorger 1955). In the 1900s, an introduced fungus led to the gradual and near extirpation of chestnuts over 90 million acres of forest, eliminating an important food resource for animals. Although the impacts of chestnut blight on wildlife were not well quantified, seven species of moth that were specific to chestnut went extinct (Opler 1978).

If disease affects species in the midtrophic level, effects may propagate up, down, and sideways through the food web. For example, myxomavirus was introduced to England and Australia with the realized hope of releasing sheep from competition with rabbits (Fenner and Ratcliffe 1965; Minchella and Scott 1991). Rinderpest (a morbillivirus related to measles) in East African ungulates also illustrates how a parasite can alter food webs (Dobson 1995; Plowright 1982; Sinclair 1979; Tompkins et al. 2001). The rinderpest epidemics of the 1800s caused mass mortality in domestic and wild artiodactyls throughout Africa (figure 9.2b). This change indirectly reduced top predators such as lions and hyenas and altered vegetation structure via grazing and fire (Plowright 1982). A vaccine was introduced into cattle, and by 1961, the ecosystem had experienced rapid recovery (Plowright 1982; Spinage 2003), including an increase in predators such as lions and hyenas, and decreases in some prey, such as gazelles, and competitors, such as wild dogs (Dublin et al. 1990; McNaughton 1992).

Figure 9.2. Interactions between infectious disease and predation. (a) Comparison of the effect of a parasite (P) on a basal taxon. The disease reduces the plant population, depleting resources for species at higher trophic levels. (b) Serengeti food web before and after rinderpest (circle). Boxes with letters indicate carnivores (C), zebra (Z), ungulates (U), and giraffe (G). (c) Comparison of the effect of a parasite on a midlevel consumer. The disease reduces the grazer population, depleting resources for species at higher trophic levels and releasing basal taxa from grazing. (d) Effect of a parasite on a predator. Reduction in predator density releases prey and affects basal taxa through a trophic cascade. (e) Effect of a parasite that cycles between predator and prey and makes prey easier for the predator to catch. Predators increase in abundance and prey decrease (many other indirect effects are possible).

Sea urchins are herbivores that can exert an enormous effect on plant communities (figure 9.2c). On tropical reefs, sea urchins, along with other grazers, help minimize the standing stock of algae. This allows invertebrates, such as corals, to dominate and form reefs. In the 1980s, an apparently infectious disease swept through sea urchin populations in the Caribbean. The near extirpation of sea urchins, coupled with overfishing of herbivorous fishes, allowed algae to grow up and choke coral reefs, adding to the worldwide decline in this ecosystem (Lessios 1988). At higher latitudes, where kelps create cathedral-like forests filled with fishes, people value algae more and urchins less. Overfishing of sea urchin predators (e.g., sea otters, spiny lobsters) increases the density of sea urchins, which then reduce kelp forests to "barrens" (Lafferty 2004). But high densities of urchins in these barrens promote epidemics of bacterial disease that can reduce urchin densities and push barren reefs back toward kelp forests (Behrens and Lafferty 2004).

Disease also affects top predators (figure 9.2d). In the temperate reefs of California, parasites of the southern sea otter are primarily from land (*Toxoplasma gondii* from cats, acanthocephalan worms from shorebirds, and terrestrial fungus) and do not have otter-to-otter transmission. Mortality rates from these parasites are high, and high mortality appears to be a main reason that otters have failed to expand their range to the south (Lafferty and Gerber 2002), releasing sea urchins from an important predator through much of the otter's former range. In a better studied example, a Scandinavian outbreak of sarcoptic mange (caused by mites) in the late 1970s through the 1980s reduced the density of red foxes. Prey (rodents, rabbits, ground birds, deer) increased as a result and then declined after the epidemic waned and fox populations recovered (Lindstrom et al. 1994).

The fate of most parasites is tied to that of their hosts. If their hosts die, this is usually a bad thing for host and parasite alike (see Holt et al., chapter 15, this volume). In nature, hosts get eaten. This puts tremendous evolutionary pressure on parasites to survive the ingestion process by establishing in the predator (Lafferty 1999). Perhaps as a result, many parasitic species have complex life cycles in which a final host must eat an intermediate host. In such life cycles, the parasite must wait for the ingestion of the intermediate host by final hosts. However, not all parasites are patient. Some parasites manipulate the behavior or appearance of the intermediate host to increase the rate at which a predator host will catch and eat it (Moore 2002). For instance, in southern California estuaries, the most common trematode, *Euhaplorchis californiensis*, encysts on the brain of killifish; the worms alter the fish's behavior, making it shimmy and swim to the surface. These fish are ten to thirty times more likely to be eaten by birds, the final host of the worm (Lafferty and Morris 1996).

In this system, the worm essentially dictates which fishes live and die. They also provide an easy snack for egrets and herons, which otherwise might have to work harder for a living. Some mathematical models indicate that such parasite-increased trophic transmission can reduce prey density; it can also increase predator density so long as the energetic costs of parasitism for the predator are not too severe (figure 9.2e) (Lafferty 1992). Other mathematical models suggest that predators may depend on parasites to supply them with easy prey (Freedman 1990).

PARASITISM

Parasites can interact with each other. Some parasites have parasites, and some compete for host resources with other parasite species. For larval trematodes, competition for resources within the snail is intense, and trematodes have special morphological and behavioral adaptations for interspecific interactions. For example, adding dominant trematode species to ponds can exclude subordinate trematode species (Lafferty 2002; Lie and Ow-Yang 1973; Nassi et al. 1979). In this example, the subordinate species are pathogenic to humans, and a consequence of this parasite-parasite interaction is improved human health (figure 9.3). Parasites can interact with the host, often via the immune system, to displace other parasites or alter their pathogenic effects on the host in various directions (Cox 2001). In shrimp, infection with one virus can reduce the effect of a second (Hedrick et al. 1994; Tang et al. 2003). In mosquitoes, filarial worms increase susceptibility to equine encephalitis virus (Vaughan et al. 1999) but decrease the development of malaria parasites (Albuquerque and Ham 1995). Despite all the potential for parasite-parasite interactions, few studies have considered what this means at the community level. Interactions between two morbilliviruses, rinderpest and canine distemper virus, are one possible example

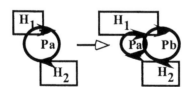

Figure 9.3 Interactions among infectious diseases. The effect of a nonvirulent parasite (Pb) on a virulent parasite (Pb) is modeled. Here, the nonvirulent parasite releases the final host from pathology through competition with the virulent parasite in the intermediate host.

of how two closely related pathogens can interact to alter communities. Anecdotal evidence suggests that carnivores develop some immunity to distemper when they feed on prey with rinderpest (Plowright 1968). This might explain the disappearance of canine distemper during the rinderpest epidemic and its resurgence in recent years (Dobson and Hudson 1986; Roelke-Parker et al. 1996). Outbreaks of distemper virus have led to periodic crashes in top predators in East Africa, with subsequent benefits for lower trophic levels.

Mutualism/Facilitation

Sometimes, by altering their host, parasites can alter communities dependent on these hosts or their actions. In one case, such manipulations can have dramatic and unexpected consequences for communities. The trematode *Curtuteria australis* reduces the ability of cockles to bury into New Zealand mudflats (perhaps this increases an infected clam's vulnerability to predation by final host birds) (Thomas et al. 1998). The shells of infected clams stick up out of the mud and provide a hard substrate for sessile invertebrates, such as limpets, that otherwise could not persist in the soft sediment (figure 9.4). Parasites can affect substrate-forming species as well, shifting communities in the opposite direction. For instance, trematodes reduce populations of a tube-building corophiid amphipod, thereby destabilizing the sediment and altering the faunal composition of a Danish mudflat (Mouritsen and Poulin 2002).

Food Web Topology

Food webs form a conceptual framework for the study of ecology, and substantial theory has developed on how the topological structure of food webs alters the flow of energy through an ecosystem and the resilience of

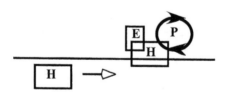

Figure 9.4 Effects of infectious disease on ecosystem engineers. The parasite makes the host sit above the surface of the sediment, where it is colonized by an epibiont (E) requiring hard substrate (which normally could not live in the habitat).

a community to change (or stability) that might occur following the addition of a new species or extirpation of an existing species. Few topological food webs (with the exception of parasitoid webs) have included parasites (Cohen et al. 1993; Polis 1991). This is because it is difficult to obtain good quantitative information on parasites in natural communities. Still, many ecologists acknowledge the potential importance of parasites in food webs and advocate their inclusion (Cohen et al. 1993; Marcogliese 2003; Marcogliese and Cone 1997; Polis 1991).

We can make some basic predictions about the effect of parasites on food webs. Parasites add links and species to food webs. This has the potential to change the linkage density or connectance of a food web. Connectance may alter the stability of a food web because it describes how strongly species are interconnected, helping to predict whether species additions or deletions will greatly alter other species. For example, Memmott et al. (2000) found that adding parasitoids to a food web decreased connectance, because most parasitoids interacted with one or a few hosts. Typical parasites are less restricted in their host ranges than are parasitoids, and we can expect that parasites that interact with many hosts will make a web more highly connected. For this reason, typical parasites are unlikely to act like parasitoids or predators in food webs. Parasites, as consumers of consumers, will tend to lengthen food-chain length (Huxham et al. 1995). In this way, they are very much like predators or parasitoids.

Existing insight into the role of typical parasites comes from estuarine food webs (see review in Sukhdeo and Hernandez 2004). In the Ythan Estuary food web, the addition of parasites slightly decreases connectance (Huxham et al. 1995). Adding parasites to a food web for a New Zealand mudflat yielded similar results (Thompson et al. 2005). By looking at the effect of each parasite species, this study found that most parasites only mildly decrease connectance. One generalist trematode, however, strongly increased connectance.

Most attempts to add parasites to food webs have considered that they operate similar to top predators. However, parasites differ from predators in several ways, the most notable being their intimate association with their prey and their relatively low biomass. Sukhdeo and Hernandez (2004) quantified the biomass of acanthocephalans in a food chain and compared this with predictions for predators. They found that the acanthocephalan population had the biomass expected from a top predator species, but that, because individual body size is very small, their abundance was much higher than expected for a top predator. Huxam et al. (1995) realized that parasites were not equivalent to top predators and predicted that including the full range of parasite links would add to connectance of food webs.

Lafferty et al. (2006) incorporated parasites into the Carpinteria salt marsh food web by using subwebs. This food web includes (1) a predator-prey subweb (this is what constitutes most published food webs) and (2) a host-parasite subweb (corresponding to previously published food webs with parasites included). In addition, a third subweb contains links where predators eat parasites. As mentioned previously, this happens in most predator-prey interactions because prey animals often have parasites in or on them. Sometimes the parasites are digested, but in a third of the links in the Carpinteria salt marsh food web, parasites can use the predator as a host. In addition, many parasites have free-living stages that may be fed on. The predator-parasite subweb contains the highest linkage density of all the subwebs in the Carpinteria Salt Marsh food web. A fourth subweb, parasite-parasite, completes the 2×2 matrix of subwebs. Including all four subwebs, connectance is three times higher with parasites than without parasites. Therefore, parasites have opposite effects on food web connectance than hitherto appreciated.

In the Carpinteria Salt Marsh food web, parasites have twice the number of hosts as predators have prey. Although top predators have few natural enemies in the predator-prey subweb, they are disproportionately attacked by parasites. For this reason, consumers at mid-trophic levels have the most natural enemies because they have a substantial number of predators and parasites compared with lower trophic levels, which have relatively few parasites, and upper trophic levels, which have few predators.

Although relatively little is known about parasites in food webs, the studies published to date indicate that parasites are likely to be worth including. They may make up most links in a food web and, at least for generalist species, may be more densely linked in webs than predators. In the Carpinteria Salt Marsh, two-thirds of the links occur in the parasite subwebs. It would seem that no food web is complete without parasites.

HUMANS

For parasites to affect communities, or even ecosystems, the parasite must alter the abundance of a host that plays an important role in the community. No host fits this better than humans. Parasites probably still affect human population densities, particularly in tropical regions. Historically, infectious diseases were a greater source of human mortality than combat. For instance, in the Spanish-American War, 10% of Spanish troop deaths occurred in battle, while 90% occurred as a result of malaria, dysentery, and other diseases (Cardona 1998). As humans spread out of Africa, they escaped some parasites and inadvertently used others as weapons against human competitors. When Native Americans colonized the New World, they found a Shangri La of abundant wildlife and, presumably, few infectious diseases; their explosive spread and growth contributed to dramatic impacts on the faunal composition of the Americas (Flannery 2001). In turn, the number of Native Americans killed by the conquistadors' muskets and swords is miniscule compared with the number killed by infectious diseases of European origin (Diamond 1997). Epidemics in native populations presumably aided European colonization and deforestation of the New World. Modern pharmaceuticals and medical science have greatly decreased the impact of infectious disease, increasing the ability of humans to dominate the globe. We are now undisputedly Earth's dominant species in our consumption, distribution, and effects on biotic and abiotic conditions. Our escape from parasites has greatly fostered this outcome.

CONCLUSIONS

Because parasites are common, it is worth considering them alongside other, more obvious consumers. The consideration of infectious agents in food webs is increasingly under way. This will allow a more complete appreciation of food webs and better inform how consumer-resource interactions affect food web stability and the evolution of consumer strategies.

There are several approaches for considering these effects, and they mirror what has been used to decipher the effects of consumer interactions in food webs. The best place to start is to identify species that have disproportionate roles in food webs—foundation species, keystone predators, and the like—and then consider which parasites might alter their population dynamics.

Additionally, one can use the food web modules presented in the figures to develop mathematical models of the population dynamics of indirect effects. Many of these exist specifically for infectious diseases (see Holt, chapter 15, this volume). A unified modeling framework applied to all the modules in the figures would permit a comparative analysis. To date, most modeling efforts use microparasite models, because these are the most likely to yield analytical solutions. To broaden the value obtained from theory, it would be worth while attempting to model the entire range of consumer strategies and then compare their effects (Lafferty and Kuris 2002). This would help determine whether infectious disease is just another type of consumer-resource interaction (see chapter 10, this volume). Obvious points of entry into understanding the differences between parasitism and other consumer strategies are the tendency toward host specialization by infectious agents and the coevolutionary responses (e.g., immune systems) specifically evolved to combat infectious pathogens.

Experimental manipulation of parasites to determine their effects on ecosystems is difficult, but the few attempts to do so have led to important insights on host population dynamics. Manipulations of parasites to investigate indirect effects have mostly been confined to laboratory investigations of apparent competition. However, because introduced species can bring parasites with them or leave them behind (see Perkins et al., chapter 8, this volume), species introductions serve as unintended experiments and have led to considerable insight into the role of infectious diseases in ecosystems (Lafferty et al. 2005). The example of the introduction of rinderpest to the Serengeti illustrates best how an infectious pathogen can dramatically change an ecosystem. We are just at the beginning of an exciting and challenging quest to uncover the role of parasites in community interactions and food webs.

LITERATURE CITED

Albuquerque, C. M. R., and P. J. Ham. 1995. Concomitant malaria (*Plasmodium gallinaceum*) and filaria (*Brugia pahangi*) infections in *Aedes aegypti*: Effect on parasite development. Parasitology 110:1–6.

Anderson, R. C. 1972. The ecological relationships of meningeal worm and native cervids in North America. Journal of Wildlife Diseases 8:304–10.

Augspurger, C. K. 1984. Pathogen mortality of tropical tree seedlings: Experimental studies of the effects of dispersal distance, seedling density, and light conditions. Oecologia 61:211–17.

Behrens, M. D., and K. D. Lafferty. 2004. Effects of marine reserves and urchin disease on southern California rocky reef communities. Marine Ecology Progress Series 279:129–39.

Bouletreau, M., P. Fouillét, and R. Allemand. 1991. Parasitoids affect competitive interactions between the sibling species, *Drosophila melanogaster* and *D. simulans*. Redia 84:171–77.

Bryce, J. 1997. Changes in the distribution of red and grey squirrels in Scotland. Mammal Review 27:171–76.

Cardona, G. 1998. A sangre y fuego. La Aventura de la Historia 1:88.

Cohen, J. E., R. A. Beaver, S. H. Cousins, D. L. Deangelis, L. Goldwasser, K. L. Heong, R. D. Holt, A. J. Kohn, J. H. Lawton, et al. 1993. Improving food webs. Ecology 74:252–58.

Connell, J. H. 1978. Diversity in tropical rain forests and coral reefs. Science 199:1302–10.

Cox, F. E. G. 2001. Concomitant infections, parasites and immune responses. Parasitology 122:S23–38.

Diamond, J. M. 1997. Guns, Germs and Steel: The Fates of Human Societies. New York: Norton.

Dobson, A. P. 1995. Rinderpest in the Serengeti ecosystem: The ecology and control of a keystone virus. *In* Proceedings of a Joint Conference of the American Association of Zoo Veterinarians, Wildlife Disease Association, and American Association of Wildlife Veterinarians, ed. R. E. Junge, 518–19. East Lansing, MI.

Dobson, A. P., and P. J. Hudson. 1986. Parasites, disease and the structure of ecological communities. Trends in Ecology & Evolution 1:11–15.

Dobson, A., K. D. Lafferty, and A. M. Kuris. 2005. Parasites and food webs. *In* Food Webs, ed. M. Pascual and J. A. Dunne, 119–35. Oxford: Oxford University Press.

Donnelly, J. S. 2001. The Great Irish Potato Famine. Gloucestershire, UK: Sutton Publishing. England.

Dublin, H. T., A. R. E. Sinclair, S. Boutin, E. Anderson, M. Jago, and P. Arcese. 1990. Does competition regulate ungulate populations? Further evidence from Serengeti, Tanzania. Oecologia 82:283–88.

Fenner, F., and F. N. Ratcliffe. 1965. Myxomatosis. Cambridge: Cambridge University Press.

Flannery, T. F. 2001. The Eternal Frontier. New York: Grove Press.

Freedman, H. I. 1990. A model of predator-prey dynamics as modified by the action of a parasite. Mathematical Biosciences 99:143–55.

Hechinger, R. F., and K. D. Lafferty. 2005. Host diversity begets parasite diversity: Bird final hosts and trematodes in snail intermediate hosts. Proceedings of the Royal Society of London. Series B, Biological Sciences 272:1059–66.

Hedrick, R. P., S. E. La Patra, S. Yun, K. A. Lauda, G. R. Jones, J. L. Congleton, and P. de Kinkelin. 1994. Induction of protection from infectious hematopoietic necrosis virus in rainbow trout *Oncorhynchus mykiss* by pre-exposure to the avirulent cutthroat trout virus (CTV). Diseases of Aquatic Organisms 20:111–18.

Holt, R. D. 1977. Predation, apparent competition and the structure of prey communities. Theoretical Population Biology 28:181–208.

Huxham, M., D. Raffaelli, and A. Pike. 1995. Parasites and food-web patterns. Journal of Animal Ecology 64:168–76.

Jensen, T., K. T. Jensen, and K. N. Mouritsen. 1998. The influence of the trematode *Microphallus claviformis* on two congeneric intermediate host species (*Corophium*): Infection characteristics and host survival. Journal of Experimental Maine Biology and Ecology 260:349–52.

Lafferty, K. D. 1992. Foraging on prey that are modified by parasites. American Naturalist 140:854–67.

Lafferty, K. D. 1999. The evolution of trophic transmission. Parasitology Today 15:111–15.

Lafferty, K. D. 2002. Interspecific interactions in trematode communities. *In* The Behavioral Ecology of Parasites, ed. E. E. Lewis, M. V. K. Sukhdeo, and J. F. Campbell, 153–69. Wallingford Oxon, UK: CAB International.

Lafferty, K. D. 2004. Fishing for lobsters indirectly increases epidemics in sea urchins. Ecological Applications 14:1566–73.

Lafferty, K. D., and L. Gerber. 2002. Good medicine for conservation biology: The intersection of epidemiology and conservation theory. Conservation Biology 16:593–604.

Lafferty, K. D., R. F. Hechinger, J. C. Shaw, K. L. Whitney, and A. M. Kuris. 2006. Food webs and parasites in a salt marsh ecosystem. *In* Disease Ecology: Community Structure and Pathogen Dynamics, ed. S. Collinge and C. Ray, 119–34. Oxford: Oxford University Press.

Lafferty, K. D., and A. K. Morris. 1996. Altered behavior of parasitized killifish increases susceptibility to predation by bird final hosts. Ecology 77: 1390–97.

Lafferty, K. D., and A. M. Kuris. 2002. Trophic strategies, animal diversity and body size. Trends in Ecology & Evolution 17:507–13.

Lafferty, K. D., K. F. Smith, M. E. Torchin, A. P. Dobson, and A. M. Kuris. 2005. The role of infectious disease in natural communities: What introduced species tell us. *In* Species Invasions: Insights into Ecology, Evolution, and Biogeography, ed. D. F. Sax, J. J. Stachowicz, and S. D. Gaines. Sunderland, MA: Sinauer.

Lessios, H. A. 1988. Mass mortality of *Diadema antillarum* in the Caribbean: What have we learned? Annual Review of Ecology and Systematics 19:371–93.

Lie, K. J., and C. K. Ow-Yang. 1973. A field trial to control *Trichobilharzia brevis* by dispersing eggs of *Echinostoma audyi*. Southeast Asian Journal of Tropical Medicine and Public Health 4:208–17.

Lindstrom, E. R., H. Andren, P. Angelstam, G. Cederlund, B. Hornfeldt, L. Jaderberg, P.-A. Lemnell, B. Martinsson, K. Skold, and J. E. Swenson. 1994. Disease reveals the predator: Sarcoptic mange, red fox predation, and prey populations. Ecology 75:1042–49.

MacKinnon, K. 1978. Competition between red and grey squirrels. Mammal Review 8:185–90.

Marcogliese, D. 2003. Food webs and biodiversity: Are parasites the missing link? Journal of Parasitology 82(S):389–99.

Marcogliese, D. J., and D. K. Cone. 1997. Food webs: A plea for parasites. Trends in Ecology & Evolution 12:320–25.

McNaughton, S. J. 1992. The propagation of disturbance in savannas through food webs. Journal of Vegetation Science 3:301–14.

Memmott, J., N. D. Martinez, and J. E. Cohen. 2000. Predators, parasitoids and pathogens: Species richness, trophic generality and body sizes in a natural food web. Journal of Animal Ecology 69:1–15.

Minchella, D. J., and M. E. Scott. 1991. Parasitism: A cryptic determinant of animal community structure. Trends in Ecology & Evolution 6:250–54.

Moore, J. 2002. Parasites and the Behavior of Animals. Oxford: Oxford University Press.

Mouritsen, K. N., and R. Poulin. 2002. Parasitism, community structure and biodiversity in intertidal ecosystems. Parasitology 124:S101–17.

Nassi, H., J. P. Pointier, and Y. J. Golvan. 1979. Bilan d'un essai de contrôle de *Biomphalaria glabrata* en Guadalupe à l'aide d'un Trématode stérilisant. Annales de Parasitologie 52:277–323.

Opler, P. A. 1978. Insects of American chestnut: Possible importance and conservation concern. *In* The American Chestnut Symposium, ed. J. McDonald, 83–85. Morgantown: West Virginia University Press.

O'Teangane, D., S. Reilly, W. I. Montgomery, and J. Rochford. 2000. Distribution and status of the red squirrel and grey squirrel in Ireland. Mammal Review 30:45–56.

Packer, A., and K. Clay. 2000. Soil pathogens and spatial patterns of seedling mortality in a temperate tree. Nature 404:278–81.

Paine, R. T. 1966. Food web complexity and species diversity. American Naturalist 100:65–75.

Park, T. 1948. Experimental studies of interspecies competition 1. Competition between populations of the flour beetles, *Tribolium confusum* Duval and *Tribolium castaneum* Herbst. Ecological Monographs 18:267–07.

Plowright, W. 1968. Rinderpest virus. Virology Monographs 3:1–94.

Plowright, W. 1982. The effects of rinderpest and rinderpest control on wildlife in Africa. Symposia of the Zoological Society of London 50:1–28.

Polis, G. A. 1991. Complex trophic interactions in deserts: An empirical critique of food-web theory. American Naturalist 138:123–55.

Roelke-Parker, M. E., L. Munson, C. Packer, R. Kock, S. Cleaveland, M. Carpenter, S. J. O'Brien, A. Pospischill, R. Hofmann-Lehmann, et al. 1996. A canine distemper virus epidemic in Serengeti lions (*Panthera leo*). Nature 379:441–45.

Roughgarden, J., and M. Feldman. 1975. Species packing and predation pressure. Ecology 56:489–92.

Schorger, A. W. 1955. The Passenger Pigeon: Its Natural History and Extinction. Madison: University of Wisconsin Press.

Sinclair, A. R. E. 1979. The eruption of the ruminants. *In* Serengeti: Dynamics of an Ecosystem, ed. A. R. E. Sinclair and M. Norton-Griffiths, 82–103. Chicago: University of Chicago Press.

Spinage, C. A. 2003. Cattle Plague: A History. New York: Kluwer.

Sukhdeo, M. V., and A. D. Hernandez. 2004. Food web patterns and the parasite perspective. *In* Parasitism and Ecosystems, ed. F. Thomas, F. Renaud, and J. Guegan, 54–67. Oxford: Oxford University Press.

Tang, K. F. J., S. V. Durand, B. L. White, R. M. Redman, L. L. Mohney, and D. V. Lightner. 2003. Induced resistance to white spot syndrome virus infection in *Penaeus stylirostris* through pre-infection with infectious hypodermal and hematopoietic necrosis virus: A preliminary study. Aquaculture 216:19–29.

Thomas, F., F. Renaud, T. de Meeüs, and R. B. Poulin. 1998. Manipulation of host behaviour by parasites: Ecosystem engineering in the intertidal zone? Proceedings of the Royal Society of London. Series B, Biological Sciences 265:1091–96.

Thomas, F., F. Renaud, F. Rousset, F. Cezilly, and T. deMeeüs. 1995. Differential mortality of two closely related host species induced by one parasite. Proceedings of the Royal Society of London. Series B, Biological Sciences 260:1091–96.

Thompson, R. M., K. N. Mouritsen, and R. Poulin. 2005. Importance of parasites and their life cycle characteristics in determining the structure of a large marine food web. Journal of Animal Ecology 74:77–85.

Tompkins, D. M., A. P. Dobson, P. Arneberg, M. E. Begon, I. M. Cattadori, J. V. Greenman, J. A. P. Heesterbeek, P. J. Hudson, D. Newborn, et al. 2001. Parasites and host population dynamics. *In* The Ecology of Wildlife Diseases, ed. P. J. Hudson, A. Rizzoli, B. T. Grenfell, H. Heesterbeek, and A. Dobson, 45–62. Oxford: University of Oxford Press.

Tompkins, D. M., J. V. Greenman, P. A. Robertson, and P. J. Hudson. 2000. The role of shared parasites in the exclusion of wildlife hosts: *Heterakis gallinarum* in the ring-necked pheasant and the grey partridge. Journal of Animal Ecology 69:829–40.

Tompkins, D. M., A. R. White, and M. Boots. 2003. Ecological replacement of native red squirrels by invasive greys driven by disease. Ecology Letters 6:189–96.

Torchin, M. E., K. D. Lafferty, A. P. Dobson, V. J. McKenzie, and A. M. Kuris. 2003. Introduced species and their missing parasites. Nature 421:628–30.

Torchin, M. E., and A. J. Mitchell. 2004. Parasites, pathogens, and invasions by plants and animals. Frontiers in Ecology and the Environment 2:183–90.

Vaughan, J. A., M. Trpis, and M. J. Turell. 1999. *Brugia malayi* microfilaria (Nematoda: Filaridae) enhance the infectivity of Venezuelan equine encephalitis virus to Aedes mosquitoes (Diptera: Culicidae). Journal of Medical Entomology 36:758–63.

Wolfe, M. S. 2000. Crop strength through diversity. Nature 406:681–82.

Wright, S. J. 2002. Plant diversity in tropical forests: A review of mechanisms of species coexistence. Oecologia 130:1–14.

Yan, G., L. Stevens, C. J. Goodnight, and J. J. Schall. 1998. Effect of a tapeworm parasite on the competition of *Tribolium* beetles. Ecology 79: 1093–03.

CHAPTER TEN

Is Infectious Disease Just Another Type
of Predator-Prey Interaction?

Spencer R. Hall, Kevin D. Lafferty, James H. Brown, Carla E. Cáceres, Jonathan M. Chase, Andrew P. Dobson, Robert D. Holt, Clive G. Jones, Sarah E. Randolph, and Pejman Rohani

SUMMARY

WHICH FACTORS FUNDAMENTALLY separate infectious disease from other types of predator-prey interactions studied by community ecologists? Could parasitism and predation be combined into a unifying model? After all, parasites and predators both convert energy and nutrients contained in their resources (hosts or prey, respectively) into new biomass and reproductive work. If these focal consumers perform similar roles, disease ecologists and community ecologists may essentially study the same problems. Therefore, they should use the same conceptual toolkits. Given this important potential for more intellectual cross-fertilization, we contemplate these questions by way of two arguments. First, we consider the case that parasitism and predation are essentially the same types of interactions, varying only quantitatively. This line of argument highlights the many similar ways in which parasites and predators interact with their resources and other species at local and macro-ecological scales. Perhaps, then, major differences between predators and parasites vary only quantitatively, as a matter of body-size scaling. Parasites are typically much smaller than their host, while predators are similarly sized or exceed the size of their prey. The second case embraces eight (or more) qualitative splits that separate different types of predators from different types of parasites (developed by Lafferty and Kuris 2002). These axes differentiate consumers that eat one versus more than one resource individual per life stage, consumers that kill or do not kill their resources, and so forth. Both lines of argument eventually lead to the same challenge, however: Just how many qualitative models must one consider to capture the range from predation to parasitism in nature? The answer almost certainly depends on the focal question (implications for population dynamics, nutrient cycling, evolutionary change, etc.) and the currency used to evaluate it.

INTRODUCTION

Nothing qualitatively distinguishes host-parasite interactions from other consumer-resource interactions. This statement provides a springboard for an interesting and challenging thought experiment contemplating fundamental similarities and differences between parasites and predators. No doubt, such a claim will immediately surprise, annoy, or enrage some students of parasitism and disease, particularly those who work to uncover the mechanistic underpinnings of disease biology in detail. After all, parasites exhibit an amazing diversity of strategies to infect their hosts, and some microbial parasites can even transfer genes horizontally (which predators cannot). Therefore, acceptance of such a claim could require much abstraction. Yet, for some questions, it might be possible and useful to aggregate predators and parasites into a single consumer strategy. Such a statement might make sense to ecologists who study interactions between free-living species (e.g., competition, predation), even if they, like many or most community and ecosystem ecologists, have not previously considered parasites in much detail. Indeed, this commonality claim, if true, might spur very useful interactions between community and disease ecologists, who already talk about similar phenomena in different languages (Mittelbach 2005). Perhaps many of the independently developed theories readily transfer among these largely separate subdisciplines.

In this chapter, we consider the question of parasites-as-predators from two viewpoints. First, we embrace a "parasites are just predators" line of reasoning to highlight the many geometric parallels between these two types of interactions (Anderson and May 1979; Earn et al. 1998; Holt et al. 2003; Morin 1999; Thomas et al. 2005). At some basic level, these geometries likely emerge because parasites, like predators, consume resources. Resources, be they humans or rabbits, contribute positively to the population growth of their consumers. Once acquired, consumers, be they the measles virus or foxes, convert some of the energy and materials concentrated in their resources into somatic and reproductive biomass and work, and dissipate the rest. Energy and materials flow through these interactions (Mitchell 2003; Polis and Strong 1996; see also Middelboe, chapter 11, and Eviner and Likens, chapter 12, this volume). Perhaps the important differences emerge, then, as a quantitative matter of scaling. The most obvious scaling dimension is variation in body size between consumers and resources: the ratio of consumer size to resource size varies greatly along a predation-parasitism gradient (Lafferty and Kuris 2002). Later we consider another, recently proposed scheme that assorts types of predators, parasites, and parasitoids along four qualitative axes (Lafferty and Kuris 2002). These axes focus on how many resource victims are attacked per life stage by the consumer, the implications of these attacks for victim fitness, whether the success of such attacks depends on attack intensity, and, in the case of parasites, whether hosts must die. Which perspective is most useful? Ultimately, the answer likely hinges on the particular question asked (e.g., concerning nutrient cycling, population growth rate and biomass, or evolutionary change) and the currency used to evaluate it.

Two Positions on Parasites as Predators

Position 1: "Parasites Are Just Predators"

At some level, disease ecologists and community ecologists have much more in common than they may realize (Mittelbach 2005). Students of both subdisciplines essentially study interspecific interactions among consumers, which attack other species to acquire their nutrients and energy, and resources, which defend or tolerate removal of acquired nutrients and energy by their attackers (Chase et al. 2000; Grover 1997). Furthermore, the net outcomes of these interactions in turn determine abundances of interactors and promote or degrade opportunities for coexistence (see chapter 7, this volume). Yet attempts to fully map webs of interactions among these consumers and resources quickly run into a "curse of dimensionality" problem. The number of links between consumers and resources explodes into temporal and spatial intractability (Raffaelli and Hall 1996; Warren 1989; Yodzis 1996). Although one might abstract properties of these complex webs using a variety of indices (Cohen 1978; Lawton and Warren 1988), we consider two more mechanistic approaches here. First, both disease and community ecologists have focused on very similar subsets of interaction "modules," starting with two species and then building up direct and indirect interactions among three to four species (Holt 1977; Holt and Pickering 1985; Holt et al. 2003; see also Lafferty, chapter 9, and Holt, chapter 15, this volume). Here we consider a few of these modules to illustrate this point. Second, many similar patterns emerge from a macroscopic look at these interactions in both disease and community ecology (Gaston and Blackburn 2000; Guégan et al. 2005). These similar patterns include diversity-latitude and diversity-area relationships and local versus regional diversity curves.

QUALITATIVE PARALLELS SEEN THROUGH THE MODULE APPROACH.
The module approach emphasizes how similar subsets of interactions arise in both predator-prey and parasite-host dynamics. The simplest module involves a binary interaction in which a consumer species, be it a

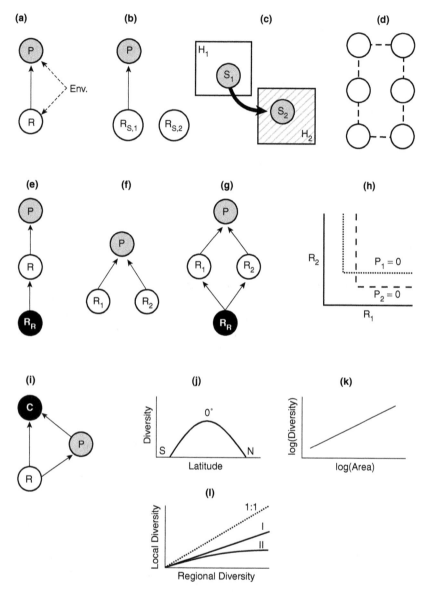

Figure 10.1 Parallel interaction geometries, life cycle complexities, spatial dependencies, and macroecological patterns between parasite-host and predator-prey interactions. (a) Environment (Env.)-dependent binary predator/parasite (*P*)-resource (*R*) interactions which can incorporate (b) differences in vulnerability of *R* to *P* with changes in stage, *S* (e.g., from S_1 to S_2), (c) stage-dependent shifts between habitats, *H*, (d) movements in space (where circles represent local populations of *R* and *P* linked by dispersal via dotted lines), and (e) cascading

predator or parasite, can potentially regulate the abundance and dynamics of its resource species (figure 10.1a). There are many examples of consumers exerting strong controls on their resources in community ecology (Morin 1999; Murdoch et al. 2003). Classic examples include predation on large zooplankton by size-selective fishes (Brooks and Dodson 1965), and the elimination of small-bodied minnows by piscivorous fishes (Mittelbach et al. 1995). Parasites can also regulate their hosts, as shown in the lab (Scott 1987; Scott and Anderson 1984) and in the field (grouse: Dobson and Hudson 1992; Hudson et al. 1992, 1998; old field grasses: Mitchell 2003; urchins: Lafferty 2004). Predators and parasites can also destabilize consumer-resource interactions (e.g., *Daphnia*-algae oscillations: McCauley et al. 1999; nematode-grouse oscillations: Hudson et al. 1998) and modify the behavior of their resource (Abrams 2004; Abrams et al. 1996; Moore 2002; Poulin 1994; Werner and Peacor 2003). Furthermore, both types of consumer-resource interactions can be strongly influenced, even synchronized, by environmental variability (Earn et al. 1998; Pascual et al. 2000; Post and Forschhammer 2002; Ranta et al. 1997; Rodó et al. 2002; Rohani et al. 1999; see also chapter 3, this volume).

These parallels suggest that two-species parasite-host and predator-prey interactions could be captured with a common, low-dimensional model (see figure 10.1a). Assuming continuous reproduction, this model might take the form

$$dP/dt = f(P, R, E) \qquad (1.a)$$
$$dR/dt = g(P, R, E), \qquad (1.b)$$

where P is predator density—or a surrogate for parasite density, using density of infected hosts to track parasites; R is resource density; and E represents environmental variability. Functions f and g capture the density-dependent and/or density-independent interplay between P, R, and E, where traits of P and R themselves may also depend on the other state variables. With work, one could specify general functional

---

indirect effects upon the resource's own resource, R_R. (f) Multiple resources, R_1 and R_2, interact with the consumer through apparent competition or (g) through a diamond arrangement. (h) Interactions among multiple consumers and resources may depend on resource ratios and stoichiometry, where nullclines of the classic resource-ratio model are shown. (i) Predators at higher trophic levels, C, may eat both P and R. Several macroecological parallels include (j) diversity peaks near the equator, (k) area-diversity relationships, and (l) proportional (type I) or saturating (type II) relationships between regional and local species richness, where the dotted line represents a 1:1 comparison.

forms of f and g that more specifically characterize this interplay. Then, to this specified model, one could also add several phenomena common to both host-parasite and predator-prey interactions, including (1) stage-specific vulnerability of the resource (e.g., size refuges: Chase 1999; recovery/immunity of infected hosts: Anderson and May 1979; figure 10.1b); (2) stage-specific (ontogenetic) switches in habitat use for consumer or resource (where multiple host species provide different habitats for various stages of a parasite's life cycle: Werner and Gilliam 1984; figure 10.1c); (3) spatially segregated populations linked by dispersal (metapopulations), where parasites disperse on at least two scales (from host to host and among host populations: Grenfell and Harwood 1997; Grenfell et al. 2001; Guégan et al. 2005; Rohani et al. 1999; figure 10.1d); and (4) "cascading" indirect effects on the prey's own resource (Carpenter and Kitchell 1993; Lafferty 2004; Power et al. 1985; figure 10.1e). Finally, such a model provides a foundation on which to add interactions of other species with both consumer and resource (see Lafferty, chapter 9, and Holt, chapter 15, this volume).

This flexibility to incorporate other species becomes particularly important for two obvious, parallel classes of interactions. The first is apparent competition (Holt 1977; Holt and Lawton 1994; Holt and Pickering 1985), in which two otherwise noninteracting resource species share a common consumer (figure 10.1f). This consumer can degrade diversity of resource species by reducing the abundance of or eliminating its competitors. In situations in which resources compete interspecifically (Grover 1997; Holt et al. 1994; Leibold 1996), these consumers can also preserve diversity by slowing or eliminating competitive exclusion (figure 10.1g; see Clay et al., chapter 7, this volume). Examples of these consumer-mediated interactions certainly occur in host-parasite systems (Holt et al. 2003; Hudson and Greenman 1998; Jaenike 1995; Tompkins et al. 2003; Yan et al. 1998) and are sometimes called "spillover" (Power and Mitchell 2004). In this case, the basic model expands to include a second resource species:

$$dP/dt = f(P, R_1, R_2, E) \qquad (2.a)$$
$$dR_1/dt = g_1(P, R_1, R_2, E) \qquad (2.b)$$
$$dR_2/dt = g_2(P, R_1, R_2, E), \qquad (2.c)$$

where the interaction between resource species R_1 and R_2 could be mediated indirectly through a shared resource of their own. With this type of model structure, one can also flip the question around: How do multiple resources, be they two separate species of hosts/prey resource

(Grover 1995; Holt et al. 2003; Begon, chapter 1, this volume) or multiple nutrients provided within the tissues of a host/prey resource (Smith and Holt 1996; Sterner and Elser 2002), influence the ability of the parasite or predator to persist? In the separate-resource case, different resource species may vary in their vulnerability to the consumer, may interfere with capture of more profitable resources, or may be completely invulnerable (Grover 1995, 1997; Holt et al. 2003; Schmidt and Ostfeld 2001; Begon, chapter 1, this volume). In this multiple nutrient–one resource species case (figure 10.1h), resources may stoichiometrically constrain their consumers (Sterner and Elser 2002), either incidentally (e.g., variable nutrient storage by plants; Hall 2004) or as part of active defense from parasites (e.g., iron withholding by hosts; Smith and Holt 1996). Amazingly, this interface between parasites, hosts, and resources of the host remains largely unexplored, despite the tremendous attention that analogous systems have garnered in community ecology.

A second type of interaction incorporates a top consumer predator that feeds on both a consumer and a resource (figure 10.1i). One label for this type of interaction is intraguild predation, a type of omnivory (Holt and Polis 1997; Polis et al. 1989; Holt, chapter 15, this volume). Notably, models of this interaction predict that the top predator (the "intraguild predator") can eliminate the consumer ("intraguild prey"), particularly at high productivity (Holt and Polis 1997). A similar interaction occurs in host-parasite systems when predators, acting as the intraguild predator, preferentially cull infected hosts (where the parasite is the intraguild prey; Ostfeld and Holt 2004; Packer et al. 2003; Holt, chapter 15, this volume). Examples include predation upon grouse infected with nematodes (Hudson et al. 1992) and predation on infected zooplankton by fishes (Duffy et al. 2005). These selective predators can potentially reduce or eliminate disease within an ecosystem, possibly through dramatic extinctions of the parasite (Duffy et al. 2005; Hall et al. 2005; Packer et al. 2003; see also Holt, chapter 15, this volume). However, it is also possible that parasites require predators to eat infected hosts in order to continue their life-cycle (Lafferty 1999). One could readily capture both scenarios by adding a top predator to the base consumer-resource model (equation 1) to create a third major food web module with blurred trophic relationships:

$$dC/dt = h(P, R, C, E) \qquad (3.\text{a})$$
$$dP/dt = f(P, R, C, E) \qquad (3.\text{b})$$
$$dR/dt = g(P, R, C, E), \qquad (3.\text{c})$$

where now the density of a top consumer (C) can shape interactions between predator and prey.

Evidence from macroecology supports the hypothesis that host-parasite and predator-prey interactions are essentially similar. Macroecology (Brown 1995; Gaston and Blackburn 2000) has yielded several interrelated, large-scale patterns that apply to both sets of interactions. These patterns include relationships between latitude and diversity, area sampled and diversity, and diversity at regional versus local scales. Each of these patterns ultimately depends on how finite living material supported in a region becomes partitioned among individuals and species. First, ecologists have documented examples of diversity peaks of parasites and predators at equatorial regions (figure 10.1j; Calvete et al. 2003; Gaston and Blackburn 2000; Guernier et al. 2004). Second, species-area relationships predict a positive but decelerating accrual of species as area of habitat increases (figure 10.1k; Gaston and Blackburn 2000; Rosenzweig 1995). Although commonly observed for a variety of species (e.g., birds: Reed 1981; Wright 1981), these relationships hold for parasites on two relevant scales (Guégan et al. 2005). The morphology of a host species provides the first scale (Guéguan and Hugueny 1994; Guéguan et al. 1992). Here, richness of parasites increases with increases in body size of host species. Habitat use by host species (Brändle and Brandl 2003; Calvete et al. 2004; Goüy de Bellocq et al. 2003; Marcogliese and Cone 1991) provides a second relevant scale. As true for the area-diversity relationships for predators (Gaston and Blackburn 2000; Rosenzweig 1995), the mechanisms driving these patterns for parasites remain a subject of debate. Viable hypotheses for both consumer types include mechanisms by which predators and parasites partition habitat and resources (Gaston and Blackburn 2000; Guégan et al. 2005).

Third, some local-to-regional diversity patterns imply a role for species interactions in both predator-prey and parasite-host systems. The diversity of species at a local site commonly scales with diversity in the region containing the site (Cornell and Lawton 1992; Gaston and Blackburn 2000; Srivistava 1999). Diversity at a local site can either increase proportionately with regional diversity (type I), or it can saturate as regional diversity becomes more rich (type II; figure 10.1l). All else being equal among locations, type I relationships typically arise if species do not interact, while type II patterns provide a signature of strong interactions at high levels of regional diversity. Most assemblages of predators and parasites show proportional increases (free-living examples: Shurin et al. 2000; Srivistava 1999; parasite examples: Calvete et al. 2003; Guégan and Kennedy 1993). This result has important implications for the management of invasive species because it suggests that local communities of newly established pest species remain undersaturated with parasites (Elton 1958; Keane and Crawley 2002; Mitchell and Power 2003; Torchin et al. 2003; see also, Perkins et al., chapter 8, and Lafferty, chapter 9, this volume). However, saturating patterns do arise in both predator and parasite systems (Calvete et al. 2004; Kennedy and Guégan 1994; Shurin 2000). If coupled with supplementary, mechanistic evidence, these patterns could signal that strong interactions may constrain membership in certain parasite and predator communities.

THE DIFFERENCES HINGE ON SCALING, AND THE CHALLENGE

Suppose that we accept the argument suggested by these parallel community and macroecological patterns, namely, the parasite-host interactions are qualitatively similar to predator-prey interactions. Perhaps, then, the main quantitative difference among them revolves around scale in general and body size in particular. The ratios in body size between consumer and resources differ over sixteen orders of magnitude (Lafferty and Kuris 2002; figure 10.2). In general, parasites are much smaller than their hosts, parasitoids and social predators are relatively similar in size to their resources, and traditional predators are larger than their prey (figure 10.2). Given these huge differences in size, small parasites and large predators perceive their environments and their resources at very different grain sizes (Gaston and Blackburn 2000). Perhaps more important these differences in size can drive wide variation in the dynamics of consumer-resource interactions, particularly in terms of the size of oscillations (Yodzis and Innes 1992). They can have large implications for the speed of turnover and recycling of materials and energy

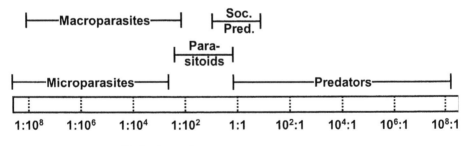

Figure 10.2. Body-size scaling in predator-prey and parasite-host interactions, drawn conceptually. Consumer-resource interactions, including those between host and parasite, scale over many orders of magnitude in relative body size between consumer and resource. Parasites tend to be smaller than their resources, parasitoids and social predators are roughly similarly sized, while predators are typically larger then their prey. (Modified from Lafferty and Kuris 2002.)

(Cyr and Pace 1993). For instance, whether they constitute small amounts of biomass in ecosystems to amounts comparable to that occupied by top predators, parasites can vastly alter biomass and nutrient flow (Polis and Strong 1996; Mitchell 2003; Sukdheo and Hernandez 2005; chapters 11 and 12, this volume).

Nonetheless, if parasite-host and predator-prey interactions mainly vary along gradients of body size, we can potentially take advantage of scaling relationships to create a unifying, overarching consumer-resource model. Several important relationships have recently emerged as functions of body size. Across many taxa, body size scales as a one-quarter or three-quarter power with basal metabolic rate, development time, and maximal population growth rates (Brown et al. 2002). Ideally, these scaling relationships can be used to predict ecological phenomena such as population density and growth rates of predators and parasites. Recent work with forest stands reveals the potential of this approach (Enquist et al. 1998, 1999). Once one controls for body size, variation in temperature (figure 10.2b) and phosphorus content might also explain variation in metabolic rates among predators and parasites (Gillooly et al. 2001, 2002). However, a comprehensive metabolic theory of ecology remains in its infancy (see Savage et al. 2004 for a recent step forward). Such a theory would connect metabolic processes at the level of the individual parasite or predator to their influence on the rate and magnitude of ecosystem-level fluxes of energy and materials. Until this theory is developed, the potential for metabolic theory to unify predator-prey and parasite-host interactions will remain unrealized.

With or without such a comprehensive metabolic theory, the ultimate challenge posed by this position ("parasites are just predators") is to write a model that captures the essence of both types of consumer-resource interactions. In other words, one must specify equation (1) to some useful degree of detail. Because size and metabolism closely associate, scaling relationships could help to solve the "plague of parameters" problem inherent in such an ambitious endeavor (Yodzis and Innes 1992). But is it possible to write such a model? If so, would the resulting structure necessarily remain so general as to offer little practical utility?

Position 2: "Predators and Parasites Differ Fundamentally"

Some might argue that well-intentioned efforts to compose such a "unifying model of consumer-resource interactions" will ultimately fail. After all, they might say, host-parasite interactions qualitatively differ among themselves (parasite, parasitoid, etc.) and from predator-prey interactions in many ways. The danger of this line of argument lies in its potential overemphasis on differences rather similarities. Are consumer-resource interactions just collections of many strategies to acquire resources? The challenge for proponents of the "predators and parasites differ fundamentally" argument is to find an intellectual middle ground. This middle ground should hinge on simple but meaningful categorization of these differences using only a few axes of separation.

A recent analysis, based on the concept of adaptive peaks, provides a very useful starting point for such an endeavor. Lafferty and Kuris (2002) outline key differences between "predators" and "parasites" and among classes of parasites (see figure 10.2). A first dichotomy immediately splits predators from parasites: how many resource victims are attacked in a life stage? Predators attack more than one individual, while parasites attack only one victim per life stage. A second axis separates parasitoids and parasitic castrators from other types of parasites: does the consumer reduce the fitness of its resource to zero? The former classes do reduce it to zero, while typical parasites do not. A third difference splits the two main categories of parasites into four. Some parasites (microparasites, macroparasites, and parasitic castrators) do not necessarily have to kill their hosts, while others (trophically transmitted parasites [Lafferty 1999] and parasitoids) require the host to die to compete their life cycle. Finally, a fourth axis can further separate some of the groups: does the effect of attacks by consumers depend on intensity (i.e., the number or density of parasites per host)? Lafferty and Kuris (2002) use this fourth axis to split the six groups forged by the first three dichotomies into ten useful ones; these ten are presently occupied by extant consumer-resource types. Here we use it to split these six into the eight classes, which are likely obvious to most readers. First, as pointed out years ago (Anderson and May 1979; May and Anderson 1979), intensity dependence cleaves macroparasitism from microparsitism because intensity and aggregation of infection by the former (but not the latter) determines pathology. This split also meaningfully divides predators into solitary and social predators. Social predators rely on other members of a group to attack relatively large-bodied resources (figure 10.3). As a result of this strategy, they can suffer Allee effects (inverse density dependence) at small population sizes (Case 1999).

While intuitively appealing, this classification scheme does not fully resolve the overarching question of this essay. For instance, it is unclear whether these eight (presented here) or ten categories (presented in Lafferty and Kuris 2002) include the essential—but not superfluous—axes on which one can array predators and parasites. Although distinct mathematical models have been written for most of these categories, do we really need eight to ten different models to characterize the array of predator-prey and parasite-host interactions? A safe answer to broad, difficult questions like this one is "it depends": the answer likely depends

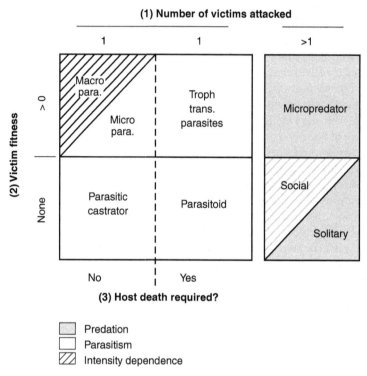

Figure 10.3. A scheme developed to qualitatively distinguish predation and parasitism (modified from Lafferty and Kuris 2002). The number of victims attacked in a lifetime (1) splits predation (right) from parasitism (left), while effects on victim fitness (2) split parasitoids and parasitic castrators (bottom) from macroparasites and trophically transferred parasites (top). This axis also splits micropredators from more typical predators. Among parasites, obligate host mortality (3) further differentiates parasitoids from parasitic castrators (bottom) and more typical parasites from trophically transmitted parasites (i.e., those requiring that their host be eaten by a predator to complete a life cycle). Finally, each of the six boxes created by 1–3 could be split again along an intensity-dependence (upper hatched triangle) or intensity independence (lower triangle) axis. Here the split is shown only to differentiate macroparasites from microparasites and social from solitary (typical) predators.

on the question examined and the currency used to address it. For instance, these characterizations outlined by Lafferty and Kuris (2002) may represent distinct, size-dependent (figure 10.2a) evolutionary strategies, or adaptive peaks, along a continuum of consumer-resource strategies. Such a hypothesis could be explored by studying trajectories of evolutionary models that use size-based scaling laws or by empirically observing consumers that are forced to switch feeding strategies. Furthermore, this classification scheme was developed based on interactions of individuals. From the perspective of individuals, these differences in behavior of consumers may matter greatly for individual fitness, and therefore they may greatly influence (co)evolutionary dynamics. However, these strategies may or may not yield different outcomes for dynamics and stability of populations or energy and material flows through food webs in ecosystems. Furthermore, they may or may not yield different outcomes for diversity maintenance or degradation when inserted into common interaction modules, such as apparent competition and intraguild predation. Definitive answers to these questions, then, likely first require specification of a currency (R_0, dynamics of populations, flow rate of energy, rate of evolutionary change, etc.) followed by systematic assessment of each of the eight to ten consumer-resource models. Perhaps some of Lafferty and Kuris's (2002) consumer-resource types can be collapsed into a subset for certain questions or currencies but not for others. Such an ambitious task presents a potentially rewarding challenge for theoretical population biologists.

Discussion

This essay approaches a difficult but fundamental question: Are parasites qualitatively different from predators? Such a pursuit presents a challenge, and we consider this essay only a start toward answering this question. In many ways, this thought experiment prompted more questions than it answered, a result that largely reflects the current stage reached in the evolution of disease ecology as a discipline. The first argument ("parasites are just predators"), even if ultimately wrong, does stress the tremendous similarities between disease ecology and more traditional community and macroecology. For whatever reason, these epidemiological and mainstream ecological subdisciplines have largely evolved separately (at least until fairly recently). This separation existed despite the fact that these subdisciplines fundamentally face very similar issues (such as introduction of exotic species, enemy release, and maintenance or degradation of diversity) even when considered with different aims in mind (eradication of parasites vs. promotion or erosion of diversity by predators). Solutions to these problems might benefit greatly from a healthy dose of conceptual cross-fertilization.

The second argument, following Lafferty and Kuris's (2002) categorization, provides a pathway to pursue a "predators and parasites differ fundamentally" position. If one posits that parasites fundamentally differ from predators, some questions still remain, however. Just how many consumer-resource types are required to satisfyingly capture the range of interactions along the parasitism-predation gradient? For some particular problems, such as nutrient cycling, one can probably combine some of Lafferty and Kuris's (2002) categories into a smaller subset, perhaps through application of size-energetic scaling relationships. After all, relative body size differences among consumers and resources appear to determine which of the eight to ten life history strategies is most profitable for consumers (see figures 10.2 and 10.3; Lafferty and Kuris 2002). Given that logic, the argument posed by the apparent dichotomy in this essay ultimately swings full circle. Perhaps, then, the ultimate answer lies somewhere between the two positions. Or perhaps the strategies used by predators and parasites (Lafferty and Kuris 2002) would emerge as stable strategies produced by the same evolutionarily explicit consumer-resource model. In this case, the two positions juxtaposed here become one and the same. We hope that a reader pursues this challenge.

ACKNOWLEDGMENTS

Research was funded in part by the National Science Foundation (OCE 02-35039) and the Institute of Ecosystem Studies. This is a contribution to the program of the Institute of Ecosystem Studies and is contribution 1185 from the W. K. Kellogg Biological Station.

LITERATURE CITED

Abrams, P. A. 2004. Trait-initiated indirect effects due to changes in consumption rates in simple food webs. Ecology 85:1029–38.
Abrams, P. A., B. A. Menge, G. G. Mittelbach, D. A. Spiller, and P. Yodzis. 1996. The role of indirect effects in food webs. In Food Webs: Integration of Patterns and Dynamics, ed. G. A. Polis and K. O. Winemiller, 371–95. New York: Chapman and Hall.
Anderson, R. M., and R. M. May. 1979. Population biology of infectious diseases. Part I. Nature 280:361–67.
Brändle, M., and R. Brandl. 2003. Species richness on trees: A comparison of parasitic fungi and insects. Evolutionary Ecology Research 5:941–52.
Brooks, J. L., and S. I. Dodson. 1965. Predation, body size, and composition of plankton. Science 150:28–35.
Brown, J. H. 1995. Macroecology. Chicago: University of Chicago Press.
Brown, J. H., V. H. Gupta, B.-L. Li, B. T. Milne, C. Restrepo, and G. B. West. 2002. The fractal nature of nature: Power laws, ecological complexity and biodiversity. Proceedings of the Royal Society of London. Series B, Biological Sciences 357:619–26.
Calvete, C., J. A. Blanco-Aguiar, E. Virgos, S. Cabezas-Diaz, and R. Villafuerte. 2004. Spatial variation in helminth community structure in the red-

legged partridge (*Alectoris rufa* L.): Effects of definitive host density. Parasitology 129:101–13.

Calvete, C., R. Estrada, J. Lucientes, A. Estrada, and I. Telletxea. 2003. Correlates of helminth community structure in the red-legged partridge (*Alectoris rufa* L.) in Spain. Journal of Parasitology 89:445–51.

Carpenter, S. R., and J. F. Kitchell. (Eds.). 1993. The Trophic Cascade in Lakes. New York: Cambridge University Press.

Case, T. J. 1999. An Illustrated Guide to Theoretical Ecology. New York: Oxford University Press.

Chase, J. M. 1999. Food web effects of prey size-refugia: Variable interactions and alternative stable equilibria. American Naturalist 154:559–70.

Chase, J. M., M. A. Leibold, and E. Simms. 2000. Plant tolerance and resistance in food webs: Community-level predictions and evolutionary implications. Evolutionary Ecology 14:289–314.

Cohen, J. E. 1978. Food Webs and Niche Space. Princeton, NJ: Princeton University Press.

Cornell, H. V., and J. H. Lawton. 1992. Species interactions, local and regional processes, and limits to the richness of ecological communities: A theoretical perspective. Journal of Animal Ecology 61:1–12.

Cyr, H., and M. L. Pace. 1993. Magnitude and patterns of herbivory in aquatic and terrestrial ecosystems. Nature 361:148–50.

Dobson, A. P., and P. J. Hudson. 1992. Regulation and stability of a free-living host-parasite system—*Trichostrongylus tenuis* in red grouse. II. Population models. Journal of Animal Ecology 61:487–98.

Duffy, M. A., S. R. Hall, A. J. Tessier, and M. Huebner. 2005. Selective predators and their parasitized prey: Top-down control of epidemics. Limnology and Oceanography 50:412–20.

Earn, D. J. D., P. Rohani, and B. T. Grenfell. 1998. Persistence, chaos, and synchrony in ecology and epidemiology. Proceedings of the Royal Society of London. Series B, Biological Sciences 265:7–10.

Elton, C. S. 1958. The ecology of invasions by animals and plants. London: Chapman and Hall.

Enquist, B. J., J. H. Brown, and G. B. West. 1998. Allometric scaling of plant energetics and population density. Nature 395:163–65.

Enquist, B. J., G. B. West, E. L. Charnov, and J. H. Brown. 1999. Allometric scaling of production and life-history variation in vascular plants. Nature 401:907–11.

Gaston, K. J., and T. M. Blackburn. 2000. Patterns and Processes in Macroecology. Oxford: Blackwell Science.

Gillooly, J. F., J. H. Brown, G. B. West, V. M. Savage, and E. L. Charnov. 2001. Effects of size and temperature on metabolic rate. Science 293:2248–51.

Gillooly, J. F., E. L. Charnov, G. B. West, V. M. Savage, and J. H. Brown. 2002. Effects of size and temperature on development time. Nature 417:70–73.

Grenfell, B. T., and J. Harwood. 1997. (Meta)population dynamics of infectious diseases. Trends in Ecology & Evolution 12:395–99.

Grenfell, B. T., O. N. Bjørnstad, and J. Kappey. 2001. Traveling waves and spatial hierarchies in measles populations. Nature 414:716–23.

Grover, J. P. 1995. Competition, herbivory, and enrichment: nutrient-based models for edible and inedible plants. American Naturalist 145:746–74.

Grover, J. P. 1997. Resource Competition. London: Chapman and Hall.

Guégan, J.-F., and B. Hugueny. 1994. A nested parasite species subset pattern in tropical fish: Host as major determinant of parasite infracommunity structure. Oecologia 100:184–189

Guégan, J.-F., and C. R. Kennedy. 1993. Maximum helminth parasite community richness in British freshwater fish: A test of the colonization time hypothesis. Parasitology 106:91–100.

Guégan, J.-F., A. Lambert, C. Lévêque, C. Combes, and L. Euzet. 1992. Can host body size explain the parasite species richness in tropical freshwater fishes? Oecologia 90:197–204.

Guégan, J.-F., S. Morand, and R. Poulin. 2005. Are there general laws in parasite community ecology? The emergence of spatial parasitology and epidemiology. In Parasitism and Ecosystems, ed. F. Thomas, F. Renaud, and J.-F. Guégan, 22–42. Oxford: Oxford University Press.

Guernier, V., M. E. Hochberg, and J.-F. Guégan. 2004. Ecology drives worldwide distribution of human infectious diseases. PloS Biology 2: 740–46.

Goüy de Bellocq, J., M. Sara, J. C. Casanova, C. Feliu, and S. Morand. 2003. A comparison of the structure of helminth communities in the woodmouse, Apodemus sylvaticus, on islands of the western Mediterranean and continental Europe. Parasitology Research 90:64–70.

Hall, S. R. 2004. Stoichiometrically-explicit competition between grazers: Species replacement, coexistence, and priority effects along resource supply gradients. American Naturalist 164:157–72.

Hall, S. R., M. A. Duffy, and C. E. Cáceres. 2005. Selective predation and productivity jointly drive complex behavior in host-parasite systems. American Naturalist 180:70–81.

Holt, R. D. 1977. Predation, apparent competition, and the structure of prey communities. Theoretical Population Biology 12:197–229.

Holt, R. D., A. P. Dobson, M. Begon, R. G. Bowers, and E. M. Schauber. 2003. Parasite establishment in host communities. Ecology Letters 6:837–42.

Holt, R. D., J. Grover, and D. Tilman. 1994. Simple rules for interspecific dominance in systems with exploitative and apparent competition. American Naturalist 144:741–71.

Holt, R. D., and J. H. Lawton. 1994. The ecological consequences of shared natural enemies. Annual Review of Ecology and Systematics 25:495–520.

Holt, R. D., and J. Pickering. 1985. Infectious disease and species coexistence: A model of Lotka-Volterra form. American Naturalist 126:196–211.

Holt, R. D., and G. A. Polis. 1997. A theoretical framework for intraguild predation. American Naturalist. 149:745–64.

Hudson, P. J., A. P. Dobson, and D. Newborn. 1998. Prevention of population cycles by parasite removal. Science 282:2256–58.

Hudson, P., and J. Greenman. 1998. Competition mediated by parasites: Biological and theoretical progress. Trends in Ecology & Evolution 13:387–90.

Hudson, P. J., D. Newborn, and A. P. Dobson. 1992. Do parasites make prey vulnerable to predation? Red grouse and parasites. Journal of Animal Ecology 61:681–92.

Jaenike, J. 1995. Interactions between mycophagous *Drosophila* and their nematode parasites: from physiological to community ecology. Oikos 72:155–60.

Keane, R. M., and M. J. Crawley. 2002. Exotic plant invasions and the enemy release hypothesis. Trends in Ecology & Evolution 17:164–170.

Kennedy, G. R., and J.-F. Guégan. 1994. Regional versus local helminth parasite richness in British freshwater fish: Saturated or unsaturated parasite communities? Parasitology 109:175–85.

Lafferty, K. D. 1999. The evolution of trophic transmission. Parasitology Today 15:111–15.

Lafferty, K. D. 2004. Fishing for lobsters indirectly increases epidemics in sea urchins. Ecological Applications 14:1566–73.

Lafferty, K. D., and A. M. Kuris. 2002. Trophic strategies, animal diversity, and body size. Trends in Ecology & Evolution. 17:507–13.

Lawton, J. H., and P. H. Warren. 1988. Static and dynamic explanations for patterns in food webs. Trends in Ecology & Evolution 3:242–45.

Leibold, M. A. 1996. A graphical model of keystone predation: Effects of productivity on abundance, incidence, and ecological diversity in communities. American Naturalist 147:784–812.

Marcogliese, D. J., and D. K. Cone. 1991. Importance of lake characteristics in structuring parasite communities of salmonids from insular Newfoundland. Canadian Journal of Zoology 69:2962–67.

May, R. M., and R. M. Anderson. 1979. Population biology of infectious diseases. Part II. Nature 280:455–61.

McCauley, E., R. M. Nisbet, W. W. Murdoch, A. M. de Roos, and W. S. C. Gurney. 1999. Large-amplitude cycles of *Daphnia* and its algal prey in enriched environments. Nature 402:653–56.

Mitchell, C. E. 2003. Trophic control of grassland production and biomass. Ecology Letters 6:147–155.

Mitchell, C. E., and A. G. Power. 2003. Release of invasive plants from fungal and viral pathogens. Nature 421:625–627.

Mittelbach, G. G. 2005. Parasites, communities, and ecosystems: conclusions and perspectives. *In* Parasitism and Ecosystems, ed. F. Thomas, F. Renaud, and J.-F. Guégan, 171–76. Oxford: Oxford University Press.

Mittelbach, G. G., A. M. Turner, D. J. Hall, J. E Rettig, and C. W. Osenberg. 1995. Perturbation and resilience in an aquatic community: A long-term study of the extinction and reintroduction of a top predator. Ecology 76:2347–60.

Moore, J. 2002. Parasites and the Behavior of Animals. Oxford: Oxford University Press.

Morin, P. J. 1999. Community Ecology. Malden, MA: Blackwell Science.

Murdoch, W. M., C. Briggs, and R. M. Nisbet. 2003. Consumer resource dynamics. Princeton, NJ: Princeton University Press.

Ostfeld, R. S., and R. D. Holt. 2004. Are predators good for your health? Evaluating evidence for top-down regulation of zoonotic disease reservoirs. Frontiers in Ecology and the Environment 2:13–20.

Packer, C., R .D. Holt, P. J. Hudson, K. D. Lafferty, and A. P. Dobson. 2003. Keeping the herds healthy and alert: Implications of predator control for infectious disease. Ecology Letters 6:797–802.

Pascual, M., X. Rodó, S. P. Ellner, R. Colwell, and M. J. Bouma. 2000. Cholera dynamics and El-Niño-Southern Oscillation. Science 289:1766–69.

Polis, G. A., C. A. Myers, and R. D. Holt. 1989. The ecology and evolution of intraguild predation: Potential competitors that eat each other. Annual Review of Ecology and Systematics 20:297–330.

Polis, G. A., and D. R. Strong. 1996. Food web complexity and community dynamics. American Naturalist 147:813–46.

Post, E., and M. C. Forschhammer. 2002. Synchronization of animal population dynamics by large-scale climate. Nature 420:168–71.

Poulin, R. 1994. Meta-analysis of parasite-induced behavioural changes. Animal Behaviour 48:137–46.

Power, M. E., W. J. Mathews, and A. J. Stewart. 1985. Grazing minnows, piscivorous bass, and stream algae: Dynamics of a strong interactions. Ecology 66:1448–56.

Power, A. G., and C. E. Mitchell. 2004. Pathogen spillover in disease epidemics. American Naturalist 164:S79–89.

Raffaelli, D. G., and S. J. Hall. 1996. Assessing the relative importance of trophic links in food webs. In Food Webs: Integration of Patterns and Dynamics, ed. G. A. Polis and K. O. Winemiller, 185–91. New York: Chapman and Hall.

Ranta, E., V. Kaitala, J. Lindström, and E. Helle. 1997. The Moran effect and synchrony in population dynamics. Oikos 78:136–42

Reed, T. 1981. The number of breeding landbird species on British islands. Journal of Animal Ecology 50:613–24.

Rodó, X., M. Pascual, G. Fuchs, and A. S. G. Faruque. 2002. ENSO and cholera: A nonstationarity link related to climate change? Proceedings of the National Academy of Sciences of the United States of America 99:12901–06.

Rohani, P., D. J. Earn, and B. T. Grenfell. 1999. Opposite patterns of synchrony in sympatric disease metapopulations. Science 286:968–71.

Rosenzweig, M. L. 1995. Species Diversity in Space and Time. Cambridge: Cambridge University Press.

Savage, V. M., J. F. Gillooly, J. H. Brown, G. B. West, and E. L. Charnov. 2004. Effects of body size and temperature on population growth. American Naturalist 163:429–41.

Schmidt, K. A., and R. S. Ostfeld. 2001. Biodiversity and the dilution effect in disease ecology. Ecology 82:609–19.

Scott, M. E. 1987. Regulation of mouse colony abundance by Heligomosomoides polygyrus (Nematoda). Parasitology 95:111–29.

Scott, M. E., and R. M. Anderson. 1984. The population dynamics of Gyrodactylus bullatarudis (Monogenea) within laboratory populations of fish host Poecilia reticulata. Parasitology 89:159–94.

Shurin, J. B. 2000. Dispersal limitation, invasion resistance, and the structure of pond zooplankton communities. Ecology 81:3074–86.

Shurin, J. B., J. E. Havel, M. A. Leibold, and B. Pinel-Alloul. 2000. Local and regional zooplankton species richness: a scale-independent test for saturation. Ecology 81:3062–73.

Smith, V. S., and R. D. Holt. 1996. Resource competition and within-disease dynamics. Trends in Ecology & Evolution 11:386–89.

Srivistava, D. S. 1999. Using local-regional richness plots to test for species saturation: Pitfalls and potentials. Journal of Animal Ecology 68:1–16.

Sterner, R. W., and J. J. Elser. 2002. Ecological stoichiometry: The biology of elements from molecules to the biosphere. Princeton, NJ: Princeton University Press.

Sukhdeo, M. V. K., and A. D. Hernandez. 2005. Food web patterns and the parasite's perspective. In Parasitism and Ecosystems, ed. F. Thomas, F. Renaud, and J.-F. Guégan, 54–67. Oxford: Oxford University Press.

Thomas, F., F. Renaud, and J.-F. Guégan. 2005. Parasitism and ecosystems. Oxford University Press, Oxford, UK.

Tompkins, D. M., A. R. White, and M. Boots. 2003. Ecological replacement of native red squirrels by invasive greys driven by disease. Ecology Letters 6:189–96.

Torchin, M. E., K. D. Lafferty, A. P. Dobson, V. J. McKenzie, and A. M. Kuris. 2003. Introduced species and their missing parasites. Nature 421:628–30

Warren, P. H. 1989. Spatial and temporal variation in the structure of a freshwater food web. Oikos 55:299–311.

Werner, E. E., and J. F. Gilliam. 1984. The ontogenetic niche and species interactions in size-structured populations. Annual Review of Ecology and Systematics 15:393–425.

Werner, E. E., and S. D. Peacor. 2003. A review of trait-mediated indirect interactions in ecological communities. Ecology 84:1083–1100.

Wright, S. J. 1981. Intra-archipelago vertebrate distributions: The slope of the species-area relation. American Naturalist 118:726–48.

Yan, G., L. Stevens, C. J. Goodnight, and J. J. Schall. 1998. Effects of a tapeworm parasite on the competition of Tribolium beetles. Ecology 79: 1093–1103.

Yodzis, P. 1996. Food webs and perturbation experiments: Theory and practice. In Food Webs: Integration of Patterns and Dynamics, ed. G. A. Polis and K. O. Winemiller, 211–17. New York: Chapman and Hall.

Yodzis, P., and S. Innes. 1992. Body size and consumer-resource dynamics. American Naturalist 139:1151–75.

CHAPTER ELEVEN
Microbial Disease in the Sea: Effects of Viruses on Carbon and Nutrient Cycling

Mathias Middelboe

SUMMARY

VIRUSES ARE DYNAMIC and integrated components of both aquatic and terrestrial ecosystems, where they are usually present in abundances of 10^6–10^8 viruses per milliliter of water or cm^3 of sediment or soil. Most of these viruses are infectious to bacteria, and they constitute a significant mortality factor for prokaryotes in all ecosystems. In the marine environment, 5%–30% of heterotrophic bacteria and cyanobacteria are infected by viruses, and approximately 5%–40% of the daily bacterial and cyanobacterial production is lost as dissolved organic matter due to viral lysis. Also, eukaryotic protists are significantly influenced by viral infections, and viruses have been shown to cause mass lysis of blooming phytoplankton populations.

In addition to the direct lethal effect of viral lysis, viral activity may significantly affect the cycling of carbon and other nutrients in ecosystems. Lysis of heterotrophic and autotrophic microorganisms by viruses liberates cellular components rich in nitrogen and phosphorus, which become available for uptake by heterotrophic bacteria. This virus-induced transformation of particulate to dissolved organic matter, the so-called viral shunt, stimulates microbial respiration and reduces the transfer of carbon to higher trophic levels and the export of particulate material to the seafloor. As agents of infectious diseases of potentially all microorganisms in the world's ecosystems, viruses therefore have a significant impact on microbial mortality and biogeochemical cycling on a local, regional, and global scale. According to recent models, as much as 25% of the total photosynthetically fixed carbon in the sea is channeled through the viral shunt, either directly by viral lysis of phytoplankton or indirectly by lysis of bacterioplankton. Since about 75% of these viral lysates are rapidly taken up and metabolized by noninfected bacteria, viral lysates obviously contribute significantly to bacterial carbon and nutrient supply. In fact, viral lysis may be a key mechanism in supplying carbon for heterotrophic bacteria, especially in oligotrophic systems, with small direct inputs of organic matter from primary producers.

INTRODUCTION

When disease ecologists get together to discuss models or experiments, inevitably they talk about diseases in plants and animals. Although pathogens that infect plants and animals are surely important, their role in basic ecosystem processes cannot rival the impacts of bacteriophages and viruses that infect unicellular organisms such as bacteria and phytoplankton (figure 11.1). This invisible world of disease ecology is characterized by true viral pathogens that basically follow the same infection patterns as are observed for viruses infecting metazoans and higher plants. These viruses are natural and integrated inhabitants of all ecosystems, and in aquatic systems they are viewed as part of the ecosystem rather than as causes of infectious diseases. In the following, however, I will consider

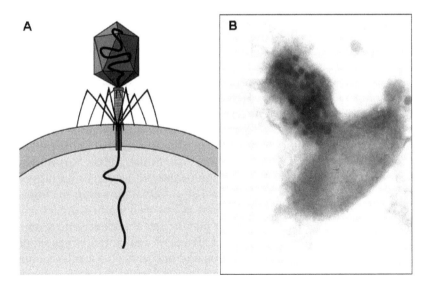

Figure 11.1. (a) Schematic diagram of a lytic virus infecting a bacterium. The virus is attached to receptors in the cell membrane and its DNA is injected into the cell. Inside the cell, the viral DNA takes control of the cell's metabolic machinery and new viruses are produced. (b) Electron micrograph of mature viruses (arrow) that are being released from an infected cell.

naturally occurring viruses that infect prokaryotic or eukaryotic micro-organisms as causes of infectious diseases of these organisms. Such viruses are found everywhere, from the deep sea to the cultivated soil and from sea ice to hot springs. They are the most abundant biological entities on the planet, usually occurring in abundances of $10^6–10^8$ viruses per milliliter of seawater or gram of soil, adding up to approximately 10^{31} viruses on a global scale (Breitbart and Rohwer 2005). Individual virus populations have been found to be distributed across large geographic areas: worldwide distribution of a virus infecting the picophytoplankton *Micromonas pusilla* was reported by Cottrell and Suttle (1991), and a particular virus that was specific for the psychrotolerant *Pseudoalteromonas* sp. was found distributed throughout the North Water Polynya (>50,000 km^2) within a two-week period (Middelboe et al. 2002). These observations indicate that viruses are efficiently spread in the environment.

Following infection, the virus may directly enter a lytic cycle (lytic viruses), where the viral genome takes control of the host cell to produce new viruses and subsequently lyse the cell, releasing the new viruses to the environment (see figure 11.1). Alternatively, the viral genetic material may be incorporated into the host genome (lysogenic viruses), where it may stay in a dormant stage (prophage) for generations until a lytic cycle is induced by some external factor. Only the lytic cycle directly affects the mortality and biogeochemical activity of microorganisms in the sea, and it is therefore the implications of lytic viral activity that are discussed here.

Lytic viral infections constitute an important loss factor for unicellular organisms. All of these organisms are potentially susceptible to viral infection by several different types of viruses, and it is now well established that viruses are responsible for 10%–50% of bacterial mortality in surface waters, and even more in environments that are unfriendly to protist grazers (e.g., low-oxygen waters) (Fuhrman 1999). Viruses have also been proposed to contribute significantly to bacterial mortality in terrestrial ecosystems (e.g., Ashelford et al. 2003); however, the impact of viruses on terrestrial environments is less well studied. Apart from having a direct regulatory influence on host cell abundance and production, there are a number of indirect effects of viral activity that significantly influence the cycling of carbon and nutrients on a local, regional, and global scale. Investigating the ecology of viruses is therefore fundamental for our ability to understand and model global nutrient cycling and microbial interactions in different ecosystems. Most of the research on these aspects of viral activity has so far been carried out in aquatic ecosystems, and the following review of the role of viruses in ecosystem functioning is primarily based on data from these systems.

Viruses in Marine Food Webs

Marine viruses have been an important issue in marine microbial research since the discovery of high oceanic viral abundances ($>10^7$ ml^{-1}; Berg et al. 1989; Proctor and Fuhrman 1990). It soon became apparent that viruses were extremely dynamic and potentially were important players everywhere in the marine environment, with a significant influence on microbial mortality, population dynamics, diversity, and nutrient fluxes. Research on marine viruses has fundamentally changed our conceptual understanding of marine food webs and element cycling (figure 11.2).

As regulators of bacterial and algal abundance and activity, viruses potentially play a key role for the regulation of microbial processes in the ocean. In plankton systems, 5–30% of heterotrophic bacteria and cyanobacteria are infected by viruses, which cause lysis of a substantial fraction of daily bacterial and cyanobacterial production (Fuhrman 1999; Wilhelm and Suttle 1999; Wommack and Colwell 2000). Recently, viruses have been found to be extremely abundant in aquatic sediments, with densities of greater than 10^8 viruses cm^{-3}. As for the water

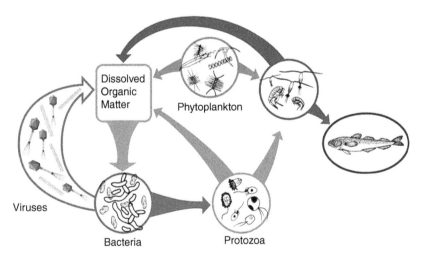

Figure 11.2. Conceptual model of the carbon flow between trophic levels in the marine food web. Part of the phytoplankton production is transported up the grazing food chain via zooplankton and fishes. From all levels in the food web, organic matter is lost as dissolved organic matter (DOM), which can be utilized by bacteria thereby forming the base of the microbial food web. By viral infection and cell lysis, part of the bacterial carbon is returned to the DOM pool, where it is again taken up by noninfected bacteria.

column, benthic viruses are dynamic components of the microbial processes in the seafloor, with population turnover times of a few hours to a few days and a close coupling to diagenetic activity.

In addition to the direct effects of viruses on microbial mortality, there are a number of side effects of virus-induced mortality that have significant impact on the dynamics and activities of marine microbial communities. Infection by a particular virus acts not on the total bacterial assemblage but rather on individual populations. By selective suppression of specific subpopulations, viruses may influence the microbial diversity and population dynamics by lysing numerically dominant strains or species (Thingstad and Lignell 1997; Wommack et al. 1999). This phenomenon has been described in the "killing the winner" theory and proposed as an important mechanism for maintaining a high diversity of bacteria and algae by preventing particular microorganisms from becoming dominant at specific growth conditions. Recent work has shown that resistant clones of bacterial and algal populations attacked by viruses are replacing the sensitive clones that are destroyed by viruses. Consequently, the influence of selective viral mortality acts not only at the population level but also at the clonal level within individual populations. So, although viruses do indeed have large effects on microbial population dynamics, the influence seems to be more complex than was previously thought (Riemann and Middelboe 2002).

The lytic cycle of infectious viruses ends by disruption of the infected cells and release of the new viruses and cell contents to the environment. These viral lysates are potentially labile compounds with an optimal nitrogen and phosphorus composition for bacteria. Consequently, viral activity converts particulate organic matter into a dissolved form, which is again available for bacterial uptake (figures 11.2 and 11.3). Viral lysis therefore affects the cycling of carbon and nutrients by increasing the regeneration of dissolved organic matter within the microbial food web. The rest of this chapter provides an update of present knowledge about the role of virus infection of bacteria and eukaryotic cells in microbial biogeochemical cycling.

BACTERIOPHAGES AND MARINE BIOGEOCHEMICAL CYCLING

Bacterial viruses, the bacteriophages, are presumed to be the dominant type of viruses in the oceans (Fuhrman 1999). As mentioned above, viral lysis of bacteria liberates cellular components rich in nitrogen and phosphorus, which become available for other, noninfected bacteria (e.g., Fuhrman 1999; Riemann and Middelboe 2002). Consequently, viral activity speeds up the recycling rate of carbon and nutrients within

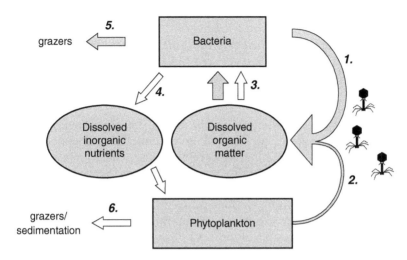

Figure 11.3. Diagram showing the impact of viruses on carbon and nutrient cycling in marine environments. Gray and white arrows illustrate virus-mediated fluxes of carbon and inorganic nutrients, respectively. (1) 5%–40 % of bacterial production and (2) 2%–10% of primary production is lost as viral lysates. (3) Viral lysates in general contribute 10%–40% of bacterial carbon supply. In oligotrophic environments, the contribution can be even higher and viral lysates may in some cases be a key mechanism for maintaining bacterial production. (4) Owing to the low C:N:P ratio of viral lysates compared with bulk DOM, bacterial utilization of viral lysates is expected to cause a net release of inorganic nutrients. (5) The export of bacterial carbon to protozoan grazers is affected by viral activity. The net effect of viral lysis of bacteria is a loss of bacterial carbon as DOM (arrow 1) and therefore a reduced output of bacterial biomass available for grazing from a given input of substrate. (6) Apart from the direct negative effect of viral lysis on primary producers (arrow 2), remineralization of inorganic nutrients as a result of viral activity (arrow 4) stimulates primary production. However, the quantitative significance of this mechanism is not known in detail.

the microbial community. Compared with a theoretical situation without viruses, the presence of viruses, therefore, stimulates bacterial carbon uptake, since a larger fraction of the organic carbon is supplied to the dissolved fraction by viral lysis. As there is a respiratory loss of carbon each time the organic carbon is metabolized by a bacterium, the end result of this organic matter recycling is the production of CO_2 and inorganic nutrients, as well as some residual fraction of refractory organic compounds. Inclusion of viruses in theoretical food web models suggested that if viral lysis of bacteria is significant and if lysates are consumed by other bacteria, then bacterial growth and respiration would be enhanced at the expense of biomass at higher trophic levels (Fuhrman 1992). In fact, the model implied that viral activity reduced the export of bacterial carbon by 37% but at the same time enhanced bacterial carbon consumption by 27%, compared to a theoretical food web without viruses (Fuhrman 1992). These theoretical considerations strongly suggested that viral lysis potentially had a significant influence on pelagic carbon cycling.

However, the quantitative significance of bacteriophage infections for biogeochemical cycling is based on a number of assumptions. First, to significantly influence marine carbon cycling, bacteriophages should constitute a significant mortality factor for bacteria. Second, viral lysates should be available for bacterial uptake, and finally, the lysates should contribute significantly as a carbon source for the heterotrophic bacteria. In addition to the stimulation of bacterial carbon cycling directly by production of lysates, the role of viruses in biogeochemical cycling also depends on the extent to which bacterial mineralization of nitrogen and phosphorus (and iron) contained in the lysates contributes to supporting phytoplankton primary production. These premises are discussed in the next section.

Bacteriophage-Induced Bacterial Mortality

Bacterial hosts are very abundant in marine ecosystems (10^5–10^6 cells per milliliter of sea water[1] and 10^7–10^9 cells per cm^3 sediment[1]), and with bacteriophages responsible for 10%–50% of bacterial mortality in both seawater (e.g., Furhman and Noble 1995) and sediments (e.g., Glud and Middelboe 2004), there is no doubt that bacteriophages in general constitute a significant controlling factor for bacterial abundance and activity. Studies comparing the impact of bacteriophage-induced bacterial mortality with other causes of mortality have often found bacteriophages and protozoan grazing to be equally important as loss factors in bacterial mortality (e.g., Fuhrman and Noble 1995; Steward et al. 1996; Weinbauer and Peduzzi 1995). Moreover, measurements of a higher frequency of infected cells in eutrophic than in oligotrophic waters have suggested that bacteriophages may have a larger impact on bacterial mortality in more productive systems (Steward et al. 1996; Weinbauer et al. 1993).

The role of bacteriophage-induced mortality in the sediment is somewhat more controversial. Recent studies have indicated that bacteriophages are equally important as agents of bacterial mortality in benthic systems, based on high rates of bacteriophage production and an estimated mortality impact of 6%–40% of bacterial production (Glud and Middelboe 2004; Mei and Danovaro 2004). However, observations of extremely low occurrence of visibly infected cells in freshwater sediments have suggested that this might not always be the case (Filippini et al., 2006). Our understanding of the role of bacteriophages in benthic ecosystems is still in its infancy, and more studies are needed to evaluate temporal and spatial variations in bacteriophage-induced bacterial mortality in benthic environments.

The Fate of Viral Lysates

An important prerequisite for the influence of viral lysates on biogeochemical cycling is that the cell contents become available for bacterial uptake following cell lysis. Zweifel and Hagström (1995) found that the majority of pelagic marine bacteria as determined by fluorescent DNA stains were in fact inactive "ghost cells" and suggested that these cells had been lysed by viruses. These results indicated that lysed cells may have long turnover times and that lysates therefore were not likely to constitute a significant substrate source for the bacteria. Other studies, however, have shown that lysed bacteria were actually efficiently converted to dissolved organic material (DOM) and that cells were not recognizable by DNA staining techniques following viral lysis (Riemann and Middelboe 2002; Shibata et al. 1997). Later, Middelboe, Riemann, et al. (2003) demonstrated a close coupling between release and subsequent uptake of viral lysates. In culture experiments, 28% of viral lysates were converted to new biomass of noninfected bacteria within two days after cell lysis, indicating that the majority of the lysates were readily available for bacterial uptake (Middelboe, Riemann, et al. 2003). In addition, the study illustrated that the result of viral lysis was not simply a recycling of organic matter but at the same time a transformation of the initial bulk DOM into labile viral lysates. This virus-induced substrate transformation provided utilizable substrate (the lysates) for a bacterial type that was unable to grow on the bulk DOM pool. The authors proposed this substrate transformation as a mechanism by which viruses may promote coexistence of diverse bacterial populations in natural communities (Middelboe et al. 2003).

Observations of colonization of visibly infected cells by noninfected bacteria suggested that the infected cells leaked DOM and functioned as nutrient point sources, which were consumed by bacteria clustering around the source (Middelboe, Riemann, et al. 2003). This suggests that lysing cells may function as "hot spots" of bioavailable substrate, contributing to the patchiness of the pelagic environment and increasing the efficiency of bacterial utilization of viral lysates. These data supported the findings by Noble et al. (1999) that heightened bacteriophage infection rates in natural bacterioplankton communities stimulated overall bacterial productivity. Studies of the fate of radioactively labeled virus particles and other lysis products have also verified that lysates are labile, with turnover times of one to four days by natural bacterioplankton (Noble and Fuhrman 1999). However, in this case, the majority of the labeled material was not accumulated into biomass, suggesting that most of the lysates were respired by the microbial community.

The bioavailability of viral lysates depends on their chemical and structural composition. Shibata et al. (1997) identified peaks of sub-micron-sized particles (0.38–0.7 µm) following viral lysis of *Vibrio algi-nolyticus* and suggested that these particles were cell debris originating from bacteria. They proposed viral lysis as an important pathway of colloid formation in the sea. Recently, Jørgensen and Middelboe (2006) found that viral lysis released significant amounts of both free and combined amino acids, in particular the D-isomers, which are important constituents of bacterial cell wall peptidoglycan. In contrast to most of the ambient DOM in the seawater, freshly released DOM from lysed cells should be labile for bacterial consumption (Amon and Benner 1996). Consequently, the available data on the fate of viral lysates to date tend to support the hypothesis that lysates are a source of labile carbon in the sea and that they are efficiently utilized by noninfected heterotrophic bacteria.

Bacteriophages and Nitrogen, Phosphorus Mineralization

Apart from supplying bacteria with labile organic carbon, lysates may constitute a significant source of phosphorus and nitrogen. Bacteria have a C:N:P ratio of approximately 30:6:1 (Vrede et al. 2002), and the product of viral lysis should have roughly the same ratio. In fact, cell contents may be rich in nitrogen and phosphorus (e.g., proteins and nucleic acids) (Weinbauer 2004), and lysis products might therefore have C:N:P ratios lower than that of intact cells, assuming that carbon is preferentially bound in the cell wall structure (Noble and Fuhrman 1999). In an experiment with phosphorus-limited bacteria, Middelboe et al. (1996) demonstrated that viral lysates supported growth of noninfected bacteria by release of organic phosphorus that was hydrolyzed by bacterial extracellular phosphatase enzymes. The phosphorus-rich lysates stimulated bacterial growth, but at the same time the bacterial growth efficiency was reduced for bacteria utilizing the viral lysates, presumably owing to increased energy requirements associated with the degradation of the lysates prior to assimilation (Middelboe et al. 1996). In a subsequent study, Noble and Fuhrman (1999) added radiolabeled lysates to natural bacterial communities and followed the fate of the material in terms of its breakdown and assimilation. The results indicated that lysates were relatively labile but that little of the bulk (^{3}H-labeled) material was taken up by bacteria in thirty-hour incubations. In contrast, experiments with ^{33}P-labeled lysates indicated that phosphorus was rapidly assimilated (more than 50% within seven hours) in phosphorus-limited seawater. These experimental data support the general view that viral lysis is an important pathway for phosphate recycling in the environment (Blackburn et al. 1996; Weinbauer 2004; Zweifel et al. 1996).

In both the above cases, the released organic phosphorus was immediately taken up and incorporated by the noninfected bacteria as a result of the phosphorus-limited growth conditions. It is reasonable, however, to assume that viral lysates often contain an excess of nitrogen and phosphorus relative to the bacterial demand. With an average bacterial C:N:P ratio as above (Vrede et al. 2002), and assuming a bacterial carbon respiration of 70% of assimilated carbon (Middelboe, Riemann, et al. 2003), the optimum C:N:P ratio of bacterial substrate should theoretically be approximately 100:6:1 in order to contain enough carbon to cover respiration demands in addition to nitrogen and phosphorus demands. Consequently, lower C:N:P ratios should result in a net mineralization of nitrogen and phosphorus. Experimental studies, however, have suggested that net bacterial release of inorganic nitrogen occurred only at substrate C:N ratios below 8–9 (i.e., C:N:P ratios of approximately 51:6:1) (Goldman and Dennet 1991), or in cases where amino acids supplied more than 80 % of bacterial carbon demand (Kirchman et al. 1989, Middelboe et al. 1995). In either case, however, viral lysates would contain nitrogen and phosphorus in excess of bacterial requirements. Therefore, in cases where bacterial growth in general is carbon limited, viral lysates should contribute to the recycling of nitrogen and phosphorus in the sea (see figure 11.3).

From this it follows that bacteriophage activity should potentially stimulate primary production in occasions with carbon-limited bacterial growth and nitrogen-/phosphorus-limited algal growth. Direct experimental evidence for such an effect of viral activity is scarce. Suttle et al. (1996) found indications that cyanobacteria growth was stimulated in the presence of bacteriophages, presumably owing to viral lysis enhancing the bacterial recycling of nutrients. In support of this, Haaber and Middelboe (unpublished) observed an accumulation of NH_4 during bacterial breakdown of algal lysates following viral lysis of the microalga *Phaeocystis pouchetii* in monocultures. In parallel cultures that also contained the microalga *Rhodomonas salina* (not exposed to viruses), the NH_4 concentration, however, remained low following lysis of *P. pouchetii* because of immediate uptake of the released ammonium by *R. salina*.

By using tracer experiments to examine the nutrient flux, Gobler et al. (1997) observed a significant release of carbon, nitrogen, phosphorus, iron, and selenium following lysis of the marine chrysophyte *Aureococcus anophagefferens*. In the same experiment, the authors demonstrated that the released nitrogen and phosphorus could alleviate both nitrogen and phosphorus limitation in other phytoplankton, either by direct algal uptake of the organic forms (DON and DOP) or indirectly by uptake of bacterially regenerated inorganic nitrogen and phosphorus. Together, these results strongly suggest that viral activity does indeed contribute to micronutrient regeneration and recycling in the marine environment, and therefore possibly also in other nutrient-limited ecosystems.

Quantitative Role of Bacteriophages in Marine Biogeochemical Cycling

It is evident from the numerous studies of bacteriophage activity and their impact on bacterial mortality and mediation of labile dissolved organic matter supply that bacteriophages potentially have significant implications for marine carbon and nutrient flow. However, we still know very little about the quantitative importance of bacteriophage-mediated element cycling and its variation in time and space.

One of the first attempts to incorporate viral activity into a budget of carbon transfer in the microbial food web (Bratbak et al. 1992) verified that viral lysates constituted a dominant substrate source for the heterotrophic bacterioplankton. However, difficulties in balancing the overall carbon budget made a complete resolution of the relative significance of the different pathways supplying bacterial carbon demand impossible.

Direct experimental evidence for the quantitative significance of viral lysis on natural bacterioplankton activity was provided by Middelboe and Lyck (2002). Similar to the theoretical approach by Fuhrman (1992), the authors used experimental conditions with reduced viral abundances to examine the effects of viruses on the carbon flow in natural microbial assemblages compared with control cultures with natural viral abundances. In agreement with Fuhrman's calculations, Middelboe and Lyck (2002) demonstrated that a reduction in viral abundance (by 62%–92%) increased bacterial net production in the cultures by approximately 40 %, although total bacterial metabolism was similar to the control cultures with natural abundance. These results confirmed the model proposed by Fuhrman and verified that viral-mediated recycling of bacterial carbon significantly influences the flux of organic carbon in the ocean.

The recycling of carbon is relevant to our understanding of carbon flow in the marine food web. Bacterial net production is considered to be roughly 20% of primary production on a volumetric basis (Cole et al. 1988), indicating that about 50% of the primary production is channeled through the bacteria. However, the estimated input of organic carbon from the primary producers cannot always account for the measured bacterial carbon demand (e.g., Strayer 1988), suggesting that bacteria themselves may contribute to bacterial carbon demand either by direct predation by other bacteria or through bacterial exudates or lysis products (Cole and Caraco 1993). Based on measurements of bacterial and viral production, Wilhelm and Suttle (2000) calculated that viral lysates potentially supplied 4%–30% of bacterial carbon demand in the Gulf of Mexico and up to 80%–95% of the carbon demand in stratified locations in the Strait of Georgia, United States. The results support the idea that viral lysis may be a key mechanism supplying substrate to the bacterioplankton and contributes to the high rates of heterotrophic bacterial production relative to primary production often observed in marine systems (e.g., del Giorgio et al. 1997; Duarte et al. 2001; Strayer 1988). The relative importance of such bacteriophage-mediated recycling for bacterial carbon demand would be expected to be highest in oligotrophic environments with little direct substrate input from the phytoplankton. Consequently, virus-mediated recycling of carbon in the ocean reduces the bacterial demand for phytoplankton carbon and represents the "missing link" between particulate and dissolved organic carbon that can explain the high bacterial-to-primary production ratios often found in oligotrophic systems.

In sediments, the role of bacteriophages seems to be different from their role in the water column. Despite observations of comparable bacteriophage-induced bacterial mortality, the impact of bacteriophages on benthic carbon cycling is moderate compared to the water column estimates (Fischer et al. 2003; Glud and Middelboe 2004). Only up to approximately 10% of benthic bacterial carbon demand is supplied by viral lysates (Fischer et al. 2003; Glud and Middelboe 2004; Middelboe et al. 2006), with an inverse relationship between the importance of lysates and bacterial activity (Middelboe et al. 2006). This suggests that the viral lysates are only a minor substrate source in the benthic environment characterized by a large benthic substrate pool, which is constantly supplied from the water column and redistributed by fauna perturbations and resuspension events (Fenchel and Glud 2000). In fact, this rationale is in line with observations that the relative importance of viral lysates in the water column decreases with increased system productivity.

PHYTOPLANKTON VIRUSES—A SHORT CIRCUIT IN THE PELAGIC CARBON FLOW

Phytoplankton constitutes the base of every pelagic food web in the ocean and is responsible for about a third of global primary production.

Viruses are important agents of disease of phytoplankton organisms and have been demonstrated to cause complete collapse of monospecific blooms of phytoplankton (Bratbak et al. 1993; Brussard 2004). The most frequently cited figures suggest that 2%–10% of total primary production is lost from the particulate fraction as a direct result of viral lysis of algal cells (see figure 11.3) (e.g., Bratbak and Heldal 2000). From a biogeochemical perspective, viral pathogens of phytoplankton introduce a short-circuit in the pelagic food web. Viral lysis of phytoplankton diverts organic matter away from metazoan grazers and the higher trophic levels in the grazing food chain and redirects the organic matter into the microbial food web. Insofar as phytoplankton primary production is approximately 49.3 Gt y^{-1} on a global scale (Ducklow and Carlson 1992), the total amount of carbon that is "lost" from the grazing food chain by viral pathogens as DOM is substantial. Like bacterial lysates, algal lysates are labile for bacterial uptake and utilized efficiently by bacteria (Bratbak et al. 1998; Gobler et al. 1997), and therefore at times constitute a carbon source for the planktonic bacteria that equals the supply from algal exudation of DOM (Bjørnsen 1988). During mass viral lysis of phytoplankton blooms, the input of algal lysates significantly alters DOM concentration and composition, with a large impact on bacterial mineralization processes (Bratbak et al. 1998; Van Hannen et al. 1999).

As important pathogens of the major primary producers in aquatic environments with major implications for the fate of primary production and the dynamics of phytoplankton communities, phytoplankton viruses play a central role in oceanic biogeochemical cycling, yet we are still on relatively thin ice when trying to quantify the impact of phytoplankton viruses on global element cycling.

CONCLUSIONS

It can be concluded that processes mediated by viral infection and subsequent lysis of microorganisms have significant influence on global carbon and nutrient cycles. Viral activity directly affects recycling and export of matter in the oceans and has indirect effects on the complex regulation of microbial community composition, which potentially has large implications for global element cycling (Riemann and Middelboe 2002).

One of the major implications is that viral lysis of cells at all trophic levels converts biomass into DOM, thus maintaining important elements such as carbon, nitrogen, phosphorus and iron in the euphotic zone (Weinbauer 2004), where they are turned over by planktonic bacteria. The consequences of this process are increased microbial respiration and mineralization of nutrients, reduced transfer of carbon to higher trophic levels, and reduced export of particulate material to the seafloor—all effects that seriously affect our general understanding of global biogeochemical cycling. These effects of viral activity are, however, to some extent counteracted by the concomitant stimulation of nitrogen and phosphorus mineralization, which is also a consequence of viral activity (see figure 11.3). Enhanced nutrient mineralization is expected to stimulate overall phytoplankton production, part of which is channeled to higher trophic levels or exported to the sediment. Obviously, such feedback mechanisms, as well as the other direct and indirect effects of viruses on microbial processes and interactions, are still not understood in sufficient detail to allow incorporation into quantitative models of marine element cycling.

In contrast to the general conception of detrimental impacts of diseases on ecosystems, viral infection of microorganisms in the sea apparently has a stimulatory effect on the overall system productivity. However, an increased understanding of the impacts of viral infections on the structure and function of the microbial food web and on biogeochemical cycles is fundamental to our ability to model the fluxes of organic matter and the regeneration of nutrients in marine environments. The speculative nature of the quantitative importance of many of the presented aspects of viral activity emphasizes the need for further studies on the role of viral diseases in biogeochemical cycling in different environments.

ACKNOWLEDGMENTS

I thank the organizers and sponsors of the Cary Conference for arranging a very constructive and interesting meeting and for financial support. Work was supported by the Danish Natural Sciences Research Council. Two anonymous reviewers provided valuable comments on the manuscript.

LITERATURE CITED

Amon, R. M. W., and R. Benner. 1996. Bacterial utilization of different size classes of dissolved organic matter. Limnology and Oceanography 41:41–51.
Ashelford, K. E., M. J. Day, and J. C. Fry. 2003. Elevated abundance of bacteriophage infecting bacteria in soil. Applied and Environmental Microbiology 69:285–89.

Berg, O., K. Y. Børsheim, G. Bratbak, and M. Heldal. 1989. High abundance of viruses found in aquatic environments. Nature 340:467–68.

Bjørnsen, P. K. 1988. Phytoplankton exudation of organic matter: Why do healthy cells do it? Limnology and Oceanography 33:151–54.

Blackburn, N., U. L. Zweifel, and Å. Hagström. 1996. Cycling of marine dissolved organic matter. II. A model analysis. Aquatic Microbial Ecology 11:79–90.

Bratbak, G., J. K. Egge, and M. Heldal. 1993. Viral mortality of the marine alga *Emiliania huxleyi* (Haptophyceae) and termination of algal blooms. Marine Ecology Progress Series 93:39–48.

Bratbak, G., and M. Heldal. 2000. Viruses rule the waves: The smallest and most abundant members of marine ecosystems. Microbiology Today 27: 171–73.

Bratbak, G., M. Heldal, T. F. Thingstad, B. Riemann, and O. H. Haslund. 1992. Incorporation of viruses into the budget of microbial C-transfer: A first approach. Marine Ecology Progress Series 83:273–80.

Bratbak, G., A. Jacobsen, and M. Heldal. 1998. Viral lysis of *Phaeocystis pouchetii* and bacterial secondary production. Aquatic Microbial Ecology 16:11–16.

Breitbart, M., and F. Rohwer. 2005. Here a virus, there a virus, everywhere the same virus? Trends in Microbiology 13:278–84.

Brussaard, C. P. D. 2004. Viral control of phytoplankton populations: A review. Journal of Eukaryotic Microbiology 51:125–38.

Cole, J. J., and N. F. Caraco. 1993. The pelagic microbial food web of oligotrophic lakes. *In* Aquatic Microbiology, ed. T. E. Ford, 101–11. Oxford: Blackwell Scientific Publications.

Cole, J. J., S. Findlay, and M. L. Pace. 1988. Bacterial production in fresh and saltwater ecosystems: a cross-system overview. Marine Ecology Progress Series 43:1–10.

Cottrell, M. T., and C. A. Suttle. 1991. Widespread occurrence of and clonal variation in viruses which cause lysis of a cosmopolitan, eukaryotic marine phytoplankton, *Micromonas pusilla*. Marine Ecology Progress Series 78:1–9.

del Giorgio, P. A., J. J. Cole, and A. Cimbleris. 1997. Respiration rates in bacteria exceed phytoplankton production in unproductive aquatic systems. Nature 385:148–51.

Duarte, C. M., S. Agusti, J. Aristegui, N. González, and R. Anadón. 2001. Evidence for a heterotrophic subtropical northeast Atlantic. Limnology and Oceanography 46:425–28.

Ducklow, H. W., and C. A. Carlson. 1992. Oceanic bacterial production. *In* Advances in Microbial Ecology, ed. K. C. Marshall, 113–81. New York: Plenum Press.

Fenchel, T., and R. N. Glud. 2000. Benthic primary production and O_2-CO_2 dynamic in a shallow water sediment: Spatial and temporal heterogeneity. Ophelia 53:159–71.

Fillipini, M., N. Buesing, Y. Bettarel, T. Sime-Ngando, and M. O. Gessner. 2006. Infection paradox: High abundance but low impact of freshwater benthic viruses. Applied and Environmental Microbiology 72:4893–98.

Fischer, U. R., C. Wieltschnig, A. K. T. Kirscher, and B. Velimirov. 2003. Does virus-induced lysis contribute to bacteria mortality in the oxygenated sediment layer of shallow Oxbow lake? Applied and Environmental Microbiology 69:5281–89.

Fuhrman, J. A. 1992. Bacterioplankton roles in cycling of organic matter: The microbial food web. In Primary Productivity and Biogeochemical Cycles in the Sea, ed. P. G. Falkowski and A. D. Woodhead, 361–82. New York: Plenum Press.

Fuhrman, J. A. 1999. Marine viruses and their biogeochemical and ecological effects. Nature 399:541–48.

Fuhrman, J. A., and R. T. Noble. 1995. Viruses and protists cause similar bacterial mortality in coastal seawater. Limnology and Oceanography 40: 1236–42.

Glud, R. N., and M. Middelboe. 2004. Virus and bacteria dynamics of a coastal sediment: Implications for benthic carbon cycling. Limnology and Oceanography 49:2073–81.

Gobler, C. J., D. A. Hutchins, N. S. Fisher, E. M. Cosper, and S. Sanudo-Wilhelmy. 1997. Release and bioavailability of C, N, P, Se, and Fe following viral lysis of a marine Chrysophyte. Limnology and Oceanography 42: 1492–504.

Goldman, J. C., and M. R. Dennet. 1991. Ammonium regeneration and carbon utilization by marine bacteria grown on mixed substrates. Marine Biology 109:369–378.

Jørgensen, N.O.G., and M. Middelboe. 2006. Occurrence and bacterial cycling of D amino acid isomers in an estuarine environment. Biogeochemistry 81: 77–94.

Kirchman, D. L., R. G. Keil, and P. A. Wheeler. 1989. The effect of amino acids on ammonium utilization and regeneration by heterotrophic bacteria in the subarctic Pacific. Deep-Sea Research 36:1763–76.

Mei, M. L., and R. Danovaro. 2004. Virus production and life strategies in aquatic sediments. Limnology and Oceanography 49:459–70.

Middelboe, M., N. H. Borch, and D. L. Kirchman. 1995. Bacterial utilization of dissolved free amino acids, dissolved combined amino acids and ammonium in the Delaware Bay estuary: Effects of carbon and nitrogen limitation. Marine Ecology Progress Series 128:109–29.

Middelboe, M., R. N. Glud, and K. Finster. 2003. Spatial distribution of viruses and relation to bacterial activity in a coastal marine sediment. Limnology and Oceanography 48:1447–56.

Middelboe, M. R. N. Glud, F. Wenzhöfer, K. Oguri, and H. Kitazato. 2006. High abundance and activity of viruses in the deep-sea sediments in Sagami Bay, Japan. Deep Sea Research I 53:1–13.

Middelboe, M., N. O. G. Jørgensen, and N. Kroer. 1996. Effects of viruses on nutrient turnover and growth efficiency of noninfected marine bacterioplankton. Applied and Environmental Microbiology 62:1991–97.

Middelboe, M., and P. G. Lyck. 2002. Regeneration of dissolved organic matter by viral lysis in marine microbial communities. Aquatic Microbial Ecology 27:187–94.

Middelboe, M., T. G. Nielsen, and P. K. Bjørnsen. 2002. Viral and bacterial production in the North Water polynya: *In situ* measurements, batch culture experiments, and characterization and distribution of a virus-host system. Deep Sea Research II 49:5063–79.

Middelboe, M., L. Riemann, G. L. Steward, V. Hansen, and O. Nybroe. 2003. Virus-induced transfer of organic carbon between marine bacteria in a model community. Aquatic Microbial Ecology 33:1–10.

Noble, R. T., and J. A. Fuhrman. 1999. Breakdown and microbial uptake of marine viruses and other lysis products. Aquatic Microbial Ecology 20:1–11.

Noble, R. T., M. Middelboe, and J. A. Fuhrman. 1999. The effects of viral enrichment on the mortality and growth of heterotrophic bacterioplankton. Aquatic Microbial Ecology 18:1–13.

Proctor, L. M., and J. A. Fuhrman. 1990. Viral mortality of marine bacteria and cyanobacteria. Nature 343:60–62

Riemann, L., and M. Middelboe. 2002. Viral lysis of marine bacterioplankton: Potential implications for organic matter cycling and bacterial clonal composition. Ophelia 56:57–68.

Shibata, A., K. Kogure, I. Koike, and K. Ohwada. 1997. Formation of submicron colloidal particles from marine bacteria by viral infection. Marine Ecology Progress Series 155:303–7.

Suttle, C. A., A. M. Chan, K. Rodda, S. M. Short, M. G. Weinbauer, D. R. Garza, and S. W. Wilhelm. 1996. The effect of cynophages on *Synechococcus* spp. during a bloom in the western Gulf of Mexico. American Geophysical Union, EOS Transactions 76:OS207–8.

Steward, G. F., D. C. Smith, and F. Azam. 1996. Abundance and production of bacteria and viruses in the Bering and Chukchi Seas. Marine Ecology Progress Series 131:287–300.

Strayer, D. L. 1988. On the limits to secondary production. Limnology and Oceanography 33:1217–20.

Thingstad, T. F., and R. Lignell. 1997. Theoretical models for the control of bacterial growth rate, abundance, diversity and carbon demand. Aquatic Microbial Ecology 13:19–27.

Van Hannen, E. H., G. Zwart, M. P. van Agtervald, H. J. Gons, J. Ebert, and H. J. Laanbrook. 1999. Changes in bacterial and eukaryotic community structure after mass lysis of filamentous cyanobacteria associated with viruses. Applied and Environmental Microbiology 65:795–801.

Vrede, K., M. Heldal, S. Norland, and G. Bratbak. 2002. Elemental composition (C, N, P) and cell volume of exponentially growing and nutrient-limited bacterioplankton. Applied and Environmental Microbiology 68:2965–71.

Weinbauer, M. G. 2004. Ecology of prokaryotic viruses. FEMS Microbiological Reviews 28:127–81.

Weinbauer, M. G., D. Fuks, and P. Peduzzi. 1993. Distribution of viruses and dissolved DNA along a trophic gradient in the Northern Adriatic Sea. Applied and Environmental Microbiology 59:4074–82.

Weinbauer, M. G., and P. Peduzzi. 1995. Significance of viruses versus heterotrophic nanoflagellates for controlling bacterial abundance in the northern Adriatic Sea. Journal of Plankton Research 17:1851–56.

Wilhelm, S. W., and C. A. Suttle. 1999. Viruses and nutrient cycles in the sea. Bioscience 49:781–88.

Wilhelm, S. W., and C. A. Suttle. 2000. Viruses as regulators of nutrient cycles in aquatic environments. *In* Proceedings of the Eighth International Symposium on Microbial Ecology, ed. C. R. Bell, M. Brylinsky, and P. Johnson-Green, 551–56. Halifax: Atlantic Canada Society for Microbial Ecology.

Wommack, K. E., and R. R. Colwell. 2000. Virioplankton: Viruses in aquatic environments. Microbiology and Molecular Biology Reviews 64:69–114.

Wommack, K. E., J. Ravel, R. T. Hill, J. Chun, and R. R. Colwell. 1999. Population dynamics of Chesapeake Bay virioplankton: Total-community analysis by pulsed-field gel analysis. Applied and Environmental Microbiology 65: 231–40.

Zweifel, U. L., N. Blackburn, and Å. Hagström. 1996. Cycling of marine dissolved organic matter. I. An experimental system. Aquatic Microbial Ecology 11:65–77.

Zweifel, U. L. and Å. Hagström. 1995. Total counts of marine bacteria include a large fraction of non-nucleoid-containing bacteria (ghosts). Applied and Environmental Microbiology 61:2180–85.

CHAPTER TWELVE
Effects of Pathogens on Terrestrial Ecosystem Function

Valerie T. Eviner and Gene E. Likens

SUMMARY

MANY STUDIES HAVE DEMONSTRATED that pathogens can have strong effects on the performance of individual organisms, population dynamics, and community interactions. A more limited set of studies suggests that pathogens can alter a wide range of ecosystem functions in terrestrial systems; however, we are lacking a framework to predict the type and magnitude of ecosystem effects that a given pathogen will have. In this chapter, we present a number of general principles that determine how pathogens influence ecosystems over time, based on the well-developed fields of disturbance ecology and the ecosystem effects of species composition. Our focus is not only on pathogens as disturbances but also as drivers of ecosystem processes, even when their presence is not readily apparent.

INTRODUCTION

Pathogens are infectious biological agents that alter the normal functioning of their hosts. Many studies have demonstrated that pathogens can have large impacts on host physiology, population dynamics, and community composition (reviewed in Burdon 1991; Gilbert 2002; Kranz 1990; see also Collinge et al., chapter 6, Clay et al., chapter 7, and Lafferty, chapter 9 this volume). Based on what is known about organisms' effects on ecosystems, many pathogen-induced changes in host populations, communities, and traits are dramatic enough that we would expect to see an accompanying change in ecosystem dynamics. However, relatively few studies have documented the consequences of pathogens on ecosystem function. These studies have demonstrated that epidemics affecting dominant plant species can alter a wide range of ecosystem functions, including net primary productivity (Kranz 1990; Spedding and Diekmahns 1972), hydrology (Bari and Ruprecht 2003; Batini et al. 1980; Hobara et al. 2001), decomposition (Cromack et al. 1991; Waring et al. 1987), nutrient cycling (Matson and Boone 1984; Waring et al. 1987), nutrient export (Hobara et al. 2001; Ohte et al. 2003), erosion

(Graniti 1998; Johnson and Wilcock 2002), and disturbance regimes (Dickman 1992). Even when there are no obvious visible signs of pathogen-induced damage or mortality, pathogens can substantially alter ecosystem processes. For example, pathogens can substantially decrease belowground production without any change in aboveground production (Agrios 2005; Mitchell 2003).

Although these studies have demonstrated that plant pathogens can alter ecosystem function, we have relatively little ability to predict the type and magnitude of ecosystem effects caused by any particular pathogen. In this chapter, we present some general principles to help account for the ecosystem impacts of pathogens. We discuss how these principles can be used to predict the ecosystem impacts of plant pathogens that act as disturbance agents and as key but invisible players in the everyday functioning of ecosystems. We also explore the ecosystem impacts of pathogens that infect organisms other than plants. Finally, we address when it is critical to consider pathogens in order to understand and manage ecosystems.

PREDICTING THE ECOSYSTEM EFFECTS OF PLANT PATHOGENS: GENERAL PRINCIPLES

The ecosystem effects of pathogens are largely mediated by the impacts of pathogens on their hosts. A number of key factors determine the ecosystem impacts of plant pathogens. These are briefly described below and discussed in more detail later in the chapter:

(1) *Pathogen impact on host survival, physiology, behavior, and reproduction.* Pathogens vary in their ecosystem impacts, depending on the host tissues they attack and whether they act as "killers, debilitators, or castrators" (Burdon 1991). The magnitude of the impact largely depends on host susceptibility, which can vary across species and across individuals within a species (Henry et al. 1992; Remold 2002).

(2) *Life stages of a host vulnerable to a pathogen.* Specific pathogens usually act on only a subset of plant life stages, or act differently on distinct life stages of a host (Agrios 2005; Castello et al. 1995; Kranz 1990). Whether a pathogen attacks its host in an immature or mature phase determines its ecosystem effects and our ability to notice the impact of pathogens on hosts and ecosystems.

(3) *Proportion of individuals/biomass infected at a site.* The proportion of individuals or biomass infected is largely a function of the susceptibility of hosts in relation to the specificity of pathogens. A highly host-specific pathogen can have large ecosystem impacts when it infects

dominant or keystone species (discussed by Collinge et al., chapter 6, this volume) but may have no ecosystem impact in a diverse host community that is largely resistant to the pathogen. In contrast, a diverse host community can be destroyed by a generalist pathogen (e.g., *Phytophthora cinnamomi* can infect many different kinds of plants in Australian jarrah forests) (Weste 2003; Weste et al. 2002).

(4) *Spatial extent and distribution of infection.* The type and intensity of ecosystem change induced by a pathogen are strongly influenced by how large an area it infects and the distribution of infection. For example, if a pathogen were to kill 30% of all trees in a watershed, the ecosystem effects would likely differ if the infected individuals were all clumped in one area versus being evenly spread throughout the watershed.

(5) *Rate of pathogen effects on hosts in relation to rate of response or recovery by hosts or individuals replacing the hosts.* Pathogens can vary greatly in the rates at which they influence hosts. Similarly, the composition of host communities can greatly influence the rate of vegetation recovery in response to pathogen-induced damage or mortality of plants. In a host community that is diverse in its susceptibility to a pathogen, a slowly progressing disease will likely have minimal short-term ecosystem impacts, since neighboring plants will rapidly fill in space made available from dying branches and roots. In contrast, a pathogen that causes rapid mortality of many overstory individuals will likely have large ecosystem impacts, since replacement of the canopy will rely on seedlings or saplings.

(6) *Functional similarity of infected individuals vs. replacements.* Many pathogens substantially alter host community composition and structure (see Clay et al., chapter 7, Collinge et al., chapter 6, and Lafferty, chapter 7, this volume). A substantial body of literature has shown that ecosystem processes can be greatly affected by shifts in plant species composition and plant diversity (reviewed in Eviner and Chapin 2003; Hooper et al. 2005), suggesting that pathogen-induced shifts in host communities can impact ecosystems. Even when there is no shift in species composition, pathogens can replace mature individuals with juveniles, and the plant traits that influence ecosystem processes can vary greatly within a species across its life stages (reviewed in Marschner 1995).

(7) *Frequency and duration of pathogen impact.* Over the long term, the ecosystem impacts of a pathogen are strongly influenced by the duration and return interval of infection. When a pathogen persists over the long term at a site, it may continuously exclude the former dominant hosts (e.g., chestnut blight, *Cryphonectria parasitica*) (Jaynes and Elliston 1982; Paillet 1988). As with other disturbances, the frequency with which a pathogen recurs at a site can have large impacts on ecosystems through its effects on host community composition and recovery of

Single disturbance

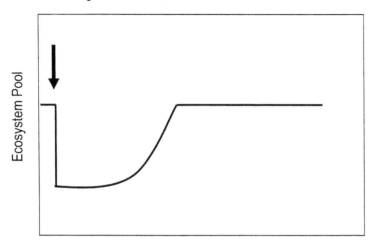

Repeated disturbance

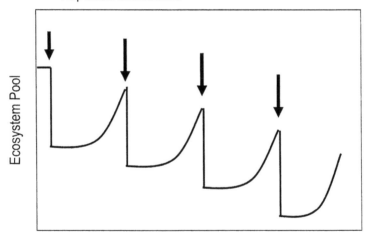

Time

Figure 12.1. Response of ecosystem pools (e.g., ecosystem carbon and nitrogen, plant biomass) to disturbance. Arrows indicate time of disturbance. (a) An ecosystem disturbance results in an immediate decrease in biogeochemical pools, which slowly recover with time after disturbance. (b) The frequency of disturbance can lead to long-term changes in ecosystem pools, particularly when the disturbance recurs before the ecosystem pool fully recovers from the previous disturbance. (Figure adapted from Vitousek and Reiners 1975, Bormann and Likens 1979, Chapin et al. 2002.)

vegetative biomass, ecosystem structure, and biogeochemical fluxes and pools (figure 12.1) (Bormann and Likens 1979; Chapin et al. 2002; Vitousek and Reiners 1975).

In this chapter, we explore how these seven factors contribute to a mechanistic understanding of the short- to long-term ecosystem effects of pathogens, focusing primarily on plant pathogens.

Plant Pathogens as Disturbances

Pathogens are often viewed as disturbance agents, organisms that disrupt the normal functioning of other organisms or ecosystems. Most studies that have investigated the ecosystem effects of pathogens have focused on large-scale epidemics that kill mature individuals, and these stand-replacing epidemics have been compared with other disturbance events such as clear-cutting (Hobara et al. 2001; Waring et al. 1987). In general, a stand-level disturbance can result in an initially large ecosystem change as a result of losses of carbon and nutrient stocks from the ecosystem. These pools return to their predisturbance baseline over time as the vegetation recovers (figures 12.1a, 12.2a) (Bormann and Likens 1979; Loucks 1970; Matson and Boone 1984; Waring et al.1987).

The key lesson to be learned from the disturbance framework is that the ecosystem effects of pathogens change over time. We outline four key stages of pathogen effects on ecosystems (figure 12.2):

- *Impact on host*. The initial impacts of a pathogen on ecosystems are largely determined by how it impacts its hosts (e.g., shifts in traits, mortality). For many pathogens, their effects on ecosystems change during this impact stage, as infection progresses from a debilitating phase to a killing phase.

- *Initial response*. The plant community usually responds during or shortly after pathogen disturbance of hosts, and the ecosystem effects of pathogens are strongly influenced by which community members respond and how rapidly they respond. This initial community response is often determined by plants already occupying the site.

- *Midterm response*. The midterm response is characterized by the second generation of establishment after pathogen disturbance.

- *Long-term response*. Whereas the three previous stages are transitory in nature, the long-term effects of pathogens are characterized by whether the ecosystem recovers to predisturbance conditions or is set on a new trajectory.

The classic disturbance curve (figure 12.2a) is based on a stand-replacing event-mortality of a large portion of the vegetation over a relatively short time. This disturbance would be followed by the rapid establishment of a plant community. Only a subset of pathogens would be expected to mirror the ecosystem effects of the classic disturbance curve (e.g., acute pathogen outbreaks that rapidly kill most plants in a given area). In many cases, the ecosystem effects of pathogens do not follow the classic disturbance curve, for a number of reasons:

1. The long-term effects of pathogens can be strongly influenced by the frequency at which they disturb plant communities. A single disturbance event can require a given amount of recovery time before biogeochemical pools and fluxes return to predisturbance levels (see figure 12.1a). If a pathogen were to return repeatedly before the ecosystem recovers, it would likely set the ecosystem on a new trajectory (figure 12.1b) (Aber and Melillo 2001; Bormann and Likens 1979; Chapin et al. 2002).

2. The short- to long-term ecosystem impacts of pathogens are largely mediated by the rate of recovery of vegetation biomass and whether this recovery involves shifts in plant species composition. When the individuals that replace the killed hosts are functionally different from the hosts, large changes in ecosystem function and structure can occur (dashed lines in figure 12.2b). For example, beech bark disease (*Nectria coccinea* or *N. galligena*) in forests of the northeastern United States can lead to gradual shifts in community composition that, in turn, alter nutrient cycling (Lovett et al. 2006). As beech (*Fagus grandifolia*)–sugar maple (*Acer saccharum*) communities shift to maple dominance, nitrification significantly increases, because sugar maple stimulates nitrification rates. Similarly, in beech-hemlock (*Tsuga canadensis*)–dominated sites, shifts from beech to hemlock lead to lower rates of nitrogen cycling and nitrogen loss (Lovett et al. 2006).

3. In some cases, there may be a time lag between plant species shifts and changes in ecosystems (dotted line in figure 12.2b). For example, in southeastern Alaska, a 3.8-fold increase in landslides occurred fifty years after large-scale mortality of yellow cedars (*Chamaecyparis nootkatensis*). Even though cedars were killed relatively rapidly, the effects of their roots on slope stability

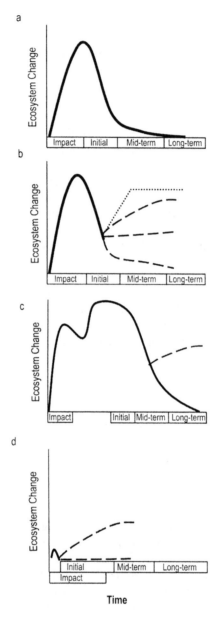

Figure 12.2. (a) The classic disturbance curve, where ecosystem processes initially change greatly in response to disturbance and then gradually recover to predisturbance levels. (b) Mid- to long-term effects of pathogens on ecosystems are determined by the composition of the plant community that reestablishes at a site after hosts are killed by pathogens (dashed lines). The effects of pathogens on ecosystems over time are also influenced by the persistence of the ecosystem effects of hosts, even after they are killed. Once these legacy effects wear off, the ecosystem can be set on a new trajectory (dotted line). (c) A lag time between

persisted for decades, until the roots finally decomposed (Johnson and Wilcock 2002).

4. Departures from the standard disturbance curve are also due to variations in the duration of each impact/response stage and the extent to which they overlap (figure 12.2). For example, *Phytophthora cinnamomi* can kill 50%–75% of overstory and understory species in Australian jarrah forests, sometimes resulting in land that is totally devoid of vegetation for a number of years (Weste 2003). In addition to the initial ecosystem change associated with disturbance, this time lag between mortality and recovery can increase the susceptibility of an ecosystem to larger changes, such as massive erosion events or continued mineralization of soil carbon and nitrogen (figure 12.2c). Midterm to long-term impacts are again largely influenced by which plant species dominate in the community over time. For example, in the most intensely affected areas of jarrah forest, forests were completely replaced by sedges and rushes (Weste 2003; Weste et al. 2002). Pathogens can determine which plant species establish over the mid- to long term by directly excluding host plants through persistent infection. Pathogen-induced ecosystem changes may also affect which plant species establish at a site. The classic disturbance curve also fails to account for the situation of pathogens gradually killing their hosts. In these cases there is often substantial overlap between the stages of pathogen impact and initial community response, resulting in minimal disturbance (figure 12.2d). For example, beech bark disease and Dutch elm disease (*Ophiostoma ulmi* or *O. novo-elmi*) often cause gradual mortality of individual trees within a mixed-species forest. As branches slowly die, neighboring overstory trees expand their own canopy, minimizing gaps (Agrios 2005). Chestnut blight causes a similarly slow die-off of individuals (Woods and Shanks 1959), but because the American chestnut was so dominant in many forests in the eastern United States, neighboring trees were often chestnuts that were also infected and could not compensate for the slow decline of neighbors, leading to small gaps in the

---

pathogen-induced host mortality and initial recovery by the community can lead to larger ecosystem changes than predicted based on the classic disturbance curve in a. Dashed lines as in b. (d) The ecosystem effects of disturbance by pathogens can be minimal when the impact stage overlaps with the initial recovery stage (e.g., when an individual gradually dies and neighbors immediately fill in the empty spaces in the canopy). Dashed lines as in b.

forest. Such gaps can increase or decrease rates of nitrogen cycling, depending on the forest structure. For example, in one study, nitrogen cycling rates increased in forests with little or no understory, similar to classic disturbance scenarios (see figure 12.2a). However, in forests with an understory, the understory plants rapidly filled in the gaps created by the death of canopy trees (figure 12.2d), so that there was no disturbance-induced stimulation of nitrogen cycling (Mladenoff 1987).

Disturbances occur over a continuous range of spatial scales, ranging from single-plant gaps to large regional disturbances. Many pathogens act at scales intermediate to these, and others move from gap agents to large patch disturbances as the disease progresses (e.g., root rots) (Agrios 2005; McCauley and Cook 1980). It is unclear whether the ecosystem effects of pathogens increase linearly with infected patch size or whether there are some thresholds below which pathogen impacts on ecosystems are minimal and above which they are substantial. Batini et al. (1980) proposed that substantial shifts in water yield occurred once pathogens killed 20% of trees in a forest. It is also critical to consider how mortality is distributed within an area. For example, for a fixed area of pathogen infection in a landscape, the magnitude of ecosystem changes will likely be much higher when the infection is in one large patch compared to infection of isolated gaps randomly distributed throughout the landscape.

The well-developed fields of disturbance ecology and species effects on ecosystems provide a strong theoretical basis that we can use to better understand the ecosystem impacts of large pathogen-induced mortality events of mature individuals. In the next section we explore the more subtle roles of pathogens in ecosystems.

Pathogens and Ecosystem Dynamics: Beyond the Disturbance Framework

Ecosystem ecologists have largely focused on pathogens as disturbance agents; mostly because the mortality of mature individuals is an obvious occurrence that can have an impact on ecosystems. However, these highly visible effects of pathogens are a small subset of the roles that pathogens play in ecosystems. It is easy to overlook the role of pathogens in regulating host population and community dynamics through pathogen-induced changes in host traits or reproduction, or when mortality occurs in immature individuals and stands. A wide range of pathogens is present but often unnoticed in natural ecosystems, as indicated by agricultural studies, which demonstrate that natural ecosystems commonly act as reservoirs for many crop diseases (reviewed in Power and Mitchell 2004). Kranz (1990) suggested that 58%–78% of disease cases in mixed plant populations go unnoticed. Is this because pathogens are present in low numbers in natural systems? Are the natural hosts less susceptible than crop hosts? Or is the presence of pathogens apparent if we were to look more closely at these systems? Undoubtedly there are individual cases that fall into any of these categories. For example, some studies have shown that pathogens are less likely to cause epidemics in natural systems (Dinoor and Eshed 1984), while others have demonstrated that pathogen impacts on natural systems can be as severe as in crop systems (Holah et al. 1993; Kranz 1990). In this section, we explore potential cases in which pathogens play a critical, but often overlooked role in the functioning of ecosystems. In contrast to considering how pathogen-induced disturbances alter ecosystems (see figure 12.2), this section explores how pathogens help to determine the *baseline* of ecosystem function, so that by removing pathogens, we would see changes in ecosystem function (figure 12.3).

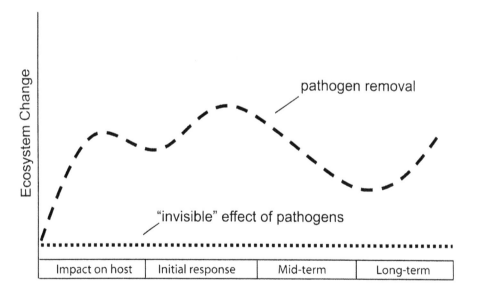

Figure 12.3. Pathogens can be critical players in the normal functioning of ecosystems. Their activity may be responsible for establishing the baseline of ecosystem function (dotted line), so that only on removal of a pathogen would a significant change in ecosystem functioning be apparent (dashed line).

Pathogens and Self-Thinning

Although we often focus on disturbances caused by pathogen-induced mortality of mature plants, most pathogen-induced plant mortality occurs at the seedling stage (Gilbert 2002) and is mediated by generalist pathogens (Kranz 1990). Self-thinning of seedlings is a critical population and community process in most plant communities (Dewar 1993; Heady 1958; Puettmann et al. 1992; Sims and Coupland 1979; Westoby 1981), and pathogens can greatly increase thinning rates in herbaceous (Spedding and Diekmahns 1972; Paul and Ayres 1986; Malmstrom et al. 2007) and woody (Castello et al. 1995) ecosystems.

Although we are not aware of any studies that have demonstrated the ecosystem effects of pathogen-induced thinning, seedling thinning plays an important role in ecosystem dynamics. Since thinning occurs at peak plant growth and competition (Heady 1958), any nutrients released from dying individuals are likely to be sequestered in growing plants. This sequestration accounts for the observations that nutrient loss is minimal and nutrient use is optimal at the self-thinning stage of stand development (Bormann and Likens 1979; Midgley 2001). In California annual grasslands, self-thinning of seedlings provides as much plant-available nitrogen as does decomposition of plant litter senesced at the end of the growing season (Eviner and Vaughn 2007). In other systems, thinning of seedlings can actually enhance the establishment of surviving seedlings (reviewed in Gilbert 2002) by initially taking up space and resources, and releasing them to surviving seedlings later in the growing season (Smith 1977).

Pathogens Alter Species Composition

Pathogens have been shown to play large roles in structuring plant community composition and diversity (reviewed in Gilbert 2002; see also Collinge et al., chapter 6, Clay et al., chapter 7, this volume). Such changes in plant species composition (reviewed in Eviner and Chapin 2003) and plant diversity (reviewed in Hooper et al. 2005) can have large effects on ecosystem processes. For example, a crown rust (*Puccinia coronata)* can change a pasture dominated by ryegrass (*Lolium perenne*) to one dominated by clover (*Trifolium repens*; Burdon and Shattock 1980), and such shifts from a grass to a legume can have a large impact on the pools and fluxes of carbon and nutrients (Eviner et al. 2006). Similarly, in a stand of *Setaria viridis* and *Bromus hordeaceus,* barley yellow dwarf virus (BYDV) decreased the fitness of *S. viridis* but had no effect on *B. hordeaceus* (Remold 2002). *Setaria* is a C4 grass and *Bromus* is a C3 grass. Since C4 grasses tend to contain higher lignin concentrations and are less digestible than C3 grasses (Trlica 2005), BYDV likely increased rates of decomposition and nitrogen cycling in this system. Such species shifts are usually attributed to competition, and the roles of pathogens in mediating these species shifts are often overlooked.

In some ecosystems, pathogens are major drivers of plant successional dynamics (Holah et al. 1997; USDA 2004). Pathogens can speed succession by increasing the competitive ability of species in the next successional stage, or delay succession by shifting competitive dominance to early successional species or by decreasing nitrogen fixers in primary succession (Dobson and Crawley 1994; Holah et al. 1997). Pathogen-induced successional shifts from herbaceous to woody species in dunes (van der Putten et al. 1993) have the potential to significantly alter erosion rates, carbon storage, water fluxes, and nutrient dynamics (reviewed in Eviner and Chapin 2003).

Although it is clear that pathogen-induced changes in plant species composition can alter ecosystems, there are few data with which to evaluate how often these shifts in plant species result in changes to ecosystem functions such as productivity and nutrient cycling. Do pathogens tend to favor species that are functionally different from the plants they displaced? Lignin, phenolics, and other secondary compounds can be responsible for plant resistance to certain pathogens (Agrios 2005), and pathogen-induced shifts to species with high concentrations of these compounds could result in large changes in ecosystems because these substances can be major drivers of plant effects on biogeochemical cycling (Fierer et al. 2001; Hattenschwiler and Vitousek 2000; Schimel et al. 1996, 1998). Although pathogens have the potential to cause such functional shifts in plant communities, it is unclear if such shifts are common in natural ecosystems.

Pathogens Alter Plant Traits

Most of the discussion in this chapter has focused on pathogens as agents of mortality; however, host plants can live for many years before succumbing to pathogens, and many pathogens do not directly cause mortality. Pathogen-induced changes in plant physiology and morphology have been well documented in plants with commercial value (reviewed in Agrios 2005) but are more difficult to detect in natural systems, owing to the large variation in plant traits across individuals in wild populations (Remold 2002). Pathogens (particularly viruses) can induce morphological changes in plants, such as stunting, dwarfing, brooming, and changes in tillering rates (Agrios 2005). Such changes in structure can influence microclimate and habitat for other key organisms in the ecosystem (reviewed in Eviner and Chapin 2003). Many pathogens cause extensive root damage or decreases in root versus shoot allocation (Agrios 2005; Holah and Alexander 1999; Mitchell 2003), without an accompanying change in aboveground biomass (Agrios 2005; Mitchell 2003). Such declines in root biomass can decrease rhizosphere stimulation of soil biogeochemical cycles, decrease root contributions to soil carbon and nutrient storage, and increase susceptibility for soil erosion and landslide events (reviewed in Eviner and Chapin 2003). Decreases in root biomass can also increase susceptibility to windthrow (Robbins 1988) and mortality by repeated grazing (Smith et al. 1986). On the flip side, pathogen-induced increases in tillering can increase the ability of grasses to withstand heavy grazing (Kranz 1990).

Pathogens can also cause large changes in plant tissue chemistry, which is a key regulator of plant effects on carbon and nutrient dynamics (Hobbie 1992; Hattenschwiler and Vitousek 2000; Fierer et al. 2001). In many plants, infection by pathogens induces secondary compounds with strong antimicrobial properties, including lignin, tannins, gums, phenolics, alkaloids, isoflavanoids, terpenoids, and proteins with antimicrobial activity (e.g., thionins, chitinase) (Belanger et al. 2003; Cao et al. 2001; Daayf et al. 2003; De Ascensao and Dubery 2000; Ishimoto et al. 2004; Krischick et al. 1991; Pegg and Ayres 1988; Ramamoorthy et al. 2002; Saunders and O'Neill 2004; Thangavelu et al. 2003; Thipyapong et al. 2004; Witzell and Shevtsova 2004; reviewed in Agrios 2005). For example, pathogens have been shown to increase lignin concentrations by 26% in alfalfa (*Medicago sativa*) (Lenssen et al. 1992), phenolics by 25%–200% in bilberry (*Vaccinium myrtillus*) (Witzell and Shevtsova 2004), and soluble phenolics by 30-fold in potato tubers (*Solanum tuberosum*) (Agrios 2005). In general, pathogens are less susceptible to these compounds than nonpathogenic fungi (Agrios 2005), suggesting the potential for large changes in microbially mediated biogeochemical cycling. These compounds can decrease feed quality, palatability, and digestibility (Clarke and Eagling 1994; Lewis et al. 1996), change soil microbial community structure (Rumberger and Marschner 2003), and alter carbon and nutrient cycling (Fierer et al. 2001; Hattenschwiler and Vitousek 2000; Schimel et al. 1996, 1998). For example, isothiocyanates decrease ammonium oxidizer populations and inhibit nitrification rates (Bending and Lincoln 2000).

Pathogens as Castrators

Pathogens can cause significant decreases in seed yield even without appreciable effects on living biomass (Spedding and Diekmahns 1972). Hudson and Dobson (1995) have hypothesized that pathogens that result in host mortality can stabilize host population density, while pathogens that reduce fecundity can result in population crashes. However, when the host species has a substantial seed bank, the composition of the community could be resilient to a temporary decrease in seed yield. When boom and bust cycles do occur, they have the potential to alter ecosystem processes when the susceptible host population is replaced by species with different functional attributes.

From Case Studies to General Principles

A number of case studies have demonstrated that pathogens can play important roles in mediating ecosystem dynamics, but pathogens are rarely included in conceptual models of ecosystem function. This absence is likely due to the lack of a generalizable, predictive framework to assess the types and magnitude of ecosystem impacts of pathogens. Some ecologists have integrated pathogens into ecosystems through their role as disturbance agents. However, it is critical to move beyond the largely artificial designation of pathogens as disturbances versus part of natural systems because we are biased in how we categorize pathogen activities. For example, we can recognize the visibly obvious signs of dieback of individual trees in forests, but tend to overlook the same process in grasslands, where it can be nearly impossible to track individual plants. The seven general principles outlined at the beginning of this chapter provide a basis to build a framework that moves beyond the artificial designation of pathogens as disturbances or intrinsic parts of ecosystems. For example, when we use the seven principles to compare pathogen-induced self-thinning to pathogens that cause disturbances (as small and large gaps) (table 12.1), the only difference is the life stage of hosts that the pathogens infect. It is likely that thinning is viewed as an invisible effect of pathogens rather than as a disturbance because ecologists don't tend to track individuals in intensely crowded juvenile stands, and tend to notice deaths only when they leave empty spaces in the canopy. The seven general principles allow for a more ecologically meaningful categorization of the activity of pathogens that will strengthen our ability to predict the types and magnitude of ecosystem effects that pathogens can have. For example, ecologists are more likely to recognize the role of pathogens in shaping community dynamics in forests than in grasslands because the signs of individual infection and mortality are far more visible in a forest. Recognizing that these processes are similar allows us to integrate our understanding of interactions between infected individuals, neighbors, and biogeochemical

TABLE 12.1

Comparison of different roles of pathogens according to the seven general principles that determine the impact of pathogens on ecosystems

	Large Gap	Small Gap	Thinning	Competition	Succession	Change in Host Traits
Impact	Mortality	Mortality	Mortality	Mortality/shifts in physiology, behavior	Mortality	Shift in physiology, behavior
Life stage	Mature	Mature	Juvenile to immature	Juvenile to mature	Immature to mature	Juvenile to mature
% Infected	High	Low	High	Low to high	Medium to high	Low to high
Spatial concentration	High	Low	Medium	Low to high	Low to high	Low to high
Timing						
Effect on host	Slow to rapid	Slow	Rapid	Slow to Rapid	Slow to rapid	Slow to rapid
Recovery	Slow	Rapid	Rapid	Rapid	Rapid	—
Likelihood of functional change	Low to high; species shift	Low to high; species shift	Low to medium; species shift	Low to high; species shift	Low to high; species shift	Low to high; shift in effect of the same species
Frequency of recurrence	Low to medium	Medium to high	Medium to high	Medium to high	Low to high	Low to high
Ecosystem functions:						
Nutrient retention	*	****	*****	Depends on species shift	Depends on species shift	Depends on trait shift
Nutrient turnover	****	**	*****			
Pathogens	Fungi, nematodes	Fungi	Fungi, bacteria	Fungi, bacteria, viruses	Fungi	Viruses, fungi, nematodes, bacteria

* indicates degree of impact.

fluxes that are easily documented in forests, with ecosystem-scale measures of yield and input/output budgets that are more feasibly studied in grasslands.

Linking the characteristics in table 12.1 to predictions of pathogen effects on the magnitude and type of ecosystem impact will require further investigation into a number of areas. These include:

• What is the relative importance of each of these factors in determining pathogen impacts on ecosystems? Do they vary by ecosystem and stage of ecosystem development? By pathogen type?

• How do interactions among these factors determine ecosystem dynamics?

• Are there tendencies for some of these factors to covary with one another?

• How often do different ecosystem processes covary versus vary independently in their response to these multiple factors? For example, the rates of nutrient retention and nutrient turnover are both very high in self-thinning stands, but in stand-replacing events, nutrient retention is low, while nutrient turnover remains high (see table 12.1).

• Do the ecosystem effects of pathogens vary depending on the type of pathogen? Most of the plant disease literature focuses on fungal pathogens, likely because 10,000 fungal species have been identified as plant diseases, compared with 100 bacterial species (Agrios 2005). But the impacts of fungi on host plants can be very different from the impacts of other pathogens such as viruses (Agrios 2005).

Understanding ecosystems also requires considering the interactions between different pathogens, and between pathogens and other biotic and abiotic factors. A large amount of plant damage and mortality is a result of a complex of diseases and their interactions with insects and abiotic disturbance agents (Castello et al. 1995; Worrall and Harrington 1988).

Ecosystem Impact of Nonplant Pathogens

Thus far we have focused on plant pathogens; however, pathogens of herbivores, predators, and decomposers can also cause large-scale changes in ecosystem function (Agrios 2005; Briggs et al. 1995; Lovett et al. 2006; Prins and van der Jeugd 1993; Read 2003; Middelboe, chapter 11, this volume). The most dramatic ecosystem effects of animal and insect pathogens occur through changes in vegetation composition. For example, the decline of impala populations due to anthrax led to shrub encroachment in northern Tanzania (Prins and van der Jeugd 1993). In the Serengeti, the decline of herbivores due to rinderpest increased the fuel load, which stimulated fires and led to the decline of the bush (reviewed in Prins and van der Jeugd 1993). Pathogens can be key regulators of insect outbreaks (reviewed in Briggs et al. 1995), allowing for the persistence of plants that are susceptible to insect attack. The decline of rabbits due to myxomatosis led to the development of an oak woodland in England (reviewed in Dobson and Crawley 1994) and significantly increased the establishment of native species in Australia (Read 2003). Rabbits not only decimated vegetation in the Australian outback, they also produced burrows, provided food for predators, increased soil turnover, and affected water penetration and runoff. Because some native Australian mammals are now rare or extinct, some activities such as soil digging are not replaced when rabbits decline due to myxomatosis or rabbit hemorrhagic disease, and ecosystem processes vary in space and time due to disease-induced fluctuations in rabbit populations (Read 2003). Decomposer organisms are also likely affected by pathogens. Pathogens can retard decomposition by limiting decomposer populations and can stimulate decomposition by inducing turnover of microbial biomass. For example, viruses that infect soil bacteria could potentially increase rates of carbon and nutrient cycling by stimulating bacterial turnover, as occurs in marine systems (Middelboe, chapter 11, this volume).

How Well Can We Predict Ecosystem Processes without Considering Pathogens?

Although pathogens may mediate thinning, succession, and competition in many systems, do we need to explicitly understand the roles that pathogens play when these population and community processes are often predictable and repeatable? Why should we further complicate ecosystem models by explicitly incorporating the role of pathogens, rather than just looking at host population and community dynamics? Global change studies have demonstrated that we must account for the response of pathogens to a changing environment in order to understand host physiological, population, and community responses. For example, the presence of pathogens influences plant community response to nitrogen additions in boreal forests (Strengbom et al. 2002) and host physiological response to elevated CO_2 (Malmstrom and Field 1997).

As is discussed in the following chapter, pathogen-induced changes in disturbance regimes, such as increased susceptibility to fire (Castello et al. 1995), windthrow (Worrall and Harrington 1988), or erosion (Wondzell 2001), could lead to state changes in the ecosystem that would not be predictable simply from host population dynamics. Some of the most significant ecosystem effects of pathogens are determined by human management of ecosystems to prevent disease and in response to pathogen outbreaks. For example, salvage logging in response to tree pathogen outbreaks accelerates soil erosion because of the increased disturbance of vegetation and soil associated with harvesting and transporting timber (Wondzell 2001). Similarly, the heavy use of antibiotics in humans and livestock has increased the prevalence of antibiotic-resistant microbes in natural systems (Burgos et al. 2005; Kolpin et al. 2002; Lateef et al. 2005; Mallin 2000). Since microbially produced antibiotics play a role in structuring microbial interactions (Burgess et al. 1999; Davelos et al. 2004; Zvenigorodskii et al. 2004), antibiotic resistance can lead to shifts in microbial community composition that can alter microbially mediated biogeochemical processes (Vaclavik et al. 2004).

We may also need to explicitly include pathogens in ecosystem models because they can represent a major pool of carbon and nutrients in many ecosystems (Mitchell 2003; Smith et al. 1992; Lafferty, chapter 9, this volume). The C and nutrients in pathogen pools may differ from other ecosystem pools in terms of how they respond to seasonal and environmental changes, their turnover times, and how labile they are. They also may drive changes in other ecosystem pools (e.g., plant or soil pools). Thus, explicitly including pathogen carbon and nutrient pools will increase our understanding of biogeochemical pools and fluxes in some ecosystems.

CONCLUSIONS

Whether their presence is readily apparent or not, pathogens play a key role in mediating many ecosystem processes. In order to integrate pathogens into our conceptual understanding of ecosystems, it is critical to consider how pathogen impacts will vary over time. Ecologists need to move beyond case studies to generalizations that can be used to predict the type and magnitude of ecosystem impacts of a pathogen. The seven key principles discussed in this chapter can guide generalizations of ecological response to disease, but ultimately, the largest ecosystem changes will arise from how humans manage ecosystems in response to disease.

ACKNOWLEDGMENTS

This is a contribution to the program of the Institute of Ecosystem Studies. We'd like to thank Kathleen Weathers, Gary Lovett, and Terry Chapin, for stimulating discussions on ecosystems and disease.

LITERATURE CITED

Aber, J. D., and J. M. Melillo. 2001. Terrestrial Ecosystems, 2nd ed. Harcourt-San Diego: Academic Press.

Agrios, George N. 2005. Plant Pathology, 5th ed. New York: Academic Press.

Bari, M. A., and J. K. Ruprecht. 2003. Water yield response to land use change in SW Western Australia. Salinity and Landuse Impact Series 31. Department of Environment, Perth, Australia.

Batini, F. E., R. E. Black, J. Byrne, and P. J. Clifford. 1980. An examination of the effects of changes in catchment condition on water yield in the Wungong catchment, Western Australia. Australian Journal of Forest Research 10: 29–38.

Belanger, R. R., N. Benhamou, and J. G. Menzies. 2003. Cytological evidence of an active role of silicon in wheat resistance to powdery mildew (*Blumeria graminis* f. sp *tritici*). Phytopathology 93:402–12.

Bending, G. D., and S. D. Lincoln. 2000. Inhibition of soil nitrifying bacteria communities and their activities by glucosinolate hydrolysis products. Soil Biology & Biochemistry 32:1261–69.

Bormann, F. H., and G. E. Likens. 1979. Pattern and Process in a Forested Ecosystem. New York: Springer-Verlag.

Briggs, C. J., R. S. Hails, N. D. Barlow, and H. C. J. Godfray. 1995. The dynamics of insect-pathogen interactions. *In* Population Biology of Infectious Diseases in Natural Populations, ed. B. T. Grenfell and A. P. Dobson, 295–326. Cambridge: Cambridge University Press.

Burdon, J. J. 1991. Fungal pathogens as selective forces in plant-populations and communities. Australian Journal of Ecology 16:423–32.

Burdon, J. J., and R. C. Shattock. 1980. Disease in plant communities. Applied Biology 5:145–219.

Burgess, J. G., E. M. Jordan, M. Bregu, A. Mearns-Spragg, and K. G. Boyd. 1999. Microbial antagonism: A neglected avenue of natural products research. Journal of Biotechnology 70:27–32.

Burgos, J. M., B. A. Ellington, and M. F. Varela. 2005. Presence of multidrug-resistant enteric bacteria in dairy farm topsoil. Journal of Dairy Science 88:1391–98.

Cao, H., L. Baldini, and L. G. Rahme. 2001. Common mechanisms for pathogens of plants and animals. Annual Review of Phytopathology 39:259–84.

Castello, J. D., D. J. Leopold, and P. J. Smallidge. 1995. Pathogens, patterns, and processes in forest ecosystems. Bioscience 45:16–24.

Chapin, F. S., III, P. A. Matson, and H. A. Mooney. 2002. Principles of Terrestrial Ecosystem Ecology. New York: Springer-Verlag.

Clarke, R. G., and D. R. Eagling. 1994. Effects of pathogens on perennial pasture grasses. New Zealand Journal of Agricultural Research 37:319–27.

Cromack, K., J. A. Entry, and T. Savage. 1991. The effect of disturbance by *Phellinus-weirii* on decomposition and nutrient mineralization in a tsuga-mertensiana forest. Biology and Fertility of Soils 11:245–49.

Daayf, F., M. El Bellaj, M. El Hassni, F. J'Aiti, and I. El Hadrami. 2003. Elicitation of soluble phenolics in date palm (*Phoenix dactylifera*) callus by *Fusarium oxysporum* f. sp *albedinis* culture medium. Environmental and Experimental Botany 49:41–47.

Davelos, A. L., L. L. Kinkel, and D. A. Samac. 2004. Spatial variation in frequency and intensity of antibiotic interactions among streptomycetes from prairie soil. Applied and Environmental Microbiology 70:1051–58.

De Ascensao, A. R. D. C., and I. A. Dubery. 2000. Panama disease: Cell wall reinforcement in banana roots in response to elicitors from *Fusarium oxysporum* f. sp *cubense* race four. Phytopathology 90:1173–80.

Dewar, R. 1993. A mechanistic analysis of self-thinning in terms of the carbon balance of trees. Annals of Botany 71:147–159.

Dickman, A. 1992. Plant pathogens and long term ecosystem changes. *In* The Fungal Community: Its Organization and Role in the Ecosystem, 2nd ed., ed. G. C. Carroll and D. T. Wicklow, 499–520. New York: Marcell Dekker.

Dinoor, A., and N. Eshed. 1984. The role and importance of pathogens in natural plant-communities. Annual Review of Phytopathology 22:443–66.

Dobson, A., and W. Crawley. 1994. Pathogens and the structure of plant-communities. Trends in Ecology & Evolution 9:393–398.

Eviner, V. T., and F. S. Chapin III. 2003. Functional matrix: A conceptual framework for predicting multiple plant effects on ecosystem processes. Annual Reviews in Ecology and Systematics 34:455–85.

Eviner, V. T., and C. E. Vaughn. 2007. Seedling thinning contributes as much N to internal N cycling as does decomposition of senesced litter. Unpublished manuscript.

Eviner, V. T., F. S Chapin III, and C. E. Vaughn. 2006. Seasonal variations in plant species effects on soil N and P dynamics. Ecology 87:974–86.

Fierer, N., J. Schimel, R. Cates, and J. Zou. 2001. Influence of balsam poplar tannin fractions on carbon and nitrogen dynamics in Alaskan taiga floodplain soils. Soil Biology & Biochemistry 33:1827–39.

Gilbert, G. S. 2002. Evolutionary ecology of plant diseases in natural ecosystems. Annual Review of Phytopathology 40:13–43.

Graniti, A. 1998. Cypress canker: A pandemic in progress. Annual Review of Phytopathology 36:91–114.

Hattenschwiler, S., and P. M. Vitousek. 2000. The role of polyphenols in terrestrial ecosystem nutrient cycling. Trends in Ecology & Evolution 15: 238–43.

Heady, H. 1958. Vegetational changes in the California annual type. Ecology 39:402–16.

Henry, M., R. I. B. Francki, and H. Wallwork. 1992. Occurrence of barley yellow dwarf virus in cereals and grasses of the low-rainfall wheat-belt of south-Australia. Plant Pathology 41:713–21.

Hobara, S., N. Tokuchi, N. Ohte, K. Koba, M. Katsuyama, S. J. Kim, and A. Nakanishi. 2001. Mechanism of nitrate loss from a forested catchment following a small-scale, natural disturbance. Canadian Journal of Forest Research–Revue Canadienne de Recherche Forestiere 31:1326–35.

Hobbie, S. E. 1992. Effects of plant-species on nutrient cycling. Trends in Ecology & Evolution 7:336–39.

Holah, J. C. and H. M. Alexander. 1999. Soil pathogenic fungi have the potential to affect the co-existence of two tallgrass prairie species. Journal of Ecology 87:598–608.

Holah, J. C., M. V. Wilson, and E. M. Hansen. 1993. Effects of a native forest pathogen, phellinus-weirii, on douglas-fir forest composition in western Oregon. Canadian Journal of Forest Research-Revue Canadienne de Recherche Forestiere 23:2473–80.

Holah, J. C., M. V. Wilson, and E. M. Hansen. 1997. Impacts of a native root-rotting pathogen on successional development of old-growth Douglas fir forests. Oecologia 111:429–33.

Hooper, D. U., F. S. Chapin III, J. J. Ewel, A. Hector, P. Inchausti, S. Lavorel, J. H. Lawton, D. M. Lodge, M. Loreau, et al. 2005. Effects of biodiversity on ecosystem functioning: A consensus of current knowledge. Ecological Monographs 75:3–35.

Hudson, P. J., and A. P. Dobson. 1995. Microparasites: Observed patterns. In Ecology of Infectious Diseases in Natural Populations, ed. B. T. Grenfell and A. P. Dobson, 144–76. Cambridge: Cambridge University Press.

Ishimoto, H., Y. Fukushi, and S. Tahara. 2004. Non-pathogenic *Fusarium* strains protect the seedlings of *Lepidium sativum* from *Pythium ultimum*. Soil Biology & Biochemistry 36:409–14.

Jaynes, R., and J. Elliston. 1982. Hypovirulent isolates of *Endothia parasitica* associated with large American chestnut trees. Plant Disease 66:769–72.

Johnson, A. C., and P. Wilcock. 2002. Association between cedar decline and hillslope stability in mountainous regions of southeast Alaska. Geomorphology 46:129–42.

Kolpin, D. W., E. T. Furlong, M. T. Meyer, E. M. Thurman, S. D. Zaugg, L. B. Barber, and H. T. Butxton. 2002. Pharmaceuticals, hormones, and other waste water contaminants in US streams, 1999–2000. A national reconnaissance. Environmental Science and Technology 36:1202–11.

Kranz, J. 1990. Fungal diseases in multispecies plant communities. New Phytologist 116:383–405.

Krischick, V. A., R. W. Goth, and P. Barbosa. 1991. Generalized plant defense: Effects on multiple species. Oecologia 85:562–71.

Lateef, A., J. K. Oloke, and E. B. Gueguimkana. 2005. The prevalence of bacterial resistance in clinical, food, water and some environmental samples in Southwest Nigeria. Environmental Monitoring and Assessment 100:59–69.

Lenssen, A. W., E. L. Sorensen, G. L. Posler, and D. L. Stuteville. 1992. Forage quality of alfalfa protected by resistance to bacterial leaf-spot. Animal Feed Science and Technology 39:61–70.

Lewis, G. C., R. H. Lavender, and T. M. Martyn. 1996. The effect of propiconazole on foliar fungal diseases, herbage yield and quality of perennial ryegrass. Crop Protection 15:91–95.

Loucks, O. L. 1970. Evolution of diversity, efficiency and community stability. Annals of Zoology 10:17–25.

Lovett, G., C. Canham, M. Arthur, K. Weathers, and R. Fitzhugh. 2006. Forest ecosystem responses to exotic pests and pathogens in eastern North America. BioScience 56:395–405.

Mallin, M. A. 2000. Impacts of industrial animal production on rivers and estuaries. American Scientist 88:26–37.

Malmstrom, C. M., C. J. Stoner, S. Brandenburg, and L. A. Newton. 2006. Grazing exert counteracting influences on survivorship of native bunchgrass seedlings competing with invasive exotics. Journal of Ecology 14:264–75.

Malmstrom, C. M., and C. B. Field. 1997. Virus-induced differences in the response of oat plants to elevated carbon dioxide. Plant Cell and Environment 20:178–88.

Marschner, H. 1995. Mineral Nutrition of Higher Plants. London: Academic Press.

Matson, P. A., and R. D. Boone. 1984. Natural disturbance and nitrogen mineralization: Wave-form dieback of mountain hemlock in the Oregon Cascades. Ecology 65:1511–16.

McCauley, K. J., and S. A. Cook. 1980. *Phellinus-weirii* infestation of 2 mountain hemlock forests in the Oregon Cascades. Forest Science 26:23–29.

Midgley, J. 2001. Do mixed-species mixed-size indigenous forests also follow the self-thinning line? Trends in Ecology & Evolution 16:661–62.

Mitchell, C. E. 2003. Trophic control of grassland production and biomass by pathogens. Ecology Letters 6:147–55.

Mladenoff, D. 1987. Dynamics of nitrogen mineralization and nitrification in hemlock and hardwood treefall gaps. Ecology 68:1171–80.

Ohte, N., N. Tokuchi, M. Katsuyama, S. Hobara, Y. Asano, and K. Koba. 2003. Episodic increases in nitrate concentrations in streamwater due to the partial dieback of a pine forest in Japan: Runoff generation processes control seasonality. Hydrological Processes 17:237–49.

Paillet, F. 1988. Character and distribution of American chestnut sprouts in southern New England woodlands. Bulletin of the Torrey Botanical Club 115:32–44.

Paul, N. D., and P. G. Ayres. 1986. Interference between healthy and rusted groundsel (senecio-vulgaris l) within mixed populations of different densities and proportions. New Phytologist 104:257–69.

Pegg, G. F., and P. G. Ayres. 1988. Fungal Infection of Plants. Cambridge: Cambridge University Press.

Power, A. G., and C. E. Mitchell. 2004. Pathogen spillover in disease epidemics. American Naturalist 164:S79–89.

Prins, H. H. T., and H. P. van der Jeugd. 1993. Herbivore population crashes and woodland structure in east-Africa. Journal of Ecology 81:305–14.

Puettmann, K., D. Hibbs, and D. Hann. 1992. The dynamics of mixed stands of *Alnus rubra* and *Pseudotsuga menziesii*: Extension of size-density analysis to species mixtures. Journal of Ecology 80:449–58.

Ramamoorthy,V., T. Raguchander, and R. Samiyappan. 2002. Induction of defense-related proteins in tomato roots treated with *Pseudomonas fluorescens* Pf1 and *Fusarium oxysporum* f. sp lycopersici. Plant and Soil 239: 55–68.

Read, J. L. 2003. Red Sand, Green Heart. Melbourne, Australia: Lothian Books.

Remold, S. K. 2002. Unapparent virus infection and host fitness in three weedy grass species. Journal of Ecology 90:967–77.

Robbins, K. 1988. Annosus root rot in eastern conifers. Forest Insect and Disease Leaflet 76. USDA Forest Service, Washington, DC.

Rumberger, A., and P. Marschner. 2003. 2-Phenylethylisothiocyanate concentration and microbial community composition in the rhizosphere of canola. Soil Biology & Biochemistry 35:445–52.

Saunders, J. A., and N. R. O'Neill. 2004. The characterization of defense responses to fungal infection in alfalfa. Biocontrol 49:715–28.

Schimel, J. P., K. Van Cleve, R. G. Cates, T. P. Clausen, and P. B. Reichardt. 1996. Effects of balsam poplar (*Populus balsamifera*) tannins and low molecular weight phenolics on microbial activity in taiga floodplain soil: Implications for changes in N cycling during succession. Canadian Journal of Botany–Revue Canadienne de Botanique 74:84–90.

Schimel, J. P., R. G. Cates, and R. Ruess. 1998. The role of balsam poplar secondary chemicals in controlling soil nutrient dynamics through succession in the Alaskan taiga. Biogeochemistry 42:221–34.

Sims, P. L., and R. T. Coupland. 1979. Producers. *In* Grassland Ecosystems of the World, ed. R.T. Coupland, 49–72. Cambridge: Cambridge University Press.

Smith, A. P. 1977. Albinism in relation to competition in bamboo *Phyllostachys bambusoides*. Nature 266:527–29.

Smith, M. A., C. E. Watson, V. H. Watson, and L. E. Trevathan. 1986. Effects of drechslera-sorokiniana infection on the yield and quality of tall fescue forage. Animal Feed Science and Technology 15:41–46.

Smith, M. L., J. N. Bruhn, and J. B. Anderson. 1992. The fungus armillaria-bulbosa is among the largest and oldest living organisms. Nature 356: 428–31.

Spedding, C., and E. Diekmahns. 1972. Pests and diseases. *In* Grasses and Legumes in British Agriculture. Slough, UK: Commonwealth Bureau of Pasture and Field Crops.

Strengbom, J., A. Nordin, T. Nasholm, and L. Ericson. 2002. Parasitic fungus mediates change in nitrogen-exposed boreal forest vegetation. Journal of Ecology 90:61–67.

Thangavelu, R., A. Palaniswami, S. Doraiswamy, and R.Velazhahan. 2003. The effect of *Pseudomonas fluorescens* and *Fusarium oxysporum* f. sp *cubense* on induction of defense enzymes and phenolics in banana. Biologia Plantarum 46:107–12.

Thipyapong, P., M. D. Hunt, and J. C. Steffens. 2004. Antisense downregulation of polyphenol oxidase results in enhanced disease susceptibility. Planta 220:105–17.

Trlica, M. J. 2005. Grass growth and response to grazing. Colorado State University Cooperative Extension. No. 6.108. Fort Collins, CO.

USDA Forest Service and State of Alaska Department of Natural Resources. 2004. Forest health conditions in Alaska 2004. USDA, Anchorage, AL.

Vaclavik, E., B. Halling-Sorensen, and F. Ingerslev. 2004. Evaluation of manometric respiration tests to assess the effects of veterinary antibiotics in soil. Chemosphere 56:667–676.

van der Putten, W. H., C. Vandijk, and B. A. M. Peters. 1993. Plant-specific soil-borne diseases contribute to succession in foredune vegetation. Nature 362:53–56.

Vitousek, P. M., and W. A. Reiners. 1975. Ecosystem succession and nutrient retention: Hypothesis. Bioscience 25:376–81.

Waring, R. H., K. Cromack, P. A. Matson, R. D. Boone, and S. G. Stafford. 1987. Responses to pathogen-induced disturbance: Decomposition, nutrient availability, and tree vigor. Forestry 60:219–27.

Weste, G. 2003. The dieback cycle in Victorian forests: A 30-year study of changes caused by Phytophthora cinnamomi in Victorian open forests, woodlands and heathlands. Australasian Plant Pathology 32:247–56.

Weste, G., K. Brown, J. Kennedy, and T. Walshe. 2002. *Phytophthora cinnamomi* infestation: A 24 year study of vegetation change in forests and woodlands of the Grampians, Western Victoria. Australian Journal of Botany 50:247–74.

Westoby, M. 1981. The place of self-thinning rule in population dynamics. American Naturalist 118:581–87.

Witzell, J., and A. Shevtsova. 2004. Nitrogen-induced changes in phenolics of Vaccinium myrtillus: Implications for interaction with a parasitic fungus. Journal of Chemical Ecology 30:1937–56.

Wondzell, S. M. 2001. The influence of forest health and protection treatments on erosion and stream sedimentation in forested watersheds of eastern Oregon and Washington. Northwest Science 75:128–40.

Woods, F. W., and R. E. Shanks. 1959. Natural replacement of chestnut by other species in the Great Smoky Mountains National Park. Ecology 40:349–61.

Worrall, J. J., and T. C. Harrington. 1988. Etiology of canopy gaps in spruce-fir forests at Crawford Notch, New Hampshire. Canadian Journal of Forest Research 18:1463–69.

Zvenigorodskii, V. I., A. I. Kuzin, E. M. Shagov, R. R. Azizbekyan, G. M. Zenova, and T. A. Voeikova. 2004. Antagonistic microbes (streptomycetes and bacilli) isolated from different soil types. Eurasian Soil Science 37: 749–54.

CHAPTER THIRTEEN
Disease Effects on Landscape and Regional Systems:
A Resilience Framework

*F. Stuart Chapin III, Valerie T. Eviner, Lee M. Talbot, Bruce A. Wilcox,
Dawn R. Magness, Carol A. Brewer, and Daniel S. Keebler*

SUMMARY

IN THIS CHAPTER WE PRESENT and evaluate a conceptual framework intended to improve predictability of the role of disease in landscape processes of socioecological systems. On a local scale, disease tends to increase the interactions among patches on a landscape, particularly through changes in the frequency and severity of disturbances and increases in the vulnerability of ecosystems to fundamental changes in state in response to such disturbances. In addition, human activities that alter landscape structure or connectivity often increase the likelihood of disease epidemics or landscape sensitivity to disease. In contrast, on global and regional scales, disease tends to reduce landscape connectivity because the social response to these ecological changes is often to restrict trade and migration. Both the *increase* in connectivity on the local scale and the *decrease* in connectivity on regional to global scales can profoundly affect ecosystem structure and functioning.

Resilience theory addresses the factors that make systems either resilient or vulnerable to radical changes in state by considering interactions between the ecological and social components of systems. This provides a framework for predicting the regional impacts of disease and reducing their social consequences. Resilience is facilitated by fostering cultural and ecological legacies that provide seeds for recovery; by maintaining biological, social, institutional, and economic diversity, which increases the options for alternative pathways of recovery; and by fostering learning and innovation, which reduces the potential impacts of disease and increases the likelihood of favorable social change.

INTRODUCTION

Many of the key drivers of landscape and regional processes are changing at unprecedented rates and in ways that are strongly influenced by human activities (Foley et al. 2005; Steffen et al. 2004; Vitousek et al.

1997). Diseases of plants, animals, and people often play a key role in these changes, both as an integral component of socioecological systems (Eviner and Likens, chapter 12, this volume) and as a trigger for changes in state. Yet disease is often viewed as a surprise that cannot be anticipated and therefore is not incorporated into predictive process-based frameworks. Earlier chapters in this book described the impacts of disease on individual organisms, communities, and ecosystems. In this chapter we extend these effects to larger spatial scales—landscapes and regions. We first assess the role of disease in landscape processes through a conceptual framework that relies on the integration of social and ecological systems. Human activities are an integral component of system dynamics on regional scales, as delineated in previous Cary Conferences (Groffman and Likens 1994; McDonnell and Pickett 1993), so we explicitly consider the social and ecological processes that characterize the reciprocal interactions of socioecological systems (Berkes et al. 2003; Machlis et al. 1997) (figure 13.1) and determine the resilience of these

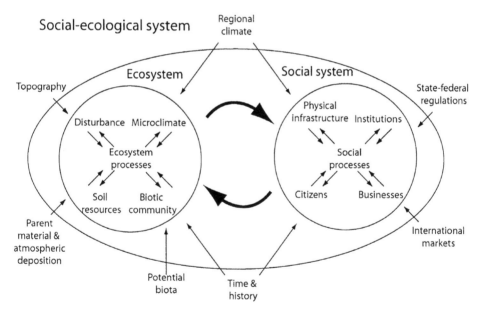

Figure 13.1. A socioecological system. The oval represents the socioecological system and the circles delineate its ecological and social subsystems. Several independent factors, such as climate, history, and international markets, determine the properties of the system; these properties are further modified by interactive controls that determine the internal dynamics of the system. (Reprinted with permission from Whiteman et al. 2004.)

systems. Resilience is the capacity of a system to absorb shocks such as disease outbreaks without changing its fundamental properties, such as its social norms, its typical economy, and the types of species it supports (e.g., grain crops or forests). Using this framework, we describe the effects of disease on landscape processes. We then use resilience theory to explore the role of disease and insect pests in causing changes in the state of socioecological systems, using bark beetles, rinderpest, malaria, and AIDS as examples.

Conceptual Framework

Pathogen and parasite interactions with their hosts are such an integral component of community dynamics that their chronic role in ecosystems is nearly invisible (Horwitz and Wilcox 2005; Lafferty, chapter 9, Eviner and Likens, chapter 12, this volume). As discussed in earlier chapters, diseases often modify competitive interactions, trophic dynamics, disturbance probabilities, and succession. Disease epidemics emerge when the pathogen-host balance shifts to be seriously deleterious to the host (Horwitz and Wilcox 2005), often creating unanticipated surprises (Gunderson 2003) that act as disturbances to alter the structure, composition, and functioning of ecosystems and landscapes (Wilcox and Gubler 2005). Many disease epidemics are triggered by human-induced environmental or biotic changes that shift the host-pathogen balance (Patz 2002; Patz et al. 2000). Here we examine disease from this perspective, considering its role in ecosystems and how ecosystem resilience influences and is influenced by disease events.

Diseases with the greatest immediate ecosystem impacts are generally those that selectively remove or suppress the dominant and keystone plant species (figure 13.2), which, by definition, are the plant species with the greatest effects on ecosystem processes (Chapin et al. 2002). Diseases can have equally dramatic impacts by selectively removing animals that control the abundance of dominant and keystone plants through trophic cascades (Polis 1999). In many aquatic ecosystems, bacteriophages (viruses that attack decomposer organisms) have a profound immediate direct effect on nutrient cycling by stimulating bacterial turnover (Middelboe, chapter 11, this volume). Over longer time scales, diseases influence ecosystem dynamics through their effects on organisms that influence disturbance regimes or species interactions (Eviner and Chapin 2003; Eviner and Likens, chapter 12, this volume).

Disease has equally profound effects on the human component of socioecological systems (figure 13.3). Disease has its greatest effects on social systems when it alters human population densities and activities,

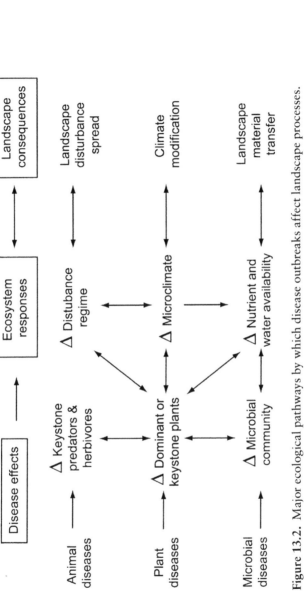

Figure 13.2. Major ecological pathways by which disease outbreaks affect landscape processes.

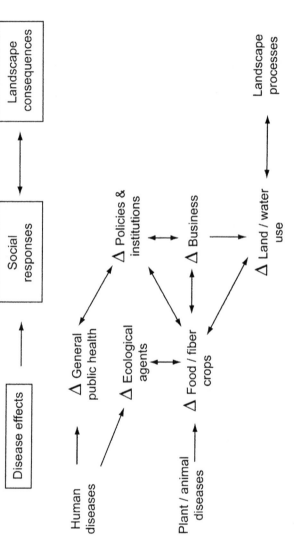

Figure 13.3. Major social pathways by which disease outbreaks affect landscape processes. The effects of disease on social responses are largely mediated by learning and education.

either by directly infecting humans (e.g., bubonic plague, *Yersinia pestis*, malaria, *Plasmodium*, tuberculosis, *Mycobacterium tuberculosis*, sleeping sickness, *Trypanosoma brucei rhodesiense* and *T. b. gambiense*) or by disrupting the food supply (e.g., the Irish famine precipitated by potato late blight, *Phytophthora infestans*, which led to the death of more than a million people and the migration of a million more) (Mc-Neil 1989). Pathogens exert many other effects, including those that alter ecosystem and landscape processes (see figure 13.2) in ways that affect the delivery of ecosystem services to society (Millennium Ecosystem Assessment 2005).

Human diseases often have differential effects on particular segments of society (Turshen 1977). Those diseases that directly affect strong ecological agents, such as farmers, foresters, or pastoralists, have disproportionately large ecological consequences. Malaria and schistosomiasis, for example, have their greatest impacts on rural populations and therefore on agricultural land use (Sachs and Malaney 2002). In addition, diseases that are perceived as posing a general health risk (particularly a risk to privileged segments of society) are more likely to influence policy than are diseases that are largely restricted to disenfranchised populations (Turshen 1977). These policies in turn often influence public health decisions, land and water management, travel, trade, and other human activities. AIDS, for example, initially received only modest research funding in the United States because it was viewed primarily as a disease of homosexuals. It received more attention by policy makers after it became recognized as a health risk for broad segments of society. In contrast, severe acute respiratory syndrome (SARS), which was immediately perceived as a health risk for the general public, particularly global travelers, received more immediate control efforts.

Plant and animal diseases also have important consequences that are mediated by socioecological feedbacks (figure 13.3). Economically important diseases (e.g., foot-and-mouth disease, *Aphthovirus*, soybean rust, *Phakopsora*, white pine blister rust, *Cronartium ribicola*) decrease the productivity of hosts that support the human food and fiber supply and the stability of economic and social systems. These diseases often trigger management actions that remove diseased organisms or alter landscape linkages in an attempt to isolate areas with the disease. An outbreak of foot-and-mouth disease in England, for example, reduced beef exports and severely restricted regional travel in 76% of the country. The restriction on the movement of cattle led to overgrazing of pastures (Fraser of Allander Institute [FAI] 2005). On the flip side, the establishment of Uruguay as a foot-and-mouth disease-free country led to a doubling of its beef exports and a corresponding expansion of pasturelands (Food and Agriculture Organization [FAO] 2001). In this way, trade barriers in one region of the world can alter land use by increasing production of that commodity in distant areas free of the disease (FAO 2001; Yuill 1991). Meanwhile, disease-prone areas change land use to accommodate another economic base. Diseases can have additional longer-term effects by altering lifestyles, policies, institutions, or business frameworks (figure 13.3). The death of livestock in Africa as a result of a short-term outbreak of Rift Valley fever led to the migration of young men from rural to urban areas, altering family structures and increasing the spread of HIV (Preslor 1999). The outbreak of Rift Valley fever in 1997–98 led to a ban on livestock products from eastern Africa by Saudi Arabia, leading to a 75% decrease in livestock exports, which constitute 90% of the foreign exchange in Somaliland. This virtually halted imports of medicine, sugar, and grains, causing the closure of many stores in urban centers (FAO 2001). These examples demonstrate the critical nature and large magnitude of change in socioecological interactions that can be triggered by disease outbreaks.

Disease Effects on Landscape Processes

Diseases affect all classes of interactions among patches on a landscape. In the absence of disease outbreaks, gravitational movement of materials such as water, nutrients, and sediments often dominate the short-term interactions among patches in a landscape (Chapin et al. 2002; Turner et al. 2001). Diseases act like other disturbances that substantially reduce the activity or biomass of dominant plants. In general, this reduces the capacity of ecosystems to retain these materials (Eviner and Likens, chapter 12, this volume), thereby increasing fluxes from one patch to another and increasing overall landscape connectivity. Dieback in Australian jarrah caused by the introduced pathogen *Phytophtora cinnamomi*, for example, increased runoff by 20% (Bari and Ruprecht 2003), and elsewhere long-term dieback caused by this disease increased water yield by 75% (Batini et al. 1980). Disease-associated diebacks can also substantially increase stream nitrate concentrations (Hobara et al. 2001; Ohte et al. 2003). In Alaska, yellow cedar decline led to a 3.8-fold increase in landslides when roots decomposed fifty years after tree death (Johnson and Wilcock 2002). Human diseases that foster migration from rural areas to cities (e.g., Rift Valley fever; FAO 2001) would also likely augment the movement of food and wastes between urban and rural areas—another increase in landscape connectivity.

Landscape processes in turn influence disease (e.g., spatial distribution, infection rates), mainly by limiting or facilitating the movement of organisms among patches (McCallum, chapter 5, this volume). Emerging infectious diseases are those that have recently expanded geographically, increased in epidemic activity, increased in virulence, or are caused by new pathogen variants or novel pathogens. About a third of these diseases are associated with natural or anthropogenic ecotones, that is, boundaries between habitat patches (Despommier et al., 2006). Ecotones result in more frequent contact among novel species, diverse selection pressures, and distinct controls over population densities, dispersal, and movements (Risser 1995). All these factors facilitate pathogen transmission, spread, and evolution and therefore the potential for disease emergence. Malaria and schistosomiasis are examples of diseases whose emergence has been facilitated by ecotonal processes (expansion of forest edge and aquatic-terrestrial zones due to deforestation and water development and dam projects). Disease can also influence landscape structure in ways that alter landscape connectivity. In Africa, for example, the introduction of rinderpest (Morbillivirus) from domestic animals to African wildlife in the nineteenth century profoundly altered the balance of shrubs and grasses, leading to changes in the structure and connectivity of both ecological and social systems. Disease can mediate interactions among landscape processes occurring on different scales. Barriers between landscape patches (e.g., roads, streams, lakes) reduce the likelihood of flea-vectored plague among prairie dog colonies (Johnson and Collinge 2004), thereby promoting the fine-scale heterogeneity created by these colonies. Conversely, policies intended to reduce disease spread can reduce the movement of people and other organisms among landscape patches.

The most dramatic impact of disease and insect pests on landscape processes is through changes in disturbance spread across landscapes. Bark beetles, for example, increase the probability of forest fires, often with longlasting and profound effects (Dickman 1992). In coastal marine ecosystems, coral bleaching in combination with diseases of sea urchins that would otherwise have grazed on reef algae led to catastrophic mortality of reef corals in the western Atlantic in the 1980s (Jackson 2001), making coastal areas more vulnerable to erosion. Kelp forests are sensitive to several impacts of disease. Sea urchins that expand in response to sea star or sea otter diseases (Conrad et al. 2005; Eckert et al. 2000) can graze down kelp (Estes and Palmisano 1974), whereas sea urchin diseases shift the balance in favor of kelp (Lafferty 2004). Kelp forests reduce current speeds, storm surges, and coastal erosion, thereby increasing the suitability of these areas for suburban development (Kay and Alder 1999). This in turn can increase public support for marine habitat protection. A shifting mosaic of disease might therefore lead to a complex matrix of changes in multiple landscape processes.

The net effect of disease outbreaks is to increase connectivity of patches in a landscape through multiple avenues of patch interactions, particularly the propagation of disturbance. If the disease spreads rapidly, the increased connectivity of landscapes might reduce landscape heterogeneity and diversity. If the disease moves slowly because of inherent dispersal limitations or disease management, it might increase landscape heterogeneity. The landscape effects of disease therefore depend on the rate and extent of disease spread relative to other process controls (Marcogliese 2004).

Disease-Induced Changes in State

Resilience Framework

Disease epidemics are disturbances that can radically alter the state of ecosystems and landscapes. Resilience theory, which addresses the factors that make systems either resilient or vulnerable to radical changes in state, therefore provides an appropriate framework for examining the role of epidemic diseases in socioecological systems (Carpenter 2003; Gunderson and Holling 2002; Walker et al. 2004). Resilience is the capacity of a system to absorb shocks such as disease outbreaks without changing its fundamental properties, for example its social norms, its typical economy, and the types of species it supports (e.g., grain crops or forests). These fundamental properties typically change slowly, but, when modified, they alter the nature of the system (Carpenter and Turner 2000; Chapin et al. 1996). Sustaining these slow variables reduces the likelihood that disease or other major perturbations will have irreversible consequences.

Most systems periodically experience severe shocks and disturbances and typically have seeds for recovery. These include propagules of important species, institutions and rules for managing crises, and memories and stories of how previous crises were managed or averted. Sustaining these legacies reduces the potential impact of disease and other disturbances and promotes recovery. Diversity of biological, social, and economic options is another mechanism for promoting recovery by increasing the number of pathways by which recovery can occur. For example, in areas where the economic base is not diversified, there is greater susceptibility to disease-induced state changes in socioeconomic systems (Department for Environment, Food and Rural Affairs [DEFRA] 2001; FAO 2002). Diversity also provides options for adapting to change when the system has been altered so substantially that recovery to the original state is unlikely.

Disease interacts with other stresses and disturbances. Systems that are already stressed are more vulnerable to disease-induced state changes. For example, industries such as agriculture that are already economically stressed and poor farmers within the agricultural sector are less resilient to disease outbreaks (DEFRA 2001; FAO 2002). The global increase in emerging infectious diseases is largely a consequence of unprecedented population growth, resulting in novel conditions that promote pathogen transmission and persistence (Wilcox and Gubler 2005). Effective management of these diseases requires a flexible approach based on an understanding of changing social and ecological conditions. When vector control programs are implemented in an inflexible manner, they may augment rather than reduce the reemergence of diseases such as malaria (Holling 1986).

In the remainder of this chapter, we highlight how resilience theory contributes to our understanding of disease-induced state changes in socioecological systems.

Social Learning to Enhance Resilience and Adaptation

Human perceptions about disease have a large effect on disease spread. How can education influence, mediate, and moderate this response? Anticipating and responding to the causes and consequences of disease epidemics requires an understanding of environmental complexity, as well as of the roles that uncertainty and variability play in environmental and social processes (Clark et al. 2001). Innovation and learning are critical to successful adaptation. This learning occurs most readily when biological, landscape, and socioeconomic diversity provide alternative possibilities for resilience or adaptation. There are at least two modes of learning that are important: (1) understanding of the dynamics of disease in socioecological systems and (2) enhancing the social and institutional capacity to respond to crises effectively.

In almost any region facing disease outbreak, a key challenge is to capture the imagination and interest of local people in a way that stimulates cooperation and appropriate action (Winch et al. 2002). It is difficult to engage the public in a meaningful dialogue about the hazards of disease outbreaks to humans and to ecosystems if policy makers and others do not have a basic understanding of the ecology and epidemiology of the disease and the environment into which it is introduced. Ideally, effective broad-scale education programs would occur before a disease reached epidemic status. Such programs would consider people's cognitive understanding of the linkages between ecology and disease, people's perceptions of the disease and health with respect to the environment, prevailing and pervasive attitudes and misconceptions, and worldviews that shape and constrain understanding of ecology and disease (Brewer et al., chapter 22, this volume). The potential to affect decision making and behavior related to disease epidemics may be enhanced, for example, by visual or other devices that promote understanding of the processes that influence disease spread, or by forecasts of how much a system might be disrupted by the disease (to this end, maps and other images are now readily available through the use of geographic information systems). Moreover, an informed public would benefit from improved communication of how disease outbreaks and epidemics are modeled, model assumptions, and how assumptions influence confidence in resulting predictions (Brewer and Gross 2003; Smith et al. 2005). The ready availability of Web-based tools to display complex data could be used to promote public understanding of disease, its influence on ecosystems, and the influence of decision-making scenarios on possible future outcomes.

When rapid changes occur as a result of disease outbreaks, most effort is applied to reducing the rates of transmission or promoting recovery. These crises, however, also represent opportunities for institutional changes that might otherwise be difficult to implement (Gunderson et al. 1995). Taking advantages of crises to implement change requires an understanding of the multiple factors that constrain human well-being and the capacity of individuals or institutions to explore innovative solutions, as described later with respect to managing the AIDS crisis in Africa.

In the following sections we provide four examples of pest and disease outbreaks that modify the thresholds of resilience in socioecological systems.

Bark Beetle

In the Kenai Peninsula of south-central Alaska, gradual climatic change led to a sudden eruption of spruce bark beetle (*Dendroctonus rufipennis*). This insect pest blocks phloem transport of sugars to roots, and its vectored bluestain fungus blocks water transport in the xylem. Massive outbreaks of this beetle-fungus combination killed the dominant tree (white spruce, *Picea glauca*) across 0.8–1.2 million ha within a decade (Berg et al. 2006). Climate warming appeared to trigger the recent outbreak by allowing the beetle to complete its life cycle in one rather than two years, suddenly altering the dynamic balance between beetles and the plant defense system. Spruce bark beetle outbreaks have occurred frequently on the Kenai, including a massive historical outbreak in the late 1870s. Outbreaks kill large-diameter trees, resulting in a growth release of pole-size trees that are less vulnerable to attack because larvae cannot develop in young trees with thin phloem. Outbreaks are limited by the availability of large-diameter trees even if favorable beetle conditions persist, causing discrete time intervals between outbreaks.

Human interactions with the ecosystem have changed since the last beetle outbreak because of the increased size and dispersion of the human population. In the past, stand-replacing fires have not occurred in beetle-kill areas (Berg et al. 2006). However, warming conditions may change fire behavior, and human settlements increase the fire ignition rate (DeWilde and Chapin 2006). The broad expanses of dry, dead trees greatly increase the risk of wildfires in what has become a wildland-urban interface. In addition, the loss of the tree canopy and the increased nutrient return to the forest floor have accelerated nutrient cycling and fostered the growth of grasses, which increase the risk of spring fires (when dry grass litter provides a ground fuel) and compete with tree seedlings, reducing the potential for stand regeneration. Widespread public concern about wildfire has led to changes in forest management policies to subsidize salvage logging to remove dead trees. In summary, the beetle outbreak substantially modified ecological and social processes on regional scales and increased the risk of disturbance spread (by fire and logging) across the landscape.

Rinderpest in Africa

Rinderpest is a highly contagious, often fatal, viral disease that infects most artiodactyls. Although it affects both domestic (e.g., cattle, water buffalo) and wild (e.g., antelope and gazelles) animals, cattle are probably its natural host. Rinderpest was apparently introduced into Africa with livestock accompanying military operations in 1884 and 1889. It rapidly became an epizootic, devastating cattle and wild ruminants throughout the continent. In East Africa, for example, it appeared in 1890, and within about two years 95% of the domestic cattle had died. In the next few years, most wild artiodactyls followed suit. Simultaneously, the loss of domestic livestock impoverished and caused famine among the pastoral Masai and subsequently among the adjacent agriculturalists both because of loss of livestock and because of resultant warfare. Weakened, the Masai and others were then hit by a smallpox epidemic, resulting in massive starvation and the consequent abandonment of large areas of the country (Branagan and Hammond 1965; Ford 1971; Sinclair and Norton-Griffiths 1979; Talbot 1963).

These ecological and social changes triggered many secondary effects. In these East African grasslands, moderate grazing and browsing combined with human-caused fires reduce woody vegetation, which will return rapidly in the absence of fires. Loss of domestic and most wild herbivores removed the grazing and browsing pressure, which, along with the cessation or reduction of fires formerly set by the people, allowed woody vegetation to take over vast areas that had been open grassland or savanna. Woody areas provide habitat for tsetse flies (*Glossina swynnertoni*), carriers of the blood parasite (*Trypanosoma*) that transmits sleeping sickness in both humans and cattle. The loss of the wildlife may have temporarily reduced the numbers of tsetse flies. However, in time, the wild ungulates increased in number. This population increase in turn provided hosts for the tsetse flies, which spread through the now extensive woodlands. The presence of the flies kept humans with their livestock out of large areas of East Africa. Control programs that reduced tsetse flies led to the return of livestock and humans, and with them came fire, which reopened much of the woodland, reexpanding savanna and grasslands. A subsequent rinderpest outbreak in the late 1950s temporarily reduced the wildebeest numbers, allowing an accumulation of grass fuel that carried hotter fires, further reducing woody vegetation. Subsequently the wildebeest numbers increased about five-fold. Their heavy grazing removed much grass fuel and thereby reduced the impact of human-ignited fires, again allowing the resurgence of some woody vegetation (Dublin 1995; Sinclair and Norton-Griffiths 1979; Talbot 1963; Talbot and Talbot 1963). These events dramatically demonstrate how disease responds to and causes complex webs of landscape-level changes in socioecological systems.

Malaria and Schistosomiasis

Malaria impacts landscape processes through its effects on people. It is the preeminent tropical parasitic disease and one of the top three killers among communicable diseases (Sachs and Malaney 2002). The incidence of malaria in the tropics is increasing rapidly for many reasons, including increasing drug and insecticide resistance, population movements into malarious regions, changing agricultural practices (including dam building and irrigation), deforestation, and weakening of public health systems in some poor countries. Malaria increases most rapidly in poor countries that have insufficient wealth for personal expenditures and government programs to reduce the incidence and provide for the treatment of disease. Malaria affects landscape processes through multiple pathways. On a local scale, malaria increases absenteeism from work and school, reducing levels of education and training. Malaria also increases the proportion of children in the population, as couples choose to have more children to compensate for anticipated malarial deaths. Together, these trends reduce agricultural productivity, savings, and human well-being. Tourism, mining, and business avoid malarial areas, further increasing the wealth disparity between malarial and nonmalarial regions (Gallup and Sachs 2001). In South Africa, malaria has caused large shifts in landscape distributions of agricultural types and people. Prior to World War II, malaria limited large-scale commercial agriculture by wealthy farmers, allowing black Africans to use this land for small-scale agriculture and herding. With increased malaria control after World War II, there was an increase in the extent of commercial farming and migration of poor white farmers onto land previously farmed by black Africans, causing black Africans to shift from small-scale agriculture to wage labor (Packard 2001).

Schistosomiasis is a water-borne disease that is increasing in the tropics, in part as a result of landscape changes associated with building of dams and expansion of irrigation. Different species of *Schistosoma* are specific to humans or cattle (in some cases both). Increases in morbidity and mortality and declines in productivity have landscape-level consequences similar to those of malaria. In Nigeria, for example, schistosomiasis reduces worker productivity, cash income, rates of land clearing, and farm size (Umeh et al. 2004).

These socioecological consequences of disease constrain the capacity of these systems to achieve economic and social changes in well-being. In this case, disease contributes to the resilience of a poverty trap that constrains changes toward a more desirable socioecological state.

AIDS

The HIV/AIDS pandemic in sub-Saharan Africa has drastically altered traditional family and community structures, leaving many orphans, whose parents died before they could pass on local knowledge of agricultural techniques best suited to local circumstances. In Zambia's Southern Province, where one-third of households care for orphans, those households not affected by HIV/AIDS can afford twice as many livestock and farming implements as female-headed households caring for orphans (FAO 2004). Potential increases in tick-borne diseases caused by a loss of local knowledge of their distribution and dynamics further reduce the availability and usefulness of livestock. In general, adult mortality has deprived families and communities of three major assets: knowledge, a strong labor force, and capital (Jayne et al. 2004).

In Zambia, aid organizations have attempted to fill this gap in the transfer of intergenerational knowledge by integrating HIV/AIDS education into agricultural education campaigns and introducing both new and traditional agricultural practices that involve labor-saving techniques. These include crop diversification (resulting in multiple harvests over the course of a year rather than a single, labor-intensive harvest) and the introduction of easier to use ploughs and tools (particularly important for children, women, and the elderly), with techniques such as minimal till-age, direct planting of seeds in pits rather than in rows, the introduction of weed-suppressant soil cover, and preparation of the soil during the dry season, which distributes labor still more evenly over the course of the year. Additional innovations include the introduction of for-profit vegetable gardens, increased husbandry of small livestock such as goats and chickens, seed multiplication, and the enhancement of "homestead gardens" that supplement staple gardens (FAO 2004). The proposed introduction of fish farming would require the digging of numerous ponds and would likely bring about significant change in local landscape processes. All of these agricultural innovations alter socioecological interactions and landscape processes.

Without these interventions, communities lose knowledge about the cultivation of traditional crops, resulting in a reduction in the planting and use of these crops (Waterhouse et al. 2004). The corresponding decrease in crop diversity and increased dependence on external seed sources reduces agricultural and household resilience. Traditional crops are especially appropriate for poorer, risk-averse households. This example highlights the importance of learning and innovation in coping with disease.

Conclusions

Disease epidemics are disturbances that alter interactions among ecological and social components of regional systems. The ecological impacts of disease are largely mediated by changes in dominant and keystone plant species or in animals that regulate these plant species. Loss of these species due to disease generally augments the fluxes of water and nutrients among patches on a landscape, increasing landscape connectivity. Diseases affect the social components of regional systems through their effects on ecological agents—farmers, foresters, and pastoralists—and the agroecosystems they manage. Over longer time scales, diseases affect policies and institutions that influence human activities and land use. Diseases have the potential to affect all pathways of interactions among patches on the landscape, but their impacts on disturbance regimes have the largest long-term consequences, sometimes leading to changes in ecosystem state. Resilience theory provides a framework for predicting, preparing for, and managing the consequences of profound changes in the state caused by disease and other perturbations. Unless we understand the feedbacks between social and ecological responses to and effects on diseases and the components of socioecological systems that foster resilience, we will fail both in predicting future disease patterns and in managing their impacts.

ACKNOWLEDGMENTS

We thank the Integrative Graduate Education and Research Training program supported by (NSF grant No. DGE-0114423 to the University of Alaska) for support of the Resilience and Adaptation program.

LITERATURE CITED

Bari, M. A., and J. K. Ruprecht. 2003. Water yield response to land use change in SW western Australia. Salinity and Landuse Impact Series 31. Perth, Australia: Department of Environment.

Batini, F. E., R. E. Black, J. Byrne, and P. J. Clifford. 1980. An examination of the effects of changes in catchment condition on water yield in the Wungong catchment, Western Australia. Australian Journal of Forest Research 10:29–38.

Berg, E. E., J. D. Henry, C. L. Fastie, A. D. De Volder, and S. Matsuoka. 2006. Spruce beetle outbreaks on the Kenai Peninsula, Alaska, and Kluane National Park and Reserve, Yukon Territory: Relationship to summer temperatures and regional differences in disturbance regimes. Forest Ecology and Management 227:219–32.

Berkes, F., J. Colding, and C. Folke. (Eds.). 2003. Navigating Social-Ecological Systems: Building Resilience for Complexity and Change. New York: Cambridge University Press.

Branagan, D., and J. A. Hammond. 1965. Rinderpest in Tanganyika: A review. Bulletin of Epizootic Diseases of Africa 13:225–46.

Brewer, C. A., and L. J. Gross. 2003. Training ecologists to think with uncertainty in mind. Ecology 84:1412–14.

Carpenter, S. R. 2003. Regime Shifts in Lake Ecosystems: Pattern and Variation. Lodendorf/Luhe, Germany: International Ecology Institute.

Carpenter, S. R., and M. G. Turner. 2000. Hares and tortoises: Interactions of fast and slow variables in ecosystems. Ecosystems 3:495–97.

Chapin, F. S., III, P. A. Matson, and H. A. Mooney. 2002. Principles of Terrestrial Ecosystem Ecology. New York: Springer-Verlag.

Chapin, F. S., III, M. S. Torn, and M. Tateno. 1996. Principles of ecosystem sustainability. American Naturalist 148:1016–37.

Clark, J. S., S. R. Carpenter, M. Barber, S. Collins, A. Dobson, J. A. Foley, D. M. Lodge, M. Pascual, R. A. Pielke, Jr., W. Pfizer, C. Pringle, et al. 2001. Ecological forecasts: An emerging imperative. Science 293:657–60.

Conrad, P. A., M. E. Grigg, C. Kreuder, E. R. James, J. Mazet, H. Dabritz, D. A. Jessup, F. Gulland, and M. A. Miller. 2005. Sea otters serve as sentinels for protozoal pathogens transmitted from the terrestrial hosts to marine mammals (abstract). *In* Cary Conference 2005: Infectious Disease Ecology, 56. Millbrook, NY: Institute of Ecosystem Studies.

Department of Environment, Food and Rural Affairs. 2001. Tackling the Impact of Foot and Mouth Disease on the Rural Economy. London: Department for Environment, Food and Rural Affairs.

Despommier, D., B. R. Ellis, and B. A. Wilcox. 2006. The role of ecotones in emerging infectious diseases. EcoHealth 3:281–89.

DeWilde, L., and F. S. Chapin III. 2006. Human impacts on the fire regime of Interior Alaska: Interactions among fuels, ignition sources, and fire suppression. Ecosystems 9:1342–53.

Dickman, A. 1992. Plant pathogens and long-term ecosystem changes. *In* The Fungal Community: Its Organization and Role in the Ecosystem, ed. G. C. Carroll and D. T. Wicklow, 499–520. New York: Marcel Dekker.

Dublin, H. T. 1995. Vegetation dynamics in the Serengeti-Mara ecosystem: The role of elephants, fire, and other factors. *In* Serengeti II, ed. A. R. E. Sinclair and P. Arcese, 71–90. Chicago: University of Chicago Press.

Eckert, G. L., J. M. Engle, and D. J. Kushner. 2000. Sea star disease and population declines at the Channel Islands. *In* Proceedings of the Fifth California Islands Symposium, ed. D. R. Browne, K. L. Mitchell, and H. W. Chaney, 390–93. Santa Barbara, CA.: Minerals Management Service.

Estes, J. A., and J. F. Palmisano. 1974. Sea otters: Their role in structuring nearshore communities. Science 185:1058–60.

Eviner, V. T., and F. S. Chapin III 2003. Functional matrix: A conceptual framework for predicting multiple plant effects on ecosystem processes. Annual Reviews of Ecology and Systematics 34:455–85.

Fraser of Allander Institute. 2005. Economic Impact of the 2001 Foot and Mouth Disease Outbreak in Scotland. Aberdeen, Scotland: Fraser of Allander Institute, University of Strathclyde, and Macauley Land Use Research Institute, Arkleton Centre for Rural Development Research, University of Aberdeen.

Food and Agriculture Organization. 2001. The State of Food and Agriculture 2001. Part III. Economic Impacts of Transboundary Plant Pests and Animal Diseases. Washington, DC: Food and Agriculture Organization.

Food and Agriculture Organization. 2002. Improved Animal Health for Poverty Reduction and Sustainable Livelihoods. FAO Animal Production and Health Paper 153. Washington, DC: Food and Agriculture Organization.

Food and Agriculture Organization. 2004. Strengthening Institutional Capacity in Mitigating HIV/AIDS Impact on the Agricultural Sector: Potential Mitigation Interventions. Lusaka, Zambia: Ministry of Agriculture and Cooperatives, Food and Agriculture Organization of the United Nations.

Foley, J. A., R. DeFries, G. P. Asner, C. Barford, G. Bonan, S. R. Carpenter, F. S. Chapin, III, M. T. Coe, G. C. Daily, et al. 2005. Global consequences of land use. Science 309:570–574.

Ford, J. 1971. The Role of the Trypanosomiases in African Ecology. Oxford: Clarendon Press.

Gallup, J. L., and J. D. Sachs. 2001. The economic burden of malaria. American Journal of Tropical Medicine and Hygiene 64:85–96.

Groffman, P. M., and G. E. Likens. (Eds.). 1994. Integrated Regional Models: Interactions between Humans and Their Environment. New York: Chapman and Hall.

Gunderson, L. H. 2003. Adaptive dancing: Interactions between social resilience and ecological crises. In Navigating Social-Ecological Systems: Building Resilience for Complexity and Change, ed. F. Berkes, J. Colding, and C. Folke, 33–52. Cambridge: Cambridge University Press.

Gunderson, L. H., and C. S. Holling. (Eds.). 2002. Panarchy: Understanding Transformations in Human and Natural Systems. Washington, DC: Island Press.

Gunderson, L. H., C. S. Holling, and S. S. Light. 1995. Barriers and Bridges to the Renewal of Ecosystems and Institutions. New York: Columbia University Press.

Hobara, S., N. Tokuchi, N. Ohte, K. Koba, M. Katsuyama, S. J. Kim, and A. Nakanishi. 2001. Mechanism of nitrate loss from a forested catchment following a small-scale, natural disturbance. Canadian Journal of Forest Research 31:1326–35.

Holling, C. S. 1986. Resilience of ecosystems: Local surprise and global change. In Sustainable Development and the Biosphere, ed. W. C. Clark and R. E. Munn, 293–317. New York: Cambridge University Press.

Horwitz, P., and B. A. Wilcox. 2005. Parasites, ecosystems, and sustainability: An ecological and complex systems perspective. International Journal of Parasitology 35:725–32.

Jackson, J. B. C. 2001. What was natural in the coastal oceans? Proceedings of the National Academy of Sciences of the United States of America 98: 5411–18.

Jayne, T. S., M. Villareal, P. Pingali, and G. Hemrich. 2004. Interactions between the agricultural sector and the HIV/AIDS pandemic: Implications for agricultural policy. ESA Working Paper 04–06. Rome: Agricultural and Development Economics Division, Food and Agriculture Organization.

Johnson, A. C., and P. Wilcock. 2002. Association between cedar decline and hillslope stability in mountainous regions of southeast Alaska. Geomorphology 46:129–42.

Johnson, W. C., and S. K. Collinge. 2004. Landscape effects on black-tailed prairie dog colonies. Biological Conservation 115:487–97.

Kay, R., and J. Alder. 1999. Coastal Planning and Management. London: Routledge.

Lafferty, K. D. 2004. Fishing for lobsters indirectly increases epidemics in sea urchins. Ecological Applications 14:1566–73.

Machlis, G. E., J. E. Force, and W. R. Burch. 1997. The human ecosystem: The human ecosystem as an organizing concept in ecosystem management. Society & Natural Resources 10:347–67.

Marcogliese, D. J. 2004. Parasites: Small players with crucial roles in the ecological theater. EcoHealth 1:151–64.

McDonnell, M. J., and S. T. A. Pickett. (Eds.). 1993. Humans as Components of Ecosystems: The Ecology of Subtle Human Effects and Populated Areas. New York: Springer-Verlag.

McNeil, W. H. 1989. Plagues and Peoples. New York: Anchor Books/Random House.

Millennium Ecosystem Assessment. 2005. Ecosystems and Human Well-being: Synthesis. Washington, DC: Island Press.

Ohte, N., N. Tokuchi, M. Katsuyama, S. Hobara, Y. Asano, and K. Koba. 2003. Episodic increases in nitrate concentrations in streamwater due to the partial dieback of a pine forest in Japan: Runoff generation processes control seasonality. Hydrological Processes 17:237–49.

Packard, R. 2001. "Malaria blocks development" revisited: The role of disease in the history of agricultural development in the Eastern and Northern Transvaal lowveld, 1890–60. Journal of Southern African Studies 27:591–612.

Patz, J. A. 2002. A human disease indicator for the effects of recent global climate change. Proceedings of the National Academy of Science of the United States of America 99:1506–8.

Patz, J., T. K. Graczyk, N. Geller, and A. Y. Yittor. 2000. Effects of environmental change on emerging parasitic diseases. International Journal of Parasitology 30:1395–405.

Polis, G. A. 1999. Why are parts of the world green? Multiple factors control productivity and the distribution of biomass. Oikos 86:3–15.

Preslor, D. B. 1999. Lessening the impact of animal diseases on developing country agriculture. In Sustainable Agriculture Solutions: Action Report. London: Noello Press.

Risser, P. 1995. The status of the science examining ecotones. BioScience 45:318–26.

Sachs, J., and P. Malaney. 2002. The economic and social burden of malaria. Nature 415:680–85.

Sinclair, A. R. E., and M. Norton-Griffiths. 1979. Dynamics of the serengeti ecosystem. In Serengeti: Dynamics of an Ecosystem, ed. A. R. E. Sinclair and M. Norton-Griffiths, 1–30. Chicago: University of Chicago Press.

Smith, K. F., A. P. Dobson, F. E. McKenzie, L. A. Real, D. L. Smith, and M. L. Wilson. 2005. Ecological theory to enhance infectious disease control and public health policy. Frontiers in Ecology and the Environment 3:29–37.

Steffen, W. L., A. Sanderson, P. D. Tyson, J. Jäger, P. A. Matson, B. Moore III, F. Oldfield, K. Richardson, H.-J. Schellnhuber, et al. (Eds.). 2004. Global Change and the Earth System: A Planet Under Pressure. New York: Springer-Verlag.

Talbot, L. M. 1963. Ecology of the Serengeti-Mara savanna of Kenya and Tanzania, East Africa. PhD dissertation, University of California–Berkeley.

Talbot, L. M., and M. H. Talbot. 1963. The wildebeest in western Masailand, East Africa. Wildlife Monographs 12:1–88.

Turner, M. G., R. H. Gardner, and R. V. O'Neill. 2001. Landscape Ecology in Theory and Practice: Pattern and Process. New York: Springer-Verlag.

Turshen, M. 1977. The political ecology of disease. Review of Radical Political Economics 9:45–60.

Umeh, J. C., O. Amali, and E. U. Umeh. 2004. The socio-economic effects of tropical diseases in Nigeria. Economics & Human Biology 2:245–63.

Vitousek, P. M., J. D. Aber, R. W. Howarth, G. E. Likens, P. A. Matson, D. W. Schindler, W. H. Schlesinger, and G. D. Tilman. 1997. Human alteration of the global nitrogen cycle: Sources and consequences. Ecological Applications 7:737–50.

Walker, B., C. S. Holling, S. R. Carpenter, and A. Kinzig. 2004. Resilience, adaptability, and transformability in social-ecological systems. Ecology and Society 9(2):5 (http://www.ecologyandsociety.org/vol9/iss2/art5).

Waterhouse, R., E. M. Devji, D. A. X. A. de Carvalho, and C. de S. P. Tinga. 2004. The Impact of HIV/AIDS on Farmers' Knowledge of Seed: Case Study of Chókwè District, Gaza Province, Mozambique. Maputo, Mozambique: International Crops Research Institute for the Semi-Arid Tropics (ICRISAT).

Whiteman, G., B. C. Forbes, J. Niemelä, and F. S. Chapin III. 2004. Bringing feedback and resilience of high-latitude ecosystems into the corporate boardroom. Ambio 33:371–76.

Wilcox, B. A., and D. J. Gubler. 2005. Disease ecology and the global emergence of zoonotic pathogens. Environmental Health and Preventive Medicine 10:263–72.

Winch, P. J., E. Leontsini, J. G. Rigau-Perez, M. Ruiz-Perez, G. G. Clark, and D. J. Gubler. 2002. Community-based dengue prevention in Puerto Rico: Impact on knowledge, behavior, and residential mosquito infestation. Journal of Tropical Medicine and Hygiene 67:363–70.

Yuill, T. M. 1991. Animal diseases affecting human welfare in developing countries: Impacts and control. World Journal of Microbiology & Biotechnology 7:157–63.

CHAPTER FOURTEEN
Research Frontiers in Ecological Systems: Evaluating the Impacts of Infectious Disease on Ecosystems

*Sharon L. Deem, Vanessa O. Ezenwa, Jessica R. Ward,
and Bruce A. Wilcox*

SUMMARY

AMONG BIOLOGISTS, THE AWARENESS OF INFECTIOUS DISEASES AND their population-, community-, and ecosystem-level impacts has increased dramatically over the past decade. This growth in interest seems to have paralleled an unprecedented global rise in the appearance of new pathogens, the spread of old pathogens, and changes in the pathology of many infectious agents in both humans and wildlife. However, the paucity of historical data makes it difficult to assess whether these current patterns represent a new or unique threat to ecosystem health. To evaluate the level of threat infectious diseases may pose to ecosystems, it is important to examine if, how, and why the impacts of diseases are changing in the modern world. Records of disease outbreaks in humans and livestock go back thousands of years, but relatively little historical information exists about infectious disease dynamics in natural ecosystems. As such, a fundamental question for disease ecologists is whether the ecosystem-level impacts of wildlife diseases are increasing on a global scale. In addition, given the mounting evidence implicating anthropogenic environmental changes in the emergence of various human infectious diseases, a second critical question is whether humans also contribute to the increased emergence and impact of infectious diseases in natural ecosystems. In this chapter we explore both questions by evaluating existing evidence from the literature and outline a research agenda for moving the field of disease ecology toward more definitive answers to these pressing questions. The research agenda we propose includes retrospective analyses of existing data on plant, animal, and human diseases to explore historical patterns of pathogen distribution, diversity, and prevalence; prospective long-term research projects to monitor infectious diseases in natural populations and communities; and experimental testing of probable drivers of disease emergence. For this research agenda to meet the ultimate objective of an increased capability to forecast disease outbreaks and to mitigate the negative impacts of these diseases, it must have multiple, integrated foci, each of which contributes to the understanding of wildlife management, ecosystem conservation, and human health in our changing world.

INTRODUCTION

Among biologists, the awareness of infectious diseases and their population-, community- and ecosystem-level impacts has increased dramatically over the past decade (Daszak et al. 2000; Deem et al. 2001; Harvell et al. 1999; Woodroffe 1999). This growth in interest is vividly reflected in an almost exponential increase in the number of publications referring to "emerging infectious diseases" over the past fifteen years (figure 14.1), and seems to have paralleled an unprecedented global rise in the appearance of new pathogens, the spread of old pathogens, and changes in the pathology of many infectious agents in both humans and wildlife. Although recent attention to high-profile disease outbreaks in wildlife such as West Nile virus (birds in North America), Ebola hemorrhagic fever (gorillas and chimpanzees in Central Africa), canine and phocine distemper (African lions and seals in northern Europe), and chytridiomycosis (amphibians worldwide) might imply that diseases are

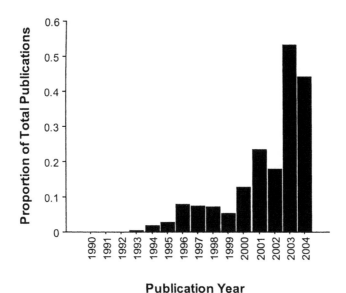

Publication Year

Figure 14.1. Proportion of publications (×1,000) cited in PubMed using the term "emerging infectious diseases" (1990–2004).

also having a larger impact on ecological systems, the paucity of historical data makes it difficult to assess whether these current patterns represent a new or unique threat to ecosystem health.

To evaluate the level of threat infectious diseases may pose to ecosystems, it is important to examine if, how, and why the impacts of diseases are changing in our modern world. Records of disease outbreaks in humans and livestock go back thousands of years (Fleming 1871, 1882; McNeill 1976), but relatively little historical information exists about infectious disease dynamics in natural ecosystems. As such, a fundamental question for disease ecologists is whether the ecosystem-level impacts of wildlife diseases are increasing on a global scale. In addition, given the mounting evidence implicating anthropogenic environmental changes in the emergence of various human infectious diseases (Institute of Medicine 2003; McMichael 2001; Patz et al. 2000, 2004), a second critical question is whether humans also contribute to the increased emergence and impact of infectious diseases in natural ecosystems. In this chapter we explore both questions by evaluating existing evidence from the literature and outline a research agenda for moving the field of disease ecology toward more definitive answers to these pressing questions. Throughout this chapter we use the term infectious disease broadly to encompass a variety of parasites and pathogens, including viruses, bacteria, protozoa, fungi, helminths, arthropods, prions, and the diseases that they cause.

ARE THE IMPACTS OF INFECTIOUS DISEASES INCREASING IN NATURAL ECOSYSTEMS?

Ecosystem properties are a function of the composition, distribution, and abundance of organisms in the system. The functional characteristics of species influence ecosystem properties through their roles as keystone species and ecological engineers, and through their interactions with other species, including infectious agents (Hooper et al. 2005). Disease impacts can occur on multiple scales, affecting individuals (fitness costs, mortality), populations (population size, gene flow), communities (shifts in dominant or abundant species, changes in species composition), and ecosystems (changes in ecosystem structure, function, and resilience). Disease can change the face of landscapes by removing keystone, abundant, and endangered species, or ecosystem engineers (Mouritsen and Poulin 2002; Collinge et al., chapter 6, this volume). The impacts of disease may also be less obvious if functional redundancy within an ecosystem allows one species to compensate for the loss of another, making it difficult to assess disease impacts at the ecosystem level. Thus, to determine whether disease impacts are increasing, a preliminary question that needs to be addressed is whether the prevalence of infectious diseases is increasing in natural systems.

The perception that disease has increased in recent years is difficult to confirm, owing to a lack of historical disease data for most systems. However, one example of historical data is a recent analysis of infectious diseases in marine organisms. Focusing on infectious disease reports in the literature between 1970 and 2001, this analysis showed that reports of disease, normalized by publication rates, increased in five of nine marine taxonomic groups, supporting the perceived increase in disease in those groups (Ward and Lafferty 2004). The analysis did not address the impact disease had on the affected populations and ecosystems, in part because disease has multiple impacts that are difficult to quantify and compare among different systems. However, several case studies suggest that when the prevalence of disease increases, the ecological impacts of disease are more severe. For example, wasting disease, caused by the slime mold *Labyrinthula zosterae* (Muehlstein et al. 1991), decimated *Zostera marina* eelgrass populations in the 1930s (Muehlstein et al. 1988). Loss of *Z. marina* precipitated the only known historical extinction of a marine invertebrate, the limpet *Lottia alveus* (Carlton et al. 1991). A more recent example is the shift from coral- to algal-dominated reefs in some areas of the Caribbean. Overfishing of herbivorous fishes, combined with disease-mediated mortality of the remaining abundant herbivore, the long-spined sea urchin, *Diadema antillarum*, released macroalgae from grazing pressure (Hughes 1994; Jackson et al. 2001; Lessios 1988). In the absence of herbivory, macroalgae increased at sites once dominated by coral (Hughes 1994).

The most noticeable impacts of disease are usually associated with epizootics (diseases that increase in incidence and prevalence suddenly and quickly) because they often have rapid, large, and unprecedented effects. The relative impact of epizootics on individuals, populations, communities, and ecosystems depends on the organisms affected (whether a consumer, a resource, or both) and on the preepizootic state of the system (i.e., how perturbed or resilient the system was). In contrast, enzootic or endemic pathogens—those with relatively constant prevalence within a population—often occur in a stable state, with less obvious effects on ecosystems. However, enzootic pathogens can act as important regulators of populations (Hudson et al. 1998; Kohler and Hoiland 2001; Tompkins and Begon 1999), and shifts in prevalence may lead to important ecosystem-level changes.

Enzootic and epizootic diseases have impacts on scales from individuals to ecosystems. The relevant question is how to assess these impacts to determine whether they are changing. This task is complicated by a lack of historical baseline data and the complexity of quantifying and comparing impacts across ecosystems. Because humans increasingly dominate the world's ecosystems (Vitousek et al. 1997) and are a major driver of changes in ecosystem processes and resilience through declines in species diversity and alteration of community composition (Chapin et al. 2000), basic ecological principles suggest that the impacts of many infectious diseases will likely increase with human alteration of natural ecosystems. However, at this point it is still unclear if and how much humans are currently influencing infectious disease dynamics in natural systems.

ARE HUMANS CONTRIBUTING TO CHANGES IN THE IMPACTS OF INFECTIOUS DISEASES ON NATURAL ECOSYSTEMS?

Many human diseases classified as emerging infectious diseases originate from human-modified ecosystems (Despommier et al. 2006). For example, the reemergence of malaria and yellow fever is associated with fragmented forests, increases in the prevalence of schistosomiasis, cholera, and leptospirosis are linked with agricultural irrigation and dam building schemes (Gubler 1998; Patz et al. 2000), and spillover of Nipah virus from its reservoir host to humans may be associated with deforestation and intensified pig farming (Chua 2002). Modified ecosystems can provide new habitat for disease vectors or reservoirs, as in the case of mosquito-borne diseases, or facilitate vector dispersal or mechanisms of pathogen transport via terrestrial or aquatic habitat corridors. Furthermore, ecotonal processes, which involve the concentration and intensification of activities and processes associated with pathogen host switching, spread, and adaptation, can also drive infectious disease emergence (see Chapin et al., chapter 13, this volume). Since the current global phenomenon of increasing emerging infectious diseases can be explained largely as a consequence of anthropogenic changes affecting ecosystems at regional and landscape scales (figure 14.2), anthropogenically driven pathogen emergence events that result in human disease are likely only a subset of similar events affecting wild species.

Human activities that have been hypothesized as drivers of disease emergence in humans and wildlife generally fall into one of four categories: (1) environmental (e.g., climate change, deforestation, habitat fragmentation) and ecological change (e.g., biodiversity loss); (2) shifts in human demography (e.g., population growth and movement, urbanization); (3) increased global travel and trade (e.g., air travel, movement of livestock, pet trade, introduced species or pathogens, ballast water); and (4) technological and agricultural practices (e.g., changes in food processing, antibiotic use, air and water pollution, intensified agriculture and animal husbandry, overfishing) (Dobson and Foufopoulos 2001; Jackson et al. 2001; McMichael 2004; reviewed in Morse 1995; Patz et al. 2000). These anthropogenic processes can directly and indirectly affect disease emergence by introducing new pathogens into naïve ecosystems, by decreasing ecosystem resilience, making ecosystems more vulnerable to pathogen invasion, or by changing enzootic host-pathogen interactions (reviewed in Kim et al. 2005). In all these cases, the ecosystem-level effects of diseases are also likely to be changing. For this reason, it is important to develop a better understanding of the associations between and among anthropogenic drivers, disease emergence, and potential ecosystem-level consequences.

A key anthropogenic driver of disease emergence is the direct introduction of pathogens into new geographic areas or hosts as a result of the movement of livestock, wildlife, plants, or pathogens (Anderson et al. 2004; Daszak et al. 2000). A classic example is the introduction of smallpox into the New World by the Spanish. The disease caused mass mortality events among Native Americans, effectively reshaping the trajectory of the human ecosystem in North America (Diamond 1997). Similarly, human activities introduced many nonhuman diseases into new areas, with important consequences for natural ecosystems. Rinderpest, introduced into Africa in the late 1800s as a result of the movement of cattle, caused massive declines of native ruminant species, with cascading effects on predator populations and woodland structure (Plowright 1982; Prins and Van der Jeugd 1993). More recently, the international trade in *Xenopus* frogs from Africa beginning in the 1930s was suggested as a mechanism for the global dissemination of chytrid fungus (Weldon et al. 2004), a pathogen implicated in amphibian population declines worldwide (Daszak et al. 1999; Retallick et al. 2004).

Large-scale disease-induced population declines and species's extinctions can have clear ecosystem-level effects ranging from the disruption of trophic interactions to the loss of functional redundancy and ecosystem resilience (Chapin et al., chapter 13, this volume). The rinderpest and chytrid fungus examples illustrate the extent to which humans can contribute to the ecosystem-level impacts of infectious disease. However, not all disease emergence events lead to massive die-offs, especially when emergence is not associated with the introduction of a novel pathogen into naïve host populations. In many cases the anthropogenic processes driving disease emergence in wildlife may simply modify existing host-pathogen relationships, leading to subtle changes in pathogen prevalence or pathogenicity that have less obvious effects at the individual, population, or community levels. Given the lack of information on historical distribution, diversity, and prevalence of endemic pathogens in

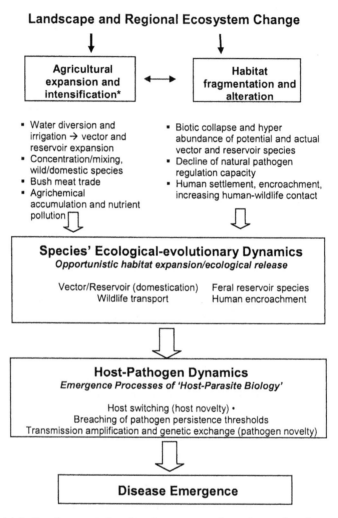

Landscape and Regional Ecosystem Change

Agricultural expansion and intensification* ↔ **Habitat fragmentation and alteration**

- Water diversion and irrigation → vector and reservoir expansion
- Concentration/mixing, wild/domestic species
- Bush meat trade
- Agrichemical accumulation and nutrient pollution

- Biotic collapse and hyper abundance of potential and actual vector and reservoir species
- Decline of natural pathogen regulation capacity
- Human settlement, encroachment, increasing human-wildlife contact

Species' Ecological-evolutionary Dynamics
Opportunistic habitat expansion/ecological release

Vector/Reservoir (domestication)   Feral reservoir species
Wildlife transport   Human encroachment

Host-Pathogen Dynamics
Emergence Processes of 'Host-Parasite Biology'

Host switching (host novelty) •
Breaching of pathogen persistence thresholds
Transmission amplification and genetic exchange (pathogen novelty)

Disease Emergence

Figure 14.2. Landscape and regional ecosystem alteration and pathogen emergence. Undisturbed natural ecosystems maintain relatively complete ecological assemblages in which component species and ecological processes are regulated and the effects of extreme environmental events are buffered by the system's inherent resilience. Human encroachment on natural habitat and wildlife encroachment on domestic habitats, along with the associated exposures, facilitates spillover and spillback of pathogens. The coalescing of human and animal hosts and of reservoir and vector species within altered ecosystems, and the movement, shifting, and mixing across the ecosystem continuum, both affect host-pathogen dynamics in a manner that facilitates disease emergence. The result is increased opportunities for host switching (pathogen novelty), the breaching of pathogen persistence thresholds (owing to unprecedented host

natural systems, how can these less obvious disease impacts be identified and assessed?

Since ecosystems were not historically monitored to detect pathogen emergence, one approach to addressing this question is to ask whether anthropogenic changes affect the conditions known to underlie or favor altered host-pathogen relationships. For example, ecological theory suggests a number of mechanisms by which human disturbances can contribute to pathogen emergence. Preeminent among these is the effect of landscape modification on the spatial distribution, composition, and abundance of species in a community. Human activities such as deforestation, the use of pesticides, and omission of various forms of pollution often result in the loss of predators. In fact, carnivorous mammals typically are the first species to disappear following forest fragmentation (Harris 1984; Laurance and Bierregaard 1997). Local extinction represents the loss of top-down natural control in ecological communities that can result in an increase in abundance of other species (Laurance and Bierregard 1997; Terborgh et al. 2001). Reduction of species diversity by whatever means (pollution, pesticides, habitat loss, nonsustainable harvesting) can contribute to the phenomenon of ecological release in remaining species, whose natural predators, competitors, or parasites are reduced in number or eliminated. Some of the released species may be reservoirs or vectors of zoonotic pathogens. If so, ecological release may result in the proliferation of released species and an increased prevalence of the pathogens they carry (Summers et al. 2003).

Climate change is another human-induced environmental change with the capacity to alter interactions between native hosts and their endemic pathogens. A major prediction regarding the effects of climate change on infectious disease is that climate warming can accelerate parasite development times and transmission rates, changing the abundance or prevalence of many endemic pathogens (Kovats et al. 2001; Harvell et al. 2002). As an example, warming temperatures in the Canadian Arctic over the last twenty years may account for the increased prevalence and intensity of the lungworm *Umingmakstrongylus pallikuukensis* in native musk oxen (Kutz et al. 2004). While *U. pallikuukensis* is thought to be an endemic parasite of musk oxen in the region, increasing parasite development rates within intermediate gastropod hosts un-

---

population densities), and transmission amplification and increased rates of pathogen adaptation and evolution of immune system detection avoidance, pathogenicity, and infectiveness (owing to increased opportunities for the interaction of endemic infection cycles and of pathogen strains, and the greater density and genetic variability of pathogen populations). (Adapted from Wilcox and Gubler 2005.)

der warmer summer conditions could be driving the recent emergence of this parasite. Selective declines in musk oxen populations infected with the parasite suggest that these changes in host-parasite dynamics may be having important population-level impacts, with as yet unknown ecosystem-level ramifications.

Although it seems likely, based on the circumstantial evidence, that humans are contributing to the effects of diseases on natural systems, we still lack direct evidence relating specific human activities to increasing prevalence, pathogenicity, or ecosystem-wide impacts of disease. In the same way, our sense that infectious diseases are increasing in prevalence and impact broadly remains largely speculative, since detection and surveillance also have increased substantially. To begin addressing these uncertainties, in the last section of this chapter we lay out a research agenda that can help guide the field of disease ecology toward answering the fundamental questions raised in this chapter.

An Integrated Research Agenda for the Future of Disease Ecology

To advance the current state of knowledge in infectious disease ecology, an interdisciplinary approach is needed in which the causes and consequences of disease emergence and responses to disease outbreaks are viewed from multiple perspectives. Key issues to be addressed include ascertaining whether infectious diseases are increasing globally, determining what the main drivers and mechanisms are for disease emergence, understanding the role of parasite and pathogen dynamics in population, community, and ecosystem ecology, and identifying associations between and among wildlife, domestic animal, ecosystem, and human health. Collaboration among professionals with unique skills is critical. Ecologists, epidemiologists, microbiologists, immunologists, and others are needed to study hosts, pathogens, and their interactions at various scales. Veterinarians, medical doctors, and public health officials are needed to explore links among wildlife, human, and domestic animal health. In addition, sociologists, economists, and politicians are required for translating scientific findings into tenable policies that mitigate the impacts of infectious diseases on natural ecosystems and human societies.

A research agenda intended to meet these objectives should have multiple, integrated foci, each of which contributes to the understanding of wildlife management, ecosystem conservation, and human health in our changing world (figure 14.3). An understanding of historical disease dynamics is essential (figure 14.3, step 1). To do this, existing data on plant, animal, and human diseases must be compiled to explore histori-

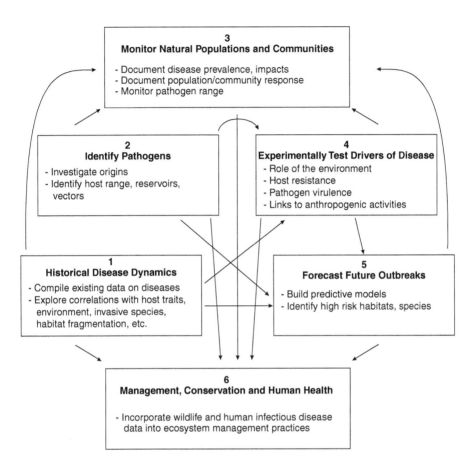

Figure 14.3. An integrated research agenda combining (1) retrospective studies of historical disease dynamics, (2) identification of pathogens and their hosts, (3) monitoring of natural populations and communities to document current disease incidence and prevalence, impacts, and range of pathogens, (4) experiments to test drivers of disease outbreaks (including the role of the environment in host resistance and pathogen virulence), and (5) theoretical ecology to build predictive models and forecast future outbreaks. This integrated research agenda allows implementation of preventive measures to mitigate the negative impacts of infectious diseases. Research clusters are interconnected, providing data and methods to further knowledge in other areas. All approaches feed into an overarching goal of management and conservation of natural resources.

cal patterns of pathogen distribution, diversity, and prevalence. Such analyses can contribute important baseline data on disease distributions in natural populations, suggest underlying processes accounting for these baseline patterns, and help reveal when and why disease patterns deviate from the baseline. Existing efforts using historical data have shown that this is a useful approach that should be expanded. Sources such as the Global Mammal Parasite Database (http://www.mammal parasites.org), a free online resource for infectious disease records in wild mammals (Nunn and Altizer 2005), illustrates the range of disease data already available in the literature and the various questions that can be addressed with such data. Also, analyses such as the one described earlier in this chapter by Ward and Lafferty (2004) show that quantitative syntheses of current data can illuminate important patterns and focus attention on future priorities.

As we collect and collate data on patterns of infectious diseases among hosts, we must remember that parasites and pathogens themselves are an integral part of ecosystems (Horwitz and Wilcox 2005). As such, it is important to document the diversity and distribution of pathogens occurring in natural systems and understand their role in shaping ecosystem processes (figure 14.3, step 2). The diversity of currently described pathogens is probably a significant underestimation of true pathogen diversity. With recent advances in molecular techniques, opportunities for identifying new parasites, exploring the phylogenetic relationships among parasites, and understanding patterns of host specificity are increasing and will be critical for understanding factors that drive disease emergence. For example, although climate change and nutrient enrichment are often hypothesized as drivers of marine diseases, experimental tests of causation are rarely possible because of the rapid and ephemeral nature of some disease outbreaks (Cerrano et al. 2000; Harvell et al. 2001) and the difficulty in isolating, identifying, and culturing the causative pathogens (Richardson 1998). Advances in parasite identification techniques will help overcome these important challenges.

Another important step is to initiate long-term research projects to monitor infectious diseases in natural populations and communities (figure 14.3, step 3). Prospective studies focused on documenting disease prevalence and incidence across hosts and habitats and tracking routes of disease spread will provide ecological and epidemiological data to predict, and possibly prevent, future disease outbreaks. Such studies would also monitor species-, community-, and ecosystem-level responses to disease, providing information on the role pathogens play in ecosystem dynamics. These monitoring projects must involve multiple disciplines to build a complete understanding of the genetic, immunological, ecological, and social implications of infectious disease. Although long-term disease monitoring is rare, platforms such as the National Ecological Observatory

Network (www.neoninc.org) pave the way for such undertakings. In addition, many existing human disease surveillance and long-term land use and cover change projects are well poised to contribute to this effort by incorporating wildlife disease surveillance into existing project protocols.

Despite the various hypotheses regarding drivers of disease emergence, obtaining direct evidence on environmental and anthropogenic factors influencing disease remains a significant hurdle for the field of disease ecology. The three steps outlined above will facilitate experimental testing of drivers of disease emergence, the next critical step in an integrated research agenda (figure 14.3, step 4). As more detailed information on historical patterns of disease distributions, parasite characteristics, and population-, community-, and ecosystem-level responses to infectious diseases becomes available, we will be able to test explicit hypotheses on the role of climate change, biodiversity loss, pollution, habitat fragmentation, and other factors in disease emergence. Experimental approaches will also provide information on pathogen virulence across hosts and levels of host resistance across species and habitats. These data, combined with long-term monitoring, will strengthen our ability to make predictions about future disease outbreaks.

An increased capability to forecast disease outbreaks is a tangible benefit of the integrated research approach described (figure 14.3, step 5). Predictive models must be built using accurate data for parameters of interest. Both retrospective and prospective studies to gather these data will be instrumental in designing these models. We will have the ability to identify species and habitats most vulnerable to the impacts of infectious disease, and the types of pathogens and parasites that pose the greatest threat. In turn, these data inform management, conservation and public health strategies to minimize the impacts of infectious diseases on wildlife, ecosystem, and human health. Ultimately, our ability to mitigate negative impacts of infectious diseases depends on interdisciplinary collaboration and the development of creative and integrative solutions. For example, it is important to recognize that pathogens are natural components of ecosystems that help shape various ecological processes. Thus, an important management challenge will be to devise strategies whereby natural interactions between pathogens and their hosts are conserved, while minimizing the effects of emerging infectious diseases.

LITERATURE CITED

Anderson, P. K., A. A. Cunningham, N. G. Patel, F. J. Morales, P. R. Epstein, and P. Daszak. 2004. Emerging infectious diseases of plants: Pathogen pollution, climate change and agrotechnology drivers. Trends in Ecology & Evolution 19:535–44.

Carlton, J. T., G. J. Vermeij, D. R. Lindberg, D. A. Carlton, and E. C. Dubley. 1991. The first historical extinction of a marine invertebrate in an ocean basin: The demise of the eelgrass limpet *Lottia alveus*. Biological Bulletin 180:72–80.

Cerrano, C., G. Bavestrello, C. N. Bianchi, R. Cattaneo-vietti, S. Bava, C. Morganti, C. Morri, P. Picco, G. Sara, et al. 2000. A catastrophic mass-mortality episode of gorgonians and other organisms in the Ligurian Sea (Northwestern Mediterranean), summer 1999. Ecology Letters 3:284–93.

Chapin III, F. S., E. S. Zavaleta, V. T. Eviner, R. L. Naylor, P. M. Vitousek, H. L. Reynolds, D. U. Hooper, S. Lavorel, O. E. Sala, et al. 2000. Consequences of changing biodiversity. Nature 405:234–41.

Chua, K. B. 2002. Nipah virus outbreak in Malaysia. Journal of Clinical Virology 26:265–75.

Daszak, P., L. Berger, A. A. Cunningham, A. D. Hyatt, D. E. Green, and R. Speare. 1999. Emerging infectious diseases and amphibian population declines. Emerging Infectious Diseases 5:735–48.

Daszak, P., A. A. Cunningham, and A. D. Hyatt. 2000. Emerging infectious diseases of wildlife: Threats to biodiversity and human health. Science 287:443–49.

Deem, S. L., W. B. Karesh, and W. Weisman. 2001. Putting theory into practice: Wildlife health in conservation. Conservation Biology 15:1224–33.

Despommier, D., B. R. Ellis, and B. A. Wilcox. 2006. The role of ecotones in emerging infectious diseases. EcoHealth 3:281–89.

Diamond, J. 1997. Guns, Germs and Steel: The Fates of Human Societies. New York: W.W. Norton.

Dobson, A., and J. Foufopoulos. 2001. Emerging infectious pathogens of wildlife. Philosophical Transactions of the Royal Society of London. Series B, Biological Sciences 356:1001–12.

Fleming, G. 1871. Animal Plagues: Their History, Nature, and Prevention. Vol. I. London: Chapman and Hall.

Fleming, G. 1882. Animal Plagues: Their History, Nature, and Prevention. Vol. II (A.D. 1800–1844). Paris: Balliere, Tindall and Cox.

Gubler, D. J. 1998. Resurgent vector-borne diseases as a global health problem. Emerging Infectious Diseases 4:442–50.

Harris, L. D. 1984. The Fragmented Forest: Island Biogeography Theory and the Preservation of Biotic Diversity. Chicago: University of Chicago Press.

Harvell, C. D., K. Kim, J. M. Burkholder, R. R. Colwell, et al. 1999. Emerging marine diseases: Climate links and anthropogenic factors. Science 285:1505–10.

Harvell, C. D., K. Kim, C. Quirolo, J. Weir, and G. Smith. 2001. Coral bleaching and disease: Contributors to 1998 mass mortality in *Briareum asbestinum* (Octocorallia, Gorgonacea). Hydrobiologia 460:97–104.

Harvell, C. D., C. E. Mitchell, J. R. Ward, S. A. Altizer, A. P. Dobson, R. S. Ostfeld, and M. D. Samuel. 2002. Climate warming and disease risks for terrestrial and marine biota. Science 296:2158–62.

Hooper, D. U., F. S. Chapin, J. J. Ewel, A. Hector, P. Inchausti, S. Lavorel, J. H. Lawton, D. M. Lodge, M. Loreau, et al. 2005. Effects of biodiversity on ecosystem functioning: A consensus of current knowledge. Ecological Monographs 75:3–35.

Horwitz, P., and B. A. Wilcox. 2005. Parasites, ecosystems and sustainability: An ecological and complex systems perspective. International Journal for Parasitology 35:725–32.

Hudson, P. J., A. P. Dobson, and D. Newborn. 1998. Prevention of population cycles by parasite removal. Science 282:2256–58.

Hughes, T. P. 1994. Catastrophes, phase shifts, and large-scale degradation of a Caribbean coral reef. Science 265:1547–51.

Institute of Medicine. 2003. Microbial Threats to Health: Emergence, Detection, and Response, ed. Smolinski, M. S., M. A. Hamburg, and J. Lederberg. Washington, DC: National Academy Press.

Jackson, J. B. C., M. X. Kirby, R. D. Berger, K. A. Bjorndal, L. W. Botsford, B. J. Bourque, R. H. Bradbury, R. Cooke, J. Erlandson, et al. 2001. Historical overfishing and the recent collapse of coastal ecosystems. Science 293:629–38.

Kim, K., A. P. Dobson, F. M. D. Gulland, and C. D. Harvell. 2005. Diseases and the conservation of marine biodiversity. In Marine Conservation Biology, ed. E. Norse and L. Crowder, 149–66. Washington, DC: Island Press.

Kohler, S. L., and W. K. Hoiland. 2001. Population regulation in an aquatic insect: The role of disease. Ecology 82:2294–305.

Kovats, R. S., D. H. Campbell-Lendrum, A. J. McMicheal, A. Woodward, and J. S. H. Cox. 2001. Early effects of climate change: Do they include changes in vector-borne disease? Philosophical Transactions of the Royal Society of London. Series B, Biological Sciences 356:1057–68.

Kutz, S. J., E. P. Hoberg, J. Nagy, L. Polley, and B. Elkin. 2004. "Emerging" parasitic infections in Arctic ungulates. Integrative and Comparative Biology 44:109–18.

Laurance, W. F., and R. O. Bierregaard. 1997. Tropical Forest Remnants: Ecology, Management, and Conservation of Fragmented Communities. Chicago: University of Chicago Press.

Lessios, H. A. 1988. Mass mortality of Diadema Antillarum in the Caribbean: What have we learned? Annual Review of Ecology and Systematics 19:371–93.

McMichael, A. J. 2004. Environmental and social influences on emerging infectious diseases: Past, present and future. Philosophical Transactions of the Royal Society of London. Series B, Biological Sciences 359:1049–58.

McMichael, T. 2001. Human Frontiers, Environments and Disease. Cambridge: Cambridge University Press.

McNeill, H. 1976. Plagues and Peoples. Garden City, NY: Anchor/Doubleday.

Morse, S. S. 1995. Factors in the emergence of infectious diseases. Emerging Infectious Diseases 1:7–15.

Mouritsen, K. N., and R. Poulin. 2002. Parasitisim, community structure and biodiversity in intertidal ecosystems. Parasitology 124:S101–17.

Muehlstein, L., D. Porter, and F. Short. 1988. Labyrinthula sp., a marine slime mold producing the symptoms of wasting disease in eelgrass, Zostera marina. Marine Biology 99:465–72.

Muehlstein, L. K., D. Porter, and F. T. Short. 1991. *Labyrinthula zosterae* sp. nov., the causative agent of wasting disease of eelgrass, *Zostera marina*. Mycologia 83:180–91.

Nunn, C. L., and S. M. Altizer. 2005. The Global Mammal Parasite Database: An online resource for infectious disease records in wild primates. Evolutionary Anthropology 14:1–2.

Patz, J. A., T. K. Graczyk, N. Geller, and A. Y. Vittor. 2000. Effects of environmental change on emerging parasitic diseases. International Journal for Parasitology 30:1395–405.

Patz, J., P. Daszak, G. M. Tabor, A. A. Aguirre, M. Pearl, J. Epstein, et al. 2004. Unhealthy landscapes: Policy recommendations on land use change and infectious disease emergence. Environmental Health Perspective 112:1092–97.

Plowright, W. 1982. The effects of rinderpest and rinderpest control on wildlife in Africa. Symposia of the Zoological Society of London 50:1–28.

Prins, H. H. T., and H. P. Van der Jeugd. 1993. Herbivore population crashes and woodland structure in East Africa. Journal of Ecology 81:305–314.

Retallick, R. W. R., H. McCallum, and R. Speare. 2004. Endemic infection of the amphibian chytrid fungus in a frog community post-decline. PLoS Biology 2:1965–71.

Richardson, L. L. 1998. Coral diseases: What is really known? Trends in Ecology & Evolution 13:438–443.

Summers, K., S. McKeon, J. Sellars, M. Keusenkothen, J. Morris, D. Gloeckner, C. Pressley, B. Price, and H. Snow. 2003. Parasitic exploitation as an engine of diversity. Biological Reviews 78:639–75.

Terborgh, J., L. Lopez, et al. 2001. Ecological meltdown in predator-free forest fragments. Science 294:1923–26.

Tompkins, D. M., and M. Begon. 1999. Parasites can regulate host populations. Parasitology Today 15:311–13.

Vitousek, P. M., H. A. Mooney, J. Lubchenco, and J. M. Melillo. 1997. Human domination of Earth's ecosystems. Science 277:494–99.

Ward, J. R., and K. D. Lafferty. 2004. The elusive baseline of marine disease: Are diseases in ocean ecosystems increasing? PLoS Biology 2:542–47.

Wilcox, B. A., and D. J. Gubler. 2005. Disease ecology and emerging zoonotic pathogens: Environmental Health and Preventive Medicine 5:263–72.

Weldon, C., L. H. du Preez, A. D. Hyatt, R. Muller, and R. Speare. 2004. Origin of the amphibian chytrid fungus. Emerging Infectious Diseases 10:2100–105.

Woodroffe, R. 1999. Managing disease threats to wild mammals. Animal Conservation 2:185–93.

Management and Applications

Introduction

Richard S. Ostfeld

THE NOTION OF PREVENTIVE MEDICINE is intuitively appealing. If disease can be prevented before the onset of symptoms, then needless suffering will be avoided, along with associated costs, economic and otherwise. For infectious diseases of humans, vaccinations are a hallmark of preventive medicine, having prevented countless cases of disease. For noninfectious diseases as well, the public health community has had many successes in identifying, for example, dietary, genetic, or lifestyle-related causes of increased disease risk and prescribing preventive measures. In a sense, preventive medicine is the acid test of epidemiology and public health. Only if the causes of variable patterns of disease are sufficiently well understood can preventive measures be prescribed and the burden of disease be avoided. Importantly, although preventive medicine is practiced at the level of individual patients, the data that allow risk factors to be understood and preventive measures to be prescribed come from analyses at the level of populations.

In recent years, a new term, *preemptive medicine*, has arisen, although its definition and implications are still under development. The former director of the U.S. National Science Foundation, Rita Colwell, has defined preemptive medicine as a holistic approach to understanding the relationships among disease, climate, and seasonality (*Discover* magazine, vol. 26, No. 2, February 2005). In contrast, the current director of the National Institutes of Health, Elias Zerhouni, describes preemptive medicine as interventions undertaken in individual patients before the onset of obvious symptoms, using knowledge of molecular precursors to disease (*AAAS News*, June 7, 2004). These conflicting characterizations of preemptive medicine are emblematic of different perspectives on the relative importance of more proximate versus more ultimate causes of disease in individuals.

Part III is largely concerned with preventive or preemptive medicine as it might be applied to humans, nonhuman animals, and plants. The major frontiers explored are where risk factors should be sought and what preventive actions might be prescribed. Infectious disease in an individual organism is the result of some pathology caused by an infectious agent, or pathogen. But to cause disease, the pathogen had to be transmitted to and then establish in the individual host. The pathogen had to occur in another host, or in a vector, or in the environment, to be transmitted to the focal host. The probability of being transmitted to the focal host depended on factors such as the abundance of the pathogen in source populations or environments, the temporal dynamics of that abundance, and the physical proximity of pathogen and focal host. These factors in turn depended on a combination of traits intrinsic to the pathogen (e.g., replication rate, longevity, antigenic profile) and extrinsic to the pathogen (e.g., mode of transmission, host physiology, host sensitivity to environmental conditions). The conflict between Colwell's and Zerhouni's definitions of preemptive medicine reflects the controversy over how much of that causal chain can and should be productively considered in trying to understand and reduce risk.

The first four chapters in this section strongly support the notion that the causes of disease in individuals and of disease emergence in populations can be traced to causally distant phenomena. In chapter 15, Holt emphasizes the potentially profound importance of the loss or reduction in predators on disease dynamics in prey and in species that interact with prey. By casting emerging infections in the context of invasion biology, Hudson et al. in chapter 16 describe the importance of interindividual variation in host behavior and physiology in influencing the probability of emergence, and recommend interventions that target "superspreaders." Garrett and Cox in chapter 17 use agricultural ecosystems to describe how disease emergence and dynamics are reduced by specific forms of genetic and species diversity of host plants, an argument with potential applications to animal, including human, diseases. Porter et al. in chapter 18 describe both the broad environmental and political contexts of a bacterial disease of corals and its management in the Florida Keys, extending the chain of causation to human institutions.

The next three chapters focus more directly on human diseases and, while embracing a broad, holistic view of causation, tend to present a more reductionist view of what would constitute preventive or preemptive action. In chapter 19, Johnson describes how the pursuit of both proximate (isolating the etiological agent) and more ultimate (determining the identity and characteristics of the reservoir) causes of a hemorrhagic disease elicited specific actions to reduce case incidence. Peters in chapter 20 describes the ecological underpinnings of several emerging diseases caused by RNA viruses, although he is less optimistic about how this knowledge could be used to control outbreaks. In chapter 21, Childs focuses on scientific and cultural distinctions between practitioners of disease ecology and those of biomedical and veterinary sciences, and makes concrete suggestions for bridging this gulf. In chapter 22, the final chapter, Brewer et al. look proactively to the future of the ecology of infectious disease by providing a detailed description of an educational agenda to integrate disease ecology into curricula from K–12 to graduate and professional training.

In its entirety, this section illustrates that, in order to use our scientific understanding as tools to anticipate and manage disease, we need to examine disease dynamics through many different lenses. Both the more holistic and the more reductionist approaches provide critical insights into disease management.

CHAPTER FIFTEEN
The Community Context of Disease Emergence: Could Changes in Predation Be a Key Driver?

Robert D. Holt

SUMMARY

ALL HOST-PATHOGEN INTERACTIONS play out against a background of other ecological drivers and the web of interactions in local communities. Parasites can infect multiple host species, and most host species are subject to infection by multiple parasites. Predators can affect host-pathogen interactions by acting as mortality factors or by facilitating disease transmission. Environmental change alters this background of species interactions and thus can indirectly modify host-pathogen dynamics in a variety of directions. Each of the many dimensions of environmental change—extirpation of endangered species, introduction of exotics, harvesting, habitat alteration and fragmentation, the addition of toxicants and pollutants, eutrophication, and climate change, not to mention the manifold synergistic interactions of all the above—can profoundly alter host-pathogen dynamics via shifts in the background of ongoing interspecific interactions.

In this chapter, I discuss several theoretical examples that illustrate effects that should be expected in anthropogenically modified ecosystems. In many natural systems, top predators are keystone species that profoundly govern system persistence and stability. Anthropogenic impacts can alter the abundance, behavior, and even the existence of top predators. Recent theoretical studies have suggested that changes in predator abundance and behavior can strongly influence disease prevalence and the potential for disease emergence across different host species, and that the quantitative magnitude and even qualitative direction of these effects depend on a suite of both epidemiological factors (e.g., the presence of acquired immunity) and ecological factors (e.g., host regulation by factors other than parasitism and predation, the presence of competing species). Here I summarize key insights from these studies and place them in the context of interactions among multiple host species.

INTRODUCTION

Ecologists increasingly recognize parasites as "hidden" but often vital drivers of the structure and dynamics of natural communities (Morand and Arias-Gonzalez 1997; Thompson et al. 2001). In epidemiology, there is likewise a growing interest in understanding how interactions among species in communities influence infectious disease emergence and dynamics (Hudson et al. 2002). Maintenance of natural host-parasite dynamics may be essential for maintaining the functioning of natural ecosystems, including the maintenance of diversity in competitive guilds and shifts in dominance during succession (Gilbert 2002). Understanding the interplay of community interactions and infectious disease is critical for many applied ecological problems of concern, such as predicting the success and impact of invasive species (Mitchell and Power 2003), conserving endangered species (Lafferty and Gerber 2002; McCallum and Dobson 1995; Torchin et al. 2003; Woodroffe 1999), unraveling the drivers of established zoonotic diseases (Ostfeld et al. 2001), and assessing the potential emergence of novel infections (Daszak et al. 2000; Woolhouse 2002). Many dimensions of human modifications of the environment could lead to shifts in species interactions and influence disease emergence and dynamics. For instance, introductions of exotics may be facilitated if the exotics harbor pathogens that harm resident competing or predatory species (as in "invasional meltdown" scenarios, D. Simberloff, personal communication). Pollutants, eutrophication, and climate change can all act as stressors on both hosts and pathogens, with a variety of effects on host-pathogen dynamics (Anderson et al. 2004; Lafferty and Holt 2003). Habitat destruction and fragmentation can lead to wholesale changes in community composition, population dynamics, and spatial connectedness, with ramifying effects on infectious diseases (e.g., Gog et al. 2000; McCallum and Dobson 2002).

Grappling with the dynamics of multispecies assemblages of interacting species is a daunting task, even without incorporating parasitism and infectious disease (Lawton 2000). One approach to the analysis of complex communities that has proven fruitful is focusing on "community modules" (Holt 1997; Holt and Hochberg 2001; Persson 1999), which in effect are multispecies extension of pairwise interactions, chosen because the module configuration involves sets of strongly interacting species that crop up in many different situations. Figure 15.1a depicts several familiar modules, and figure 15.1b shows analogous modules with pathogens. Here I consider some generalizations that emerge from analyses of models of several such modules, illuminating how changes

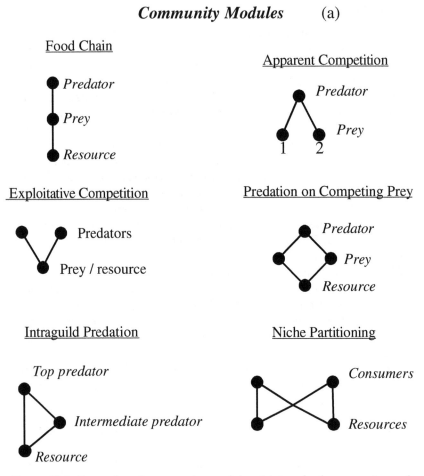

Figure 15.1. Examples of community modules. (a) Standard community ecology. (b) Modules involving infectious disease dynamics.

in interspecific interactions can influence infectious disease emergence and dynamics. I start by considering the interplay of predation and infectious disease emergence and dynamics, reviewing ideas that are presented in more detail elsewhere (Holt and Roy 2007; Ostfeld and Holt 2004; Packer et al. 2003). I then discuss how altered predation can lead to changes in host community composition, with feedback effects on infectious disease.

Community Modules in Epidemiology (b)

<u>Parasite Chain</u>

<u>Shared Parasitism</u>

Hyperpathogen

Pathogen

Host

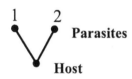

Pathogen

Host

1   2

<u>Competition among parasites</u>

<u>Predation on Host-pathogen System</u>

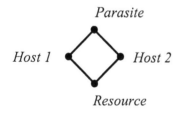

1   2
Parasites

Host

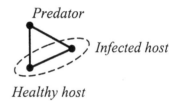

Predator

Infected host

Healthy host

<u>Keystone Parasitism</u>

<u>Niche Partitioning</u>

Parasite

Host 1   *Host 2*

Resource

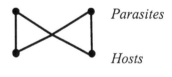

Parasites

Hosts

Figure 15.1. *(continued)*

THE INTERPLAY OF PREDATION AND HOST-PATHOGEN INTERACTIONS

A fundamental attribute of natural communities is that species are tied together in intricate webs of trophic interactions, including predator-prey and plant-herbivore interactions (Cohen 1978; Polis and Winemiller 1996). A common feature of food webs (though poorly documented in detail) is that many species are simultaneously hosts for pathogens and prey for predators (Dobson et al. in press; Marcogliese and Cone 1997; Memmott et al. 2000). This blending of interactions can potentially arise at any trophic level. For instance, consumption of plants by herbivores can influence the dynamics of fungal plant pathogens. A likely example was noted in the famed manipulative experiments in deserts of the Southwest U.S.A. initiated by James Brown. Following rodent removal, a pathogen outbreak occurred in a desert annual (whose seeds had been kept low in abundance by rodent predation), likely because of the increased abundance of the host plant (Inouye 1981). Predation has several distinct consequences for infectious disease ecology as well as population dynamics. First, predators are often intermediate hosts for macroparasites with complex life cycles (Brown et al. 2001; Dobson 1988; Lafferty 1992; Parker et al. 2003; Poulin 1998). Second, even if predators are not directly involved in disease transmission, by imposing mortality or modifying prey behavior, predators can influence disease dynamics (e.g., Bjornstad et al. 2001; Dwyer et al. 2004; Grenfell 1992). My focus is on exploring the latter population dynamic effect, but this by no means implies the former is unimportant.

Some species, by virtue of their life histories, ecological requirements, and relevance to human concerns, are particularly vulnerable to environmental change. Vertebrate predators seem particularly sensitive to human disturbance (Terborgh et al. 2001; Turner 1996). Humans often view predators with fear and loathing, because of imagined or real threats to humans themselves, their livestock, and game animals; worldwide, there has been persecution of large carnivores. Predator control has been a routine facet of wildlife management (Murie 1940). Even if humans do not deliberately kill predators, ecosystem disruption can be disproportionately felt at high trophic levels. Bioaccumulation of pesticides led to precipitous declines in predatory birds during the twentieth century. Top predators tend to occur in low abundance, have large home range requirements, and have low reproductive rates, and so are vulnerable to disturbance. Habitat destruction and fragmentation can lead to truncated food chains (Holt et al. 1999). Changes in predation on a focal host species can then alter levels of infection in it, which in turn influences the likelihood of spillover infection onto alternative hosts.

Systems with a mixture of predation and parasitism broadly parallel the community module of "intraguild predation" (Holt and Polis 1997; see figure 15.1); predators compete exploitatively with parasites by preying on healthy hosts, and predators can directly consume parasites when they attack infected prey. Several articles analyzing models with this general structure have indicated that one broad effect of predator reduction may be the unleashing of host-pathogen interactions within prey populations (Dobson and Hudson 1992; Hall et al. 2004; Ostfeld and

Holt 2004; Packer et al. 2003). Recent studies suggest a richer array of possibilities. Here I briefly summarize results presented more fully elsewhere (Holt and Roy 2007) that suggest that the effect of predator reduction or removal on disease dynamics may vary with factors such as host regulation, immunity, and the predator's pattern of prey selectivity.

PREDATORS AS MORTALITY FACTORS

It is useful to reflect on systems in which parasites are not responsible for host regulation (Holt and Dobson 2007). Such a system emerges as a limiting case of models in which hosts can be regulated by a mixture of factors (e.g., food availability, space). Understanding this limiting case sheds light on the behavior of a variety of more complex systems (Holt and Roy 2007; Roy et al. 2007). So let us now imagine the host to be regulated by mechanisms such as territoriality to a carrying capacity K (as often assumed in theoretical treatments in human epidemiology; e.g., Kermack and McKendrick 1927). With this assumption about strong compensatory density dependence, the following simple model (analyzed in Holt and Roy 2007) describes the dynamics of the infection (there is no separate equation for susceptible individuals, because host numbers are fixed at carrying capacity, $K = S + I + R$, where S, I, and R are respectively the densities of susceptible, infected, and recovered and immune hosts):

$$\frac{dI}{dt} = \beta(K - I - R)I - [\gamma + m_I(K,I,R,C)]I$$
$$\frac{dR}{dt} = \gamma I - m_R(K,I,R,C)R. \tag{1}$$

Here, β is the transmission coefficient for density-dependent infection and γ is the rate at which infected individuals recover to become immune. The quantities m_i are mortality rates, which in general depend on host numbers, level of infection, and predator abundance (C). For the moment, we assume that mortality is a function only of predator abundance (we relax this assumption below) and increases with increasing predator numbers, and that predator numbers are regulated independently of the host.

Analysis of equilibrial prevalence in equation 1 (Holt and Roy 2007) reveals several interesting features. First, the minimum host carrying capacity below which the pathogen disappears declines as predator abundance decreases. Thus, predator reduction can indirectly permit establishment of novel infections that previously could not persist. This

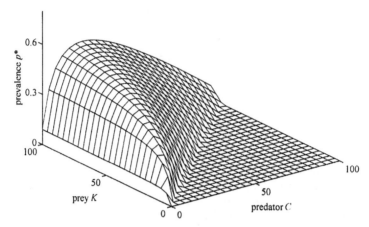

Figure 15.2. Prevalence as a function of predator and host abundance, for a model in which the pathogen is nonregulatory (i.e., total host numbers are fixed at carrying capacity) and host mortality scales linearly with predator numbers. The model is given by equation 1 in the text, and the prevalence is given by a formula shown in Holt and Roy (2007). The parameters in the example shown are as follows: $\beta = \gamma = 1$, $m_o = 0.1$, $a_I = 0.9$, and $a_R = 0.1$. Note that the hump-shaped pattern is quite strong, even though predation is sharply focused on infected individuals. (For analyses and other numerical examples, see Holt and Roy 2007.)

result is quite general, and in particular should not depend on the assumption of density-dependent disease transmission. For a pathogen to persist in a single host population, its R_0 must exceed unity, so each infection on average gives rise to one or more secondary infections. If predators impose mortality on infected hosts, the expected life span of infected individuals should increase if there are fewer predators, so more secondary infections can be produced per primary infection. Disease emergence into novel hosts should thus generically be facilitated by predator reduction.

One limiting case of equation 1 is the absence of acquired immunity. In this case, prevalence increases as mortality declines, so predator reduction should increase the level of infection in a host population (this is the effect explored in Hudson et al. 1992; Packer et al. 2003; Ostfeld and Holt 2004; and Hall et al. 2004). By contrast, with acquired immunity, prevalence (and absolute density) of infecteds can at times actually *increase* with increasing mortality of recovered hosts. Figure 15.2 shows a numerical example (for other examples, see Holt and Roy 2007). Overall, prevalence does tend to decrease with increasing predation, but going from zero to low predator abundance one sees a sharp increase in the prevalence (and absolute density) of infection. The reason is that because of our assumption of compensatory density dependence, predation that reduces the abundance of recovered individuals increases the recruitment of fresh susceptible hosts into the population (thus increasing infection rates). But with further increases in mortality, few host individuals reach the immune class, and any further effect of predators is largely via impacts on infected individuals (decreasing prevalence). The net pattern is a non-monotonic relationship between predator abundance and the prevalence of the infection in the host population, with a peak at rather low predator abundance. Overall, there is a broad trend for increasing predation to depress disease incidence (as emphasized in Packer et al. 2003), but one might also expect to see counterexamples, particularly when the pathogen is nonregulatory and predation is initially low. Holt and Roy (2007) show that comparable effects can emerge in models where the host is regulated solely by the pathogen, but for different reasons.

SATURATING FUNCTIONAL RESPONSES

So far, we have made the simplifying assumption that the per capita mortality rates of the host are functions solely of predator density. So predation affects the equilibrium, and may influence whether or not equilibrium is reached, but it does not fundamentally alter the basic dynamics of the system, because it does not generate novel feedbacks. But in more realistic models of predator-prey interactions, prey density can also affect per capita capture rates by the predator; for instance, high prey numbers can satiate the predator's functional response. Hall et al. (2004) recently showed that incorporating a saturating functional response (but no numerical response) into an SI model (with a mixture of direct density dependence and regulation by parasitism) could lead to a variety of dynamical effects. For instance, because of the positive density dependence emerging from the saturating functional response, one can observe abrupt transitions between alternative states, with and without the pathogen (see also Hethcote et al. 2004).

We can illustrate this particular effect quite simply with an SI model along the lines of model 1 above, in which total host numbers are regulated by factors other than the pathogen (thus $S + I = K$), and the predator is only able to attack infected prey. Assume that the per capita mortality rate of infected individuals is equal to a constant, plus mortality due to predation, as follows:

$$m(I, C) = m_0 + C\left[\frac{a}{1 + abI}\right].$$

The mortality due to predation is the consumption rate per predator (which we assume is given by the familiar disk equation, where a is the instantaneous attack rate during search and h is handling time) times the density of predators.

We assume that in the absence of mortality due to predation, the pathogen can persist. The dynamics of the infection are given by

$$\frac{dI}{dt} = \beta I(K - I) - m(I,C)I,$$

where the first term is the rate of production of new infections (assuming density-dependent disease transmission) and the second term is the total rate of mortality of infected individuals. When the infection is rare, I is small, and the infection disappears if $C > (BK-m_0)/a$ But for a range of predator density C above this quantity, an unstable equilibrium generally exists, such that if the initial value for I is greater than this amount, the infection will persist and equilibrate at a stable level of prevalence. There are three equilibria; the stable equilibria are at zero infectives and at a level less than the carrying capacity, with an unstable equilibrium at an intermediate level of infection. Thus, there is a threshold density of infectives, below which the infection tends to disappear, and above which it will be maintained. A decrease in the abundance of predators lowers the threshold density of infected individuals required for disease establishment.

In effect, if predators selectively consume infected prey and exhibit saturating responses, there is an Allee effect (positive density dependence) in one demographic parameter. It is well known that positive density dependence can generate alternative stable states and contribute to sustained unstable dynamics, and these generalizations carry over to host-pathogen-predator systems.

A slight but often realistic complication in the model is to assume that there is a second, reservoir host with I' infectives and a perpetual small trickle of infections emanating from that species, say $\beta'I'(K - I)$. The model can now have two non-zero stable equilibria, at low and high density of infectives. There are now several distinct effects of a decrease in predator density. If the predator attacks the alternative host, then for the reasons given above, as those attacks decline, the quantity I' should increase, and it becomes more likely the disease can become established in the focal host species (and persist even when the alternative host is absent). Even if this is not true, as C decreases, the lower stable equilibrium rises and the unstable equilibrium shrinks. There comes a point where the only equilibrium remaining is one with a high density of infected individuals. Gradual predator reduction could then lead to abrupt

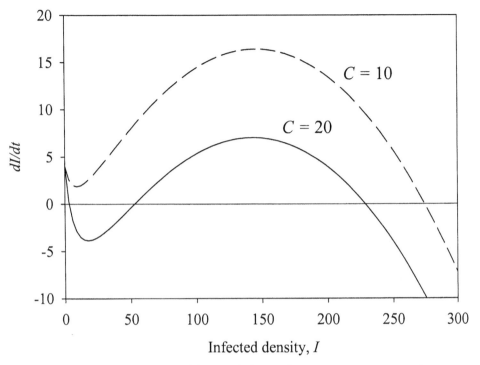

Figure 15.3. Alternative equilibria and abrupt disease emergence across host species. Growth rates of infected individuals are plotted against abundance of infecteds in a focal host, using model 1 in the text; mortality is due to a saturating predator response. It is assumed that predation is focused on infected hosts, and that an alternative reservoir host provides a low rate of cross-species transmission (see text for details). Because of the interplay of spillover infection and positive density dependence in host mortality, over some range of predator abundance alternative equilibria exist. At high predator numbers (not shown), only a low equilibrium maintained by "spillover" infection exists. At intermediate predation there is an alternative equilibrium, with a high level of infection (where the infection has escaped control by the predator). A reduction in predator abundance (from a high C, indicated by solid line, to a lower C, shown by dashed line) can lead to an upsurge of infection that is initially maintained at a low equilibrium by an alternative reservoir host. This equilibrium will be maintained where predator numbers return to their initial level. Hence, even a transient reduction in predation can lead to the emergence of a previously minor infectious disease in the focal host.

shifts in disease incidence in focal host populations (see figure 15.3 for an example).

In addition to such shifts between alternative states of high versus low or no infection, incorporating a saturating functional response can lead to sustained oscillatory behaviors—in effect, recurrent epidemics. Changes in predator density can then lead to shifts in the frequency and amplitude of such epidemics. These dynamic impacts of predation may be more important than effects on mean abundance of infectives or prevalence. These effects arise even in the simplest SI model (Hall et al. 2004) and also occur with SIR dynamics (Roy et al. 2007).

To summarize, I have argued that predator removal can facilitate initial establishment of a novel parasite in a host species. Moreover, given hosts regulated by an already established pathogen, predator removal may often lead to an upsurge in the density of infected host individuals, making spillover infection onto other species more likely. With such spillover, there can be abrupt shifts between states of low and high prevalence within a focal host species as predator numbers gradually decline. With nonregulatory pathogens, counterintuitive cases arise in which the abundance of infected individuals *declines* along with predator abundance. Finally, the interplay of predation and parasitism can change host dynamics. If the host is regulated largely by the parasite, and if a principal effect of parasitism is increased risk of predation, a reduction in predator numbers may lead to explosive population growth. Even if the host stays regulated, one may observe shifts in dynamical behavior arising from changes in predation. The particular examples I have presented here are for density-dependent disease transmission. A significant task for future work will be to examine a broader spectrum of transmission modes (Antonovics et al. 1995; Hochberg 1993; McCallum et al. 2001) and assumptions about host regulation (Roy et al. 2007).

Predation, Multiple Host Species, and Pathogen Persistence

Although I have focused so far on how shifts in predation alter the abundance and dynamics of single host populations, with consequences for emergence in novel hosts, in many natural systems multiple host species could be involved in the initial system. Multiple-host systems are not rare. Woolhouse (2002) states that "most pathogens can infect multiple hosts, and three-quarters of emerging human pathogens are zoonotic." Cleaveland et al. (2001) found that of 1,415 human pathogens, 62% have known alternative hosts; of 616 pathogens of livestock, 77% have known alternative hosts; and of 374 pathogens of domestic carnivores (cats and dogs), a whopping 90% have known alternative hosts. Many pathogens of corals infect multiple species from taxonomically distant genera (Harvell et al. 2004). Williams and Jones (1994) showed that for nematodes and digeneans in fish, over half the parasite species infect two or more host species; many macroparasite systems involve multiple host species differing in abundance and habitat use (Morgan et al. 2004). Many plant pathogens exploit multiple host species. For instance, sudden oak death (caused by the fungus *Phytophora* sp.) is causing mass mortality of three oak species in California (Myers and Bazely 2003:151). In sclerophyll vegetation in Australia, cinnamon fungus (*Phytophthora cinnamomi*) "has been found to infect all native plant species with which it comes into contact" (Wilson et al. 2003), with more than 60% of the plant species present at a site wiped out after infection, due to adult death and depressed seedling recruitment. For many emerging diseases, alternative hosts are often implicated yet unknown (e.g., the Ebola and Marburg viruses in humans must be drawn from an as yet unknown reservoir species). These many examples suggest that cross-species infection may be the norm, not the exception, in host-pathogen systems.

Observing a wide host range for a pathogen does not always imply that alternative hosts are strongly dynamically coupled, because host ecology and behavior may lead to most infections being focused within each host separately (e.g., Begon et al. 1999). However, there are a growing number of examples for which shared parasitism is believed to be a strong driver of host community dynamics (Dobson 2004; Gilbert et al. 2001; Hudson and Greenman 1998; Power and Mitchell 2004). An increasing body of theory aims at understanding the consequences of cross-species infection for disease dynamics and community composition (e.g., Begon et al. 1992; Bowers and Begon 1991; Bowers and Turner 1997; Dobson 2004; Dobson and Foufopoulos 2001; Greenman and Hudson 2000; Holt and Pickering 1985; Holt et al. 2003; Norman et al. 1999; for a review, see Begon, chapter 1, this volume). Here I note some likely consequences of predation in these multihost systems.

Given density-dependent transmission (and even frequency-dependent transmission systems often prove on close inspection to have a density-dependent component; see Antonovics et al. 1995), the concept of a threshold host population size required for establishment of a pathogen is of great importance. With multiple host species, the appropriate generalization of threshold abundance is the zero-growth isocline of an infection, when it is rare; this describes the minimal community configurations that permit establishment. Holt et al. (2003; see also Holt and Dobson 2007) describe how one can construct isoclines from multispecies host-pathogen models. The curvature of the isocline reflects the relative strength of within versus between-species infection.

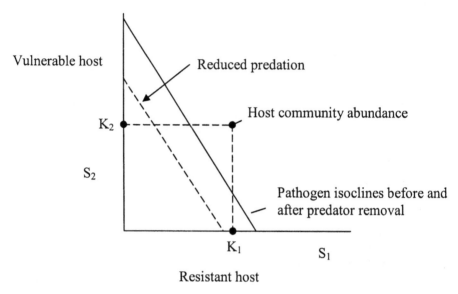

Figure 15.4. Predator removal can lead to apparent competition between hosts. The solid line is the parasite's zero-growth isocline in the presence of a predator. We assume that the hosts do not directly compete, so that in the absence of the parasite each equilibrates at its respective carrying capacity (middle dot). As drawn, neither host on its own can sustain the parasite. Thus, exclusion due to apparent competition will not occur. Reduced abundance of a predator reduces mortality and so displaces the isocline toward the origin (dashed line). Now, host species 1 can sustain the parasite on its own, and the second host may be vulnerable to exclusion because of heightened parasitism.

If predators impose mortality on infected hosts, then, as noted above, this in effect increases the threshold host density required for pathogen establishment. With multiple hosts, the impact of predators is to shift the isocline away from the origin. Conversely, a reduction in predation should push the isocline toward the origin. This not only makes it easier for a pathogen to get established in the first place, it may also intensify apparent competition between alternative host species, via shared parasitism, that otherwise are not interacting. Figure 15.4 depicts a parasite that initially can persist only if both host species are present, because each species alone at its carrying capacity is below the parasite's threshold host density. Even if apparent competition occurs in this case, it is not expected to cause extinction of either host species; if one grows rare, the pathogen would disappear, permitting the host to increase again. Assume that in the initial community a generalist predator is present attacking these hosts; this imposed mortality helps determine the position of the parasite isocline. If the predator is now removed, say by anthropogenic disturbance, the threshold host densities required for parasite persistence should shrink (the dashed diagonal line in the figure). In the example shown, after predator removal, the parasite can persist on host 1 alone; should this host be sufficiently productive, host 2 could be excluded by apparent competition (Begon and Bowers 1992; Holt and Pickering 1985; Tompkins et al. 2000, 2001; the inverse of this effect is the dilution effect: Ostfeld and Keesing 2000; for reviews, see Begon, chapter 1, this volume, and Keesing et al. 2006). So, predator removal can lead to an upsurge in infections in the trophic level below and enhance the potential for cascading extinctions due to parasite-mediated competition.

Disease Emergence due to Host Species Loss

As a final example of how host-pathogen dynamics can shift because of a change in interspecific interactions, in figure 15.5 we assume the shared parasite has greater within-species than between-species transmission, and that the two hosts compete such that their summed densities are a constant (e.g., this occurs in competition for space). The assumption about disease transmission implies that the parasite isocline bows out away from the origin (Holt et al. 2003). In the example shown, when the two hosts are present together and equally abundant (the middle dot), the parasite cannot persist (unless, of course, there is spillover from other hosts or habitats). But when one host is removed, the other host increases in abundance, and now the parasite can persist on this single host. This is one plausible mechanism that could produce the dilution effect of increasing disease levels observed along decreasing gradients in species richness. Keesing et al. (2006; see also Begon, chapter 1, this volume) explore in more detail different mechanisms that can produce dilution, and anti-dilution effects; many of these involve factors that can lead to nontraditional parasite isoclines with nonnegative slopes.

Keystone Parasitism

We have largely focused on the consequences of removal or reduction of top predators from communities. But it is also likely that human impacts on the environment can lead to increased predation on some prey species. For instance, many predators are introduced along with humans, either

Competitive hosts
with asymmetric transmission

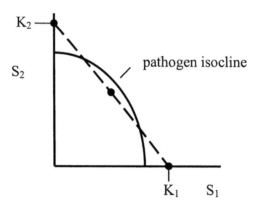

Figure 15.5. Host species removal can facilitate pathogen invasion. Unlike fig-ure 15.4, here we assume that the two potential host species compete, so that each has a lower abundance when the other is present (the dashed line). We as-sume that they equilibrate at equal abundances (the middle dot). The pathogen experiences greater within-species than between-species transmission, so the pathogen has a zero-growth isocline that bows away from the origin (Holt et al. 2003). Removal of one competitor allows the other competitor to increase in density, which in turn permits the pathogen to invade.

accidentally or deliberately (e.g., mongoose on oceanic islands, feral cats worldwide). Moreover, many communities can exhibit trophic cascades in which the removal of a top predator can lead to more intense preda-tion at lower trophic levels as a result of upsurges in the abundances of intermediate predators. This then can disrupt preexisting mechanisms of coexistence, including coexistence of hosts mediated by native parasites. If direct density dependence or competition occurs between as well as within hosts, a parasite can lead to a "keystone parasitism" effect (figure 15.1b), analogous to the familiar keystone predation effect in community ecology (Holt et al. 1994; Leibold 1996). Bowers and Turner (1997) spliced the Holt-Pickering (1985) model with the standard Lotka-Volterra model for direct competition and demonstrated that parasite-mediated coexistence of host species could occur; a necessary condition is that within-species infection rates are stronger than between-species infec-tion rates. In effect, competing species can coexist because they partition "enemy-free space" (Holt and Lawton 1994). Thus, pathogens can mediate preexisting competitive interaction between species and promote co-existence. Jaenike and Perlman (2002) suggest that keystone parasitism may help explain biogeographical patterns in multispecies assemblages of mycophagous *Drosophila* (which compete as larvae for their mushroom resources); because nematode aggregation and transmission increase with host density, competitively dominant fruit fly species experience higher levels of parasite-mediated control.

Theoretical studies of keystone predation have shown that it tends to occur at intermediate levels along gradients in productivity or disturbance (Holt et al. 1994). Although it has not been addressed specifically in the literature, it is very likely that increases in mortality due to release from top predators can disrupt coexistence mediated by native parasites. The reason is that, following the logic presented above, an increase in the abundance of these mesopredators should lead to a substantial decline in the incidence of parasitism and thus weaken the impact of keystone parasitism relative to forces such as competition for shared resources.

More broadly, entire guilds of natural enemies may foster coexistence in natural prey communities if those natural enemies act in a species-specific manner. Specialist predators can prevent any single prey species from dominating competition for a shared resource (Grover 1997). This is the basic idea of the Janzen-Connell hypothesis for explaining high plant species diversity, for which an increasing body of evidence is emerging (e.g., Augspurger 1983, 1989; Packer and Clay 2000). Both the ecological and evolutionary dynamics of parasites are likely to lead to keystone parasitism; if hosts evolve to become more abundant because of an increase in competitive dominance, they are also more likely to be able to sustain specialist pathogens, which tends to check their potential for driving competitive exclusion. One suggestive line of evidence for the likely ubiquity of keystone parasitism effects comes from invasion biology. Many introduced species in their novel range carry fewer parasites than in their native ranges and often explode to greater abundances than observed in their native ranges, with severe impacts on other species (Mitchell and Power 2003; Torchin et al. 2003). Mesopredator release could lead to a cascade of additional extinctions because of the diminution in the importance of native parasites playing keystone roles in guilds of competing host species.

Given the potential for rapid evolutionary responses in parasites (e.g., Antonovics et al. 2002), and the potential for trade-offs in utilization of alternative hosts, cross-species infection may rapidly generate an ensemble of related parasite species linking the members of a host community, rather than just one. It is likely that this will weaken apparent competition, and also require one to consider competition among parasites

(Dobson 1985, 1990; Roberts and Dobson 1995; Hochberg and Holt 1990; Holt and Dobson 2007; Holt and Hochberg 1998). A task for future work is to put together the treatment of coexistence in parasites sketched above with these models of apparent competition among hosts. In general resource-consumer models of exploitative competition, it is rather easy to generate systems in which two consumers stably coexist on two distinct prey species, but if one species is removed at either trophic level, there are cascading extinctions at the other trophic level (Holt, unpublished results). An important challenge is to understand the interwoven problems of the maintenance of diversity in host communities and in parasite guilds, and how both can be modulated more broadly by food web interactions. This problem may even pertain to what appear to be one-parasite, many-host systems. Because parasites can rapidly evolve, what appear to be generalist parasites on taxonomic grounds may actually consist of multiple, genetically distinct strains with different dynamical properties; indeed, host switching often involves genetic change (Woolhouse et al. 2001). What appear to be one-parasite, multiple-host species systems may under the surface actually be multiparasite, multihost systems.

CONCLUSIONS

I have barely scratched the surface of the manifold ways human alterations to interspecific interactions can influence infectious disease emergence and dynamics. A particularly important issue is to examine shifts in interspecific interactions in a spatial context. Habitat destruction and fragmentation are dominant themes of global change, and many shifts in community interactions are driven by alterations in the spatial context in which those interactions play out. In fragmented landscapes, there is more potential for contact between species that were originally in different communities, which could enhance chances for cross-species transmission. Moreover, during the course of anthropogenic landscape transformation, one may observe transient dynamics that have long-term impacts. If landscape change occurs rapidly, and if species are mobile and respond to a worsening local environment by leaving and seeking out remnant patches of original, undisturbed habitat, there can be a transient spike in numbers as individuals crowd into those remnants. If the species harbors an infectious disease and if disease transmission increases with crowding, then this reshuffling of "landscape refugees" may cause transient spikes of infections, leading both to a greater likelihood of transmission across host species and to potential local extinctions, either for the host itself or its pathogen. There are many challenges remaining in clarifying the consequences of such interaction webs for infectious disease systems, but it is likely that this issue must be addressed if we are to fully gauge the likely impact of human modifications to the environment on infectious disease ecology.

ACKNOWLEDGMENTS

I thank the University of Florida Foundation and the National Institutes of Health for support and Michael Barfield and Manojit Roy for assistance on our collaborative projects.

LITERATURE CITED

Anderson, P. K., A. A. Cunningham, N. G. Patel, F. J. Morales, P. R. Epstein, and P. Daszak. 2004. Emerging infectious diseases of plants: Pathogen pollution, climate change and agrotechnology drivers. Trends in Ecology & Evolution 19:535–44.

Antonovics, J., M. Hood and J. Partain. 2002. The ecology and genetics of a host shift: *Microbotryum* as a model system. American Naturalist 160:S40–53.

Antonovics, J., Y. Iwasa and M. P. Hassell. 1995. A generalized model of parasitoid, venereal, and vector-based transmission processes. American Naturalist 145:661–75.

Augspurger, C. K. 1983. Seed dispersal by the tropical tree, *Platypodium elegans*, and the escape of its seedlings from fungal pathogens. Journal of Ecology 71:759–71.

Augspurger, C. K. 1989. Impact of pathogens on natural plant populations. *In* Plant Population Ecology, ed. A. J. Davy, M. J. Hutchings and A. R. Watkinson, 413–433. Oxford: Blackwell Scientific Publications.

Begon, M. and R. G. Bowers. 1992. Beyond host-pathogen dynamics. Ecology of Infectious Diseases in Natural Populations, ed. B. T. Grenfell and A. P. Dobson, 478–509. Cambridge: Cambridge University Press.

Begon, M., R. G. Bowers, N. Kadianakis, and D. E. Hodgkinson. 1992. Disease and community structure: the importance of host self-regulation in a host-host-pathogen model. American Naturalist 139:1131–50.

Begon, M., S. M. Hazel, D. Baxby, K. Bown, R. Cavanagh, J. Chantrey, T. Jones, and M. Bennett. 1999. Transmission dynamics of a zoonotic pathogen within and between wildlife host species. Proceedings of the Royal Society of London. Series B, Biological Sciences 266:1939–45.

Bjornstad, O. N., S. M. Sait, N. C. Stenseth, D. J. Thompson, and M. Begon. 2001. The impact of specialized enemies on the dimensionality of host dynamics. Nature 409:1001–6.

Bowers, R. G., and M. Begon. 1991. A host-host-pathogen model with free-living infective stages, applicable to microbial control. Journal of Theoretical Biology 148:305–29.

Bowers, R. G., and J. Turner. 1997. Community structure and the interplay between interspecific infection and competition. Journal of Theoretical Biology 187:95–109.

Brown, S. P., F. Renaud, J. F. Guegan, and F. L. Thomas. 2001. Evolution of trophic transmission in parasites: The need to reach a mating place? Journal of Evolutionary Biology 14:815–20.

Cleaveland, S. M. K. Laurenson, and L. H. Taylor. 2001. Diseases of humans and their domestic mammals: Pathogen characteristics, host range and the risk of emergence. Philosophical Transactions of the Royal Society of London. Series B, Biological Sciences 356:991–99.

Cohen, J. E. 1978. Food Webs and Niche Space. Princeton, NJ: Princeton University Press.

Daszak, P., A. A. Cunningham, and A. D. Hyatt. 2000. Emerging infectious diseases of wildlife: Threats to biodiversity and human health. Science 287: 443–49.

Dobson, A. P. 1985. The population dynamics of competition between parasites. Parasitology 91:317–47.

Dobson, A. P. 1988. The population biology of parasite-induced changes in host behavior. Quarterly Review of Biology 63:139–65.

Dobson, A. P. 1990. Models for multi-species parasite-host communities. In The Structure of Parasite Communities, ed. G. Esch, C. R. Kennedy, and J. Aho, 261–88. London: Chapman and Hall.

Dobson, A. P. 2004. Population dynamics of pathogens with multiple host species. American Naturalist (Supplement) 164:S64–78.

Dobson, A., and J. Foufopoulos. 2001. Emerging infectious pathogens of wildlife. Philosophical Transactions of the Royal Society of London. Series B, Biological Sciences 356:1001–12.

Dobson, A. P., and P. J. Hudson. 1992. Regulation and stability of free-living host-parasite system Trechostrongylus tenuis in red grouse: Population models. Journal of Animal Ecology 61:487–500.

Dobson, A. P., K. D. Lafferty, A. M. Kuris, and C. Packer. In press. Parasites and food webs. In Parasite Ecology, ed. J. Dunne and M. Pascual. Oxford: Oxford University Press.

Dwyer, G., J. Dushoff, and S. H. Yee. 2004. The combined effects of pathogens and predators on insect outbreaks. Nature 430:341–45.

Gilbert, G. S. 2002. Evolutionary ecology of plant diseases in natural ecosystems. Annual Review of Phytopathology 40:13–43.

Gilbert, L., R. Norman, K. M. Laurenson, H. W. Reid, and P. J. Hudson. 2001. Disease persistence and apparent competition in a three-host community: An empirical and analytical study of large-scale, wild populations. Journal of Animal Ecology 70:1053–61.

Gog, J., R. Woodroffe, and J. Swinton. 2000. Disease in endangered metapopulations: The importance of alternative hosts. Proceedings of the Royal Society of London. Series B, Biological Sciences 269:671–76.

Greenman, J. V., and P. J. Hudson. 2000. Parasite-mediated and direct competition in a two-host shared macroparasite system. Theoretical Population Biology 57:13–34.

Grenfell, B. T. 1992. Parasitism and the dynamics of ungulate grazing systems. American Naturalist 139:907–29.

Grover, J. 1997. Resource Competition. London: Chapman and Hall.

Hall, S. R., M. A. Duffy, and C. E. Caceres. 2004. Selective predation and productivity jointly drive complex behavior in host-parasite systems. American Naturalist 165:70–81.

Harvell, C. D., R. Aronson, N. Baron, J. Connel, A. Dobson, S. Ellner, L. Gerber, et al. 2004. The rising tide of ocean diseases: Unsolved problems and research priorities. Frontiers in Ecology and the Environment 2:375–82.

Hethcote, H. W., W. Wang, L. Han, and Z. Ma. 2004. A predator-prey model with infected prey. Theoretical Population Biology 66:259–68.

Hochberg, M. E. 1993. Nonlinear transmission rates and the dynamics of infectious diseases. Journal of Theoretical Biology 153:301–21.

Hochberg, M. E., and R. D. Holt. 1990. The coexistence of competing parasites. I. The role of cross-species infection. American Naturalist 136:517–41.

Holt, R. D. 1997. Community modules. In Multitrophic Interactions in Terrestrial Ecosystems, ed. A. C. Gange and V. K. Brown, 333–49. Oxford: Blackwell.

Holt, R. D., and A. P. Dobson. 2007. Extending the principles of community ecology to address the epidemiology of host-pathogen systems. In ed. S.K. Collinge and C. Ray, 6–27. Disease ecology, Oxford. Oxford Univ. Press. Unpublished manuscript.

Holt, R. D., A. P. Dobson, M. Begon, R. G. Bowers, and E. Schauber. 2003. Parasite establishment and persistence in multi-host-species systems. Ecology Letters 6:837–42.

Holt, R. D., J. Grover, and D. Tilman. 1994. Simple rules for interspecific dominance in systems with exploitative and apparent competition. American Naturalist 144:741–77.

Holt, R. D., and M. E. Hochberg. 1998. The coexistence of competing parasites. II. Hyperparasitism and food chain dynamics. Journal of Theoretical Biology 193:485–495.

Holt, R. D., and J. H. Lawton. 1994. The ecological consequences of shared natural enemies. Annual Review of Ecology and Systematics 25:495–520.

Holt, R. D, J. H. Lawton, G. A. Polis, and N. Martinez. 1999. Trophic rank and the species-area relationship. Ecology 80:1495–504.

Holt, R. D., and J. Pickering. 1985. Infectious disease and species coexistence: A model of Lotka-Volterra form. American Naturalist 126:196–211.

Holt, R. D., and G. A. Polis. 1997. A theoretical framework for intraguild predation. American Naturalist 149:745–64.

Holt, R. D., and M. Roy. 2007. Predation can increase the prevalence of infectious disease. American Naturalist 169:690–99.

Hudson, P. J., A. P. Dobson, and D. Newborn. 1992. Do parasites make prey vulnerable to predation? Red grouse and parasites. Journal of Animal Ecology 61:681–92.

Hudson, P. J. and J. Greenman. 1998. Competition mediated by parasites: Biological and theoretical progress. Trends in Ecology & Evolution 13:387–90.

Hudson, P. J., A. Rizzoli, B. T. Grenfell, H. Heesterbeek, A. P. Dobson (Eds.). 2002. The Ecology of Wildlife Diseases. Oxford: Oxford University Press.

Inouye, R. S. 1981. Interactions among unrelated species: Granivorous rodents, a parasitic fungus, and a shared prey species. Oecologia 49:425–27.

Jaenike, J., and S. J. Perlman. 2002. Ecology and evolution of host-parasite associations: Mycophagous *Drosophila* and their parasitic nematodes. American Naturalist 160:S23–39.

Keesing, F., R. D. Holt, and R. S. Ostfeld. 2006. Effects of species diversity on disease risk. Ecology Letters 9: 485–98.

Kermack, W. O., and A. G. McKendrick. 1927. Contributions to the mathematical theory of epidemics, part I. Proceedings of the Royal Society of London A 115:700–721.

Lafferty, K. D. 1992. Foraging on prey that are modified by parasites. American Naturalist 140:854–67.

Lafferty, K. D., and L. R. Gerber. 2002. Good medicine for conservation biology: The intersection of epidemiology and conservation theory. Conservation Biology 16:593–604.

Lafferty, K. D., and R. D. Holt. 2003. How does environmental stress affect the population dynamics of disease? Ecology Letters 6:654–64.

Lawton, J. H. 2000. Community Ecology in a Changing World. Oldendorf/Luhe, Germany: Ecology Institute.

Leibold, M. A. 1996. A graphical model of keystone predators in food webs: Trophic regulation of abundance, incidence and diversity patterns in communities. American Naturalist 147:784–812.

Marcogliese, D. J., and D .K. Cone. 1997. Food webs: A plea for parasites. Trends in Ecology & Evolution 12:320–25.

McCallum, H., N. Barlow, and J. Hone. 2001. How should parasite transmission be modelled? Trends in Ecology & Evolution 16:295–300.

McCallum, H., and A. Dobson. 1995. Detecting disease and parasite threats to endangered species and ecosystems. Trends in Ecology & Evolution 10: 190–94.

McCallum, H., and A. Dobson. 2002. Disease, habitat fragmentation and conservation. Proceedings of the Royal Society of London. *Series B*, Biological Sciences 269:2041–49.

Memmott, J., N. D. Martinez, and J. E. Cohen. 2000. Predators, parasitoids, and pathogens: Species richness, trophic generality, and body sizes in a natural food web. Journal of Animal Ecology 69:1–15.

Mitchell, C. E., and A. G. Power. 2003. Release of invasive plants from fungal and viral pathogens. Nature 421:625–27.

Morand, S., and E. Arias-Gonzalez. 1997. Is parasitism a missing ingredient in model ecosystems? Ecological Modelling 95:61–74.

Morgan, E. R., E. J. Milner-Gulland, P. R. Torgerson, and G. F. Medley. 2004. Ruminating on complexity: Macroparasites of wildlife and livestock. Trends in Ecology & Evolution 19:181–88.

Murie, A. 1940. Ecology of the Coyote in the Yellowstone. Fauna Series, No. 4, U.S. Department of the Interior, National Park Service.

Myers, J. H., and D. R. Bazely. 2003. Ecology and Control of Introduced Plants. Cambridge: Cambridge University Press.

Norman, R., R. G. Bowers, M. Begon, and P. J. Hudson. 1999. Persistence of tick-borne virus in the presence of multiple host species: Tick reservoirs and parasite mediated competition. Journal of Theoretical Biology 200:111–18.

Ostfeld, R. S., and R. D. Holt. 2004. Are predators good for your health? Evaluating evidence for top-down regulation of zoonotic disease reservoirs. Frontiers in Ecology and the Environment 2:13–20.

Ostfeld, R. S., and F. Keesing. 2000. The function of biodiversity in the ecology of vector-borne zoonotic diseases. Canadian Journal of Zoology 78: 2061–78.

Ostfeld, R. S., E. M. Schauber, C. D. Canham, F. Keesing, C. G. Jones, and J. O. Wolff. 2001. Effects of acorn production and mouse abundance on abundance and Borrelia burgdorferi infection prevalence of nymphal Ixodes scapularis ticks. Vector-Borne and Zoonotic Diseases 1:55–63.

Packer, A., and K. Clay. 2000. Soil pathogens and spatial patterns of seedling mortality in a temperate tree. Nature 404:278–85.

Packer, C., R. D. Holt, A. Dobson, and P. Hudson. 2003. Keeping the herds healthy and alert: Impacts of predation upon prey with specialist pathogens. Ecology Letters 6:797–802.

Parker, G. A., J. C. Chub, M. A. Ball, and G. N. Roberts. 2003. Evolution of complex life cycles in helminth parasites. Nature 425:480–84.

Persson, L. 1999. Trophic cascades: Abiding heterogeneity and the trophic level concept at the end of the road. Oikos 85:385–97.

Polis, G. A., and K. O. Winemiller. (Eds.). 1996. Food Webs: Integration of Patterns and Dynamics. New York: Chapman and Hall.

Poulin, R. 1998. Evolutionary Ecology of Parasites: From Individuals to Communities. London: Chapman and Hall.

Power, A. G., and C. E. Mitchell. 2004. Pathogen spillover in disease epidemics. American Naturalist 164:S79–89.

Roberts, M. G., and A. P. Dobson. 1995. The population dynamics of communities of parasitic helminths. Mathematical Biosciences 126:191–214.

Roy, M., and R. D. Holt. 2007. Effects of predation on host-pathogen dynamics in SIR models. Unpublished manuscript.

Terborgh, J., L. Lopez, P. V. Nunez, et al. 2001. Ecological meltdown in predator-free forest fragments. Science 294:1923–26.

Thompson, J. N., O. J. Reichman, P. J. Morin, G. A. Polis, M. E. Power, R. W. Sterner, C. A. Couch, L. Gough, R. D. Holt, et al. 2001. Frontiers of ecology. Bioscience 51:15–24.

Tompkins, D. M., J. V. Greenman, and P. J. Hudson. 2001. Differential impact of a shared nematode parasite on two gamebird hosts: Implications for apparent competition. Parasitology 122:187–93.

Tompkins, D. M., J. V. Greenman, P. A. Robertson, and P. J. Hudson. 2000. The role of shared parasites in the exclusion of wildlife hosts: Heterakis *gallinarum* in the ring-necked pheasant and the grey partridge. Journal of Animal Ecology 69:829–40.

Torchin, M. E., K. D. Lafferty, A. P. Dobson, V. J. McKenzie, and A. M. Kuris. 2003. Introduced species and their missing parasites. Nature 421:628–30.

Turner, I. M. 1996. Species loss in fragments of tropical rain forest: A review of the evidence. Journal of Applied Ecology 33:200–209.

Williams, H. H., and A. Jones. 1994. Parasitic Worms of Fish. London: Taylor and Francis. UK.

Wilson, B. A., A. Lewis, and J. Aberton. 2003. Spatial model for predicting the presence of cinnamon fungus (*Phytophthora cinnamomi*) in sclerophyll vegetation communities in southeastern Australia. Austral Ecology 28:108–15.

Woodroffe, R. 1999. Managing disease threats to wild mammals. Animal Conservation 2:185–93.

Woolhouse, M. E. J. 2002. Population biology of emerging and re-emerging pathogens. Trends in Microbiology 10(Suppl.):S3–7.

Woolhouse, M. E. J., L. H. Taylor, and D. T. Haydon. 2001. Population biology of multihost pathogens. Science 292:1109–12.

CHAPTER SIXTEEN
The Emergence of Wildlife Disease and the Application of Ecology

Peter J. Hudson, Sarah E. Perkins, and Isabella M. Cattadori

SUMMARY

SINCE MOST EMERGING DISEASES are zoonotic infections derived from wildlife, they can be considered invasive species that exploit a new habitat. In this paper we consider emerging diseases as interspecific transmission events, but the fitness of the invading parasite is determined by the success of subsequent dispersal (transmission) to new resource-rich patches (susceptible hosts) and sustaining the chain of transmission. Successful invasion does not depend simply on invading a closely related host species but on the number of times invasion is attempted and the size of the individual dose. Successful invasion also depends on the presence of competitors and release from host immunity. In epidemiology, the likelihood of invasion is usually measured as R_0, the average number of infected individuals established within the new host species when an infectious individual is introduced into a population of susceptible hosts. Likelihood of persistence, on average, will fall with R_0, although a longer period of infectiousness will increase persistence even if R_0 remains the same. Estimates of R_0 generally make assumptions about the average case, yet there is increasing evidence that in many parasitic diseases, infectiousness and susceptibility may covary, such that just a few individuals may be responsible for much of the transmission; these individuals are sometimes referred to as superspreaders. In the control of emerging infections in wildlife, stopping interspecific spillover through culling or vaccination may be useful in some situations but can also lead to increased rates of transmission. We suggest that the identification and treatment of superspreaders and other features that influence R_0 can provide a novel way of controlling spillover events and preventing epidemics of emerging diseases.

INTRODUCTION

In general terms, emerging infectious diseases are considered to be parasitic infections of humans that are increasing in incidence (Woolhouse and Dye 2001). This definition encompasses not only the newly recognized diseases such as Nipah viral infection, SARS, Ebola hemorrhagic fever, and HIV but also the reemerging diseases, such as malaria and tuberculosis. Typically, emerging diseases of wildlife are overlooked in this definition, but interestingly, the majority of human emerging diseases are zoonotic in origin and are often directly transmitted by RNA viruses that have jumped from wildlife hosts and infected humans (Taylor et al. 2001). The narrow definition of emerging disease as simply human infections ignores many of the other important and instructive systems in which parasites spill over from one host species to a new host species. For example, we know that human pathogens can invade wild animal species (e.g., *Mycobacterium tuberculosis,* measles virus), that infections pass between wildlife and domestic animals (West Nile, louping ill and rabies viruses), and that parasites spill over from one wildlife species to another (phocine distemper virus). Parasites are introduced for the control of pest species (myxomavirus in rabbits), and introduced hosts and their parasites can have impacts on resident species (Hudson and Greenman 1998).

The ecology of emerging disease in both humans and wildlife is more than just the risk of infection; it also includes the likelihood of exposure, the conditions for establishment, and the special conditions that lead to successful transmission. Novel insights can be gained by taking a broad and interdisciplinary view of disease emergence, by applying ecological principles, and by drawing parallels with invasion ecology (see Perkins et al., chapter 8, this volume). In this chapter we consider emerging diseases as interspecific transmission events—in effect, invasion of a new habitat—but the fitness of the invading parasite is determined by the success of subsequent dispersal (transmission) to new resource-rich patches (susceptible hosts) and sustaining the chain of transmission. A critical point here is that the ecology of invasion and the ecology of persistence are not the same biological processes. This allows us to draw the important distinction between emerging diseases that spill over and infect a new host species but have no onward transmission and emerging diseases with onward transmission that become established in the host population.

Rarely is there a disease with both successful invasion and persistence within a new host species. A qualitative view of the papers in the journal *Emerging Infectious Diseases* shows that by far the majority of pathogens that invade a new host species (usually humans) exhibit little or no onward transmission. For example, West Nile virus and H5N1, the virus causing bird flu, transmit to humans from a sylvatic reservoir but probably have no onward human-to-human transmission, except in special cases such as blood transfusion (West Nile virus: Sampathkumar 2003; H5N1: Buxton Bridges et al. 2000), while the viruses responsible for SARS and Ebola hemorrhagic fever do not persist because of host mortality and quarantine. The recent exceptions where pathogens have invaded a new species and then have persisted are few and fascinating: dengue fever in humans, a vector-borne flavivirus infection of wild primates, HIV1 in humans, originally a viral infection of chimpanzees, and myxomavirus, a vector-transmitted poxvirus that was intentionally introduced into wild European rabbits.

Since emerging diseases are generally multiple-host infections, we focus on between-species transmission events and move from reviewing information on the emerging diseases of humans to the dynamics of cross-species infections in wildlife. We have used an ecological approach to describe emergence by examining the characteristics of invasive parasites, in much the same way as ecologists have considered the characteristics of invasive plants and animals (Kolar and Lodge 2001; Shea and Chesson 2002). Many of the disease invasion events are driven by ecological conditions, but we may also expect molecular mechanisms such as mutation, recombination, and reassortment to provide opportunities for disease invasion to occur (Webby et al. 2004). Of course, much of the human concern providing impetus to disease emergence studies is that a rare and virulent pathogen will cause high human mortality or the loss of an enigmatic wildlife species. Therefore the underlying tenet of most epidemiological research is to identify effective means of disease control, and we highlight these efforts where applicable. Throughout we use the term *parasite* in the broadest definition to include all parasitic organisms that cause disease symptoms, ranging from viruses, bacteria, protozoans through to helminths and vectors, but we use the term *pathogen* as synonymous with the microparasitic infections, particularly the disease-causing viruses and bacteria.

INTERSPECIFIC PARASITE INVASION

Parasite invasion of a new host species involves a series of steps:

Step 1: Interspecific exposure. Exposure risk is the probability of a host being exposed to an infectious parasite. Although the level of risk may depend on the occasional rare contact event, environmental drivers and anthropogenic factors often play a major role. For example, there is evidence that the heavy rainfall in New Mexico in 1993 led to increased rodent density and an increased prevalence of hantavirus in rodents, which resulted in increased transmission to humans when the rodents entered houses (Parmenter et al. 1993). Interestingly, as this case illustrates, disease emergence is often preceded by an increase in prevalence in the zoonotic wildlife host, which implies that emergence can be caused by an increase in the basic reproduction ratio (R_0) in the zoonotic host species. In this instance, R_0 is increased through an increase in the size of the host population, but R_0 can be increased by changes in the transmission process. Parasites that are transmitted by mosquito or tick vectors usually bite a broad range of hosts, and so we may expect interspecific exposure to occur more frequently through vectors than with directly transmitted parasites. For example, increased exposure to infected nymph ticks has been an important driving force in the emergence of Lyme disease where the decline of agriculture, the reestablishment of woodland, and changes in tick host biodiversity have increased nymph infection prevalence and the risk of humans being exposed (LoGiudice et al. 2003).

Step 2: Scaling the host barriers. After exposure, the parasite needs to overcome the physical and molecular barriers of the host's defense system. No doubt mucus, other secretions (saliva, gastric juices), and the mucociliary escalator in the lungs remove most of these parasitic agents. However, even if they do scale these barriers, they are then subject to immediate attack by the innate immune response and the actions of phagocytes, localized inflammatory responses, and activation of complement, which together probably prevent many infectious particles from ever establishing in a novel host.

Step 3: Establishment within the host. Once inside the host, the parasite must establish within a suitable cell or tissue. For viruses, which require specific cells, this is a critical step in the invasion process, since if both host and virus do not express the right receptors, the virus will fail to invade, and the host can be deemed noncompetent. Many invading parasites probably fail to establish at this stage simply because of receptor incompatibility or because the immune system rapidly clears the infection (Webby et al. 2004).

Step 4: Production of transmission stages. Once in a new cell or tissue the challenge now is to replicate, or grow so that new infective stages are produced. At this stage the acquired immune response often responds and prevents any further multiplication or growth inside the host (Webby et al. 2004). Consequently, hosts may become infected but the production of transmission stages fail.

Step 5: Successful transmission and persistence within the invaded host population. The probability of infecting a susceptible host species depends on the contact rate between the host and infective stage, and the infectious period of the parasite. Some pathogens, including many of the DNA viruses, have evolved persistent mechanisms that allow them to have a long infectious period. For example the Varicella zoster virus which causes chickenpox, generates a short-lived infection in children but sets up latent infections in the dorsal root ganglia of the spinal cord that can be reactivated to cause the disease shingles in later life and lead to infection of susceptible hosts. Similarly, herpesvirus, another DNA virus that produces fever blisters around the lips and genitals, can reemerge. In fact, while the virus lytically infects epidermal and mucosal cells, it also latently infects neurons and disappears within the host until the neurons are stimulated and the host once again becomes infectious. In contrast, the majority of RNA viruses multiply rapidly within the host and generate a crisis that leads either to host death or to the development of an adaptive immune response that curtails the infection so that the infectious host is effectively removed from the population. In this instance the pathogens are cleared from the body, and the long-term persistence of the pathogens relies on host-to-host transmission or some other reservoir of infection; otherwise the infection fades out. Interestingly, high rates of transmission and a high value of R_0 may well lead to a rapid epidemic in which the infection spreads through the invaded population faster than new susceptible hosts are recruited, so the epidemic fades out rapidly. For example, the two outbreaks of phocine distemper virus (PDV) in North Sea harbor seals have been short-lived, and it is assumed that the virus is circulating in harp seals (a reservoir) and occasionally jumps to harbor seals (Dietz et al. 1989). Similarly, Ebola outbreaks usually occur when an African woodland worker becomes infected, then hemorrhagic symptoms result in a rapid transmission within the human community, but, as a consequence of mortality rates in excess of 95%, sustained transmission is prevented. The pathogen may remain persistent because it is circulating in a wildlife host in the rainforest, although at the current time we do not know the identity of this reservoir host. Interestingly, the majority of emerging diseases follow this pattern and are little more than occasional spillovers that infect just a small proportion of the host population.

Conclusions and Consequences for Controlling Invasive Infections

Since the majority of emerging diseases are zoonotic, the likelihood of a parasite invading a new host is a function of the unlikely events of exposure and susceptibility coinciding. In broad terms it seems probable that humans are exposed to a constant rain of parasites, and the cumulative probability of progressing through each of the five stages to disease emergence becomes increasingly less likely. In particular, once invasion has occurred and a chain of transmission established, this is unlikely to be sustained. As such, many emerging diseases are a function of increased rates of interspecific transmission. This constant emergence and fade-out of an infection has been called "virus chatter" by Don Burke and refers to the regular passage of viruses from wildlife to bushmeat hunters in central Africa (Wolfe et al. 2005).

Emerging Infectious Diseases as an Ecological Invasion Process

Because interspecific transmission is essentially an invasion process and because the ecology of invasion of new habitats by novel species has been studied in detail, we now consider the characteristics of ecological invasions and compare these with the characteristics of disease emergence.

Habitat Similarity and Invasion

Most ecological invasions are a consequence of anthropogenic drivers; indeed, the majority of large vertebrate invasions have been a direct consequence of intentional introductions, usually by acclimation societies attempting to make their new habitat like their original home (Vander Zanden 2005). In contrast, most disease invasions are not intentional (cf. myxoma). One might expect the success of ecological invasion to be determined primarily by habitat or climate matching between source and novel habitats, where habitat structure, seasonality, and resource availability are similar between the two habitats. Quite surprisingly, this is not the case. Crawley (1987) found no evidence for climate matching in insect introductions used to control weed species. Similarly, Moulton and Pimm (1981) found that temperate bird species introduced into Hawaii were just as likely to succeed as tropical species.

Turning to invasion by parasites, we can consider the habitat of a parasite as the specific tissue it infects in a specific host species so that we could expect parasites to be more likely to invade the same tissues in closely related than distantly related host species. Interestingly, the patterns observed are similar to those for ecological invasions. In the broadest terms, plant parasites, be they nematodes or viruses, rarely infect animal species, and vice versa. Moreover, vertebrate parasites do not tend to spill over into invertebrates, although there is a big group of arboviruses and helminths that use invertebrates as vectors or intermediate hosts. However, at lower taxonomic levels it is surprising to see that microparasites frequently infect species that are not phylogenetically close (e.g., the tick-borne encephalitis virus complex infects and causes viremias in humans, grouse, sheep, and a small proportion of vole species). This is an interesting observation and means that making predictions about spillover and the emergence of new diseases is not a simple consequence of phylogenetic distance. Of course, we are examining host susceptibility, and we cannot ignore the possibility that many closely related species may simply not be exposed to infectious particles. In this respect, preliminary evidence suggests that viruses that utilize taxonomically distant host species are those that use the highly conserved cell receptors on the host. Woolhouse (2002) provides some preliminary analysis of data from Genbank that is most tantalizing. Based on conserved protein receptors, where there was more than 85% homology between human and mouse amino acid sequences, he showed that viruses with a host range that encompassed hosts from different taxonomic orders were significantly more likely to use conserved receptors than viruses with a narrow host range. Cell receptors have been identified for only eighty-eight viruses, so he correctly points out that the data are incomplete and may be phylogenetically confounded. This is an exciting hypothesis, and with additional evidence we may be able to examine which cell receptors each virus uses, the ecological risks of exposure, and the likelihood of disease emergence. For example, the rabies virus invades a wide range of host species that are not often in the same geographic location but utilizes the integrin acetylocholine receptor that is conserved across many carnivores, chiroptera, and humans (Baranowski et al. 2001; Woolhouse et al. 2005).

Invasion Effort

The primary predictor of successful ecological invasion is the invasion effort, defined as the number of times invasion was attempted and the number of individuals in each invasion incident (Kolar and Lodge 2001; Shea and Chesson 2002). Introductions of just a small number of individuals fail simply because of the stochastic problems related to small populations. Data on 424 game bird releases showed that the initial propagule size was immensely important, but still 85% of releases failed to establish, and success ultimately depended on both the number of individuals and the number of releases attempted (Pimm 1991).

In epidemiology, the invasion effort is equivalent to the force of infection (βI): the number of infectious individuals of one species (I) and the likelihood of a successful transmission event (β). The value of β depends on a number of factors, but an examination of the microbe literature indicates that workers frequently consider that a threshold number of parasites is needed to establish an infection in a single host. For example, 10^5 bacteria are needed for *Bordetella bronchiseptica* infection of mice (Harvell, personal communication), a viremia of 10^4 is needed to ensure a tick becomes infected with louping ill virus, while six entomopathogenic nematodes are needed to infect an insect host. The dose required for invasion may also vary with host and virus strain. For example, Blancou et al. (1997) record that a million times more rabies viruses taken from a fox are needed to infect a dog or cat than another fox.

Competition and Predation in the Invaded Community

From an ecological point of view, once an invading species finds a host resource it can exploit, the niche may well be occupied by a competing species. A generalist invader may be able to invade, but only a specialist may be able to survive and persist. Success will depend on the structure of the food web; a close competitor may simply outcompete the invader for resources but could also fail if it becomes vulnerable prey to the predator of the resident species. Alternatively, the invading species may benefit from predator release in the new community and do well because they have effectively escaped from predators and parasites in their native habitat (Mitchell and Power 2003; Torchin et al. 2003; Perkins et al., chapter 8, this volume).

In a similar manner, invading parasites face a community of established parasites in the resident host that they may have to compete with. Comparative analysis of the communities of parasites have tended to conclude that communities are little more than random assemblages of species (Poulin 2001). If this is the case, then an invasive species may well find a niche to invade. In contrast, laboratory-based studies find evidence of very strong competition between species, often mediated by the host immune response, and in this instance invasion may not succeed because of the effects of direct or cross-immunity, where the cross-immunity effectively acts as a predator (Adams et al. 1989; Behnke et al. 2001; Christensen et al. 1987; Cox 2001).

Conclusions and Consequences for Controlling Infections

Many of the characteristics of parasite invasion of a new host species are reflected in the invasive animal or plant emergence in a novel habitat. Successful invasion does not depend simply on invading a closely related host species but on the number of times invasion is attempted and the size of the individual dose. Successful invasion also depends on the presence of competitors and release from host immunity.

INVASION, PERSISTENCE, AND R_0

The likelihood of invasion is often measured in epidemiology as R_0, the average number of infected individuals established within the new host species when an infectious individual is introduced into a population of susceptible hosts. Factors that act to increase the value of R_0, by reducing parasite mortality—increasing the period of infectiousness or the rate of infection—have a major influence on the success of invasion. If

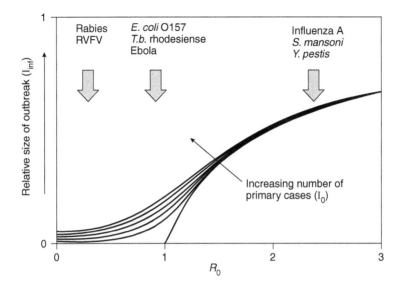

Figure 16.1. Determinants of outbreak size. Relationship between expected final outbreak size (I_{inf}, as fraction of total population) and basic reproduction ratio (R_0) for increasing numbers of primary cases (I_0, increasing from 0% to 5% total population, shown by arrow). The model is the recursive equation $I_{inf} = 1 - (1 - I_0)\exp[-R_0 I_{inf}]$. (Adapted from Kermack and McKendrick 1927.) The limiting cases are (1) $R_0 \ll 1$, where outbreak size is determined largely by the number of primary cases, and (2) $R_0 \gg 1$, where outbreak size is determined largely by the size of the susceptible host population. In the range $R_0 \approx 1$, outbreak size is very sensitive to changes in either the number of primary cases or the basic reproduction ratio. Suggested examples of zoonotic pathogens whose dynamics lie in different parts of this spectrum are shown. Species names: RVFV, Rift Valley fever virus; *T. b. rhodesiense*, *Trypanosoma brucei rhodesiense*; *S. mansoni*, *Schistosoma mansoni*; *Y. pestis*, *Yersinia pestis*. (After Woolhouse 2002.)

there is an established parasite with cross-immunity to the invading parasite, then a general rule is that the species with the greater R_0 will win this competition (more details in Hudson et al. 2002).

In a foundation publication on the dynamics of infectious diseases, Kermack and McKendrick (1927; later developed by Woolhouse and Dye 2001) examined the relative size of an outbreak in relation to the value of R_0. When R_0 is less than unity, invasion is unlikely, although the initial size of the outbreak will simply be determined by the number

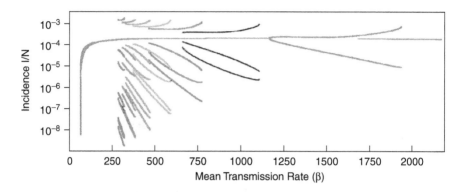

Figure 16.2. The bifurcation diagram for a seasonally forced SEIR model (susceptible, exposed, infected, recovered) showing incidence on 1 January, normalized by (constant) population size. In this example the control parameter is the mean transmission rate β, but this equates also to host birth rate or the proportion vaccinated. Each attractor is identified with a different color. When transmission rate or birth rate is high, then there is a unique annual attractor, but as this is reduced, then we observe increasing cycle periods until the annual attractor is extinguished. (After Earn et al. 2000.)

of infectious hosts that arrive in the population. As R_0 increases, the likelihood of successful invasion increases and the relative size of the subsequent outbreak becomes larger (figure 16.1). Hence an epidemic in which $R_0 = 10$ is more easily generated from a single invading infectious individual and will produce a larger outbreak than an outbreak in which $R_0 = 2$. Moreover, the epidemic curve will be different: a high value of R_0 not only burns through the host population faster but also results in a larger trough after the initial epidemic, and thus is less likely to persist than an invading disease with a moderate value of R_0.

This last point is immensely important and one that is frequently misunderstood in the epidemiological literature. In effect, while the likelihood of invasion increases with R_0, persistence is greatest when R_0 is of moderate size for any given host birth rate, and so R_0 does not provide a suitable means of estimating persistence. Indeed, R_0 is essentially the outcome between the tension of parasite burn out and the input of susceptibles, determined by the host birth rate (figure 16.2). In other words, not all R_0 values are the same. The actual dynamics and likelihood of persistence depend on host birth rate, transmission rate, and period of infectiousness. This may explain why none of the emerging diseases in humans that persist have a direct transmission route that is density dependent. The situation differs when there is a reservoir of infection, usually a reservoir species with regular spillover, or a vector that effectively acts as the reservoir of infection and ensures persistence when there are few susceptibles available for infection. This would also be different for macroparasites, where R_0, defined as the average number of successful infections that arise from each worm, increases the mean parasite intensity.

Conclusions and Consequences for Controlling Infections

This section had identified two important epidemiological observations that should be included in any general model of parasite invasion: the likelihood of invasion is determined by R_0, but persistence is not, and similar R_0 values can have very different dynamics.

WHAT DETERMINES R_0?

In essence, R_0 is the average ratio of parasite birth to death rates in the host population such that when births are greater than deaths, the parasite can successfully invade a population. Infections of hosts by microparasites are usually transient and generate either host mortality or strong immunity, and the overall loss of parasites through these various mortality factors determines the period that the host remains infectious. Anderson and May (1991) point out that R_0 can be estimated from the average age of infection, A, where $(R_0 \sim 1\text{-}(1/A))$. In many respects, this means that estimating R_0 can be straightforward in well-studied epidemics but may not be applicable for a new host species where new data on incidence (e.g., PDV) or contact tracing between infectious and susceptible hosts (e.g., SARS) are required. Even so, if we use a coarse estimate of R_0, we need to assume that many of the parameters are independent, that the population-level outcome is assumed to be a consequence of the average case, and that the variance is equivalent to the mean. In this respect, these assumptions ignore many of the heterogeneities in disease transmission that can generate nonlinear effects and have a large influence on the value of R_0. Large variation between hosts in contact rates and periods of infectiousness and covariation in susceptibility and transmission increase R_0. Identifying not only the nonlinearities but also how they influence the value of R_0 is a major challenge.

SUPERSPREADERS

Superspreaders are infectious individuals that infect a relatively large number of susceptible hosts and consequently have a disproportionately high contribution to R_0. One of the most notorious superspreaders was Typhoid Mary, an Irish cook who was an asymptomatic carrier of typhoid fever. Between 1900 and 1907 she initiated twenty-eight outbreaks of typhoid fever simply as a consequence of her unhygienic habits and a predilection of undercooking the food she prepared (Bourdain 2001); one outbreak in Ithaca allegedly led to 1,400 people being infected. Another well-known superspreader was Gaetan Dugas, considered patient zero for the emergence of HIV in North America; he was a promiscuous homosexual with an estimated 250 partners per annum who remained sexually active until he died at thirty-two years of age. This high contact rate resulted in at least 248 known infections, including all nine of the men in the Los Angeles sexual network where HIV was first recorded. In all probability, HIV1 spilled over from chimpanzees into the human population on numerous occasions in rural areas of Africa, but it needed a superspreader in the form of Dugas to invade the Western world. Of course, not all homosexual people have such a large number of partners. For example, a 1996 survey in the United Kingdom found that on average, homosexual men reported 7.6 partners per year, but some had more than fifty, such that the variance was twenty-three times greater than the mean (Anderson and May 1991). The point is that while spillover events may occur, there can be huge variations in the likelihood of a chain of transmission starting, but a superspreader greatly increases the success of an initial parasite invasion (figure 16.3). One interesting characteristic of superspreaders is that they may be asymptomatic hosts, shedding parasites but not exhibiting disease symptoms. When this feature varies between hosts within a population, it can lead to rapid disease transmission and difficulty in isolating or quarantining hosts (Fraser et al. 2004).

In human infections, contact tracing may allow us to work backward through an epidemic and identify the initial host that introduced the infection, but can we predict who the superspreaders will be? With hindsight, we could predict that a sexually promiscuous individual with many partners would act as a superspreader if that person was infected with HIV; however, the behavior of the individual is probably not sufficient evidence, since in this instance a superspreader who was asymptomatic would have a much greater impact than one who developed disease symptoms and was shunned by susceptible hosts. But do superspreaders occur in other parasite systems, and can we expect superspreaders in wild animal populations? We now examine some case studies.

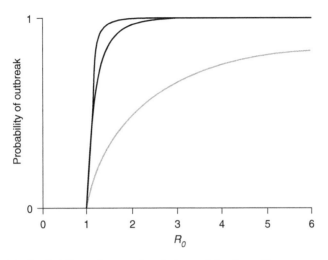

Figure 16.3. Probability of an outbreak in an infectious disease according to how the initial case is embedded in the host population. A superspreader is more likely to initiate an outbreak than a host with few contacts, even when R_0 is the same. The probability of an outbreak is estimated as $P = (1 - 1/R_0)^n$.

Tick-Borne Encephalitis in the Yellow-Necked Mouse

Tick-borne encephalitis (TBE) is an emerging disease in humans, who become infected with the virus when bitten by an infectious tick, but this is a spillover infection in humans and naturally circulates in rodents, usually the yellow-necked mouse (*Apodemus flavicollis*). Interestingly, there is no detectable viremia within the mouse; instead, transmission of TBEv occurs between cofeeding ticks, adjacently feeding infected and susceptible ticks between which the virus is passed. As such, TBEv transmission relies on cofeeding tick aggregations, and so superspreaders of TBEv can be identified as those individuals carrying the majority of the cofeeding groups. Perkins et al. (2003) examined more than two thousand rodent captures and found that the hosts that carried large numbers of larvae were also more likely to have nymphs and cofeeding groups. The authors estimated that 93% of the potential transmission of TBEv was accounted for by the top 20% most cofeeding-tick-infested rodents in the host population (figure 16.4). They also showed that these were most likely to be the large-body-mass, sexually active males. In other words, they were able to predict the individuals responsible for the TBEv transmission, the superspreaders. In general terms, targeting control efforts toward this 20% would effectively

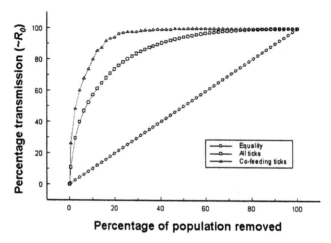

Figure 16.4. The reduction in transmission potential associated with sequential removal of hosts from the population as illustrated by tick-borne encephalitis cofeeding groups of ticks on mice. The degree of curvature indicates how the distributions differ from equality. For all ticks observed on hosts, 20% of the population was accountable for 74% of transmission potential; for ticks observed only in cofeeding aggregations, 20% of the population was responsible for 94% of transmission potential. If the population were homogeneous and biting rates were equal, then 20% of the hosts would account for exactly 20% of the transmission potential. (After Perkins et al. 2003.)

eliminate the disease, and this focused treatment would be a fraction of the cost of a more widespread treatment.

Pathogen-Induced Transmission of Gastrointestinal Worms

Myxoma is a poxvirus infection of wild Brazilian tropical forest rabbits (*Sylvilagus brasiliensis*) that was intentionally introduced into European rabbits as a biological control (Fenner and Fantini 1999). Interestingly, the virus is immunosuppressive and downmodulates expression of MHC2, in this way reducing the humoral response against macroparasite infections. One consequence of this reduced immunity is that rabbits infected with myxomavirus have greater species richness of gastrointestinal nematodes (Boag et al. 2001) and higher worm intensities of the helminths to which they exhibit a strong acquired immune response (e.g., *Trichostrongylus retortaeformis*), but not higher intensities in species where the host exhibits little acquired immunity (*Graphidium strigosum*). In other words, myxomavirus infection makes individual rabbits more susceptible to worm infection, and these individuals then produce more infective stages. The myxomavirus in effect turns the average individual rabbit into a superspreader. Thus, the immunosuppressive effects of one parasite can turn a host into a superspreader of a second parasite.

Conclusions and Consequences of Superspreaders

In many parasite-host systems, it seems that the variance in the number of hosts infected per infectious host is greater than the average number of cases, such that a few individuals are responsible for most of the transmission events. These hosts may also be more susceptible to infection, and this will lead to nonlinearities and increase R_0. Indeed, one could define the superspreaders as the 20% of individuals accounting for at least 80% of the transmission, a definition in line with Woolhouse et al. (1997). This is an easy and workable definition that has the advantage of focusing not just on the individual Typhoid Marys but on a proportion of the population. Superspreader individuals could also play an important role in initiating successful invasion. Clearly, there is a need to look more closely at the variation between individual hosts in the period of infectiousness, observe whether these covary with susceptibility, and thus identify the nonlinearities that may have a dramatic influence on R_0.

CONTROLLING EMERGING DISEASES IN WILDLIFE

The control of emerging diseases must focus on two events: preventing the initial spillover by reducing exposure to the infectious stage, and, when a spillover event occurs, breaking the chain of transmission by either culling infected individuals or vaccinating susceptible hosts. Identifying and culling the superspreaders could have a major influence in disease control.

Preventing Spillover Events: Culling or Vaccinating the Reservoir

Logically, preventing spillover by reducing contact rates between host species should be the primary method for stopping invasive parasites emerging. In this respect, if superspreaders are both more susceptible and more infectious, they could play a major role in receiving the spillover infection and initiating an epidemic in the target species. Disease spillover has been controlled in wild animal populations in a number of scenarios.

Louping ill virus is a tick-borne pathogen, closely related to the TBE virus, that causes 80% mortality in red grouse (*Lagopus lagopus scoticus*) and is sustained through nonviremic transmission in mountain hares (*Lepus timidis*) (Hudson et al. 1995). Ticks require a large mammalian host to feed the adult stages, so the hares not only permit nonviremic transmission but also help maintain the life cycle of the ticks (Gilbert et al. 2001). Laurenson et al. (2004) undertook a large-scale experiment in which they reduced hare density and then recorded a fall in seroprevalence to louping ill in grouse. This case illustrates how a reservoir host can play a key role in sustaining infections, and more specifically how culling the reservoir can help reduce infection in a target host.

In the United Kingdom, the European badger (*Meles meles)* is believed to be the reservoir host for bovine tuberculosis (TB). Badgers excrete bacteria that can infect cattle sharing the same pasture, but when and how spillover from badgers to cattle occurs is not known. The accepted view was that spillover to cattle came from badgers, so the management strategy once bovine TB was detected in a cattle herd was to kill all the infected badgers on the targeted farm, in an attempt to remove the assumed source of infection. A large-scale experiment that aimed to investigate the feasibility of this management strategy has recently identified that this strategy increases rather than decreases the rate of spillover (Donnelly et al. 2003). One explanation for this outcome is that culling disrupts the social structure of the badger, with the result that surviving members of the infected group disperse and increase spillover on other farms. Interestingly, in this case, culling the reservoir host leads to increased rates of spillover, counter to what is expected. In the meantime, the full experiment has yet to reach completion, and the trial continues with proactive culling of badgers, the removal of all badgers irrespective of infection status, and comparing the rate of cattle herd breakdown in experimental areas with that in areas of reactive culling and no culling.

Domestic dogs act as an important reservoir of infection of both rabies virus and canine distemper virus (CDV), which has placed a number of endangered canid species at risk of extinction. At the current time, rabies and CDV threaten to seriously reduce populations of Ethiopian wolves and wild dogs in Africa (Randall et al. 2004). The question is whether vaccination of local domestic dogs could be sufficient to reduce the risk of spillover to an acceptable level. Logic may suppose that vaccinating the endangered species would be the optimum strategy to protect the threatened hosts; however, previous attempts at doing this with wild dogs in the Serengeti coincided with the loss of wild dogs and brought into question whether handling and vaccination may have exacerbated the decline (Woodroffe and Ginsberg 1999). Cleaveland et al. (2003) showed that 70% vaccination of domestic dogs reduced the rate of rabies outbreak in the Serengeti population of domestic dogs and hence should have reduced spillover rates into wild canid populations. However, 70% vaccination of domestic dogs in Ethiopia did not prevent an itinerant dog introducing an infection into the threatened Ethiopian wolf in 2003, when at least thirty-eight wolves died. Modeling indicates that about 30% vaccination of the wolves could provide sufficient protection to prevent a serious epidemic in Ethiopian wolves (Haydon et al. 2002; Randall et al. 2004), and now a combined strategy of vaccination is used.

Targeted Management: Controlling the Superspreaders

We have highlighted the relative importance of superspreaders in introducing infections into populations. The identification and even removal of these hosts should be an effective means of controlling disease outbreaks. We now examine how effective this strategy would be, then consider the use of ecological processes to reduce rates of infection.

The Effectiveness of Controlling Superspreaders

Heterogeneities in the transmission rates of parasites and pathogens influence R_0. The greater the number of subgroups, all experiencing different transmission rates, the higher R_0 is expected to be (Woolhouse et al. 1997). A manifestation of these nonlinearities is the aggregated distribution that is universal for the majority of host-macroparasite distributions (Shaw et al. 1998) or the differences in an individuals contact rate within a population. An obvious consequence of these aggregated distributions of infection is that certain individuals in the tail of the distribution are superspreaders, who contribute disproportionately to the value of R_0.

Woolhouse (1997) and colleagues examined host-parasite contact rates for ten data sets and found that for all, the removal of the 20% of hosts most heavily infected reduced R_0 by 80%, a pattern termed the 20/80 rule. Therefore, removal of this relatively small proportion of the population would reduce to unity any disease for which $1 < R_0 \leq 5$. This simple relationship gives us a general rule that will successfully reduce R_0 by targeting a relatively small proportion of the population and highlights the importance of identifying the superspreaders within a population.

The real challenge for the future is to try and identify the superspreaders in a population before an outbreak so that we can selectively focus control techniques on these individuals, either by prioritizing vaccination to prevent infection or by administering priority treatment after infection to stop shedding.

CONCLUSIONS

Emerging diseases are primarily multiple-host systems in which a pathogen moves from one species and invades a novel host species. The process of disease invasion is very similar to the process of animal and plant invasion into novel habitats, and the two show interestingly similar characteristics. In all likelihood, exposure to novel parasites is not uncommon, but infection and onward transmission are rare events. Although epidemiologists use the basic reproduction ratio, R_0 as a means of measuring the ability to invade of a new host population, R_0 of the same value are not necessarily the same, and there can be variation in the likelihood of subsequent establishment within the host population, depending on the period of infectiousness and host birth rate. Furthermore, estimates of R_0 generally make assumptions about the mean case, yet there is increasing evidence that in many parasitic disease, there are large variations in infections, such that just a few individuals may be responsible for much of the transmission. These individuals are often referred to as superspreaders. In the control of emerging infections in wildlife, stopping interspecific spillover through culling or vaccination may be useful in some situations but can also lead to increased rates of transmission. We suggest that the identification and treatment of superspreaders can provide a novel way of controlling spillover events and preventing epidemics.

LITERATURE CITED

Adams D. B., B. H. Anderson, and R. G. Windon. 1989. Cross-immunity between *Haemonchus contortus* and *Trichostrongylus colubriformis* in sheep. International Journal for Parasitology 19:717–22.

Anderson, R. M., and R. M. May. 1991. Infectious Disease of Humans: Dynamics and Control. Oxford: Oxford University Press.

Baranowski, E., C. M. Ruiz-Jarabo, and E. Domingo. 2001. Evolution of cell recognition by viruses. Science 292:1102–05.

Behnke, J. M., A. Bajer, E. Sinski, and D. Wakelin. 2001. Interactions involving intestinal nematodes of rodents: Experimental and field studies. Parasitology 122:S39–51.

Blancou J., M. F. A. Aubert, M. G. Blancher, M. R. Nordmann, M. A. Rerat, and M. C. Laroche. 1997. Transmission of the rabies virus: Importance of the species barrier. Bulletin de l'Acadamie Nationale de Medecine 181:301–12.

Boag, B., J. Lello, A. Fenton, D. M. Tompkins, and P. J. Hudson. 2001. Patterns of parasite aggregation in the wild European rabbit *Oryctolgaus cuniculus*. International Journal of Parasitology 31:1421–28.

Bourdain, A. 2001. Typhoid Mary. New York: Bloomsbury.

Buxton Bridges, C., J. M. Katz, W. H. Seto, P. K. Chan, D. Tsang, W. Ho, K. H. Mak, W. Lim, J. S. Tam, et al. 2000. Risk of influenza A (H5N1) infection among health care workers exposed to patients with influenza A (H5N1), Hong Kong. Journal of Infectious Diseases 181:344–48.

Christensen, N. O., P. Nansen, B. O. Fagbemi, and J. Monrad. 1987. Heterologous antagonistic and synergistic interactions between helminths and between helminths and protozoans in concurrent experimental infection of mammalian hosts. Parasitology Research. 73:387–410.

Cleaveland, S., M. Kaare, P. Tiringa, T. Mlengeya, and J. Barrat. 2003. A dog rabies vaccination campaign in rural Africa: Impact on the incidence of dog rabies and human dog-bite injuries. Vaccine 21:1965–73.

Cox, F. E. G. 2001. Concomitant infections, parasites and immune responses. Parasitology 122:S23–38.

Crawley, M. J. 1987. What makes a community invasible? *In* Colonization, Succession and Stability, ed. M. J. Crawley, P. J. Edwards, and A. J. Gray, 429–54. Oxford : Blackwell Scientific Publications.

Dietz, R., M. P. Heidejorgensen, and T. Harkonen. 1989. Mass deaths of harbor seals (*Phoca vitulina*) in Europe. Ambio 18:258–64.

Donnelly, C. A., R. Woodroffe, D. R. Cox, J. Bourne, G. Gettinby, A. M. Le Fevre, J. P. McInerney, and W. I. Morrison. 2003. Impact of localized badger culling on tuberculosis incidence in British cattle. Nature 426:834–37.

Earn, D. J. D., P. Rohani, B. M. Bolker and B. T. Grenfell. 2000. A simple model for complex dynamical transitions in epidemics. Science 287:667–70.

Fenner, F. and B. Fantini. 1999. Biological Control of Vertebrate Pests: The history of Myxomatosis. An Experiment in Evolution. Wallingford: Cabi Publishing.

Fraser, C., S. Riley, R. M. Anderson, and N. M. Ferguson. 2004. Factors that make an infectious disease outbreak controllable. Proceedings of the National Academy of Sciences of the United States of America 101:6146–51.

Gilbert, L., R. Norman, K. M. Laurenson, H. W. Reid, and P. J. Hudson. 2001. Disease persistence and apparent competition in a three-host community: An empirical and analytical study of large-scale wild populations. Journal of Animal Ecology 70:1053–61.

Haydon, D. T., M. K. Laurenson, and C. Sillero-Zubiri. 2002. Integrating epidemiology into population viability analysis: Managing the risk posed by rabies and canine distemper to the Ethiopian wolf. Conservation Biology 16: 1372–85.

Hudson, P. J., R. Norman, M. K. Laurenson, D. Newborn, M. Gaunt, H. Reid. E. Gould, R. Bowers, and A. P. Dobson. 1995. Persistence and transmission of tick-borne viruses: *Ixodes ricinus* and louping-ill virus in red grouse populations. Parasitology 111:S49–58.

Hudson, P. J., and J. Greenman. 1998. Competition mediated by parasites: Biological and theoretical progress. Trends in Ecology & Evolution 13:387–90.

Hudson, P. J., A. P. Rizzoli, B. T. Grenfell, H. Heesterbeek, and A. P. Dobson. 2002. The Ecology of Wildlife Diseases. Oxford: Oxford University Press.

Kermack, W. O., and A. G. McKendrick. 1927. A contribution to the mathematical theory of epidemics. Proceedings of the Royal society of London. Series B, Biological Sciences 115:700–21.

Kolar, C. S., and D. M. Lodge. 2001. Progress in invasion biology: Predicting invaders. Trends in Ecology & Evolution 16:199–204.

Laurenson, K. M., L. Gilbert, R. A. Norman, H. W. Reid, and P. J. Hudson. 2004. Identifying disease reservoirs in complex system: Mountain hares as reservoirs of ticks and louping ill virus, pathogens of red grouse. Journal of Animal Ecology 72:177–86.

LoGiudice, K., R. S. Ostfeld, K. A. Schmidt, and F. Keesing. 2003. The ecology of infectious disease: Effects of host diversity and community composition on Lyme disease risk. Proceedings of the National Academy of Sciences of the United States of America 100:567–71.

Mitchell, C. E., and A. G. Power. 2003. Release of invasive plants from fungal and viral pathogens. Nature 421:625–27.

Moulton, M. P., and S. L. Pimm. 1981. Species introductions to Hawaii. *In* Ecology of Biological Invasions of North America and Hawaii, ed. H. Monney and J. A. Drake, 231–49. Berlin: Springer-Verlag.

Parmenter, R. R., J. W. Brunt, D. I. Moore, and M. S. Ernest. 1993. The Hantavirus epidemic in the Southwest: Rodent population dynamics and the implications for transmission of Hantavirus-associated adult respiratory distress syndrome (HARDS) in the Four Corners Region. *In* Report to the Federal Centers for Disease Control and Prevention, Atlanta, Georgia,1–45. Sevilleta LTER Publication No. 41. Albuquerque: Sevilleta Long-Term Ecological Research Program, Department of Biology, University of New Mexico.

Perkins, S. E., I. M. Cattadori, V. Tagliapietra, A. P. Rizzoli, and P. J. Hudson. 2003. Empirical evidence for key hosts in persistence of a tick borne disease. International Journal of Parasitology 33:909–17.

Pimm, S. L. 1991. The Balance of Nature. Chicago: University of Chicago Press.

Poulin, R. 2001. Interactions between species and the structure of helminth communities. Parasitology 122:S3–11.

Randall, D. A., S. D. Williams, I. V. Kuzmin, C. E. Rupprecht, L. A. Tallents, Z. Tefera, K. Argaw, F. Shiferaw, D. L. Knobel, et al. 2004. Rabies in endangered Ethiopian wolves. Emerging Infectious Diseases 10:2214–17.

Sampathkumar, P. 2003. West Nile virus: Epidemiology, clinical presentation, diagnosis, and prevention. Mayo Clinic proceedings 78:1137–44.

Shea, K., and P. Chesson. 2002. Community ecology theory as a framework for biological invasions. Trends in Ecology & Evolution 17:170–176.

Shaw, D. J., B .T. Grenfell, and A. P. Dobson. 1998. Patterns of macroparasite aggregation in wildlife host populations. Parasitology 117:597–610.

Taylor, L. H., S. M. Latham, and M. E. J. Woolhouse. 2001. Risk factors for human disease emergence. Philosophical Transactions of the Royal Society of London. Series B, Biological Sciences 356:983–89.

Torchin, M. E., K. D. Lafferty, A. P. Dobson, V. J. McKenzie, and A. M. Kuris. 2003. Introduced species and their missing parasites. Nature 421:628–30.

Vander Zanden, M. J. 2005. The success of animal invaders. Proceedings of the National Academy of Sciences of the United States of America 102:7055–56.

Webby, R., E. Hoffmann, and R. Webster. 2004. Molecular constraints to interspecies transmission of viral pathogens. Nature Medicine 10:S77–81.

Wolfe, N. D., W. Heneine, J. K. Carr, A. D. Garcia, V. Shanmugam, U. Tamoufe, J. N. Torimiro, A. T. Prosser, M. LeBreton, et al. 2005. Emergence of unique primate T-lymphotropic viruses among central African bushmeat hunters. Proceedings of the National Academy of Sciences of the United States of America 102:7994–99.

Woodroffe, R., and J. R. Ginsberg. 1999. Conserving the African wild dog *Lycaon pictus*. I. Diagnosing and treating causes of decline. Oryx 33:132–42.

Woolhouse, M. E. J. 2002. Population biology of emerging and re-emerging pathogens. Trends in Microbiology 10:S3–7.

Woolhouse, M. E. J., and C. Dye. 2001. Population biology of emerging and re-emerging pathogens: Preface. Philosophical transactions of the Royal Society of London. Series B, Biological Sciences 356:981–82.

Woolhouse, M. E. J., C. Dye, J. F. Etard, T. Smith, J. D. Charlwood, G. P. Garnett, P. Hagan, J. L. K. Hii, P. D. Ndhlovu, et al. 1997. Heterogeneities in the transmission of infectious agents: Implications for the design of control programs. Proceedings of the National Academy of Sciences of the United States of America 94:338–42.

Woolhouse, M. E. J., D. T. Haydon, and R. Antia. 2005. Emerging pathogens: The epidemiology and evolution of species jumps. Trends in Ecology & Evolution 20:238–44.

CHAPTER SEVENTEEN
Applied Biodiversity Science: Managing Emerging Diseases in Agriculture and Linked Natural Systems Using Ecological Principles

K. A. Garrett and C. M. Cox

SUMMARY

PATHOGEN REPRODUCTION TENDS to be host frequency dependent, so that disease may be more problematic when particular crop species or genotypes are very common. Nonetheless, production agriculture is dominated by extensive monocultures. This situation is partly an artifact of agricultural policy and decision making, but it also reflects the real challenges of understanding and manipulating the ecological genomics of a single crop genotype, not to mention multiple species and genotypes. There are trade-offs in investing agricultural research in many versus only a few agricultural species. Agricultural diseases may emerge or reemerge for a number of reasons, including new pathogen introductions, new adaptation of pathogens to previously effective resistance genes, new types of host homogeneity (such as the widespread deployment of Texas male sterile cytoplasm in maize), trading policies that increase the economic impact of uncommon infections, and changes in the abiotic environment. Technological abilities in ecological genomics that are needed to support the management of emerging and long-term diseases include the ability to manipulate disease resistance genes in crops, the ability to devise crop plant communities on multiple spatial scales that are optimal for a range of agronomic traits, the ability to manipulate microbial communities for disease-suppressive characteristics, and the ability to minimize undesired impacts on ecosystems surrounding agricultural systems. The construction of crop variety mixtures is an example of a technology that draws heavily on ecological ideas and has also contributed greatly to our understanding of disease ecology through experiments examining the effects of patterns of host variability on disease through time and space. New forms of agricultural systems, such as perennial grains, may also offer environmental benefits, such as reduced erosion and nutrient leaching, but will also challenge ecological genomics to provide effective methods of disease management, since pathogens may more easily accumulate in long-term agricultural stands. Although agricultural systems have typically grown less diverse over the past century, advances in ecological genomics are likely to make it feasible for systems to incorporate higher levels of diversity in the future as agricultural scientists are better able to influence and manage complex agricultural communities for reduced disease.

INTRODUCTION

Reproduction in plant pathogens tends to be host frequency dependent, and many pathogens are specific to one or a few crop species or genotypes. As a result, when susceptible genotypes of a particular species are present at higher frequency, covering a higher proportion of agricultural land, losses to disease for that species will tend to be higher (Garrett and Mundt 1999). Also, greater crop host abundance may lead to more rapid adaptation of pathogen populations to resistance genes, so that the genes are useful for shorter periods of time (McDonald and Linde 2002). Yet a small number of species and genotypes dominate agricultural production (Strange and Scott 2005). It is not that careful evaluation has led agricultural scientists to the conclusion that the best human strategy is to deploy only a relatively small number of agricultural species and varieties. Rather, many factors together lead to lower diversity in agricultural plant communities (table 17.1). On each scale of decision making, economic and political factors push agriculture toward lower diversity, whether or not lower diversity is the best long-term strategy.

Any of the strategies for managing diseases proposed by other authors in this book might also be usefully applied in some agricultural contexts. But, in contrast to the management of diseases of humans and natural systems, agricultural genotypes and individuals can, in theory, be completely replaced to increase disease resistance, with the main ethical concern being preservation of traditional varieties. Thus, agricultural scientists have emphasized manipulating the genetic composition of agricultural species for disease management, often trying to perfect a single crop genotype. Improving even a single crop species has, in fact, often been a substantial challenge.

Ecologists have long been intrigued by the relative fitness of specialist and generalist organisms, and the question of how to partition agricultural research effort can be framed similarly: Are agricultural specialists or agricultural generalists better adapted? (Or, since agriculture is embedded in a complex economic system that requires the cooperation of large numbers of people, the question might also be posed as whether agricultural societies are better off as generalists or specialists.) Many agricultural specialists work with only a single crop species. Investing

TABLE 17.1
Decision making that determines the composition of agricultural plant communities

Decision Makers	Forms of Choices	Pressures away from Agricultural Diversity
Policy makers	Which farming systems to subsidize through farm bills	Lobbyists representing producers of major crop species are strongest.
Agribusiness administrators	Which crop varieties, biocontrol agents, and chemicals to produce and offer for sale	Specialization in particular crop species may be most profitable in the short run.
Public research and extension scientists	Which crop species to support with research and breeding programs	Researchers typically are hired to specialize in one or two crop species.
Farmers	Which crop species and varieties to grow	Limited availability of crop species with needed varietal and strategic development.
Food processors	Which crop species and varieties to purchase	Specialization in particular crop species may be most profitable in the short run.
Consumers	Which crop species and varieties to consume	Limited knowledge about and cultural support for other options.

a great deal of effort in a single agricultural species makes it possible to understand all the major pathogens of the species fairly well and makes it easier to develop more or less effective disease resistance and other strategies for managing diseases. By contrast, an agricultural generalist might develop many crop types, though each type might be less modified and characterized than would otherwise have been possible. These decisions could be viewed through the lens of economic portfolios, but the typical approach of agricultural researchers might more aptly be termed "insider trading" in that, rather than simply observing and predicting the characteristics of crop genotypes and systems, crop genotypes are directly manipulated in ways that may make their performance easier to predict. And the process of gaining knowledge can itself be accelerated, so that more resources are available or the same resources can be used to produce more information. For example, genetic markers can now be used to rapidly screen seedling trees to determine whether they have disease resistance genes that are expressed only in adults (Collard et al. 2005).

A goal of this chapter is to demonstrate both how agricultural research and agricultural systems have contributed to disease ecology and how new insights into community ecology could greatly benefit agricultural disease management. Agricultural plant pathology has pioneered the study of the effects of biological diversity on disease ecology and agricultural systems. Though production agriculture is often characterized by strikingly low biological diversity, agriculture offers excellent model systems for studying the effects of biodiversity on ecological processes, since agricultural pathogens are relatively well understood and techniques for handling them in studies of spatial and temporal diversity have been developed. Agricultural research is also making extensive contributions to the field of ecological genomics, including characterizations of how plant and pathogen genomes and gene expression influence the community ecology of host-pathogen interactions (Garrett, Hulbert, et al. 2006). Here we discuss forms of human technology available for managing disease, how agricultural homogeneity and other factors may contribute to the emergence of disease, how crop diversity can be manipulated to manage disease, how the composition of soil microbial communities may be manipulated to suppress disease, and the effects of agricultural adaptations to disease on surrounding systems. We emphasize plant-based agricultural systems, which are most readily manipulated through breeding programs and management strategies.

HUMAN AGRICULTURAL TECHNOLOGY AND INFORMATION BASE

Human agriculture differs from, for example, the fungus gardening of leaf-cutter ants in the huge role of intentional information accumulation and transfer. This information takes both genetic forms, such as improved varieties, and strategic (memetic) forms, such as strategies for reducing inoculum loads. Even ant agriculturalists may distribute biocontrol agents (Currie et al. 1999), thus transmitting genetic information for disease management.

Components of current human technological ability include the ability to manipulate genetic disease resistance, to manipulate nonpathogen microbial communities to enhance the disease suppressiveness of soils, to manipulate acquired and induced disease resistance (through chemical agents or microbial communities), to develop pesticides and biocontrol agents (Fravel 2005), to develop cultural practices such as sanitation (removal of diseased tissues) to reduce inoculum levels, and to develop uses of crop biodiversity such as rotation schemes, intercropping, and variety mixtures. Boudreau and Mundt (1997) have reviewed the principles of many disease management practices that have an ecological basis, including modifications to plant density and microclimate. Some of these technologies currently are marginal, and all can be improved on through better understanding of community ecology. Genetic engineering is a new tool that can potentially bring in forms of genetic information that are completely new to crop systems.

Information is a particular challenge for agricultural disease management in developing countries. Loss of traditional information concerning traditional varieties and land races, wild crop relatives (which may be important sources of resistance genes for breeding programs), and traditional management methods is a common risk (e.g., Hijmans et al. 2000). Useful information is particularly needed for orphan crops (Nelson et al. 2004), an example of which is finger millet, a crop that is of particular importance to resource-poor farmers but has seen little recent genetic or strategic investment.

Information requirements for good management are even greater at the interface between agricultural and natural (or unmanaged) systems. Some native systems, like tallgrass prairie, do require management choices in human landscapes about factors such as burning frequency. These choices could be adjusted to take into account effects on adjacent agricultural systems if, for example, more frequent burning reduced pathogen emigration to agricultural fields from prairie, as long as such decisions did not affect the integrity of the natural system. Conversely, agricultural systems should be managed to minimize spread of pathogens to natural systems. This can be challenging even for well-publicized diseases such as sudden oak death, which has been widely distributed via nursery stock (Rizzo et al. 2005). Data about the nature of exchanges of plant pathogens between managed and unmanaged systems, are generally rare.

CAUSES OF EMERGENCE OR REEMERGENCE OF
AGRICULTURAL DISEASES

Agricultural diseases may emerge through several well-known though not necessarily easily predictable mechanisms. Exotic invasive pathogens may be introduced, such as the the soybean rust fungus, which was introduced into the United States in 2004, probably by way of a hurricane (Stokstad 2004). Familiar pathogens may reemerge in the classic boom-and-bust cycle that often characterizes the deployment of major resistance genes in agriculture, with analogies in coevolution in natural systems but with the added potential for human strategy to delay evolution of pathogens that can overcome resistance (Clay and Kover 1996; Leach et al. 2001; Clay et al., chapter 7, this volume). New forms of host population homogeneity may unexpectedly support rapid pathogen reproduction. For example, Texas male sterile cytoplasm was widely used in U.S. maize breeding programs and had become common throughout U.S. maize varieties in the early 1970s. This cytoplasm surprisingly conferred susceptibility to a new form of the southern corn leaf blight pathogen, causing widespread losses in maize production (Ullstrup 1972). Trading policies can result in emergent diseases even when yield losses are minimal. For example, though Karnal bunt is a minor wheat disease, many countries refuse wheat imports from regions where the disease is present, so the introduction of this pathogen into new areas has important economic impacts (Rush et al. 2005). Changes in the abiotic environment may also be associated with the emergence of diseases (Garrett, Dendy, et al. 2006). In an intriguing study of dried wheat specimens sampled across 160 years, Bearchell et al. (2005) found that the ratio of abundance of two wheat pathogens was closely correlated with environmental SO_2 levels.

AGRICULTURAL DIVERSITY FOR DISEASE MANAGEMENT

Agricultural biodiversity can have important effects on diseases on multiple scales. In field agriculture, the use of crop variety mixtures or intercropping can be used to reduce disease. Diverse landscapes of agricultural and natural systems are also assembled, though with less intentionality, by communities of land managers. On a larger scale, a mixture of agricultural and unmanaged fields has the potential to dilute available host tissue for specific pathogens or to provide additional hosts and "green bridges" through time and space for more host-generalist pathogens.

Mitchell et al. (2002) demonstrated that decreased plant species diversity in grassland plant communities increased the pathogen load for the overwhelming majority of foliar fungal diseases evaluated. Indeed, Harper (1977, 1990) concluded, "host specific pathogens appear to penalize a population that becomes dominated by a single species," and therefore "diversity in plant communities may reflect the failure of pure stands."

The effects of diversity in nature are often multilateral. Increased diversity at one trophic level often results in increased diversity at other levels (Armbrecht et al. 2004). For example, an increased number of plant species in a community frequently correlates with an increased number of insect species (Andow 1991; Armbrecht et al. 2004; Murdoch et al. 1972). More diverse pathogen populations or communities are found in natural ecosystems where plant diversity is high than in monocultures in conventional agroecosystems (Browning 1974; Gilbert 2002; Mundt 2002).

In an agricultural field, perhaps the simplest way to increase plant diversity is through mixing different cultivated varieties (cultivars) within a crop species. Mixtures of at least two crop cultivars increases the genetic diversity, and this strategy has proved effective in reducing disease and pest severity, increasing yield stabilities, and strengthening resilience to physiological stresses (e.g., Bowden et al. 2001; Cox et al. 2004; Mundt 2002; Power 1999). Kansas wheat farmers are increasingly planting seed blends typically composed of three cultivars (Bowden et al. 2001). According to the Kansas Agricultural Statistics Service, cultivar mixtures currently cover 15.2% of the wheat acreage in Kansas, covering more land than any single unblended wheat cultivar other than the extremely popular wheat cultivar Jagger. Wheat mixtures are also commonly grown in the Pacific Northwest (Mundt 2002). It makes sense to decrease the dependence on one cultivar, since even a superior cultivar has its flaws. Combining cultivars that have complementary characteristics reduces the risks of crop failure and increases stability.

An example of striking success of cultivar mixtures is the use of mixtures of high-value rice varieties that are susceptible to the disease rice blast with lower-value resistant varieties (Zhu et al. 2000). Use of this strategy has dramatically reduced fungicide use through a large area of China. The success of this system for disease management may be due in part to the unusually large scale at which it is deployed, since the success of smaller-scale mixtures may be reduced if high levels of inoculum are supplied from adjacent plots or fields (Garrett et al. 2001).

Although mixture effectiveness against residue- and soil-borne diseases remains less predictable, there is little doubt that mixtures of small grains can substantially reduce the severity of foliar diseases caused by polycyclic, specialized, wind-dispersed pathogens, such as those causing rusts, powdery mildews, and rice blast (Browning 1974, 1988; Cox et al. 2004; Garrett and Mundt 2000; Mundt 2002; Mahmood et al. 1991; McDonald et al. 1988; Wolfe 1985). Infection by viruses may also be reduced in grass mixtures if the pathogen population is partitioned between grass types or if vector behavior markedly changes. For example, barley yellow dwarf virus (BYDV) infects three major prairie grasses, but the dominant virus strains appear to differ among these species and also to differ from the dominant wheat virus strain (Garrett et al. 2004). Further, aphid vectors of BYDV exhibited shorter feeding times in oat mixtures than in oat monocultures, with associated lower rates of virus transmission (Power 1999). Some of the mechanisms proposed for reduced disease severity include dilution of inoculum through greater spatial distance between susceptible genotypes (Burdon and Chilvers 1977; Chin and Wolfe 1984; Wolfe 1985), induced resistance (Lannou et al. 1995), and compensation by the resistant cultivar through increased tiller number (Finckh and Mundt 1992).

Mixture effectiveness largely depends on the plant host and life history of the pathogen. Garrett and Mundt (1999) suggest that the effects of host diversity tend to be greatest when the host genotype unit area is small (defined as the area occupied by an independent unit of host tissue of the same genotype; Mundt and Browning 1985), there is strong host-pathogen specialization, the pathogen's dispersal gradient is shallow (i.e., inoculum levels drop off after relatively long distances versus a steep gradient in which inoculum levels drop off after relatively short distances), characteristic lesion sizes are small, and the number of pathogen generations over the course of an epidemic is large (i.e., several cycles of inoculum are produced). With these criteria, diseases can be compared according to the magnitude of host diversity effect predicted (Cox et al. 2004; Cox, Garrett, and Bockus 2005; Garrett and Mundt 1999).

As a test of this hypothesis, Cox et al. (2004) directly compared the relative effectiveness of wheat cultivar mixing for two wheat diseases, tan spot and leaf rust, in a field experiment over two growing seasons at two different locations in Kansas. In contrast to leaf rust, caused by a highly specialized, polycyclic, wind-borne pathogen with a shallow dispersal gradient, the tan spot pathogen survives in plant residue and has a steep dispersal gradient (Sone et al. 1994). Two annual wheat cultivars, one susceptible to leaf rust (Jagger) and the other susceptible to tan spot (2145), were mixed in different proportions and inoculated with each pathogen alone and in combination. For both tan spot and leaf rust, disease severity decreased substantially on the susceptible cultivar as the proportion of that cultivar decreased in mixture. However, as predicted, mixtures were significantly ($P < 0.0001$) more effective at reducing the severity of leaf rust compared with tan spot (figure 17.1).

Determining the necessary level of plant diversity required (Main 1999) and the manner in which mixtures are best deployed for effectively reducing disease is a complex task. Small increases in host diversity in annual wheat populations, such as two cultivar mixtures in which one cultivar is resistant to disease, have substantially reduced disease severity (e.g., Cox et al. 2004). Even though these systems are much simpler than most natural plant communities, there are still many complicated competitive interactions and host-microbe interactions that need to be understood to optimize use of agricultural diversity. For example, wheat mixtures developed to decrease stripe rust severity were more effective at intermediate than at higher or lower planting densities

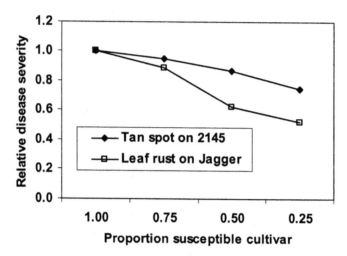

Figure 17.1. Relative effects of wheat cultivated variety (cultivar) mixtures for reducing two diseases. Mixtures were more effective at reducing the severity of leaf rust (on the leaf rust–susceptible cultivar Jagger) than at reducing the severity of tan spot (on the tan spot–susceptible cultivar 2145), as indicated by the significantly steeper slope for leaf rust ($P < 0.0001$). The relative severity in each mixture is expressed as a proportion of the severity in a monoculture of the susceptible cultivar. (Reproduced from Cox et al. 2004.)

(Garrett and Mundt 2000), though there is no clear reason for this result. Higher levels of diversity, such as polycultures of different plant species, may also be desirable for reasons that go beyond the scope of plant pathology, such as more efficient moisture and nutrient management, especially for perennial crops (Crews 2005; Main 1999).

Diverse landscapes of agricultural and natural systems are also assembled, though with less intentionality, by communities of land managers. The spread of pathogens between these systems has received little attention because of the difficulty of conducting such large studies. BYDV is one pathogen that moves between agricultural and natural systems such as tallgrass prairie (Garrett et al. 2004); invasive grasses play a role in increasing its abundance and therefore its spillover effects on other species (Malmstrom et al. 2005). Invasive plant species may also have impacts during agricultural epidemics. For example, kudzu may play an important role as an additional host in the new epidemic of soybean rust in the United States. The combination of widespread soybean plantings and invasive kudzu stands may also have significant effects on native legumes susceptible to the soybean rust fungus.

Definitions of biological control range from strict to inclusive. For the most conservative definition, the introduction of one or a few organisms to control a targeted plant pathogen, commercial success is still a challenge (Fravel 2005; Weller 1988). As an anonymous observer has put it, "it's like releasing poodles on the Serengeti." Perhaps a more enlightened ecological approach relies not on a single or a few introduced antagonists for disease suppression, but rather on enhancing the resident soil microbial community (Mazzola 2004).

Disease-suppressive soils in which indigenous microflora collectively protect plants against soil-borne pathogens have been described for many soil-borne pathogens (reviewed in Garbeva et al. 2004; Mazzola 2004; Weller et al. 2002). General suppression refers to the total soil microbial biomass, which limits the fitness of a plant pathogen through competition for resources or through more direct forms of antagonism (Weller et al. 2002). Suppression of this type is often enhanced by practices that increase soil microbial activity and diversity, such as the addition of organic amendments, crop rotation, cover crops, organic farming, reduction of or abstinence from pesticides, and the buildup of soil fertility (Garbeva et al. 2004; Mazzola 2004; Weller et al. 2002). Indeed, indicators for soil health may be indicators of a soil's ability to suppress disease caused by soil-borne pathogens (van Bruggen and Semenov 2000).

Plant type is also one of the major determinants of microbial community structure in soil (Garbeva et al. 2004). Plant species and genotypes culture microflora differentially within their rhizosperes and surrounding soil (Berg et al. 2002; da Mota et al. 2002; Grayston et al. 1998; Kaiser et al. 2001; Mazzola and Gu 2002). Monocultures of certain crops and cultivars have been indicated as building up soil supressiveness either to the continuing crop over time (e.g., take-all disease of wheat; Weller et al. 2002) or to the following crop as part of a rotation (Mazzola and Gu 2002).

Diversity at one trophic level often results in diversity at other levels. For example, higher numbers and greater complexity of antagonist isolates against the soil-borne pathogen *Rhizoctonia solani* AG 3 were found in species-rich permanent grassland than in arable land under rotation or monoculture of maize (van Elsas et al. 2002). Moreover, higher densities of particular antibiotic-producing genes were detected in the grasslands, whereas these genes were present in low densities or absent from the arable land under rotation.

Effects of Agriculture on Other Systems and the Challenge of Developing Perennial Agriculture

Because of the unique degree of control humans potentially have over their food web, the effects of diseases on agroecosystems include their effects on human decision making. Decisions about management often involve trade-offs: use of minimum tillage systems, and even more dramatically, perennial systems, reduces erosion but tends to increase disease risks for pathogens that survive in senesced plant material (Bockus and Shroyer 1998; Cox et al. 2005). The common use of annuals in agriculture has resulted in increased erosion and the potential for movement of nitrogen and other agricultural inputs to neighboring natural systems.

Much of the ongoing degradation of fifteen of the twenty-four ecosystem services examined in the United Nations–sponsored Millennium Ecosystem Assessment (MEA), in particular sudden changes in water quality, the creation of "dead zones" along the coasts, and shifts in regional climate, are linked to agriculture. The MEA report has labeled agriculture the "largest threat to biodiversity and ecosystem function of any single human activity." Because natural and agricultural ecosystems merge seamlessly into one another, conservation efforts that ignore agriculture will lack impact.

The development of herbaceous perennial grains for human consumption would reduce the agricultural impacts outlined in the MEA report. Crop species currently being developed include wheat, rice, corn, sorghum, sunflower (oil crops), and legumes (Glover 2005; Moffat 1996; Pimm 1997). Although decidedly controversial and still in the toddler stage, the development of perennial grains suitable for agriculture is well underway and prospects are excellent (C. M. Cox et al. 2002; T. S. Cox et al. 2002; DeHaan 2005; Scheinost 2001). The challenges of and potential for disease management of perennial grain crops were reviewed by Cox, Garrett, and Bockus (2005).

A limited number of studies have compared annual versus perennial life histories in terms of pathogen defense. In response to six maize viruses, three annual teosinte (*Zea*) lines were susceptible to all of the viruses, whereas perennial diploid and tetraploid teosintes were each susceptible to only one of the six viruses (Nault et al. 1982). Wheatgrasses (*Thinopyrum* spp.), the perennial relatives of modern wheat (*Triticum aestivum*), and perennial hybrids of wheatgrass × wheat show high levels of resistance to many common wheat diseases, including *Cephalosporium* stripe, wheat streak mosaic (and its wheat curl mite vector), stripe and leaf rust, BYDV, eyespot (C. M. Cox et al. 2002; Friebe et al. 1996; Jones et al. 1995; Juahar and Peterson 1996), and tan spot (Cox, Garrett, Cox, et al. 2005). These differences in susceptibility probably developed because perennial plants have evolved in the context of longer-term exposure to pests and pathogens in natural ecosystems than have annual plants. Despite this greater exposure, perennials dominate most native landscapes and constitute roughly 80% of North America's native flora (Hart 1977).

Conclusions

Changes in current policies could support management of disease in agriculture. Artificial incentives could be removed from farm policies that support lower agricultural diversity. Interestingly, current political pressures for reduced trade protectionism may act to reduce farm subsidies that support production of only a small number of specific crop species. More research in agriculture should be dedicated to developing theory about how best to allocate resources for agricultural research. Although a great deal of research effort is invested in understanding and manipulating the genomes of a small number of crop species, more investment is needed in understanding the broader context of optimal crop plant communities and microbial communities for disease management. A more diverse agricultural system is likely to be more stable in the face of new or newly important diseases.

Ecological theory, including ecological genomics, can contribute greatly to the development of better agricultural systems. A new area of ecology could focus on how to adapt theory in community ecology to incorporate the role of human decision making about the use of genetic and strategic information for disease management. Agroecology would benefit from an increased ability to predict the outcomes of interactions between and among plants, microbes, and pathogen vectors. Ecological genomics could contribute a better understanding of the genetic basis of these interactions so that better crop mixtures can be developed, for disease management as well as other desirable outcomes. In microbial ecology, better theory is needed for the prediction and manipulation of microbial interactions to facilitate soils that inhibit disease.

We close this discussion with two forms of technology optimism, as illustrated in figure 17.2. First, there is continuing great potential for humans to develop technological abilities that will support better disease management as well as enable transitions to new types of cropping

Degree of agricultural stability and productivity ─────────

Number of agricultural species/genotypes   - - - - - - - -
that optimizes stability and productivity

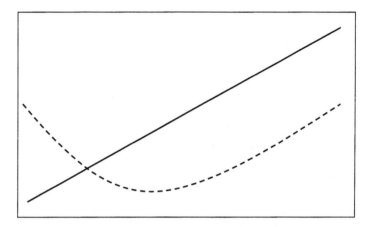

Human technological ability in ecological genomics

Figure 17.2. Proposed relationship between human technological ability in ecological genomics and the number of agricultural species that maximizes productivity and stability of agricultural systems. Here increments in human technological ability are defined as those that increase agricultural productivity and stability. That is, technological advances are defined as those that increase desirable agricultural outcomes so the relationship is defined as a straight line. The relationship between technological ability and optimal number of agricultural species or types is speculative, based on the idea that (1) at low levels of technological ability, many species or genotypes are required to optimize within a heterogeneous environment; (2) at intermediate levels of technological ability, a limited number of species or genotypes can be understood and manipulated successfully; and (3) at higher levels of technological ability, many species or genotypes and their interactions can be understood and manipulated.

systems such as perennial grains. This is shown in figure 17.2 as a straight line, indicating that technological abilities are defined here as those that lead to desirable outcomes such as productivity and stability, as well as reduced undesirable impacts on surrounding ecosystems. Second, as humans gain new technological abilities in ecological genomics, they will be able to effectively influence and manage a wider range of agricultural species and genotypes (collectively, "crop types") to enhance agricultural productivity and stability. Previously, a diverse portfolio of crop types was probably necessary to optimize agricultural systems. We may currently be at a low point in deployed agricultural diversity in temperate regions and approaching a low point in tropical regions, as agriculture industrializes and research focuses on a small number of crop types. This may be a useful short-term approach, as great strides are made in understanding the ecological genomics of some particularly important crop species and developing tools that may be applied to other systems. Deployment of variety mixtures and intercropping is often limited by an incomplete understanding of the genomic basis for interactions such as competition between crop types and for differential responses to climatic heterogeneity that may result in undersirable differences in maturity rates for different crop types. Similarly, more complete understanding is needed for optimal deployment of crop types to exert desired selection pressures on pathogens and beneficial microbes. Also, on larger scales, deployment of greater agricultural diversity is often limited by an inability to simultaneously develop many crop types to a high level of performance and economic competitiveness. However, as it becomes easier to develop and implement strategies for the use of agricultural diversity through multiple crop types, rather than continuing toward a single "supergenotype," the benefits of agricultural systems may be optimized by making use of the complementary traits of many crop types.

ACKNOWLEDGMENTS

Work was supported by NSF grants DEB-0130692 and DEB-0516046, by NSF grant EF-0525712 as part of the joint NSF-NIH Ecology of Infectious Disease program, by NSF grant EPS-0236913 with matching funds from the Kansas Technology Enterprise Corporation, by the NSF Long Term Ecological Research Program at Konza Prairie, by the U.S. Agency for International Development for the Sustainable Agriculture and Natural Resources Management Collaborative Research Support Program (SANREM CRSP) under terms of Cooperative Agreement Award No. EPP-A-00-04-00013-00 to the Office of International Research and Development at Viginia Tech and for the IPM CRSP, by USDA grant 2002-34103-11746, and by the Office of Science (PER), U.S. Department of Energy, grant DE-FG02-04ER63892. This is Kansas State Experiment Station Contribution No. 06-131-B. We thank two anonymous reviewers for comments that improved the manuscript, and also R. L. Bowden, B. Flinchbaugh, P. Garfinkel, and F. White.

LITERATURE CITED

Andow, D. A. 1991. Vegetational diversity and arthropod population response. Annual Review of Entomology 36:561–86.

Armbrecht, I., I. Perfecto, and J. Vandermeer. 2004. Enigmatic biodiversity correlations: Ant diversity responds to diverse resources. Science 304:284–86.

Bearchell, S. J., B. A. Fraaije, M. W. Shaw, and B. D. L. Fitt. 2005. Wheat archive links long-term fungal pathogen population dynamics to air pollution. Proceedings of the National Academy of Sciences of the United States of America 102:5438–42.

Berg, G., N. Roskot, A. Steidle, L. Eberl, A. Zock, et al. 2002. Plant-dependent genotypic and phenotypic diversity of antagonistic rhizobacteria isolated from different *Verticillium* host plants. Applied and Environmental Microbiology 68:3328–38.

Bockus, W. W., and J. P. Shroyer. 1998. The impact of reduced tillage on soilborne plant pathogens. Annual Review of Phytopathology 36:485–500.

Boudreau, M. A., and C. C. Mundt. 1997. Ecological approaches to disease control. *In* Environmentally Safe Approaches to Crop Disease Control, ed. N. A. Rechcigl and J. E. Rechcigl, 33–92. Boca Raton, FL: Lewis Publishers.

Bowden, R., J. Shroyer, K. Roozeboom, M. Claassen, P. Evans, B. Gordon, B. Heer, K. Janssen, J. Long, et al. 2001. Performance of Wheat Variety Blends in Kansas. Kansas State University Agricultural Extension Bulletin No. 128. Kansas State University Agriculture Experiment Station and Cooperative Extension Service, Manhattan, KS.

Browning, J. A. 1974. Relevance of knowledge about natural ecosystems to development of pest management programs for agroecosystems. Proceedings of the American Phytopathological Society 1:191–99.

Browning, J. A. 1988. Current thinking on the use of diversity to buffer small grains against high epidemic and variable foliar pathogens: Problems and future prospects. *In* Breeding Strategies for Resistance to the Rusts of Wheat, ed. N. W. Simmonds and S. Rajaram, 76–90. Mexico, DF: CIMMYT.

Burdon, J. J., and G. A. Chilvers. 1977. Controlled environment experiments on epidemic rates of barley mildew in different mixtures of barley and wheat. Oecologia 28:141–46.

Clay, K., and P. X. Kover. 1996. The Red Queen hypothesis and plant/pathogen interactions. Annual Review of Phytopathology 34:29–50.

Chin, K. M., and M. S. Wolfe. 1984. Selection on *Erysiphe graminis* in pure and mixed stands of barley. Plant Pathology 33:535–46.

Collard, B. C. Y., M. Z. Z. Jahufer, J. B. Brouwer, and E. C. K. Pang. 2005. An introduction to markers, quantitative trait loci (QTL) mapping and marker-assisted selection for crop improvement: The basic concepts. Euphytica 142:169–96.

Cox, C. M., K. A. Garrett, and W. W. Bockus. 2005. Meeting the challenge of disease management in perennial grain cropping systems. Renewable Agriculture and Food Systems 20:15–24.

Cox, C. M., K. A. Garrett, R. L. Bowden, A. K. Fritz, S. P. Dendy, and W. F. Heer. 2004. Cultivar mixtures for the simultaneous management of multiple diseases: Tan spot and leaf rust of wheat. Phytopathology 94:961–69.

Cox, C. M., K. A. Garrett, T. S. Cox, W. W. Bockus, and T. Peters. 2005. Reactions of perennial grain accessions to four major cereal pathogens of the Great Plains. Plant Disease 89:1235–40.

Cox, C. M., T. D. Murray, and S. S. Jones. 2002. Perennial wheat germplasm lines resistant to eyespot, *Cephalosporium* stripe, and wheat streak mosaic. Plant Disease 86:1043–48.

Cox, T. S., M. Bender, C. Picone, D. L. Van Tassel, J. B. Holland, E. C. Brummer, B. E. Zoeller, A. H. Paterson, and W. Jackson. 2002. Breeding perennial grain crops. Critical Reviews in Crop Science 21:59–91.

Crews, T. E. 2005. Perennial crops and endogenous nutrient supplies. Renewable Agriculture and Food Systems 20:25–37.

Currie, C. R., J. A. Scott, R. C. Summerbell, and D. Malloch. 1999. Fungus-growing ants use antibiotic-producing bacteria to control garden parasites. Nature 398:701–704.

da Mota, F. F., A. Nóbrega, I. E. Marriel, E. Paiva, and L. Seldin. 2002. Genetic diversity of *Paenibacillus polymyxa* populations isolated from the rhizosphere of four cultivars of maize (*Zea mays*) planted in Cerrado soil. Applied Soil Ecology 20:119–32.

DeHaan, L. R., D. L. Van Tassel, and T. S. Cox. 2005. Perennial grain crops: A synthesis of ecology and plant breeding. Renewable Agriculture and Food Systems 20:5–14.

Finckh, M. R., and C. C. Mundt. 1992. Plant competition and disease in genetically diverse wheat populations. Oecologia 91:82–92.

Fravel, D. R. 2005. Commercialization and implementation of biocontrol. Annual Review of Phytopathology 43:337–59.

Friebe, B., K. S. Gill, N. A. Tuleen, and B. S. Gill. 1996. Transfer of wheat streak mosaic virus resistance from *Agropyron intermedium* into wheat. Crop Science 36:857–61.

Garbeva, P., J. A. van Veen, and J. D. van Elsas. 2004. Microbial diversity in soil: Selection of microbial populations by plant and soil type and implications for disease suppressiveness. Annual Review of Phytopathology 42:243–70.

Garrett, K. A., S. P. Dendy, E. E. Frank, M. N. Rouse, and S. E. Travers. 2006. Climate change effects on plant disease: Genomes to ecosystems. Annual Review of Phytopathology 44:489–509.

Garrett, K. A., S. P. Dendy, A. G. Power, G. K. Blaisdell, H. M. Alexander, and J. K. McCarron. 2004. Barley yellow dwarf disease in natural populations of dominant tallgrass prairie species in Kansas. Plant Disease 88:574.

Garrett, K. A., S. H. Hulbert, J. E. Leach, and S. E. Travers. 2006. Ecological genomics and epidemiology. European Journal of Plant Pathology 115:35–51.

Garrett, K. A., and C. C. Mundt. 1999. Epidemiology in mixed host populations. Phytopathology 89:984–90.

Garrett, K. A., and C. C. Mundt. 2000. Effects of planting density and the composition of cultivar mixtures on stripe rust: An analysis taking into account limits to the replication of controls. Phytopathology 90:1313–21.

Garrett, K. A., R. J. Nelson, C. C. Mundt, G. Chacón, R. E. Jaramillo, and G. A. Forbes. 2001. The effects of host diversity and other management components on epidemics of potato late blight in the humid highland tropics. Phytopathology 91:993–1000.

Gilbert, G. S. 2002. Evolutionary ecology of plant diseases in natural ecosystems. Annual Review of Phytopathology 40:13–43.

Glover, J. D. 2005. The necessity and possibility of perennial grain production systems. Renewable Agriculture and Food Systems 20:1–4.

Grayston, S. J., S. Wang, C. D. Campbell, and A. C. Edwards. 1998. Selective influence of plant species on microbial diversity in the rhizosphere. Soil Biology and Biochemistry 30:369–78.

Harper, J. L. 1977. Population Biology of Plants. London: Academic Press.

Harper, J. L. 1990. Pests, pathogens, and plant communities: An introduction. In Pests, Pathogens, and Plant Communities, ed. J. J. Burdon and S. R. Leather, 3–14. Oxford: Blackwell Scientific Publications.

Hart, R. 1977. Why are biennials so few? American Naturalist 111:792–99.

Hijmans, R. J., K. A. Garrett, Z. Huamán, D. P. Zhang, M. Schreuder, and M. Bonierbale. 2000. Assessing the geographic representativeness of genebank collections: The case of Bolivian wild potatoes. Conservation Biology 14:1755–65.

Jones, S. S., T. D. Murray, and R. E. Allan. 1995. Use of alien genes for the development of disease resistance in wheat. Annual Review of Phytopathology 33:429–43.

Jauhar, P. P., and T. S. Peterson. 1996. *Thinopyron* and *Lophopyrum* as sources of genes for wheat improvement. Cereal Research Communications 24:15–21.

Kaiser, O., A. Puhler, and W. Selbitschka. 2001. Phylogenetic analysis of microbial diversity in the rhizoplane of oilseed rape (*Brassica napus* cv. Westar) employing cultivation-dependent and cultivation-independent approaches. Microbial Ecology 42:136–49.

Lannou, C., C. de Vallavieille-Pope, and H. Goyeau. 1995. Induced resistance in host mixtures and its effects on disease control in computer-simulated epidemics. Plant Pathology 44:478–89.

Leach, J. E., C. M. Vera Cruz, J. Bai, and H. Leung. 2001. Pathogen fitness penalty as a predictor of durability of disease resistance genes. Annual Review of Phytopathology 39:187–224.

Mahmood, T., D. Marshall, and M. E. McDaniel. 1991. Effect of winter wheat cultivar mixtures on leaf rust severity and grain yield. Phytopathology 81:470–74.

Main, A. R. 1999. How much biodiversity is enough? In Agriculture as a Mimic of Natural Ecosystems, ed. E. C. Lefroy, R. J. Hobbs, M. H. O'Connor and J. S. Pate, 23–41. Dordrecht, The Netherlands: Kluwer.

Malmstrom, C. M., A. J. McCullough, H. A. Johnson, L. A. Newton, and E. T. Borer. 2005. Invasive annual grasses indirectly increase virus incidence in California native perennial bunchgrasses. Oecologia 145:153–64.

Mazzola, M. 2004. Assessment and management of soil microbial community structure for disease suppression. Annual Review of Phytopathology 42:35–59.

Mazzola, M., and Y. H. Gu. 2002. Wheat genotype-specific induction of soil microbial communities suppressive to disease incited by *Rhizoctonia solani* anastomosis group (AG)-5 and AG-8. Phytopathology 92:1300–307.

McDonald, B. A., R. W. Allard, and R. K. Webster. 1988. Responses of two-, three-, and four-component barley mixtures to a variable pathogen population. Crop Science 28:447–52.

McDonald, B. A., and C. Linde. 2002. Pathogen population genetics, evolutionary potential, and durable resistance. Annual Review of Phytopathology 40:349–79.

Mitchell, C. E., D. Tilman, and J. V. Groth. 2002. Effects of grassland plant species diversity, abundance, and composition on foliar fungal disease. Ecology 83:1713–26.

Moffat, A. S. 1996. Higher yielding perennials point the way to new crops. Science 274:1469–1470.

Mundt, C. C. 2002. Use of multiline cultivars and cultivar mixtures for disease management. Annual Review of Phytopathology 40:381–410.

Mundt, C. C., and J. A. Browning. 1985. Development of crown rust epidemics in genetically diverse oat populations: Effect of genotype unit area. Phytopathology 75:607–10.

Murdoch, W. W., F. C. Evans, and C. H. Peterson. 1972. Diversity and pattern in plants and insects. Ecology 53:819–29.

Nault, L. R., D. T. Gordon, V. D. Damsteegt, and H. H. Iltis. 1982. Response of annual and perennial teosintes (*Zea*) to six maize viruses. Plant Disease 66:61–62.

Nelson, R. J., R. L. Naylor, and M. M. Jahn. 2004. The role of genomics research in improvement of "orphan" crops. Crop Science 44:1901–4.

Pimm, S. L. 1997. In search of perennial solutions. Nature 389:126–27.

Power, A. G. 1991. Virus spread and vector dynamics in genetically diverse plantpopulations. Ecology 72:232–41.

Rizzo, D. M., M. Garbelotto, and E. M. Hansen. 2005. *Phytophthora ramorum:* Integrative research and management of an emerging pathogen in California and Oregon forests. Annual Review of Phytopathology 43:309–35.

Rush, C. M., J. M. Stein, R. L. Bowden, R. Riemenschneider, T. Boratynski, and M. H. Royer. 2005. Status of Karnal bunt of wheat in the United States 1996–2004. Plant Disease 89:212–23.

Scheinost, P. L., D. L. Lammer, X. Cai, T. D. Murray, and S. S. Jones. 2001. Perennial wheat: The development of a sustainable cropping system for the U.S. Pacific Northwest. American Journal of Alternative Agriculture 16:147–51.

Sone, J., W. W. Bockus, and M. M. Claassen. 1994. Gradients of tan spot of winter wheat from a small-area source of *Pyrenophora tritici-repentis*. Plant Disease 78:622–27.

Stokstad, E. 2004. Plant pathologists gear up for battle with dread fungus. Science 306:1672–73.

Strange, R. N., and P. R. Scott. 2005. Plant disease: A threat to global food security. Annual Review of Phytopathology 43:83–116.

Ullstrup, A. J. 1972. The impacts of the southern corn leaf blight epidemics of 1970–1971. Annual Review of Phytopathology 10:37–50.

van Bruggen, A. H. C. and A. M. Semenov. 2000. In search of biological indicators for soil health and disease suppression. Applied Soil Ecology 15:13–24.

van Elsas J. D., P. Garbeva, and J. Salles. 2002. Effects of agronomical measures on the microbial diversity of soils as related to the suppression of soilborne plant pathogens. Biodegradation 13:29–40.

Weller, D. M. 1988. Biological control of soilborne plant pathogens in the rhizosphere with bacteria. Annual Review of Phytopathology 26:379–407.

Weller, D. M., J. M. Raaijmakers, B. B. McSpadden-Gardener, and Thomashow, L. S. 2002. Microbial populations responsible for specific soil suppressiveness to plant pathogens. Annual Review of Phytopathology 40:309–48.

Wolfe, M. S. 1985. The current status and prospects of multiline cultivars and variety mixtures for disease resistance. Annual Review of Phytopathology 23:251–73.

Zhu, Y., H. Chen, J. Fan, Y. Wang, Y. Li, J. Chen, J. X. Fan, S. Yang, L. Hu, et al. 2000. Genetic diversity and disease control in rice. Nature 406:718–22.

CHAPTER EIGHTEEN
The Ecology of an Infectious Coral Disease in the Florida Keys: From Pathogens to Politics

James W. Porter, Erin K. Lipp,
Kathryn P. Sutherland, and Erich Mueller

SUMMARY

CORAL REEFS ARE IN SEVERE DECLINE. Between 1996 and 2000, 38% loss of living coral cover was recorded in the Florida Keys. By far the greatest loss, in terms of both absolute abundance and percent loss, occurred in the elkhorn coral, *Acropora palmata*. Once the most common coral in the Caribbean, this species had lost more than 90% of its surface area in the Florida Keys. All elkhorn populations observed to decline in the Florida Keys exhibited signs of white pox disease prior to demise. The pathogen that causes white pox is *Serratia marcescens*, a common fecal enteric bacterium found in the gut of humans and other terrestrial animals. To date, this pathogen has not been found in any other fully marine invertebrate or nonestuarine fish. Even though the origin of the bacterium is not definitively established, data collected to date are consistent with the hypothesis that it is of human origin.

Water-related tourism in the Florida Keys generates more than U.S. $3.1 billion annually. By producing carefully written print and broadcast news press releases, and by tying the issues to both environmental and economic health, we were able to influence the public discourse about the ecology of infectious disease in a manner that promoted environmental quality and economic security. This information spurred local, state, and federal officials to propose significant improvements to both wastewater and storm water treatment. These infrastructural improvements carry with them substantial costs, which are being met by significant tax increases supported by the public and by politicians of all political parties.

THE DILEMMA

Research on the ecology of infectious disease exists at the interface between pure and applied science. As such, these investigations have the potential to bridge the historical chasm between the laboratory and the ballot box. Particularly when the study of wildlife disease has direct impacts on the livelihood of the voting public, there is no way for scientists involved to avoid politics. By acknowledging this, and by providing timely and relevant information to the public debate, it is possible to responsibly influence public perception and thereby public policy. Communication is an essential part of the scientific process and one that need not stop at the laboratory door. If scientific evidence is to influence public policy and foster political action, then scientists must share their knowledge to make this happen.

The Problem

Coral reefs are declining rapidly. The most reliable estimates suggest that worldwide, 27% have already been lost, with another 16% at serious risk for loss (Wilkinson 2002). Within the Caribbean, populations of the once most common reef-building coral, *Acropora palmata,* are being decimated, with losses of living cover in the Florida Keys (figure 18.1) averaging 87% (figures 18.2) or greater (Bruckner 2003; Dustan 1999; Dustan and Halas 1987; Gardner et al. 2003; Hughes et al. 2003; Miller, Bourgue,

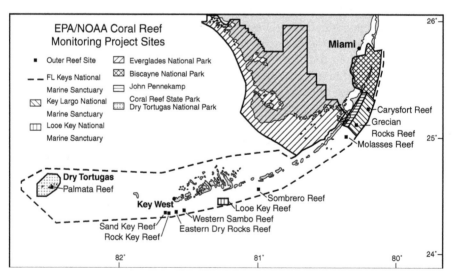

Figure 18.1. Locations of *Acropora palmata* monitoring stations in the Florida Keys. Coral reef monitoring stations were chosen using a stratified random selection procedure (E-MAP) throughout the Florida Keys (Porter et al., 2002) and have been monitored annually since 1996.

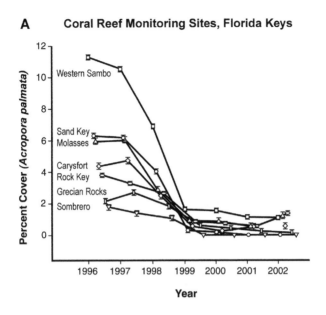

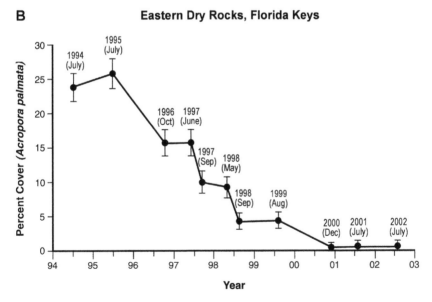

Figure 18.2. (a) Percent cover of *Acropora palmata* at seven reef sites in the Florida Keys National Marine Sanctuary, 1996–2002 (see figure 18.1). By 1999, the percent cover of this once abundant species had decreased at every site, for a Keys-wide average loss of 87%. (b) Ninety-eight percent of all living *A. palmata* at Eastern Dry Rock Reef, Key West, Florida, died between 1994 and 2002.

et al. 2002, Patterson et al. 2002; Sutherland and Ritchie 2004). Severe population declines of *A. palmata* in Florida (Dustan 1999; Dustan and Halas 1987; Miller, Baums, et al. 2002, Patterson et al. 2002; Sutherland and Ritchie 2004) and elsewhere in the Caribbean (Bruckner 2003; Gardner et al. 2003; Hughes et al. 2003) have led to the listing of this species under the Endangered Species Act (Bruckner 2003; Diaz-Soltero 1999; Miller, Baums, et al. 2002; Precht et al. 2002).

Hurricanes (Bythell et al. 2000; Fong and Lirman 1995; Hughes 1994; Lirman 2000a,b, Lirman and Fong 1997), bleaching (Harvell et al. 1999, Hoegh-Guldberg 1999), predation (Miller 2001, Miller, Bourgue, et al. 2002), disease (Aronson and Precht 2001; Gladfelter 1982; Porter and Meier 1992; Rodriguez-Martinez et al. 2001), and human disturbances (Richmond 1993; Rogers et al. 1988; Tomasick and Sander 1987) have contributed to declines in elkhorn populations throughout the Caribbean. However, the majority of recent losses in the Florida Keys are associated with white pox disease (Patterson et al. 2002; Sutherland and Ritchie 2004). One year after the first documentation of the disease (Holden 1996), white pox was found at all surveyed reefs in the Florida Keys (Patterson et al. 2002; Porter et al. 2002). Signs of active white pox disease (figure 18.3) were observed at Eastern Dry Rocks Reef, Florida, every year between 1996 and 2000 (Patterson et al. 2002) and at each of the other seven monitored reefs with living cover of *A. palmata*

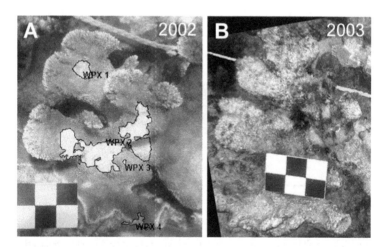

Figure 18.3. Paired images show infection by white pox disease on *Acropora palmata* colonies (a, from July 16, 2002) and almost complete colony mortality a year later (B, from August 19, 2003) from the Coral Reef Monitoring Project Value-Added Site on Grecian Rocks, Key Largo, Florida.

every year between 1997 and 2002 (figure 18.2). Observations of white band disease, the only other disease known to kill *A. palmata*, were rare at monitored reefs in the Florida Keys between 1996 and 2002 (Sutherland and Ritchie 2004).

Since *Acropora palmata* reproduces primarily by fragmentation rather than by sexual reproduction (Aronson and Precht 2001; Bruckner and Bruckner 2001; Knowlton et al. 1990; Szmant 1986), disease is especially devastating not only to adult survival, but also to juvenile recruitment (Patterson et al. 2002). Remnant patches of *A. palmata* are also subject to heavy losses from the predatory snail, *Coralliophila abbreviata*, which preferentially feeds on this coral species (Miller 2001). In combination, these factors will prevent rapid recovery of *A. palmata* in the Florida Keys.

THE PATHOGEN

Serratia marcescens is a cause of white pox disease. In 2002, we reported at least one definitive agent of WPD, confirmed through fulfillment of Koch's postulates, as the bacterium *Serratia marcescens* (Patterson et al. 2002). Since *S. marcescens* cannot be cultured from all apparent white pox lesions, however, we proposed that when lesions are confirmed to harbor this bacterium that the disease be termed acroporid serratiosis to reflect its etiology. *S. marcescens* is a common gram-negative bacterium classified as a coliform and a member of the Enterobacteriaceae family. It is found in the intestines of humans, insects, and other animals and in freshwater, soil, and plants (Grimont and Grimont 1994). It is a pathogen of humans, cows, goats, chickens, fishes, insects, and plants (Grimont and Grimont 1994). In humans, the bacterium is an opportunistic pathogen associated with waterborne infections in tropical waters (Hazen 1988) and hospital-acquired diseases, including meningitis, pneumonia, endocarditis, urinary tract infection, wound infections, and septicemia (Grimont and Grimont 1994; Hejazi and Falkiner 1997; Miranda et al. 1996; Shi et al. 1997). Environmental isolates of *S. marcescens* are often characterized by the production of a red pigment, prodigiosin. Like human clinical strains, which are typically nonpigmented, the known white pox pathogen (isolate PDL100) lacks prodigiosin (Patterson et al. 2002).

White pox is one of eighteen coral diseases documented worldwide (Sutherland et al., 2004), but *S. marcescens* is only the fifth pathogen to be confirmed as a coral disease agent through fulfillment of Koch's postulates (Ben-Haim and Rosenberg 2002; Ben-Haim et al. 2003; Denner et al. 2003; Geiser et al. 1998; Kushmaro et al. 1996, 1997, 1998, 2001; Richardson, Goldberg, Carlten, et al. 1998; Richardson, Goldberg, Kuta, et al. 1998; Rosenberg et al. 1998; Smith et al. 1996) and the first agent with a possible link to human sewage pollution. The identification of *S. marcescens* as a coral pathogen marked the first time that a common member of the human gut microbiota was shown to be a marine invertebrate pathogen (Patterson et al. 2002). Concurrent studies also show that human sewage markers (e.g., human enteric viruses) are prevalent among nearshore corals and environments of the Florida Keys (Griffin et al. 1999; Lipp et al. 2002). The potential significance of both studies bears directly on coral reef management and strongly suggests that land-based activities affect reef coral health and coral reef survival.

Serratia marcescens is ubiquitous in terrestrial and freshwater ecosystems, and in the guts of both terrestrial and freshwater vertebrates and invertebrates. Very little is known of its prevalence and ecology in the marine environment. To date, the bacterium has not been found in any fully marine invertebrate (other than *A. palmata*) or nonestuarine fish. To assess the prevalence of *S. marcescens* in the marine coastal waters of the Florida Keys, we screened a variety of potential sources, including (1) offshore coral reefs, (2) nearshore waters from beaches, navigable channels, and residential boat canals, and (3) human wastewater (table 18.1). Genetic typing of *S. marcescens* isolates from these environments may help to identify sources and potential modes of disease transmission. Once the dominant transmission mechanism has been identified, it can be incorporated into mathematical models (Grenfell and Dobson 1995; Rohani et al. 1999) in order to predict and evaluate the impact of white pox on *A. palmata*.

All isolates were identified to species level using PCR techniques directed at 16S rDNA. To date, we have identified 292 *S. marcescens* isolates from sources in the Florida Keys (table 18.1). These studies reveal that *S. marcescens* is uncommon in offshore reef environments. Nonpigmented strains of this bacterium are common in untreated human sewage and in the canal waters of the Florida Keys (table 18.1; Sutherland et al. 2007). These spatial data point to, but are not definitive proof of, a human origin for white pox.

Because *S. marcescens* may be derived from a variety of environments, it is currently unknown what the source may be for the coral pathogen. To address this question, environmental isolates were compared for analysis of genetic similarity and variability using pulsed-field gel electrophoresis (PFGE). This procedure has been applied extensively in human clinical epidemiological studies on *S. marcescens* and has allowed the determination of sources of infection in a variety of diseases for both hospital-acquired and environmental infections (e.g. Miranda et al. 1996; Shi et al. 1997). Although this method has been applied to source tracking of fecal indicator bacteria (e.g., *E. coli* and *Enterococcus* spp.) for

TABLE 18.1.
Summary of detection and isolation of *Serratia marcescens* from various environmental sources in the Florida Keys, September, 2003–August, 2004

	N	No. of Serratia marcescens		
		Pigmented	Nonpigmented	Total
A. *palmata* mucus (apparently healthy)	108	0	0	0
A. *palmata* mucus (predation lesions)	82	0	0	0
Reef water	16	0	0	0
Coralliophilia abbreviata (snails)	64	0	0	0
Hermodice carunculata (fireworms)	3	0	4	4
Parrotfish feces	48	0	0	0
Seabird guano	51	0	23	23
Beach water	5	0	22	22
Canal water	17	10	81	91
Advanced wastewater treatment influent	4	5	150	155
Advanced wastewater treatment effluent	2	1	0	1
Total	400	16	276	292

Note: All isolates were confirmed by PCR (Sutherland et. al., 2002). The known white pox pathogen, *S. marcescens* PDL 100, is nonpigmented.

determining host origin (Scott et al. 2002), it has not been used to its full potential for marine invertebrate disease epidemiology.

Using PFGE and digital band pattern analysis, we have compared the genetic fingerprints of the coral pathogenic strain, *S. marcescens* PDL100 (Pox Disease Looe Key, isolate 100; Patterson et al. 2002), and environmental strains from potential source areas (see figure 18.1). Preliminary application of this technique has not yet identified a source of the pathogen but reveals that the pathogenic strain, PDL100, clusters with isolates from wastewater and contaminated canals.

THE POLITICS

Coral loss in the Florida Keys has been headline news for more than a decade (Dorfman 1992; Keating 1991; Porter and Meier 1992). Subsequent studies, over longer time scales and broader spatial scales, have confirmed this downward trajectory (Porter et al. 2002). Although many South Florida scientists were slow to accept these findings and their significance (Keating 1991), commercial fishermen and most local inhabitants of the Keys suspected from their own experience that these results were correct and the profound implications that these findings had for their annual $3.1 billion tourist economy (Associated Press 1994).

An initial working hypothesis was that multiple stressors were responsible for the decline (Porter et al. 1999). This diffuse hypothesis, however, did not identify specific management actions required to reverse the loss. Systematic observations on Floridian coral reefs revealed that infectious disease was a major contributor to coral decline (Porter et al. 2001). As with the discovery of coral loss, the discovery of a new coral disease of unknown origin afflicting elkhorn coral was widely reported (Associated Press 1996; Holden 1996). *Acropora palmata* is the icon species of the Florida reef tract, and even before the identification of the cause of this wildlife disease, local people were aware of its potential threat to their livelihoods (Madigan 1996).

Identification of the causative agent of white pox as a fecal enteric bacterium caused a firestorm of press coverage both in print (e.g., Morgan 2002) and on television (ABC Corporation 2002; CBS Corporation 2002; NBC Corporation 2002). Before the release date, we authored a carefully prepared press statement to ensure accurate reporting. Specifically, we noted that although *S. marcescens* is a human enteric bacterium, it is also found in other organisms. This important qualifier needed to be reported because although *S. marcescens* is found in human waste, it had not been proved that the coral pathogenic strain came from humans. This "controlled news release" strategy was successful. Even newspaper headlines presented the actual findings of the research (Morgan 2002) rather than presenting, as if they were conclusions, potential implications of the study. As the press coverage continued, careful qualifiers enhanced credibility both in scientific circles and in the political arena.

With the discovery of white pox, the ecology of infectious disease had been inserted into South Florida politics. By returning telephone calls, granting interviews, distributing reprints, and giving talks at multiple venues in the Florida Keys, the authors of the study (Patterson et al. 2002) made the findings widely available to taxpayers, voters, and elected officials. When Keys-wide elections for county commissioner occurred in November 2002, the two leading candidates made affordable wastewater treatment the core of their campaigns (Fusaro 2002). A major issue was how to pay for the proposed infrastructural upgrades. As property values rise, so do tax revenues. The two leading candidates (both Republicans) chose to reverse a decades-old strategy of rolling back mileage rates to hold the tax digest constant. Instead, both ran on platforms of increasing taxes (Matley 2002).

Following his successful campaign, David Rice issued the following statement:

> There is a link between environmental quality and economic prosperity in the Florida Keys. Tourism is our economy. If we don't preserve this, we have no product. Voters are well aware that wastewater treatment issues need to be addressed in the Florida Keys. Including this "environmental plank" in my campaign platform was not a risky election strategy. We are going to AWT [Advanced Wastewater Treatment Standards], and people are insisting that it be AWT. The arguments now are about how to do it, not whether it should be done. While it may have been difficult for politicians to talk about these issues in the past, it certainly is not now.

This is clearly a real-life application of the precautionary principle, since at no time had a 100% similarity between human and white pox *S. marcescens* strains been proved or implied. In addition to white pox, research revealed a high prevalence of sewage contamination and infectious viruses along Keys' canals and beaches (e.g., Griffin et al. 1999). Furthermore, high levels of fecal indicator bacteria resulted in the closure of beaches around Key West at the height of the summer tourist season (Bolen 1999), 30% of swimmers in an annual "Swim Around Key West" reported infections after swimming in these waters (Nobles et al. 2000). Florida Keys residents decided that, for several environmental reasons, costly improvements to water quality made financial sense. The seriousness of the issue for residents of the Florida Keys drove the electoral process to link good government with environmental protection.

No one underestimates the amount left to be done or the magnitude of the problem (Pandolfi et al. 2005). However, as Causey and Andrews (2005) point out, local, state, and federal governments have enacted stringent legislation mandating significant water quality improvements in the Florida Keys, and, just as significantly, have combined forces to pay for these improvements. Of the $438 million that is needed to upgrade wastewater treatment throughout the Keys (Monroe County 2000), more than $97 million was obligated from local county property taxes in 2003 (Matley 2002). In addition, the City of Key West spent almost $10 million to upgrade its sewage treatment plant to AWT

standards by 2004, six years before state and federal clean water stat-
utes take effect (Kruczynski 2005). The proof that this system is working is visible in a comparison of the number of *S. marcescens* strains
isolated from influent and effluent wastewater of the Key West AWT
plant (table 18.1). Nonpigmented *S. marcescens* isolates (the white pox
isolate is also nonpigmented) were never found in treated wastewater
effluent (table 18.1).

The Future

The *S. marcescens* isolates collected from environments in the Florida
Keys need to be tested for pathogenicity against *A. palmata*. Fulfillment
of Koch's postulates with any of these isolates will identify an origin of
the coral pathogen in the Florida Keys. However, experimentation with
pathogenic strains of *S. marcescens* poses an acute ethical dilemma. As a
listed threatened species, inoculation of *A. palmata* in the field is inap-
propriate and could contribute to new disease outbreaks if not properly
contained. To overcome this test-subject problem, we have propagated
clonal lines of *A. palmata* (Becker and Mueller 2001). *Acropora palmata*
is one of the most difficult corals to maintain in artificial environments
(Borneman 2001). A pilot project at the Perry Institute for Marine Science on Lee Stocking Island in the Bahamas, however, has demonstrated
the feasibility of culturing *A. palmata* in an open tank system for almost
two years (figure 18.4; Becker and Mueller 2001). In this new system, *A.
palmata* survivorship has been 75%. Ten confirmed genotypes of *A. palmata* are now in culture and available for study of disease resistance,
lethality, and mechanisms of pathogenesis.

Our ability to culture elkhorn coral in the laboratory will facilitate
the search for disease-resistant strains of *A. palmata*. Identification of
disease-resistant *A. palmata* phenotypes will also lead to the identification of the gene or genes that confer resistance, and could provide valuable information regarding mechanisms of pathogenesis. The production
of disease-resistant clones of this listed threatened species could have
special significance to coral reef restoration efforts.

Future elkhorn coral recovery will depend on at least three interlac-
ing factors: (1) the evolution of resistant strains of *A. palmata*, (2)
coastal zone management of the discharge of wastewater into nearshore
environments, and (3) mitigation of water quality conditions that may
have favored the establishment of *S. marcescens* in the first place. We already have anecdotal evidence from observational studies in the Florida
Keys that natural selection may be producing disease-resistant elkhorn
colonies. Despite the fact that juvenile colonies can be afflicted with

white pox (Patterson et al. 2002), we have seen sexual recruits of *A. palmata* without signs of white pox growing in fields of elkhorn coral widely affected by the disease. We recommend rigorous enforcement of existing state and federal no-discharge regulations and the installation of readily accessible boat pump-out stations at all marinas throughout the Florida Keys. Even in the absence of proof that it is a human strain of *S. marcescens* that kills coral, a reduction of the infusion of *S. marcescens* into coastal waters is prudent. *Serratia marcescens* is not common in the marine environment (table 18.1). Therefore, even if the transmission of white pox is now fully marine, a reduction in the input of new *S. marcescens* strains may benefit the evolution of disease resistant strains of *A. palmata*.

Finally, we need to address the water quality conditions that may have favored the establishment and survival of *S. marcescens* in the marine environment. Elevated sea surface temperatures from global warming (Harvell et al. 1999, 2004) and elevated nutrients (both nitrogen and phosphorus) from terrestrial runoff or aerial deposition (Boyer and Jones 2002) threaten costal ecosystems everywhere, but especially in tropical waters, where reef organisms already live close to their upper lethal temperatures and thrive best under low-nutrient conditions (Porter and Tougas 2000). A reduction in conditions favoring microbial growth will almost certainly promote the settlement and growth of juvenile corals

Figure 18.4. (a) *Acropora palmata* colonies are grown in culture at the Perry Institute for Marine Science's Lee Stocking Island facility in the Bahamas. (b) *A. palmata* genetic propagules flourish at the laboratory and exhibit rapid growth, as in this example from August 2003 to May 2004 in the L.S.I. coral culture facility.

and contribute to the post-settlement survival of adult colonies. Water quality improvements usually focus on reducing nutrient inputs. We recommend adding to this list of specific improvements pathogens, and a better understanding of the ecology of infectious disease.

PUBLIC AWARENESS AND CIVIC RESPONSIBILITY

The elements that allowed us to effectively introduce science into the public dialogue in the Florida Keys included (1) a voting public that recreates and works out-of-doors and that (2) already suspected environmental degradation was occurring. As boaters and fishermen, many also understood the link between environmental quality and economic prosperity. In addition, (3) we decided from the beginning that our environmental news story was important, and therefore we wrote the press release as carefully as we wrote the scientific paper. To us, environmental protection matters; we had a chance to foster this value, and we took it. Finally, (4) we decided to work with broadcast news agencies to reach the largest number of eligible voters. As the environmental dimensions of our work became apparent, we pulled together video of both the discovery and the investigatory process. Students and institutional media specialists were of immeasurable help in this regard. We left 99% of this footage on the cutting room floor and shared the remaining video freely with local and national broadcasting agencies. In the sound bite world of modern broadcast news, we stuck to the essential elements of the story. We respected the viewing audience by telling them what the data meant, not just what the data were. In broadcast news, content comes from context, not the other way around.

We fully acknowledge the potential risks involved in speaking in public forums. A well-crafted press release is the best defense against being misconstrued, misquoted, or misrepresented. The rewards are worth the risk because a sustainable future is everybody's responsibility.

LITERATURE CITED

ABC Corporation, 2002. ABC World News with Peter Jennings. June 17.

Aronson, R. B., and W. F. Precht. 2001. White-band diseases and the changing face of Caribbean coral reefs. Hydrobiologia 460:25–38.

Associated Press. 1994. Florida coral reef declining rapidly. New York Times, August 9.

Associated Press. 1996. Diseases ravaging some Keys reefs: Yet unknown, whether nature or man is to blame. Miami Herald, December 18.

Becker, L., and E. Mueller. 2001. The culture, transplantation and storage of *Montastraea faveolata, Acropora cervicornis* and *A. palmata*: What we have learned so far. Bulletin of Marine Science 69:881–96.

Ben-Haim, Y., and E. Rosenberg. 2002. A novel *Vibrio* sp. pathogen of the coral *Pocillopora damicornis*. Marine Biology 141:47–55.

Ben-Haim, Y., F. L. Thompson, C. C. Thompson, M. C. Cnockaert, B. Hoste, J. Swings, and E. Rosenberg. 2003. *Vibrio coralliilyticus* sp. nov., a temperature-dependent pathogen of the coral *Pocillopora damicornis*. International Journal of Systematic and Evolutionary Microbiology 53:309–15.

Bolen, M. 1999. Water problems empty beaches: Tourists cutting vacations short as bacteria spreads. The Key West Citizen, June 17.

Borneman, E. H. 2001. Aquarium Corals: Selection, Husbandry, and Natural History. Charlotte, VT: T. E. H. Publications.

Boyer, J. N., and R. D. Jones. 2002. A view from the bridge: External and internal forces affecting the ambient water quality of the Florida Keys National Marine Sanctuary. *In* The Everglades, Florida Bay, and Coral Reefs of the Florida Keys, ed. J. W. Porter and K. G. Porter, 609–28. Boca Raton, FL: CRC Press.

Bruckner, A. W. 2003. Proceedings of the Caribbean *Acropora* Workshop: Potential Application of the U.S. Endangered Species Act as a Conservation Strategy. NOAA Technical Memorandum NMFS-OPR-24. Washington, DC: NOAA.

Bruckner, A. W., and R. J. Bruckner. 2001. Condition of restored *Acropora palmata* fragments off Mona Island, two years after the *Fortuna Reefer* ship grounding. Coral Reefs 20:235–43.

Bythell, J. C., Z. M. Hillis-Starr, and C. S. Rogers. 2000. Local variability but landscape stability in coral reef communities following repeated hurricane impacts. Marine Ecology Progress Series 204:93–100.

Causey, B. D., and K. Andrews. 2005. Reassessing U.S. Coral Reefs. Science 308:1740–41. CBS Corporation. 2002. CBS 4 Sunday Morning with Eliot Rodriguez. June 23.

Denner, E. B. M., G. Smith, H. J. Busse, P. Schumann, T. Narzt, S. W. Polson, W. Lubitz, and L. L. Richardson. 2003. *Aurantimonas coralicida* gen. nov., sp. nov., the causative agent of white plague type II on Caribbean scleractinian corals. International Journal of Systematic and Evolutionary Microbiology 53:1115–22.

Diaz-Soltero, H. 1999. Endangered and threatened species: A revision of candidate species list under the Endangered Species Act. Federal Register 64: 33466–68.

Dorfman, A. 1992. Summit to save the earth: The world's next trouble spots. Time 139:64–65.

Dustan, P. 1999. Coral reefs under stress: Sources of mortality in the Florida Keys. Natural Resource Forum 23:147–55.

Dustan, P., and J. C. Halas. 1987. Changes in the reef-coral community of Carysfort Reef, Key Largo, Florida: 1972–1982. Coral Reefs 6:91–106.

Fong, P., and D. Lirman. 1995. Hurricane caused population expansion of the branching coral *Acropora palmata* (Scleractinia): Wound healing and growth patterns of asexual recruits. Marine Ecological Progress Series 16:317–35.

Fusaro, S. 2002. Commission candidates offer voters clear choices: Afford-able wastewater treatment is at the core. Florida Keys Keynoter, August 24, 1–5.

Gardner, T. A., I. M. Côté, J. A. Gill, A. Grant, and A. R. Watkinson. 2003. Long-term region-wide declines in Caribbean corals. Science 301:958–60.

Geiser, D. M., J. W. Taylor, K. B. Ritchie, and G. W. Smith. 1998. Cause of sea fan death in the West Indies. Nature 394:137–38.

Gladfelter, W.B. 1982. White-band disease in *Acropora palmata*: Implications for the structure and growth of shallow reefs. Bulletin of Marine Science 32:639–43.

Grenfell, B. T., and A. P. Dobson. (Eds.). 1995. Ecology of Infectious Diseases in Natural Populations. Cambridge: Cambridge University Press.

Griffin, D. W., C. J. Gibson III, E. K. Lipp, K. Riley, J. H. Paul and J. B. Rose. 1999. Detection of viral pathogens by reverse transcriptase PCR and of mi-crobial indicators by standard methods in the canals of the Florida Keys. Ap-plied Environmental Microbiology 65:4118–25.

Grimont, P. A., and F. Grimont. 1994. Genus VIII *Serratia* Bizio, 1823. *In* Ber-gey's Manual of Determinative Bacteriology, Vol. 4, ed. J. G. Holt, N. R. Kreig, P. H. A. Sneath, J. T. Staley, and S. T. Williams, 477–84. Baltimore: Williams and Wilkins.

Harvell, C. D., R. Aronson, N. Baron, J. Connell, A. Dobson, S. Ellner, L. Gerber, K. Kim, A. Kuris, et al. 2004. The rising tide of ocean diseases: Unsolved problems and research priorities. Frontiers in Ecological Environ-ment 2:375–82.

Harvell, C. D., K. Kim, J. M. Burkholder, R. R. Colwell, P. R. Epstein, J. Grimes, E. E. Hofmann, E. Lipp, A.D.M.E. Osterhaus, et al. 1999. Emerging marine diseases: Climate links and anthropogenic factors. Sci-ence 285:1505–10.

Hazen, T. C. 1988. Fecal coliforms as indicators in tropical waters: A review. Toxicity Assessment International Journal 3:461–77.

Hejazi, A., and F. R. Falkiner. 1997. *Serratia marcescens*. Journal of Medical Microbiology 46:903–12.

Hoegh-Guldberg, O. 1999. Climate change, coral bleaching and the future of the world's coral reefs. Marine and Freshwater Research 50:839–66.

Holden, C. 1996. Coral disease hot spot in the Florida Keys. Science 274: 2017.

Hughes, T. P. 1994. Catastrophes phase shifts and large-scale degradation of a Caribbean coral reef. Science 265:1547–51.

Hughes, T. P., A. H. Baird, D. R. Bellwood, M. Card, S. R. Connolly, and C. Folke. 2003. Climate change, human impacts, and the resilience of coral reefs. Science 301:929–33.

Keating, D. 1991. Florida's barrier reef seen doomed by 2000; Some scientists dispute finding of study. Washington Post, December 31.

Knowlton, N., J. C. Lang, B. D. Keller. 1990. Case study of natural population collapse: Post-hurricane predation on Jamaican staghorn corals. Smithsonian Contributions to Marine Sciences 31:1–25.

Kruczynski, W. 2005. Reassessing U.S. coral reefs. Science 308:1741.

Kushmaro, A., E. Banin, Y. Loya, E. Stackebrandt, and E. Rosenberg. 2001. *Vibrio shiloi* sp. nov., the causative agent of bleaching of the coral *Oculina patagonica*. International Journal of Systematics and Evolutionary Microbiology 51:1383–88.

Kushmaro, A., Y. Loya, M. Fine, and E. Rosenberg. 1996. Bacterial infection and coral bleaching. Nature 380:396.

Kushmaro, A., E. Rosenberg, M. Fine, Y. Ben-Haim, and Y. Loya. 1998. Effect of temperature on bleaching of the coral *Oculina patagonica* by *Vibrio shiloi* AK-1. Marine Ecology Progress Series 171:131–37.

Kushmaro, A., E. Rosenberg, M. Fine, and Y. Loya. 1997. Bleaching of the coral *Oculina patagonica* by *Vibrio* AK-1. Marine Ecology Progress Series 147:159–65.

Lipp, E. K., J. L. Jarrell, D. W. Griffin, J. Jacukiewicz, J. Lukasik, and J. B. Rose. 2002. Preliminary evidence for human fecal contamination in corals of the Florida Keys, U.S.A. Marine Pollution Bulletin 44:666–70.

Lirman, D. 2000a. Lesion regeneration in the branching coral *Acropora palmata*: Effects of colonization, colony size, lesion size and lesion shape. Marine Ecology Progress Series 197:209–15.

Lirman, D. 2000b. Fragmentation in the branching coral *Acropora palmata* (Lamark): Growth, survivorship, and reproduction of colonies and fragments. Journal of Experimental Marine Biology and Ecology 251:41–57.

Lirman, D. 2003. A simulation model of the population dynamics of the branching coral *Acropora palmata*: Effects of storm intensity and frequency. Ecological Modeling 161:167–80.

Lirman, D., and P. Fong. 1997. Patterns of damage to the branching coral *Acropora palmata* following Hurricane Andrew: Damage and survivorship of hurricane-generated asexual recruits. Journal of Coastal Research 13:67–72.

Madigan, N. 1996. Key West reefs dying of mysterious disease. New York Times, December 24, B-11.

Matley, A. 2002. Higher taxes likely for virtually everyone in the Keys. Florida Keys Keynoter, August 24, 8–9.

Miller, M. W. 2001. Corallivorous snail removal: Evaluation of impact on *Acropora palmata*. Coral Reefs 19:293–95.

Miller, M. W., I. B. Baums, D. E. Williams, and A. M. Szmant. 2002. Status of candidate coral, *Acropora palmata*, and its snail predator in the upper Florida Keys National Marine Sanctuary: 1998–2001. NOAA Technical Memorandum NMFS-SEFSC-479. Washington, DC: NOAA.

Miller, M. W., A. S. Bourque, J. A. Bohnsack. 2002. An analysis of the loss of acroporid corals at Looe Key, Florida USA: 1983–2000. Coral Reefs 21: 179–82.

Miranda, G., C. Kelly, F. Solorzano, B. Leanos, R. Coria and J. E. Patterson. 1996. Use of pulsed-field gel electrophoresis typing to study an outbreak of infection due to *Serratia marcescens* in a neonatal intensive care unit. Journal of Clinical Microbiology 34:3138–41.

Monroe County. 2000. Sanitary Wastewater Master Plan. Key West, FL: Monroe County Commission.

Morgan, C. 2002. Bacteria killing a major coral in the Keys; Organism found in human waste. Miami Herald, June 18.

NBC Corporation. 2002. NBC 6 Evening News with Hank Tester. June 18.

Nobles, R. E., P. Brown, J. B. Rose, and E. K. Lipp. 2000. The investigation and analysis of swimming-associated-illness using the fecal indicator enterococcus in southern Florida's marine waters. Florida Journal of Environmental Health 169:15–19.

Pandolfi, J. M., J. B. C. Jackson, N. Baron, R. H. Bradbury, H. M. Guzman, T. P. Hughes, C. V. Kappel, J. C. Ogden, H. P. Possingham, and E. Sala. 2005. Are U.S. coral reefs on the slippery slope to slime? Science 307:1725–26.

Patterson, K. L., J. W. Porter, K. B. Ritchie, S. W. Polson, E. Mueller, E. C. Peters, D. L. Santavy, and G. W. Smith. 2002. The etiology of white pox, a lethal disease of the Caribbean elkhorn coral, *Acropora palmata*. Proceedings of the National Academy of Sciences of the United States of America 99:8725–8730.

Porter, J. W., P. Dustan, W. C. Jaap, K. L. Patterson, V. Kosmynin, O. W. Meier, M. E. Patterson, and M. Parsons. 2001. Patterns of spread of coral disease in the Florida Keys. Hydrobiologia 460:1–24.

Porter, J. W., V. Kosmynin, K. L. Patterson, K. G. Porter, W. C. Jaap, J. L. Wheaton, K. Hackett, M. Lybolt, C. P. Tsokos, et al. 2002. Detection of coral reef change by the Florida Keys Coral Reef Monitoring Project. *In* The Everglades, Florida Bay, and Coral Reefs of the Florida Keys, ed. J. W. Porter and K. G. Porter, 749–69. Boca Raton: CRC Press.

Porter, J. W., S. K. Lewis, and K. G. Porter. 1999. The effect of multiple stressors on the Florida Keys coral reef ecosystem: A landscape hypothesis and a physiological test. Limnology and Oceanography 44:941–49.

Porter, J. W., and O. W. Meier. 1992. Quantification of loss and change in Floridian reef coral populations. American Zoology 32:625–40.

Porter, J. W., and J. I. Tougas. 2000. Reef ecosystems: Threats to their biodiversity. Encyclopedia of Biodiversity 5:73–95.

Precht, W. F., A. W. Bruckner, R. B. Aronson, and R. J. Bruckner. 2002. Endangered acroporid coral of the Caribbean. Coral Reefs 21:41–42.

Richardson, L. L., W. M. Goldberg, R. G. Carlton, and J. C. Halas. 1998. Coral disease outbreak in the Florida Keys: Plague Type II. Revista Biología Tropical 46:187–98.

Richardson, L. L., W. M. Goldberg, K. G. Kuta, R. B. Aronson, G. W. Smith, K. B. Ritchie, J. C. Halas, J. S. Feingold, and S. L. Miller. 1998. Florida's mystery coral killer identified. Nature 392:557–58.

Richmond, R. H. 1993. Coral reefs: Present problems and future concerns resulting from anthropogenic disturbances. American Zoology 33:524–36.

Rodriguez-Martinez, R. E., A. T. Banaszak, and E. Jordan-Dahlgren. 2001. Necrotic patches affect *Acropora palmata* (Scleractinia: Acroporidae) in the Mexican Caribbean. Diseases of Aquatic Organisms 47:229–34.

Rogers, C. S., L. McLain, and E. Zullo. 1988. Damage to coral reefs in Virgin Islands National Park and Biosphere Reserve from recreational activities. *In* Proceedings of the Sixth International Coral Reef Symposium 2:405–10. Townsville, Queensland, Australia: James Cook University Press.

Rohani, P., D. J. D. Earn, and B. T. Grenfell. 1999. Opposite patterns of synchrony in sympatric disease metapopulations. Science 286:968–71.

Rosenberg, E., Y. Ben-Haim, A. Toren, E. Banin, A. Kushmaro, M. Fine, and Y. Loya. 1998. Effect of temperature on bacterial bleaching of corals. *In* Current Perspectives in Microbial Ecology, ed. E. Rosenberg, 242–54. Washington, DC: ASM Press.

Scott, T. M., J. B. Rose, T. M. Jenkins, S. R. Farrah, and J. Lukasik. 2002. Microbial source tracking: Current methodology and future directions. Applied Environmental Microbiology 68:5796–803.

Shi, Z.-Y., P Y.-F. Liu, Y.-H. Lin, and B.-S. Hu. 1997. Use of pulsed-field gel electrophoresis to investigate an outbreak of *Serratia marcescens*. Journal of Clinical Microbiology 35:325–27.

Smith, G. W., L. D. Ives, I. A. Nagelkerken, and K. B. Ritchie. 1996. Caribbean sea fan mortalities. Nature 383:487.

Sutherland, K. P., J. W. Porter, and C. Torres. 2004. Disease and immunity in Caribbean and Indo-Pacific zooxanthellate corals. Marine Ecology Progress Series 266:273–302.

Sutherland, K. P., J. W. Porter, J. W. Turner, J. C. Venton, K. M. Caspary, and E. K. Lipp. 2007. Prevalence and diversity of the coral pathogen, *Serratia marcescens*, in wastewater and marine environments in the Florida Keys, USA. Unpublished manuscript.

Sutherland, K. P., and K. B. Ritchie. 2004. White pox disease of the Caribbean elkhorn coral *Acropora palmate*. *In* Coral Health and Disease, ed. E. Rosenberg and Y. Loya, 289–97. Heidelberg: Springer-Verlag.

Szmant, A. M. 1986. Reproductive ecology of Caribbean reef corals. Coral Reefs 5:43–54.

Tomasick, T., and F. Sander. 1987. Effects of eutrophication on reef-building corals. II. Structure of scleractinian coral communities on fringing reefs, Barbados, West Indies. Marine Biology 94:53–75.

Wilkinson, C. (Ed.). 2002. Status of Coral Reefs of the World: 2002. Townesville, Australia: Australian Institute of Marine Sciences.

CHAPTER NINETEEN
Infection and Ecology: *Calomys callosus,* Machupo Virus, and Acute Hemorrhagic Fever

Karl M. Johnson

INTRODUCTION

ATTEMPTS TO EXPLAIN THE ORIGIN and transmission of pathogens and how they variously cause disease certainly require knowledge of pathogen biology and biomedical science. But time and again, ecological methods have proved critical for understanding the interactions that can lead to prevention or control of infection, as well as for appreciating evolutionary history relevant to the origins and dissemination of given pathogens.

This chapter is a personal account of such discovery and how my team and I came to partially understand the underlying ecology of a particularly deadly zoonotic virus that causes severe human disease.

BACKGROUND

Beginning in 1959, residents of two small villages and many ranches in the eastern Department of Beni, Bolivia, began to suffer an acute febrile disease that often led to gastrointestinal hemorrhage, a peculiar neurological syndrome, clinical shock, and a frightening 30% death rate. The epicenters of these ongoing endoepidemics were a tiny town of less than 300 persons near Magdalena and a larger village of about 2,500 persons, San Joaquin, located some seventy miles to the west along the Machupo River (figure 19.1; Mackenzie et al. 1964).

Hispanics had settled the Beni for more than 200 years, but none in the affected region could recall a clinical disease that resembled this apparent new plague. It erupted in a portion of the department that is a vast tropical grassland savanna at about 15°S and some 800 feet in altitude. Rainfall averages almost sixty inches annually, and there is a pronounced dry season from May through September. The atmosphere fills with smoke during August and September, when the dry grass is systematically burned to allow new growth when rains return in October. Many Amazon River tributaries that flow north and east from the steep Atlantic slope of the Andes dissect the savanna. There also are numerous small *alturas* in the plain, forested islands of better soil raised thirteen to thirty

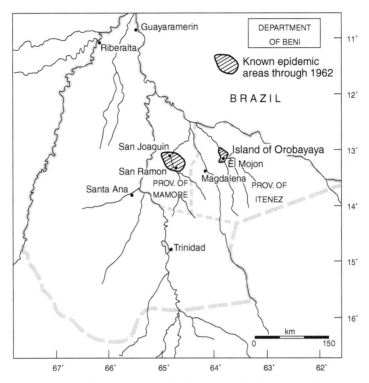

Figure 19.1. Map of the portion of Bolivia in which the initially recognized hemorrhagic fever epidemic occurred.

feet above the surrounding lateritic grasslands. The alturas provide principal sites for villages and large ranches because they afford some protection against huge snowmelt- and rain-induced floods in May or June that sometimes cover much of the savanna. In 1962 there were fewer than 125,000 humans in the department, and perhaps twenty times that number of cattle. These stringy animals continue to provide a significant supply of beef for human consumption on the high Alto Plano of central Bolivia. Carcasses were hauled up to La Paz at more than 13,00 feet in four-motor World War II bombers, principally the B-17. There were no roads in the affected portion of the Beni; access was only by boat on the extensive river system or by airplane.

In 1962 Dr. Ronald Mackenzie, a recent graduate of UC Berkeley's master's of public health program in epidemiology, visited Bolivia as part of a nutrition team organized by the Department of Defense to estimate the health status of young men in rural areas of the country.

After the failed Bay of Pigs invasion in 1961, the U.S. military was concerned that Che Guevara fully intended to mount a guerilla invasion somewhere in mainland South America. Response and defense were deemed to rely heavily on local resources, and it was felt important to obtain a reliable health baseline, as well as to establish contacts with sympathetic individuals and communities along the extensive borders with Peru, Brazil, Argentina, and Paraguay. Mackenzie had recently joined the National Institute of Allergy and Infectious Diseases (NIAID) laboratory in the Panama Canal Zone, the Middle America Research Unit (MARU). The lab was established in 1958 as a listening post for sylvatic yellow fever, which occasionally rolled up from Colombia through the eastern Darien Province of Panama to cross the canal and continue into southern Mexico. Beyond this passive function, MARU was enjoined to look for "new opportunities" in zoonotic viral diseases.

When Mackenzie arrived in La Paz in May 1962, shivering and oxygen-starved at the highest commercial airport in the world, he was quickly taken to see the minister of health and within three days was flown to Magdalena in one of the meat planes to see "El Tifu Negro," the name given the new disease by Bolivian physicians. Events proved these doctors and scientists half-right; the reservoir was a rodent, but the pathogen turned out to be a virus new to science, not a rickettsial relative of typhus. Mackenzie made two trips to the epidemic region in the summer of 1962 and managed to collect acute and convalescent serum samples from three patients who survived the illness (Mackenzie et al. 1964)

He also learned that Argentine workers had recovered a virus in 1958 from patients on the fertile pampas northwest of Buenos Aires who had suffered a disease that was remarkably similar to that seen in the Beni. They also reported that a similar virus was identified from several species of field mice, principally of the *Calomys* genus, newly erected by Philip Hershkovitz at the Field Museum in Chicago (Hershkovitz 1962). Finally, these biologists claimed that ectoparasitic mites taken from infected mice were virus positive, thus closing a putative basic life cycle for this zoonotic pathogen. Just before Christmas, workers at the main laboratories of NIAID in Bethesda, Maryland, showed that the three Bolivian hemorrhagic fever survivors had developed antibodies reactive with the Argentine virus, named Junin, using a now superceded method called complement fixation (CF) (Wiebenga et al. 1964). It was well recognized in those days that CF was not a species-specific test; it measured antigens shared by members of immunologically related groups of viruses. So the first question was whether the Bolivian syndrome was caused by Junin virus or by an unrecognized relative. If Junin was the culprit, a major issue would be to reconcile the 700-mile gap between two endemic foci, as well as the differences in tropical versus temperate grassland biomes. If the virus proved to be a new relative of Junin, at least there might be vertebrate hosts and arthropod vectors also related to those of the pampas to the south.

INVESTIGATION—1963

I joined the staff at MARU in September 1962, just weeks before the Cuban missile crisis. After mutual atomic destruction between the United States and the Soviet Union was narrowly avoided, the U.S. Department of Defense rapidly moved headquarters for Latin America from Puerto Rico to the Canal Zone. For decades, our first priority had been to defend the canal; now it was time to confront Castro and Che Guevara more directly. President Kennedy had promised not to invade Cuba as part of the settlement with the Soviet Union, and it was widely known that Guevara had publicly championed guerilla invasions on the South American mainland. A four-star army general led the newly named Southern Command with assistance from a three-star air force officer. Our government also decided that Bolivia was the most likely initial target for an operation led by Che. That judgment proved correct. Thus, beginning in 1963, our embassy and military assistance group in La Paz was soon the largest in South America. There was feverish activity to train and equip Bolivian military forces. There was intense effort and much expenditure to mount highly visible "civic action" programs in health and agriculture designed to win the hearts and minds of the population. Bolivia at that time had a socialist government and had recently undergone a major reshuffling of land ownership. Many were undoubtedly sympathetic to Che.

Our laboratory was located about 500 yards from Southern Command headquarters in an area called Quarry Heights. We also had a small research group from Walter Reed Army Research Institute in Washington, so news of the disease on the eastern Bolivian frontier soon reached command top brass. As part of the campaign to fight potential guerilla attacks, highly trained military special operations teams were sent to the Canal Zone. The 8th Special Forces unit had physicians and cross-trained Hispanic technicians who could move at a moment's notice. The air force organized a squadron of fliers, who had World War II planes taken from mothballs in the Arizona desert. Decals on these planes could be removed in minutes so that they could operate in counterinsurgency emergencies without overt recognition. These were the air commandos. They sported blue neckerchiefs and wore Australian-style hats, usually with both brims turned up. We soon had firm offers from Quarry Heights for transportation to and from the Beni, with funds to manage local expenses and renovate space for a field laboratory at any site we chose, and for Special Forces medical teams to help with clinical management of the disease. We found those inducements irresistible.

Mackenzie learned in January 1963 that disease activity was concentrated in the region of San Joaquin, about 100 miles north of Trinidad, the department capitol. In March, he and our newly arrived ecologist, Merle Kuns, flew with the air commandos to San Joaquin (a three-day voyage with stops in Lima and La Paz) to evaluate the status of the epidemic there. They found several patients in a poorly supplied hospital, discovered that there were no motorized vehicles, and electricity for only four hours each day after sundown. With two excellent Bolivian physicians running interference, they selected space for a laboratory, arranged to have furniture made by hand, mapped the town and placed new numbers on each house, performed five autopsies, the first ever done in the village, and began to collect wildlife specimens for taxonomic orientation. Almost immediately, Kuns found that the most common rodent in homes and gardens was almost certainly a member of the new *Calomys* genus.

During this time I was preoccupied with issues surrounding the need for maintenance of a cold chain that had temperatures of dry ice or lower. If we were to recover the causative virus in the village where the disease raged, we had to have uninterrupted supplies of dry ice or liquid nitrogen. Neither of these materials was available in Bolivia. We built Styrofoam-lined plywood boxes that just accommodated fifty-pound blocks of dry ice and buried them in shallow graves outside MARU. If not disturbed, we learned, the ice would last for fifteen days. So we went to Quarry Heights and asked for air commando trips to San Joaquin every two weeks. "For how long?" we were asked. "For the rest of the year, if needed." We received that promise.

I also was concerned that virus isolation work in the field might be dangerous for field technicians and laboratorians, including myself. We knew only that this disease had significant mortality and affected both sexes and all age groups. The mode of transmission was unknown. Was there person-to person transmission? Could aerosols be involved? To deal with this problem, I explained what we did not know and asked that persons interested in joining the work team do so only as volunteers. There would be no prejudice toward anyone who opted out. Everyone with skills required, including our Panamanian technicians, volunteered. Now the question was how to create a safe workspace for virus isolation attempts in a field lab. I visited the army biological laboratories at Fort Detrick, Maryland, and was soon rewarded with a lightweight aluminum and plastic isolator specially designed for our needs. It had two pairs of glove ports and an exhaust motor to produce negative air pressure inside the chamber, with filter material placed at the exhaust so that contaminated air could not be discharged into the room. It functioned in a manner almost identical to that of large heavy stainless steel cabinets that were the first versions of what today provide Biosafety Level 4 (BSL-4), or maximum containment of agents that are highly virulent, infectious as aerosols, and for which no specific therapy is available. We took the world's first portable BSL-4 lab to frontier Bolivia.

Our research team arrived in San Joaquin on May 12, 1963, with two tons of gear. Laboratories for field and virus studies were almost ready. Hamster and mouse colonies started in March were now producing litters of newborns for use in isolation attempts. On May 18 we dedicated the virus lab in an emotional ceremony organized by the town's leaders. That evening, a small boy was brought in from a tiny hamlet fifteen miles distant. He was feverish, extremely dehydrated, and comatose after eight days of illness, during which time minor bleeding from nose and mouth had been noted by the family. He died within two hours, and we were permitted to perform an autopsy. Tissue samples from several organs were processed in the plastic isolator through the night and inoculated into newborn mice and hamsters. Seven days later one baby hamster was missing from a litter inoculated with a suspension of human spleen cells, and the next day all other animals were overtly ill. We harvested tissues from this group of hamsters, combined brain samples from each individual, and did quantitative titration of the presumptively infected brains. After another week of incubation all animals became ill, including those that received brain diluted 1 million-fold. This meant that there were at least 10 million hamster-lethal units of infectivity in each gram of original hamster tissue. In less than three weeks we had isolated the virus of the disease we now called Bolivian hemorrhagic fever (BHF). Within another ten days the original isolate had been freeze-dried and sent back to MARU and on to the Bethesda laboratories (Johnson, Wiebenga, et al. 1965). Eventually it was found to be related to, but immunologically distinct from, Junin virus of Argentina (Tauraso et al. 1964). We named our new pathogen Machupo in honor of the river that flowed just more than one-half mile from the town.

Mackenzie, meanwhile, was busy with epidemiological investigation of the disease. He quickly confirmed that the virus was active both in San Joaquin and in small ranches and tiny villages within a radius of about twenty-five miles from the town. Most rural cases came from farms to the south of the village, and Mackenzie soon had solid data to show that almost 90% of San Joaquin infections since October 1962 had occurred in the southern half of the village. Merle Kuns worked hard in urban and rural ecological settings and discovered two principal rodents: *Calomys*, which he thought matched Hershkovitz's description of *C. callossus*, and the spiny rat, *Proechimys guyannensis*, which he found in each of four localities sampled. The former species was by far the most numerous rodent in urban settings, where it occupied the niche commonly observed for *Mus musculus* in many regions of the world. The spiny rat never entered houses but was common in brushy areas little disturbed by humans. Kuns also captured small numbers of the arboreal rice rat, *Orizomys bicolor*, in thatch and tile roofs of the houses in San Joaquin. The only other candidate vertebrates that fit the observed pattern of human disease occurrence were bats. Animals of several genera, most prominently *Eumops*, *Molossus*, *Myotis*, and *Noctilio*, were present in the village and in human outposts in the countryside.

Several days after the hamsters inoculated with human spleen tissue sickened, a young boy brought in a *Calomys* mouse bleeding from the anus and convulsing. We leapt at this clue, suspicious that this rodent also had hemorrhagic fever. Sure enough, Machupo virus was recovered from its brain (Johnson et al. 1966). We began to test ectoparasites of this species, hoping to identify an arthropod vector for transmission of infection to humans. Unfortunately, we encountered no further hemorrhagic *Calomys*, despite offering significant reward for any that could be delivered to the lab. Another puzzle was that there were approximately equal numbers of *Calomys* in both northern and southern halves of San Joaquin, and that Conrad Yunker, an expert acarologist from our sister NIH field lab in Hamilton, Montana, could not find differences in species or prevalence of these parasites on rodents from either portion of the village. At the height of the dry season in late September, grasslands ablaze to prepare pastures for imminent rains, and with human disease at a seasonal low ebb (Johnson et al. 1967), the fatigued and frustrated research team returned to MARU to test thousands more vertebrate and arthropod samples in a forlorn attempt to elucidate both reservoir and vector of this virus new to science and disease new to medicine. Suspicious of *Calomys*, I had carried six pairs of adults back to the Canal Zone from houses in the northern part of town that had not experienced human BHF cases. My goal was to start a breeding colony and eventually to carry out experiments to elucidate the behavior of Machupo virus in this mouse.

Investigation—1964

By the end of 1963 we had no further clue to the ecological puzzle of Machupo virus. Virus was not recovered from more than 28,000 ectoparasites (Kuns 1965). Unfortunately, we had not preserved many samples of *Calomys* tissue because we waited for further mice with hemorrhage to be encountered. The *Calomys* colony was doing nicely (Justines and Johnson 1970), but it would be several months before there were enough animals to mount the battery of experiments already planned. But now we were under immense pressure from the government of Bolivia, from our embassy in La Paz, and from our supporters in Quarry Heights to "do something"—something obvious and dramatic to demonstrate that we were not just ivory tower scientists doing studies and writing for our specialty journals. We decided to concentrate field effort on bats, and to organize an intense rodent control experiment in San Joaquin—target, *Calomys*. Mackenzie made a new census of the village and collected a random set of serum samples to help estimate numbers of still susceptible persons in different sections. He noted in March that new BHF cases occurred almost equally in northern and southern portions of the village (Mackenzie 1965).

Eastern and western portions of the town were defined such that estimated numbers of susceptible individuals were almost equal (392 and 454), with a central north-south buffer zone that had 187 susceptible individuals. Cumulative weekly numbers of new cases of BHF in eastern and western zones during the month of April were remarkably similar (figure 19.2; Johnson et al. 1973). On May 1, fifteen to twenty mouse

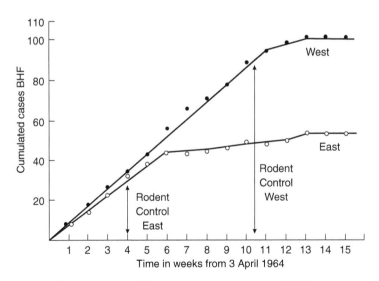

Figure 19.2. Cumulative weekly numbers of new cases of BHF in eastern and western zones of San Joaquin, Bolivia, beginning April 3, 1964.

snap traps were placed in houses and gardens in the eastern zone. As shown in the figure, new disease in that zone abruptly declined after two weeks. Four weeks later traps were added to the western and buffer zones, and after another two weeks the last case of an eighteen-month epidemic in San Joaquin was admitted to hospital on June 28 (Kuns 1965). The toll was more than 700 clinical infections and 122 deaths. This spectacular result was far more dramatic than we had imagined possible. But it also gave us insight into, and new questions regarding, the ecology of Machupo virus infection.

The abrupt virtual cessation of new human disease almost exactly two weeks after trapping control began in each sector of the village clearly demonstrated that person-to-person transmission of virus did not occur. It also suggested that the human incubation period was about seven to ten days. That is because the vast majority of *Calomys* removed from each sector were captured during the first ten days of trapping, and most of these in turn were taken in the initial three nights. We live-trapped seventeen *Calomys* in the western sector before the kill traps were deployed and were surprised to find that thirteen of these were infected with Machupo virus. Urine samples were obtained from thirteen of these mice, and five of these also were virus-positive (Johnson, Mackenzie, et al 1965). Parenthetically, we traced the source of the *Calomys* brought to us in 1963, and learned that it had almost certainly been poisoned by warfarin used by the owner of the home from which that mouse had been taken. Altogether, it appeared that *Calomys* mice suffered chronic, clinically silent infection by Machupo virus and that continuous excretion of virus in urine provided environmental contamination that led to human infection and disease. Acting on this information, a team of BHF-immune men and boys was organized in San Joaquin to mount an ongoing *Calomys* control campaign in towns and ranches of the region.

Even before the rodent control miracle, another mystery nagged at the back of our minds. A village similar in size to San Joaquin and located a dozen miles upstream on the Machupo River was little affected by BHF. The only conspicuous difference was that San Ramon was located right on the riverbank. Merle Kuns investigated rodents in the village, and Rose Navarro, our Hispanic Peace Corps volunteer who provided epidemiological eyes and ears throughout the field studies, interviewed citizens and obtained serum samples from a randomly selected fraction of the population. San Ramon domiciles and shops were infested with *Rattus norvegicus,* not *Calomys*. Rose learned that there had been a few cases of illness clinically compatible with BHF among persons resident at the edge of town farthest from the river, and antibodies were subsequently documented in a few individuals from this section of the village. Did *Rattus* exclude *Calomys*? Were the rats also infected but unable to chronically excrete virus into the environment? There were many unanswered questions.

ELUCIDATION—1965–72

Experimental investigation of Machupo virus infection in *Calomys* could now begin. The MARU colony flourished, and we generated a large aliquoted pool of the virus that had been passaged only twice in baby hamsters to use in all work. We also had found a continuous monkey cell line that allowed development of visible holes of dead cells in lawns of these cells in small dishes when Machupo virus suspensions were added to the medium. Thus, accurate, quantitative measurement of virus concentrations in specimens was now possible. Specific virus-neutralizing antibodies also were measurable with this method in which a fixed, countable number of plaques were reacted with dilutions of sera to determine the absence, presence, and power of such antibodies (Webb et al. 1969).

Initial studies revealed that no amount of virus evoked detectable acute illness in mice of either sex or any age, including newborns. This result was independent of route of administration, including direct intracranial inoculation. We saw no bleeding and no convulsions—final evidence that the clinical disease observed in that first virus-positive *Calomys* in San Joaquin was a red herring.

We soon had another fundamental surprise. All mice less than nine days of age developed persistent infections marked by huge amounts of virus in blood, lymphatic organs, salivary glands, and brain. These mice failed to develop virus-neutralizing antibodies and continually excreted about 500,000 infectious units of virus/ml in urine. In contrast, we discovered that mice older than eight days exhibited a split response to infection. Half of them behaved exactly as those less than nine days old. However, an equal number initially showed viremia and viruria but after one to three months produced good antibodies and virtually eliminated detectable virus from organs and body fluids (table 19.1). The chronically viremic response was an example of immune tolerance (Justines and Johnson 1969) that we termed type A. Animals that developed antibodies exhibited a diminished form of chronic infection. Small amounts of virus were recovered from saliva and urine (Justines and Johnson 1968). This pattern was designated type B.

It seemed evident that Type A mice were the main contributors to contamination of the environment that led to human infection with Machupo virus. Virus deposited on unprotected food or aerosolized from urine on dirt floors of houses that were swept each morning by women

TABLE 19.1

Age of *Calomys* rodents when inoculated with Machupo virus, and resultant type of infection (viremia but no detectable antibody, or detectable antibodies but no viremia)

Age at Inoculation	Number Inoculated	Type of Infection at 4–5 Months	
		Viremia, No Antibodies	Antibodies, No Viremia
0–2 Days	53	53	0
4–8 days	9	9	0
9 days	7	4	3
11–14 days	17	8	9
20–22 days	16	7	9
8 weeks	27	14	13
6 months	26	13	13
Total, ≥ 9 Days	93	46	47

of the town was likely ingested or inhaled by those who had type A-infected *Calomys* in cohabitation. I was much pleased that we had used the plastic isolator for work in San Joaquin. We found that virus in urine of type A mice was rapidly inactivated, regardless of the extreme range of pH encountered with different diets (figure 19.3). But addition of protein to such urine increased survival time by several hours.

Gustavo Justines, the senior Panamanian technician at MARU, who later earned a doctorate at the University of Wisconsin, did an interesting study with these chronically infected mice. He found that food containing large amounts of water, such as bananas and yucca, caused the animals to excrete almost ten times the daily volume of urine that was produced when mice were fed dry corn and rice and provided with water in bottles. We surmised that banana consumption was greater than for corn because mice had similar caloric need, regardless of the source of such calories. Justines learned that the concentration of Machupo virus in urine of mice on each diet did not differ, despite the fact that bananas resulted in urinary pH of more than 8, while corn produced slightly acid urine (figure 19.3). So wet food caused a tenfold daily increase in virus excretion. This pattern reflected the incidence of human BHF in San Joaquin between rainy and dry seasons. Other things equal—and we had no direct data regarding seasonal *Calomys* densities in the town—inherent ecological factors appear to have played a role beyond mouse and virus in the observed pattern of human infection.

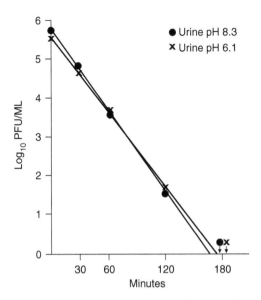

Figure 19.3. Decline in virus concentration in *Calomys* urine at two different pH levels. The pH extremes had no detectable effect on virus survival in urine.

Our working idea regarding chronic viruric infection of *Calomys* mice as the single mechanism for human disease still had some holes. Did the other rodents found in San Joaquin (and San Ramon) also experience chronic infection? That answer was no. *Proechimys, Orizomys,* and *Rattus* adults exhibited no disease, failed to exhibit viremia or viruria, and simply developed antibodies to Machupo virus as memories of silent infection (Webb et al. 1975). We did not test animals of these species less than a week of age, but we were unable to find virus or neutralizing antibodies to it in tissues or sera of individuals of these species captured in endemic areas of Bolivia. The single-virus, single-host idea seemed more probable. If true, however, the crucial question was, how did it work? How did intergenerational transmission occur? Fighting among chronically viremic mice might provide a mechanism for horizontal transmission. Was this the whole story?

We dug deeper, undertaking an experiment to assess the role of types A and B infection in reproduction. Results of the type A part of that work are shown in table 19.2. If the male was infected, there was significant reduction in both frequency of pregnancy and number of offspring. But if a chronically infected female was mated with a normal male, there were virtually no new mice. We decided quickly that the experiment with

TABLE 19.2

Effect of tolerant (type A) Machupo virus infection on the number of litters and average litter size of virgin *Calomys callosus* mice

Infection Status	No. of Pairs	Litters No. (%)	Litter Size, Average
Female normal, male normal	48	32 (67)	5.7
Female normal, male positive	37	14 (38)	3.2
Female positive, male normal	44	3 (7)	1.7

type A-infected animals of both sexes was not necessary. Type B infectious matings produced offspring in numbers not significantly different from those of normal mice (data not shown).

Patricia Webb and Gus Justines next carried out time-defined examinations of the influence of virus on *Calomys* reproduction. They found that all normal females that became pregnant after exposure to type A males had a litter of live offspring. These females were highly viremic at the time of parturition and had large amounts of virus in breast milk that immediately infected each newborn mouse. All of these mice, in turn, exhibited pernicious type A infection. When they were weaned and bred to normal males, almost no live births occurred. In a series of serial sacrifice studies, Webb and Justines demonstrated that these females became pregnant as often as normal paired mice, but that fetuses generally died during the last three days of the twenty-two- to twenty-six-day gestation period. Most of these fetuses contained Machupo virus, but there was no histological evidence of viral damage (table 19.3). If the initial dam eventually developed a type B response with antiviral antibodies, subsequent fecundity was radically different. These mice thenceforth had a reproductive experience no different from that of uninfected sex partners.

TABLE 19.3

Numbers of viable and of dead fetuses of type A (sensitive) infected and uninfected (normal) *Calomys* females paired with normal males

Gestation (days)	No. of Females	Viable Fetuses Virus	Viable Fetuses No Virus	Dead Fetuses Virus	Dead Fetuses No Virus
Infected 18–26*	18	18	0	18	12
Normal 17–21	17	0	119	0	1

*Normal gestation is 21–22 days. Infected animals examined after that interval had only resorbing dead fetuses.

Breast milk from type B dams provided Machupo antibodies instead of virus to their litters, thus protecting the next generation from infection against low-level chronic maternal shedding of virus until the young mice were weaned and the passive maternal antibodies disappeared. These mice were now fully susceptible to venereal as well as horizontal blood-borne infection from bites and scratches.

An immunofluorescent method disclosed several important aspects of this venereal infectious process (Peters et al. 1973). In this test, very thin frozen sections of organs of interest are placed on slides and reacted with antibodies to Machupo virus. If Machupo viral antigens are present in the tissues, primary viral antibodies react with fluorescein-labeled antihost antibodies. Thus, a positive result shows bright green color in the anatomical locations where virus has replicated. We found that type A male mice had persistent infection of the epithelial lining of the epididymis, but not the spermatozoa. We could not find viral antigen in ova of type A female *Calomys,* but the epithelial lining of the uterus stained spectacularly green. Based on data from *Mus musculus,* a species that has a gestational interval similar to that of *Calomys,* we surmised that the placenta of a pregnant animal might close at about four to five days after insemination. To examine the possible effect of this event, we bred normal mice and infected females the next morning and three, five, and seven days after conception. The results (table 19.4) showed clearly that by five days after successful breeding, virus did not reach any fetus and all animals were born normally (Webb et al. 1975). Once the placenta was closed, virus could not gain access to fetuses. We concluded that because it requires about five days for Machupo infection to become completely systemic, one cycle of successful reproduction occurred in all females, with half of them destined to be virtually sterile thereafter.

TABLE 19.4

Machupo virus infection and viability of embryos in pregnant female *Calomys callosus* infected intraperitoneally at varying times after mating with uninfected males

Virus Inoculated (days after mating)	No. of Females	No. with Dead or Infected Embryos	Viable Embryos		Dead or Resorbing Embryos	
			Virus Positive	Virus Negative	Virus Positive	Virus Negative
0	3	3	2	5	9	1
3	6	4	0	28	2	1
5–7	8	0	0	46	0	0

Note: Animals were autopsied at 17–19 days of gestation.

There was one more major dilemma. If half of *Calomys* rodents that are sexually infected with Machupo virus become partial or almost complete reproductive cripples, how does the species survive? We wondered whether host genetics were in play, but realized that answers would be very tedious to obtain. Nevertheless, Webb and Justines embarked on three years of breeding experiments linked to testing of weaned animals for phenotypic type A or B outcomes. Thirty-seven pairs of *Calomys* were mated and up to six successive F_1 brother-sister pairings as well as backcrosses were carried out. Adult animals were eventually infected with virus, and phenotypic A versus B outcomes were determined ninety days later. As shown in table 19.5, F_1 mice of homozygous parents appeared to function as though a single dominant gene determined outcome. When successive offspring of these "homozygous" pairings were tabulated, however, it was evident that genetic control of response to infection was due to more than a single gene and that the type A pattern was phenotypically dominant (Webb et al. 1975). We concluded that if the experimental data were representative of nature, Machupo virus might well serve as a servo-regulator of *Calomys* populations. As population increased in distinct inbred deems, genetically type A animals would increasingly dominate. Finally, when interdeemic contact became common as populations peaked, a venereal epizootic would race through the species, giving rise to a generation of predominantly type A-infected females that became functionally sterile. Populations would crash, to be temporarily dominated by small numbers of type B individuals.

TABLE 19.5
Results of genetic crosses between type A and type B *Calomys callosus* individuals

	No. of Animals Showing Type A or Type B Response		Ratio A/B
	A*	B†	
Total F_1 of 37 matings	167	187	1:1
F_1 offspring of parents:			
A-A	71	25	3:1
B-B	27	79	1:3
All offspring of parents:			
A-A	348	75	5:1
B-B	213	247	1:1

* A = viremia, no N antibody.
† B = N antibody, no viremia.

We could never test this hypothesis directly, because our field resources were constantly demanded in rodent control activities and because wherever *Calomys* were present, local authorities had only one goal—elimination. Whether virus infection or variation in primary plant production linked to global pacific temperature cycles is the main driving force for population control of *Calomys callosus* remains unknown. But the question continues to tantalize, as infection has not been clearly linked to regulation of any other wild vertebrate species that suffers population crash as a recurrent phenomenon.

In any event, our experimental ecology led to one other very practical result. We discovered that mice that suffered type A infection, particularly at birth, also developed acute hemolytic anemia within two weeks. They gained weight more slowly than normal juveniles and developed enlarged spleens. This extramedullary hematopoesis partially alleviated the anemia, but splenomegaly in these animals was as chronic as the tolerant infection. Hemoglobin and spleen weights of normal mice and those infected through ingestion of maternal milk are shown in figure 19.4 (Johnson et al. 1973). Early rodent control work disclosed that the prevalence of virus infection in *Calomys* was almost seven times greater in sites where human disease was current or had occurred within six months than in localities that had been without human disease for that interval or longer. Where virus infection was high, *Calomys* spleens longer than 18 millimeters (about 0.20 g) were commonly encountered. Bolivia in the 1960s had no virological laboratory, but the control program that kept the Beni virtually free of BHF for the next twenty-five years was both efficient and economical (Mercado 1975). A small team of virus-immune individuals visited villages and ranches, set traps for three nights, autopsied all *Calomys*, and measured spleens with a plastic metric ruler. If even one large spleen was discovered, a second larger team was brought to the site and a campaign to eliminate the carrier species was maintained for up to one month. The rulers were the cheapest virological tools ever deployed as ecological weapons.

One final codicil deserves mention. In 1993, Ron Mackenzie and I visited San Joaquin to celebrate thirty years since discovery of Machupo virus. The town was prosperous, twice the size since our last visit. Jeeps and pickup trucks were everywhere, and there was now a dry-season dirt road to Trinidad. We visited San Joaquin's new, larger, better-equipped hospital and interviewed the last two rodent control men still based in the town. There had been no BHF in San Joaquin for many years. But these gentlemen showed us skins of rodents that they said now controlled domiciliary environments in the village. They were all *Rattus norvegicus*. Thus, rats had finally made the journey from the tiny Machupo River port into the town itself. They had supplanted the

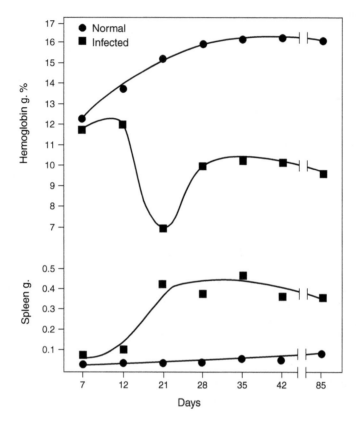

Figure 19.4. Hemoglobin levels and spleen masses in uninfected *Calomys callosus* and those infected from mother's milk.

unique host of the pathogen that had caused so much grief decades before. Mackenzie and I also realized one other important point. *Rattus* did not ask for any budget to maintain effective biological control of human infection. All they needed was space for shelter and scraps of food unintentionally left by those who would no longer suffer at the hands of a local rodent. Ecology, as every other scientific discipline, is no stranger to serendipity.

LITERATURE CITED

Hershkovitz, P. 1962. Evolution of Neotropical Rodents (Muridae) with Special Reference to the Phyllotine Group. Fieldiana: Zoology 46:1–524.

Johnson, K. M., S. B. Halstead, and S. N. Cohen. 1967. Hemorrhagic fevers of Southeast Asia and South America: A comparative appraisal. Progress in Medical Virology 9:105–58.

Johnson, K. M., M. L. Kuns, R. B. Mackenzie, P. A. Webb, and C. E. Yunker. 1966. Isolation of Machupo virus from wild rodent *Calomys callosus*. American Journal of Tropical Medicine and Hygiene 15:103–6.

Johnson, K. M., R. B. Mackenzie, P. A. Webb, and M. L. Kuns. 1965. Chronic infection of rodents by Machupo virus. Science 150:1618–19.

Johnson, K. M., P. A. Webb, and G. Justines. 1973. Biology of Tacaribe-complex viruses. *In* Lymphocytic Choriomeningitis Virus and Other Arenaviruses, ed. F. Lehmann-Grube, 241–58. New York: Springer-Verlag.

Johnson, K. M., N. H. Wiebenga, R. B. Mackenzie, M. L. Kuns, N. M. Tauraso, A. Shelokov, P. A. Webb, G. Justines, and H. K. Beye. 1965. Virus isolations from human cases of hemorrhagic fever in Bolivia. Proceedings of the Society for Experimental Biology and Medicine 118:113–18.

Justines, G., and K. M. Johnson. 1968. Use of oral swabs for detection of Machupo-virus infection in rodents. American Journal of Tropical Medicine and Hygiene 17:788–90.

Justines, G., and K. M. Johnson. 1969. Immune tolerance in *Calomys callosus* infected with Machupo virus. Nature 222:1090–91.

Justines, G., and K. M. Johnson. 1970. Observations on the laboratory breeding of the cricetine rodent *Calomys callosus*. Laboratory Animal Care 20: 57–60.

Kuns, M. L. 1965. Epidemiology of Machupo virus infection. II. Ecological and control studies of hemorrhagic fever. American Journal of Tropical Medicine and Hygiene 14:813–16.

Mackenzie, R. B. 1965. Epidemiology of Machupo virus infection. I. Pattern of human infection, San Joaquin, Bolivia, 1962–1964. American Journal of Tropical Medicine & Hygiene 14:808–13.

Mackenzie, R. B., H. K. Beye, L. Valverde Ch., and H. Garron. 1964. Epidemic hemorrhagic fever in Bolivia. I. A preliminary report of the epidemiologic and clinical findings in a new epidemic area in South America. American Journal of Tropical Medicine and Hygiene 13:620–25.

Mercado, R. R. 1975. Rodent control programs in areas affected by Bolivian hemorrhagic fever. Bulletin of the World Health Organization 52:691–96.

Peters, C. J., P. A. Webb, and K. M. Johnson. 1973. Measurement of antibodies to Machupo virus by the indirect fluorescent technique. Proceedings of the Society for Experimental Biology and Medicine 142:526–31.

Tauraso, N. M., N. H. Wiebenga, and A. Shelokov. 1964. Plaque neutralization studies of Bolivian hemorrhagic fever virus. Bacteriological Proceedings, 122. American Society for Microbiology (abstracts of annual meeting). Washington, DC.

Webb, P. A., K. M. Johnson, and R. B. Mackenzie. 1969. The measurement of specific antibodies in Bolivian hemorrhagic fever by neutralization of virus plaques. Proceedings of the Society for Experimental Biology and Medicine 130:1013–19.

Webb, P. A., G. Justines, and K. M. Johnson. 1975. Infection of wild and laboratory animals with Machupo and Latino viruses. Bulletin of the World Health Organization 52:493–99.

Wiebenga, N. H., A. Shelokov, C. J. Gibbs, and R. B. Mackenzie. 1964. Epidemic hemorrhagic fever in Bolivia. II. Demonstration of complement-fixing antibody in patients' sera with Junin virus antigen. American Journal of Tropical Medicine and Hygiene 13:626–28.

CHAPTER TWENTY

Resolved: Emerging Infections of Humans Can Be Controlled by Ecological Interventions

C. J. Peters

SUMMARY

IT HAS BECOME VERY APPARENT over the past decade that new and emerging viruses will plague us for the foreseeable future. The causes were addressed in detail by a recent Institute of Medicine report, and the conclusions were that underlying factors would only worsen in an accelerated fashion and that interactions among these factors would exacerbate the situation. Equally alarming is the fact that new vaccines, anti-infectives, and measures directed against vectors and reservoirs are not being developed by commercial or public sector researchers. This chapter analyzes ten recent serious emerging viral infections and concludes that ecological factors drive most. An additional conclusion is that current understanding of the ecology of these emergences is totally inadequate to provide serious measures for ecologically based control strategies. Even when we comprehend the underlying factors partially, it is unlikely that ecological manipulations would be possible or would be economically and socially accepted. The lesson is clear: we need to understand more about the ecological determinants behind transmission of these diseases, and we need to find practical ways to exert leverage, with the goal of control.

INTRODUCTION

This chapter addresses the proposition of its title, and I will assert that the answer is no. This conclusion, which is as speculative as the affirmative would be, is not based on the idea that ecology in the narrow sense or the broad sense cannot contribute to our understanding of disease emergence but rather on the recognition that current understanding of the variables leading to infectious disease emergence and our ability to control these variables are insufficiently developed. I support this assertion with the analysis of several examples drawn from RNA viral diseases coming from nature and with which I have some personal familiarity.

The sources of human disease emergence were recently described in an important and widely regarded study (Institute of Medicine [U.S.] Committee on Emerging Microbial Threats to Health et al. 1992) and then reexamined a decade later with essentially similar conclusions (Committee on Emerging Microbial Threats to Health in the 21st Century 2003). I have placed "Microbial adaptation and change" as the first entry in each list. The two studies identified general factors that lead to changes in disease incidence and appearance of "new" diseases; and their conclusions were generally in agreement. The second study had more emphasis on social underpinnings of changes, classic ecological factors, and the interactions of different factors (the coming together of disparate influences to result in "the perfect storm"). These publications were very important in identifying general factors in disease emergence and in establishing this phenomenon as a real part of today's global civilizations. However, they did not critically examine the detailed mechanisms of these events because the data were not available: Why now? Why this microbe? What were the relative contributions of the different factors? What was the driving force? In the absence of the rate-limiting step in emergence, would the disease have occurred?

First IOM report: Classification of factors in human disease emergence

- Microbial adaptation and change

- Human demographics and behavior

- Technology and industry

- Economic development and land use

- International travel and commerce

- Breakdown or lack of public health measures

Second IOM report: Classification of factors in human disease emergence

- Microbial adaptation and change

- Human demographics and behavior

- Technology and industry

- Economic development and land use

- International travel and commerce

- Breakdown of public health measures

- Human susceptibility to infection

- Climate and weather

- Changing ecosystems

- Poverty and social inequality

- War and famine

- Lack of political will

- Intent to harm

Note that the first line of IOM reports 1 and 2 is the only part that is dependent on the microbe and includes an essentially passive connotation, "adaptation," and an active element, "change." The ability of microbes with their short generation time and, in the case of RNA viruses, their error-prone replication mechanisms to adapt to changing circumstances is clearly important, but how often does the microbial change truly drive the emergence of a disease, rather than being indifferent or perhaps merely adapting to a niche created by the receptive milieu? In fact, genetic change and selection, if taken in the broad sense, can explain all the emergences we see. The factors listed in and after "Microbial adaptation and change" are simply ways to enumerate some of the selective factors underlying emergence. In this chapter I argue that, although we lack carefully collected data to definitively support the conclusion, the selection factors are usually the culprit that drive emergence. Genetic factors tend to be adaptive, at least with the limited analysis we have. There is, of course, one major exception: influenza A, in which genetic changes can result in a "new" virus that may be capable of causing pandemic influenza. Although influenza is not included here, it is worth noting that these genetic changes may arise through ecologically driven forces.

A caution about use of the word *ecology* in this chapter. I will use ecology mainly in the sense of "the environment as it relates to living organisms; 'it changed the ecology of the island.'"*

From time to time, I may use ecology in a broader context. Ecology (from the Greek *oikos* = house) has different contexts if we talk of HIV/ AIDS, West Nile virus, or rheumatic fever. A virus lurking within a seminal ejaculate is only indirectly affected by climate and weather, but its emergence, epidemiology, and global spread are much more closely allied to human demographics and behavior, economic development and land use, and international travel and commerce.

---

* WordNet® 2.0, © 2003 Princeton University, accessed August 5, 2005.

I have chosen ten diseases to discuss the proposition that ecology will be helpful in controlling emerging diseases. Many of the viruses have been discovered since the 1970s, and they are all small RNA viruses. These diseases should be ideal for ecological control because they are transmitted in nature and spill over into human beings only incidentally. I will try to identify the factors that governed their emergence, the relative contribution of ecological change and virus mutation, and how ecological modification could be used to control them.

CASE STUDIES: ARENAVIRIDAE

The first identified arenavirus was lymphocytic choriomengitis virus (LCMV), isolated in 1933 and identified as a cause of aseptic meningitis in humans. It was found to be a chronic, vertically transmitted infection of *Mus musculus* and then linked to the geographic range of at least two subspecies of *Mus* (reviewed in Peters 1995). Later arenaviruses have all conformed to the general paradigm: each is a chronic infection of a single rodent species, the spread of virus to humans is direct (often by aerosol), and human risk is directly related to the rodent-human interaction. The biology of LCMV has been very important in the development of modern immunology (Zinkernagel 2002), but its ecology has been surprisingly poorly studied. The distribution has been linked to that of its reservoir host, and the autumn influx of house mice into human dwellings correlates with the autumn peak of cases in rural communities. However, only a few studies have examined the determinants of the local prevalence and behavior of the last relevant to virus infection.

In the 1950s, Arribalzaga (1955) reported an unusual and very clinically characteristic disease occurring among corn harvest workers in the humid pampas of Argentina. This disease was christened Argentine hemorrhagic fever (AHF), and the virus was named Junin for one of the towns in the most affected area. The distribution of the disease has spread progressively since then, and as the disease moves into new areas, often the older afflicted zones have a marked decrease in incidence of disease (Enria et al. 2005; Maiztegui et al. 1986). The increased incidence of the disease in new areas of involvement is associated with an increased prevalence of infection in the reservoir rodent, *Calomys musculinus* (Mills et al. 1994). Several workers in the field have proposed ecological explanations, such as land use, defoliant use for weed control, methods of harvest (hand vs. machine), and rodent-virus genetic explanations for the spread, but none of these has held up to critical scrutiny, and the explanations are conflicting. Without any clear-cut reason for the spread of the disease, it is hard to imagine how an ecological strategy can lead to its control. Furthermore, the locus of maximum risk is in the rich agricultural regions of the country, and any control measures would have to take into account the impact on the gross national product and the need for workers to plant, tend, and harvest the crops. One important finding has been the proclivity of the natural reservoir rodent to reside along fence lines, roadsides, and other linear habitats rather than in the fields themselves. This does provide one ecologically based strategy to ameliorate risk, provided brush can be cleared from these habitats and workers can be trained to avoid them (Enria et al. 2005; Mills et al. 1994). Fortunately, there is an effective vaccine that is now being produced in Argentina to protect those at highest risk (Enria and Barrera Oro 2002; Maiztegui et al. 1998; D. A. Enria, personal communication, 2005).

Another severe arenavirus disease emerged as Bolivian hemorrhagic fever (BHF) in the tropical areas of Bolivia between the Andes and the Brazilian border. This disease was first noted as a sporadic and severe "black typhus" but later became a serious epidemic disease within small towns in the Beni Department of Bolivia. The isolation of the virus and its identification with the disease showed a third arenavirus associated with human disease. In addition, it provided the lever for control. The reservoir rodent, *Calomys callosus*, is a common savanna and cropland rodent, but it has a proclivity to enter towns and human dwellings (Johnson et al. 1966). It was selectively eliminated from the town of San Joaquin, resulting in control of the disease and definitive proof of the rodent's importance as the source of human infection (Johnson et al. 1967, Johnson, chapter nineteen, this volume).

Was BHF an example of how ecological studies helped in the control of a human disease? I would suggest that it was only indirectly so. It is not a far reach to eliminate a rodent that carries a virus from a town with an epidemic of that virus. It is true that there are questions about why *C. callosus* enters towns and dwellings rather than staying in its natural grassland habitat as many other rodents do, and it is true that we do not understand the interactions of cats and other rodents in exclusion of *C. callosus* from towns (Mackenzie 1965), but the fact remains that assiduous trapping of the reservoir rodent eliminates disease in the towns of the Beni (Mercado 1975). Disease still occurs from time to time in persons who work in agricultural habitats in the area where control of *C. callosus* is impractical with our current state of knowledge, and there is also a risk of person-to-person transmission that has been subsequently revealed (Kilgore et al. 1995; Peters et al. 1974), but the pattern of a rodent reservoir still holds and still provides the basis for efficient control efforts within towns. Interestingly, the actual reservoir is probably not *C. callosus* but rather a cryptic *Calomys* species (Salazar-Bravo et al. 2002).

Lassa fever was first defined as a biomedical entity in 1969 (Buckley et al. 1970; Frame et al. 1970) but subsequently was shown to be a common disease in certain areas of West Africa (McCormick and Fisher-Hoch 2002). As with other arenavirus diseases, Lassa virus infection is a chronic infection of rodents. Unlike most arenavirus diseases, it seems to be a chronic infection of two closely related species, *Mastomys hubertii* and *M. erythroleucus* (Enria et al. 2005). However, the taxonomic situation of *Mastomys* species is in disarray, and the reservoir status of this very important virus is unclear. Most infections appear to be acquired peridomestically and from *Mastomys* invading homes in African villages. The ecological conditions that have led to intensive home infestation include the local habitat (coastal vs. savanna vs. forest), social differences (diamond mining areas, house construction, disruption by war), and less well-understood factors (Bausch et al. 2001; McCormick and Fisher-Hoch 2002). *Mastomys* is among the most numerous rodents in African villages of the endemic area and is usually absent only when *Rattus* has migrated from coastal introduction sites and dominates (Demby et al. 2001).

Infection is associated with large numbers of infected rodents in the homes (McCormick and Fisher-Hoch 2002). Rodent infection is relatively focal in villages, and this focality even extends to the individual house (Demby et al. 2001). This might provide leverage for an ecological strategy for control, but the houses themselves do not lend themselves to rodent-proofing, and elimination of *Mastomys* from the environment is impossible. Well-designed approaches to rodent control applied to high-incidence Lassa fever villages have resulted in significant decreases in disease incidence but are neither generally applicable nor sustainable over time (McCormick and Fisher-Hoch 2002; McCormick et al. 1987).

Because the arenaviruses are regularly transmitted among rodents and infected rodents are such a strongly associated risk factor, it appears that viral variation is not a factor in human infection. Genetic analysis of individual viruses from donor rodent to sick human have not been extensively carried out, so it is impossible to be sure that point mutations do not occur as viruses spread through the human body. There are clearly major ecological factors that favor human infection, but economic and cultural issues, combined with our ignorance of the complex ecological situations, make it difficult to recommend a comprehensive amelioration strategy beyond simply trapping in homes and reducing their permeability to rodents. In the case of Lassa virus, the African village homes themselves do not lend themselves to such modification.

CASE STUDIES: FILOVIRIDAE

Marburg virus appeared in Europe in 1967, when it was imported with African green monkeys from Uganda and infected laboratory workers and others. When Ebola appeared in Zaire and Sudan in 1976, it provided additional members of this new virus family, christened Filoviridae (Wahl-Jensen et al. 2006). Once in human populations, these viruses are moderately transmissible and are spread because of the poor hygienic conditions that are present, particularly in African hospitals, a problem in human ecology that is best subsumed under poverty and behavior (Peters 2005). However, the underlying distribution and circulation of virus continue to be a puzzle. Filoviruses are the only virus family whose basic survival strategy remains unknown, although they clearly are not primate viruses.

Ebola virus most likely has a mammalian reservoir, but it has been very difficult to determine which species this might be (Murphy and Peters 1998). Multiple laboratory studies and testing of more than 7,000 vertebrates and 30,000 invertebrates captured in areas in which the virus has recently been active have failed to yield virus or even important evidence of antibody to indicate virus circulation (reviewed in Pourret et al. 2005). This is the classic approach for detecting the natural cycle of a virus that circulates in nature, and it clearly failed. It is probably not extraordinary that these studies did not yield the reservoir because of a variety of factors, including the complexity of the tropical ecosystem (Leirs et al. 1999; Murphy and Peters 1998). For example, during the investigations of the 1995 Kikwit, Zaire, Ebola epidemic, almost 3,000 vertebrates were captured, but no one expert in the local fauna was available to help in identification and processing. Samples were taken for subsequent DNA typing, formalin-fixed animals were sent to research museums around the world for definitive identification, and indeed, four possible new species of insectivore were discovered.

Ebola virus was initially detected in Sudan and Zaire between 1976 and 1979, but there were a limited number of sites, and the only suggestive generalization that emerged was that the initial human Ebola infection generally occurred in forest or nearby gallery forest settings during rainy seasons. Marburg virus seemed to have a somewhat different ecologic distribution (Peters et al. 1997). Then, from 1994 to 1997 and from 2002 to 2004, multiple outbreaks in humans and apes were recognized (Peterson et al. 2004; Pourrut et al. 2005). In fact, these outbreaks often had a disproportionate effect on chimpanzees and gorillas, with human involvement often being secondary to harvesting dead apes for culinary purposes (Leroy et al. 2004). This still did not provide a careful measure of filovirus activity, but it did give more geographic and temporal information as to virus transmission. In addition, Marburg virus became more active during this time (Pourrut et al. 2005).

Two studies used this information to approach the circulation of filoviruses from different perspectives using ecological tools. In one, satellite remote sensing used advanced very high-resolution radiometer (AVHRR) data and relied on established correlations with vegetation and rainfall (Pinzon et al. 2004). Ebola outbreaks correlated with markedly dryer conditions at the end of rainy seasons, and their geographic sites were correctly predicted.

Another approach used ecological niche modeling (Peterson et al. 2004) to examine where Ebola virus might be distributed and what ecological zones were involved. The conclusions were that the reservoirs would be present in tropical broadleaf forests, particularly in the Congo basin and selected other habitats. Although these results may not at first seem to be overwhelming, the long odds against the investigators and the strong correlations observed were in fact quite encouraging and left open the opportunity to continue to use these methods to predict future epidemics and validate or falsify the approaches. They can also be used to narrow the search for the reservoir.

Careful genetic analysis of the viruses is also useful in understanding how the viruses relate to ecological factors. One of the characteristics of filoviruses is that each virus species has relatively little genetic variability, but it is not known how this relates to the nature of the reservoir, although one might speculate that the homogeneity implies limited genetic variation of the host or spatial movement and mixing such as occurs with chiropterans in the tropics. We are reasonably confident that each of the human epidemics follows a single introduction because the viruses within the epidemic are virtually identical and remain so until virus transmission is curtailed (Rodriguez et al. 1999). There are exceptions: epidemiologically unconnected Ebola (Zaire) epidemics in Gabon (Georges-Courbot et al. 1997; Leroy et al. 2002), multiple introductions of Marburg virus in a single mining area of the Democratic Republic of Congo (Zeller 2000), and different independently infected apes in Gabon (Leroy et al. 2004) all show small but consistent sequence variations suggesting independent origins for the viruses. Marked phylogenetic differences among some Ebola virus isolates show they make up four species (Sanchez et al. 1996, 1998), and thus we are presumably searching for four reservoirs, and a fifth for Marburg.

CASE STUDIES: A PHLEBOVIRUS (RIFT VALLEY FEVER VIRUS)

Rift Valley fever virus (RVFV) was first isolated in Kenya in the 1930s and was quickly recognized as an important epizootic disease of sheep and cattle, with extensive human infection occurring during the epidemics (for reviews, see Peters and Linthicum 1994; Peters and Meegan 1981; Swanepoel et al. 1994). In 1975, the occurrence of Marburg virus disease in Johannesburg prompted an increased surveillance for viral hemorrhagic fever (VHF), and for the first time, virologists detected the VHF syndrome as a minority among the milder clinical cases resulting from RVFV infection.

The reservoirs and mechanisms of persistence in nature remained a mystery until studies in the 1980s revealed that part of the virus's strategy was transovarial transmission in floodwater *Aedes* (*Ochlerotatus*) mosquitoes. Eggs from these mosquitoes must be thoroughly saturated with water before hatching occurs. Thus, the larvae emerge and develop into infected female mosquitoes only after heavy rains saturate the depressions where these vectors breed. The mosquito may then feed on a vertebrate host and may infect that host. However, the mosquitoes may oviposit in other similar ecological sites, presumably transmitting the virus transovarially to maintain the virus in nature. This has been verified by rearing larval mosquitoes from specialized depressions in East Africa, which have been shown to emerge already infected (Linthicum et al. 1985). There are still many unknowns. The mosquitoes are difficult to colonize, and thus the cycle has not been reproduced in the laboratory. Virtually all systems of this type require reintroduction of virus into the arthropod, but without the ability to study the cycle in the laboratory, it is difficult to reach a deeper understanding of the process with RVFV. Another major riddle is the natural vertebrate reservoir for the virus prior to the introduction of sheep and cattle ca. 3,000 years ago.

Endemic transmission occurs frequently, but epidemics are associated with unusually heavy rainfall. We hypothesize that the sequence of events begins with flooding of the habitats of the transovarially infected vector and the emergence of infected mosquitoes. As they bite susceptible sheep and cattle, viremia occurs, and the vectors, which are present in increased numbers with extraordinary precipitation, become infected while feeding on infected livestock and are thought to be the major transmitters (Linthicum et al. 1999; Peters 1997; Swanepoel et al. 1994). In selected areas of Kenya it was possible to predict RVFV circulation by advanced high-resolution radiometry based on vegetation (and topography?).

CASE STUDIES: HANTAVIRUSES IN THE AMERICAS

Hantavirus diseases have been known for a very long time because of their characteristic Eurasian syndrome, hemorrhagic fever with renal syndrome (HFRS). The causative agent of the disease in much of Asia was finally isolated in Korea in 1976 (Lee et al. 1978). However, it was not until 1993 that hantaviruses were linked to human disease in the Americas, and the disease was one new to science, hantavirus pulmonary syndrome (HPS) (Nichol et al. 1993). It was soon found that there were many American hantaviruses and that a substantial proportion, if not all, could cause HPS (Peters 2002). Each virus was linked with one rodent, and the phylogenetics of the virus correlated with that of the rodent, indicating coevolution from a predecessor that crossed the Behring Strait 30–50 million years ago (Hughes and Friedman 2000; Monroe et al. 1999). Rodent contact is the major risk factor for human disease (Childs et al. 1995). In Paraguay, Chile, and Panama, the epidemic was related to unusual ecological conditions that had resulted in fluctuations of rodent populations (Ruedas et al. 2004; Toro et al. 1998; Williams et al. 1997); the situation in Argentina and Brazil was more complex and involved multiple rodent and virus species.

Longitudinal rodent studies were undertaken to try to link the transmission of the virus most prevalent in the southwestern United States, Sin Nombre virus (SNV), to the population density and infection prevalence of the reservoir *Peromyscus maniculatus* (Mills et al. 1999). The original 1993–94 human epidemic and a smaller one in 1998–99 followed El Niño/Southern Oscillation events. The connection seemed to map most closely to winter and spring precipitation that resulted in proliferation of the reservoir rodent (Yates et al. 2002). The increased number of rodents was important, but equally or more influential was the higher proportion of infected rodents the next year. This phenomenon, referred to as delayed density dependence of infection, is an interesting and important phenomenon in hantavirus infections (Niklasson et al. 1995; Yates et al. 2002).

Another approach has been taken to identify signatures from Landsat, local precipitation, and altitude data. Data from initial set of HPS patients in the southwestern United States were used to construct an algorithm to identify high-risk areas. Then, data from a later set of patients were used to verify the value of the predictive algorithm (Glass et al. 2000). The same algorithm can predict rodent seroprevalence and thus risk in later years; interestingly, the high-risk areas change from year to year (Glass et al. 2002).

It has been extremely difficult to study the links of human and rodent infection because of the low incidence of human disease, even in epidemic years. One possible explanation for the disparity is that animals recently infected with hantaviruses have up to a 10^4 higher titer of virus shed in urine than animals that remain infected for life (Hutchinson et al. 1998). Thus, one chronically infected rodent in the early phases of infection may be as dangerous as 1,000–10,000 rodents later in their infection. This, of course, makes us turn to rodent studies as an imperfect marker for human risk when we evaluate methods such as rodent exclusion to prevent human disease (Glass et al. 1997). It also makes it very difficult to definitively evaluate provocative findings, such as the higher prevalence of hantavirus antibody in *P. maniculatus* trapped inside structures than in those that do not enter, and the increased traffic of deer mice into such structures after trapping out resident mice (Douglass et al. 2003). It has been noted that hantavirus (and perhaps arenavirus) rodent-borne diseases are associated with areas of low species diversity, but it is difficult to know exactly how to manipulate this finding in the southwestern United States to lower population levels of an ecological generalist such as *P. maniculatus*. Even simple measures of rodent-proofing homes and peridomestic habitat modification are difficult to achieve against a background of rural living, poverty, and cultural differences. Perhaps one element is the low incidence of hantavirus pulmonary syndrome compared with the situation with Bolivian hemorrhagic fever, in which the disease incidence was high and rodent trapping has been sustained after a dramatic demonstration of efficacy forty years ago.

CASE STUDIES: HENNIPAVIRUSES

In 1994, a previously unrecognized virus named Hendra caused a handful of horse deaths and secondarily three human cases. The widely separated foci suggested that this was not a single emergence event, and then the virus recurred in a horse in 1999. In 1998, a related virus named Nipah emerged to cause extensive disease in pigs, with transmission to humans resulting in 285 recognized cases of human encephalitis, with an almost 40% case fatality (Wong et al. 2002). Later, Nipah or a related virus emerged to cause multiple foci of cases in Bangladesh, possibly with person-to-person transmission (Hsu et al. 2004). These viruses formed a new genus within the well-established virus family Paramyxoviridae (Chua et al. 2000)—perhaps surprising, in light of the amount of work that has been done on this virus family over the years. These viruses appear to chronically infect *Pteropus* bats. If the pattern seen with some other viruses holds, there may be a group of viruses that correlate in their phylogenetics with the structure of the megachiropterans that are their hosts. These fruit-eating bats are limited in their geographic distribution, but it seems likely that we can look forward to continuing reemergences over time, possibly because of ecological disturbances affecting the bats. Clearly, the intensification of swine farming in Malaysia is a candidate factor in the emergence of Nipah, as pigs can become infected from bats and can amplify and transmit virus to humans. Unfortunately, the real basis for the appearance of these diseases has not been elucidated, and we lack an understanding of the bat-virus relationships, viral genetic changes, and other factors that led to their emergence.

THE OTHER SIDE OF THE DEBATE

The arguments presented so far obviously are slanted to one person's experience and would also look different if more vector-borne diseases were under consideration. For example, Western equine encephalitis virus has been subjected to intensive field and laboratory study to devise integrated management strategies, with some success (Reeves 1990). Interestingly, much of the decrease in human disease may be a consequence of changing ways of living. Exposure to crepuscular feeding by *Culex tarsalis* has been greatly diminished with the rise of television and air conditioning replacing a cool evening on the front porch for many people (Gahlinger et al. 1986).

The importation of West Nile virus into the United States in 1999 showed the gaps in our understanding of the factors that regulate arbovirus transmission. Arguments raged among experts as to whether the virus could persist in its new environment or whether it would die out. In spite of intensive application of intuitive approaches to control and eradication we now have a nationwide distribution of virus, with more than 200 cases of encephalitis. West Nile epidemiology in the Americas is strongly influenced by ecological considerations, much of which we still do not understand. If we had a better understanding of these factors, we might have been able to influence the spread, or in the future control transmission in a more effective way.

It is also fair to say that study of the ecology of the diseases in table 20.1 has contributed to our understanding of their emergence and has suggested approaches to ecological prediction and control. We know that ecological variables drive the incidence of most of these diseases, so ecological understanding could be enormously helpful in anticipating or preventing emergence and in control. But we still do not have a deep understanding of how the factors interact to modulate virus transmission. We will be successful in achieving ecologically based control strategies only when we have in-depth multidisciplinary studies, and even then we will be hampered by factors such as the expense of manipulating the en-

TABLE 20.1
Emerging virus diseases under consideration

Virus	Disease	Emergence
Family Arenaviridae		
Junin	Argentine hemorrhagic fever	1950s
Machupo	Bolivian hemorrhagic fever	1960s
Lassa	Lassa fever	1969
Family Filoviridae		
Ebola	Ebola hemorrhagic fever	1976
Marburg	Marburg hemorrhagic fever	1967
Family Bunyaviridae, genus Phlebovirus		
Rift Valley fever	Rift Valley fever	1930
Family Bunyaviridae, genus Hantavirus		
Sin Nombre	Hantavirus pulmonary syndrome	1993
Andes	Hantavirus pulmonary syndrome	1996
Family Paramyxoviridae, genus Hennipavirus		
Hendra	Hendra infection	1995
Nipah	Nipah encephalitis	1997

vironment and measuring the effects, climate and weather, which do not lend themselves to purposeful human intervention, the widespread distribution of some of the reservoirs and vectors, and finally the difficulties in altering human behavior.

CONCLUSIONS

The viruses discussed in this chapter exhibit several characteristics when they infect their disease target, including a high impact on health without the participation of secondary factors such as nutrition, other parasites, and similar variables often highly relevant in the ecological study of animal and plant diseases. In many cases, the pathway to ecological manipulation is not alterable, either because of its nature (climate, El Niño) or for cultural and economic reasons (rodent-proofing houses, stopping bush meat harvest, eliminating pig farming). Some of these diseases may possibly be ameliorated through improvements in food production, alleviation of poverty, and development in general. Nevertheless, we should strive to understand the impact of ecological changes on the system of circulation of viruses outside humans, because this gives us greater predictive power and may eventually suggest more effective ecological strategies.

It is not apparent that any of the diseases in table 20.1 are driven by viral mutation. Viral change seems to be more important in adaptation to take advantage of the ecological niches that open up through shifts in the environment and through travel and transport of reservoirs and vectors.

Future studies should be multidisciplinary and should take advantage of the classic approaches of ecological science and medical epidemiology, but they also need to incorporate phylogenetic and genetic analysis of the microbes and the reservoir or vectors, the pathogenesis of virus infection, quantitative measurements of virus shedding, and the genetics of adaptation to new hosts. We need to work out the mechanisms of virus emergence for established examples such as those in table 20.1, and we must be ready to study future occasions from their initial stages.

Literature Cited

Arribalzaga, R. 1955. Una nueva enfermedad epidemica a germen desconocido: Hipertermia, nefrotoxica, leucopenica y enantematica. Dia Medico 27:1204–10.

Bausch, D. G., A. H. Demby, M. Coulibaly, J. Kanu, A. Goba, A. Bah, N. Conde, H. L. Wurtzel, K. F. Cavallaro, et al. 2001. Lassa fever in Guinea. I. Epidemiology of human disease and clinical observations. Vector Borne and Zoonotic Diseases 1:269–81.

Buckley, S. M., J. Casals, and W. G. Downs. 1970. Isolation and antigenic characterization of Lassa virus. Nature 227:174.

Childs, J. E., J. W. Krebs, T. G. Ksiazek, G. O. Maupin, K. L. Gage, P. E. Rollin, P. S. Zeitz, J. Sarisky, R. E. Enscore, and J. C. Butler. 1995. A household-based, case-control study of environmental factors associated with hantavirus pulmonary syndrome in the southwestern United States. American Journal of Tropical Medicine and Hygiene 52:393–97.

Chua, K. B., W. J. Bellini, P. A. Rota, B. H. Harcourt, A. Tamin, S. K. Lam, T. G. Ksiazek, P. E. Rollin, S. R. Zaki, et al. 2000. Nipah virus: A recently emergent deadly paramyxovirus. Science 288:1432–35.

Committee on Emerging Microbial Threats to Health in the 21st Century. 2003. Microbial Threats to Health, Emergence, Detection, and Response. Washington, DC: National Academy Press.

Demby, A. H., A. Inapogui, K. Kargbo, J. Koninga, K. Kourouma, J. Kanu, M. Coulibaly, K. D. Wagoner, T. G. Ksiazek, et al. 2001. Lassa fever in Guinea. II. Distribution and prevalence of Lassa virus infection in small mammals. Vector Borne and Zoonotic Diseases 1:283–97.

Douglass, R. J., A. J. Kuenzi, C. Y. Williams, S. J. Douglass, and J. N. Mills. 2003. Removing deer mice from buildings and the risk for human exposure to Sin Nombre virus. Emerging Infectious Diseases 9:390–92.

Enria, D. A., and J. G. Barrera Oro. 2002. Junin virus vaccines. Current Topics in Microbiology and Immunology 263:239–61.

Enria , D. A., J. N. Mills, R. Flick, M. D. Bowen, D. Bausch, W. J. Shieh, and C. J. Peters. 2005. Arenavirus infections. *In* Tropical Infectious Diseases: Principles, Pathogens, and Practice, 2nd ed., R. L. Guerrant, D. H. Walker, and P. F. Weller. New York: Elsevier.

Frame, J. D., J. M. Baldwin, Jr., D. J. Gocke, and J. M. Troup. 1970. Lassa fever, a new virus disease of man from West Africa. I. Clinical description and pathological findings. American Journal of Tropical Medicine and Hygiene 19:670–76.

Gahlinger, P. M., W. C. Reeves, and M. M. Milby. 1986. Air conditioning and television as protective factors in arboviral encephalitis risk. American Journal of Tropical Medicine and Hygiene 35:601–10.

Georges-Courbot, M. C., A. Sanchez, C. Y. Lu, S. Baize, E. Leroy, J. Lansout-Soukate, C. Tevi-Benissan, A. J. Georges, S. G. Trappier, et al. 1997. Isolation and phylogenetic characterization of Ebola viruses causing different outbreaks in Gabon. Emerging Infectious Diseases 3:59–62.

Glass, G. E., J. E. Cheek, J. A. Patz, T. M. Shields, T. J. Doyle, D. A. Thoroughman, D. K. Hunt, R. E. Enscore, K. L. Gage, et al. 2000. Using remotely sensed data to identify areas at risk for hantavirus pulmonary syndrome. Emerging Infectious Diseases 6:238–47.

Glass, G. E., J. S. Johnson, G. A. Hodenbach, C. L. Disalvo, C. J. Peters, J. E. Childs, and J. N. Mills. 1997. Experimental evaluation of rodent exclusion methods to reduce hantavirus transmission to humans in rural housing. American Journal of Tropical Medicine and Hygiene 56:359–364.

Glass, G. E., T. L. Yates, J. B. Fine, T. M. Shields, J. B. Kendall, A. G. Hope, C. A. Parmenter, C. J. Peters, T. G. Ksiazek, et al. 2002. Satellite imagery characterizes local animal reservoir populations of Sin Nombre virus in the southwestern United States. Proceedings of the National Academy of Sciences of the United States of America 99:16817–22.

Hsu, V. P., M. J. Hossain, U. D. Parashar, M. M. Ali, T. G. Ksiazek, I. Kuzmin, M. Niezgoda, C. Rupprecht, J. Bresee, and R. F. Breiman. 2004. Nipah virus encephalitis reemergence, Bangladesh. Emerging Infectious Diseases 10:2082–87.

Hughes, A. L., and R. Friedman. 2000. Evolutionary diversification of protein-coding genes of hantaviruses. Molecular Biology and Evolution 17:1558–68.

Hutchinson, K. L., P. E. Rollin, and C. J. Peters. 1998. Pathogenesis of a North American hantavirus, Black Creek Canal virus, in experimentally infected *Sigmodon hispidus*. American Journal of Tropical Medicine and Hygiene 59:58–65.

Institute of Medicine (U.S.) Committee on Emerging Microbial Threats to Health, J. Lederberg, R. E. Shope, and S. C. Oaks. 1992. Emerging Infections: Microbial Threats to Health in the United States. Washington, DC: National Academy Press.

Johnson, K., S. Halstead, and S. Cohen. 1967. Hemorrhagic fevers of Southeast Asia and South America: A comparative appraisal. Progress in Medical Virology 9:105–58.

Johnson, K. M., M. L. Kuns, R. B. Mackenzie, P. A. Webb, and C. E. Yunker. 1966. Isolation of Machupo virus from wild rodent *Calomys callosus*. American Journal of Tropical Medicine and Hygiene 15:103–6.

Kilgore, P. E., C. J. Peters, J. N. Mills, P. E. Rollin, L. Armstrong, A. S. Khan, and T. G. Ksiazek. 1995. Prospects for the control of Bolivian hemorrhagic fever. Emerging Infectious Diseases 1:97–100.

Lee, H. W., P. W. Lee, and K. M. Johnson. 1978. Isolation of the etiologic agent of Korean hemorrhagic fever. Journal of Infectious Diseases 137:298–308.

Leirs, H., J. N. Mills, J. W. Krebs, J. E. Childs, D. Akaibe, N. Woollen, G. Ludwig, C. J. Peters, and T. G. Ksiazek. 1999. Search for the Ebola virus reservoir in Kikwit, Democratic Republic of the Congo: Reflections on a vertebrate collection. Journal of Infectious Diseases 179(Suppl. 1):S155–63.

Leroy, E. M., S. Baize, E. Mavoungou, and C. Apetrei. 2002. Sequence analysis of the GP, NP, VP40 and VP24 genes of Ebola virus isolated from deceased, surviving and asymptomatically infected individuals during the 1996 outbreak in Gabon: Comparative studies and phylogenetic characterization. Journal of General Virology 83:67–73.

Leroy, E. M., P. Rouquet, P. Formenty, S. Souquiere, A. Kilbourne, J. M. Froment, M. Bermejo, S. Smit, W. Karesh, et al. 2004. Multiple Ebola virus transmission events and rapid decline of central African wildlife. Science 303:387–90.

Linthicum, K. J., A. Anyamba, C. J. Tucker, P. W. Kelley, M. F. Myers, and C. J. Peters. 1999. Climate and satellite indicators to forecast Rift Valley fever epidemics in Kenya. Science 285:397–400.

Linthicum, K. J., F. G. Davies, A. Kairo, and C. L. Bailey. 1985. Rift Valley fever virus (family Bunyaviridae, genus Phlebovirus). Isolations from Diptera collected during an inter-epizootic period in Kenya. Journal of Hygiene 95:197–209.

Mackenzie, R. B. 1965. Epidemiology of Machupo virus infection. I. Pattern of human infection, San Joaquin, Bolivia, 1962–1964. American Journal of Tropical Medicine and Hygiene 14:808–13.

Maiztegui, J., M. Feuillade, and A. Briggiler. 1986. Progressive extension of the endemic area and changing incidence of Argentine hemorrhagic fever. Medical Microbiology and Immunology 175:149–52.

Maiztegui, J. I., K. T. McKee, Jr., J. G. Barrera Oro, L. H. Harrison, P. H. Gibbs, M. R. Feuillade, D. A. Enria, A. M. Briggiler, S. C. Levis, et al. 1998. Protective efficacy of a live attenuated vaccine against Argentine hemorrhagic fever. AHF Study Group. Journal of Infectious Diseases 177:277–83.

McCormick, J. B., and S. P. Fisher-Hoch. 2002. Lassa fever. Current Topics in Microbiology and Immunology 262:75–109.

McCormick, J. B., P. A. Webb, J. W. Krebs, K. M. Johnson, and E. S. Smith. 1987. A prospective study of the epidemiology and ecology of Lassa fever. Journal of Infectious Diseases 155:437–44.

Mercado, R. 1975. Rodent control programmes in areas affected by Bolivian haemorrhagic fever. Bulletin of the World Health Organization 52:691–96.

Mills, J. N., B. A. Ellis, J. E. Childs, K. T. McKee, Jr., J. I. Maiztegui, C. J. Peters, T. G. Ksiazek, and P. B. Jahrling. 1994. Prevalence of infection with Junin virus in rodent populations in the epidemic area of Argentine hemorrhagic fever. American Journal of Tropical Medicine and Hygiene 51:554–62.

Mills, J. N., T. L. Yates, T. G. Ksiazek, C. J. Peters, and J. E. Childs. 1999. Long-term studies of hantavirus reservoir populations in the southwestern

United States: Rationale, potential, and methods. Emerging Infectious Diseases 5:95–101.

Monroe, M. C., S. P. Morzunov, A. M. Johnson, M. D. Bowen, H. Artsob, T. Yates, C. J. Peters, P. E. Rollin, T. G. Ksiazek, and S. T. Nichol. 1999. Genetic diversity and distribution of *Peromyscus*-borne hantaviruses in North America. Emerging Infectious Diseases 5:75–86.

Murphy, F. A., and C. J. Peters. 1998. Ebola virus: Where does it come from and where is it going? *In* Emerging Infections, ed. R. M. Krause, 375. New York: Academic Press.

Nichol, S. T., C. F. Spiropoulou, S. Morzunov, P. E. Rollin, T. G. Ksiazek, H. Feldmann, A. Sanchez, J. Childs, S. Zaki, et al. 1993. Genetic identification of a hantavirus associated with an outbreak of acute respiratory illness. Science 262:914–17.

Niklasson, B., B. Hornfeldt, A. Lundkvist, S. Bjorsten, and J. LeDuc. 1995. Temporal dynamics of Puumala virus antibody prevalence in voles and of nephropathia epidemica incidence in humans. American Journal of Tropical Medicine and Hygiene 53:134–40.

Peters, C. 1995. Arenavirus Diseases. *In* Kass Handbook of Infectious Diseases: Exotic Viral Infections, ed. J. Porterfield. New York: Chapman and Hall.

Peters, C. J. 1997. Emergence of Rift Valley fever. *In* Factors in the Emergence of Arbovirus Diseases, ed. J. F. Saluzzo and B. Dodet, 253–64. Paris: Elsevier.

Peters, C. J. 2002. Human infection with arenaviruses in the Americas. Current Topics in Microbiology and Immunology 262:65–74.

Peters, C. 2005. Marburg and Ebola: Arming ourselves against the deadly filoviruses. New England Journal of Medicine 352:2571–73.

Peters, C., R. Kuehne, R. Mercado, R. Le Bow, R. Spertzel, and P. A. Webb. 1974. Hemorrhagic fever in Cochabama, Bolivia, 1971. American Journal of Epidemiology 99:425–33.

Peters C. J., and K. J. Linthicum. 1994. Rift Valley fever. *In* Handbook of Zoonoses, ed. G. W. Beran, 125–38. Boca Raton, FL: CRC Press.

Peters, C. J., and J. Meegan. 1981. Rift Valley fever. *In* Viral Zoonoses, ed. G. W. Beran, and J. H. Steele, 403–20. Boca Raton, FL: CRC Press.

Peters, C., S. Zaki, and P. Rollin. 1997. Viral hemorrhagic fevers. *In*. Atlas of Infectious Diseases: External Manifestations of Systemic Infections, ed. G. Mandell, 10.1–10.26. Philadelphia: Current Medicine.

Peterson, A. T., J. T. Bauer, and J. N. Mills. 2004. Ecologic and geographic distribution of filovirus disease. Emerging Infectious Disease 10:40–47.

Pinzon, J. E., J. M. Wilson, C. J. Tucker, R. Arthur, P. B. Jahrling, and P. Formenty. 2004. Trigger events: Enviroclimatic coupling of Ebola hemorrhagic fever outbreaks. American Journal of Tropical Medicine and Hygiene 71:664–74.

Pourrut, X., B. Kumulungui, T. Wittmann, G. Moussavou, A. Delicat, P. Yaba, D. Nkoghe, J. P. Gonzalez, and E. M. Leroy. 2005. The natural history of Ebola virus in Africa. Microbes and Infection 7:1005–14.

Reeves W. C. 1990. Epidemiology and Control of Mosquito-Borne Arboviruses in California, 1943–1987. Sacramento: California Mosquito and Vector Control Association.

Rodriguez, L. L., A. De Roo, Y. Guimard, S. G. Trappier, A. Sanchez, D. Bressler, A. J. Williams, A. K. Rowe, J. Bertolli, et al. 1999. Persistence and genetic stability of Ebola virus during the outbreak in Kikwit, Democratic Republic of the Congo, 1995. Journal of Infectious Diseases 179(Suppl. 1):S170-76.

Ruedas, L. A., J. Salazar-Bravo, D. S. Tinnin, B. Armien, L. Caceres, A. Garcia, M. A. Diaz, F. Gracia, G. Suzan, et al. 2004. Community ecology of small mammal populations in Panama following an outbreak of hantavirus pulmonary syndrome. Journal of Vector Ecology 29:177–91.

Salazar-Bravo, J., J. W. Dragoo, M. D. Bowen, C. J. Peters, T. G. Ksiazek, and T. L. Yates. 2002. Natural nidality in Bolivian hemorrhagic fever and the systematics of the reservoir species. Infection, Genetics and Evolution 1:191–99.

Sanchez, A., S. G. Trappier, B. W. Mahy, C. J. Peters, and S. T. Nichol. 1996. The virion glycoproteins of Ebola viruses are encoded in two reading frames and are expressed through transcriptional editing. Proceedings of the National Academy of Sciences of the United States of America 93:3602–7.

Sanchez, A., S. G. Trappier, U. Stroher, S. T. Nichol, M. D. Bowen, and H. Feldmann. 1998. Variation in the glycoprotein and VP35 genes of Marburg virus strains. Virology 240:138–46.

Swanepoel, R. C. J. 1994. Rift Valley Fever. In Infectious Diseases of Livestock, with Special Reference to Southern Africa, ed. J. A. W. Coetzer, G. R. Thomson, and R. C. Tustin, 688–717. New York: Oxford University Press.

Toro, J., J. D. Vega, A. S. Khan, J. N. Mills, P. Padula, W. Terry, Z. Yadon, R. Valderrama, B. A. Ellis, et al. 1998. An outbreak of hantavirus pulmonary syndrome, Chile, 1997. Emerging Infectious Diseases 4:687–94.

Wahl-Jensen, V., H. Feldmann, A. Sanchez, S. Zaki, P. Rollin, and C. J. Peters. 2006. Filovirus Infections. In Tropical Infectious Diseases: Principles, Pathogens and practice, ed. R. L. Guerrant, D. H. Walker, and P. F. Weller. Philadelphia: Elsevier Churchill Livingstone.

Williams, R. J., R. T. Bryan, J. N. Mills, R. E. Palma, I. Vera, F. De Velasquez, E. Baez, W. E. Schmidt, R. E. Figueroa, et al. 1997. An outbreak of hantavirus pulmonary syndrome in western Paraguay. American Journal of Tropical Medicine and Hygiene 57:274–82.

Wong, K. T., W. J. Shieh, S. Kumar, K. Norain, W. Abdullah, J. Guarner, C. S. Goldsmith, K. B. Chua, S. K. Lam, et al. 2002. Nipah virus infection: Pathology and pathogenesis of an emerging paramyxoviral zoonosis. American Journal of Pathology 161:2153–67.

Yates, T. L., J. N. Mills, C. A. Parmenter, T. G. Ksiazek, R. R. Parmenter, J. R. Vande Castle, C. H. Calisher, S. T. Nichol, K. D. Abbott, et al. 2002. The ecology and evolutionary history of an emergent disease: Hantavirus pulmonary syndrome. Bioscience 52:989–98.

Zeller, H. 2000. Lessons from the Marburg virus epidemic in Durba, Democratic Republic of the Congo (1998–2000). Medecine Tropicale 60:23–26.

Zinkernagel, R. M. 2002. Lymphocytic choriomeningitis virus and immunology. Current Topics in Microbiology and Immunology 263:1–5.

CHAPTER TWENTY-ONE
From Ecological Theory and Knowledge to Application

James E. Childs

Summary

This chapter considers means to increase communication and understanding between practitioners of infectious disease ecology (IDE) and mainstream public and veterinary health practitioners (PVHP). Successful integration of the remarkable achievements and advances in IDE into standard health practice would improve human, animal, and ecosystem health.

The discussion of IDE focuses on academically based scientists who are developing theoretical constructs of the interaction of pathogen and host populations mediated by disease and immunity (Anderson and May 1991; Nowak and May 2000), scientists who are developing mathematical models or simulations of specific disease processes or epidemiological patterns of disease spread (Coyne et al. 1989; Russell et al. 2005), and scientists who are developing integrative but more epidemiologically traditional statistical models to predict or forecast disease outbreaks that are environmentally driven (Allen and Cormier 1996; Glass et al. 2002; Jones et al. 1998). This latter approach capitalizes on novel technologies to inform epidemiological models of risk analysis. Common examples use remotely sensed satellite data linked with structured layers of georeferenced data (e.g., demographic, physiogeographic, disease incidence) through a geographic information system (Brownstein et al. 2003; Glass et al. 2002) and employ spatial statistics developed for health applications (Waller and Gotway 2005).

To illustrate the need to develop new strategies for integrating the IDE and PVHP communities, I describe two contrasting perspectives on the utility of IDE in a specific health intervention. I argue that as end users become more familiar with the array of tools IDE makes available for complementing and bolstering traditional disease prevention and control practices, the obvious advantages of integrating IDE approaches should become self-evident and self-sustaining. For now, however, IDE is a long way from being an integral part of health practice.

DIFFERENT PERSPECTIVES ON THE UTILITY OF IDE
IN HEALTH PRACTICE

The following quotations were taken from two recently published papers. The authors are identifiable as influential members of the IDE and PVHP communities, respectively. Both publications were conceived and published after reviewing and assessing efforts to control the same disease outbreak, the foot-and-mouth disease (FMD) outbreak among ruminants in the United Kingdom in 2001. The quotations emphasize how different views and opinions can arise among the IDE and PVHP communities evaluating a common experience.

In commenting on the role played by the IDE community in helping guide interventions and control efforts in the FMD outbreak, K. F. Smith, A. P. Dobson, F. E. McKenzie, L. A. Real, D. L. Smith, and M. L. Wilson (2005) describe a set of policies (stringent movement restrictions and ringed culling policies around infected farms) recommended by the IDE community. They then state that

> These policies . . . were in place a month after the epidemic started and provide a remarkable testimony to the ability of the ecological epidemiology community to respond to a crisis. . . .
>
> The UK foot and mouth outbreak provides some salutary lessons for public policy and future infectious disease outbreaks. Retrospective calculations of the time course of the epidemic suggest that the initial 2-week delay may have eventually led to a doubling of cases and of the number of cattle and sheep herds culled. All this suggests that the ecologists should have been brought in sooner.

A different point of view of involvement of the IDE community in the United Kingdom's experience with the 2001 FMD is offered by M. D. Salman (2004), an influential member of the veterinary epidemiological community:

> I feel that modeling was a major issue in persuading decision-makers to adopt specific control strategies. . . . Now, the following points about modeling and its influence in the last European/UK FMD outbreak are strictly my opinion (and do not necessarily reflect the opinions of the other participants or speakers).
>
> • Most of the models published prior to this outbreak were not used because they were not validated prior to the outbreak. Many of these published models were academic exercises rather than aids in the decision making process.

• There were at least two models published after the outbreak—but they were too late to help in this particular outbreak (e.g. Keeling et al. 2001, Ferguson et al. 2001).

• Modelers were not involved in actually implementing disease control during the outbreaks in the UK and the Netherlands. Therefore, the modelers did not appreciate the practical complexity of thse system and the difficulty in collecting relevant data.

My hope is that modelers have learned from this experience and will attempt to become more acquainted with existing animal-production systems prior to the epidemics of disease. Furthermore, we should strive to use all available resources (such as manpower and laboratory support) to validate and test assumptions before they are needed in an emergency.

From these excerpts, there is obviously a difference of opinion between the IDE and PVHP communities about the usefulness and applicability of the models and products derived from IDEs in contributing to the successful control of the FMD outbreak. Of note, each set of authors suggests or concludes that adherents of the "other" scientific community have much to learn from "the facts" apparent on reviewing the response to the FMD outbreak, and each proffers suggestions for the other community as an object lesson to help guide future responses to similar conditions.

From Salman (2004), the object lesson for IDEs was that academically grounded IDEs should learn from this experience with FMD and become involved in field-based implementation of control procedures to gain firsthand experience of the process involved in implementing outbreak control and to familiarize themselves with the limitations or opportunities to collect and use all available data. In addition, IDEs must field-validate models for real-time implementation; otherwise, they can be dismissed as little more than academic exercises. Based on the previous quotations from Smith et al. (2005) and the comment that "disease prevention and control has depended upon public health officials applying knowledge gained from the work of population ecologists," it is fair to conclude that these authors conclude that IDEs ecologists should be recruited early in the strategic planning for disease control and that PVHPs already base many control decisions on the work generated by IDES.

The role of each community would appear to be complementary, and the skills and experience each brings to bear on infectious disease control should be inextricably linked by a mutual interest in understanding the natural history of infectious diseases and advancing the science and practice of intervention and control of disease outbreaks. Yet clearly

TABLE 21.1

Suggestions and model systems for promoting better communication and integration of infectious disease ecology (IDE) with health practice

Approaches for Advancing Communication Between the IDE and PVHP Communities	Strengths and Products	Limitations
White paper panel (e.g., as done with bioterrorism)	Collects a diverse group of professionals with appreciation for a common problem, a shared desire to develop a collaborative strategy, and willingness to produce a written report outlining potential directions for addressing needs.	Potentially low visibility and funding.
Joining or attending professional society meetings	Ability to directly communicate with an audience of active, practicing professionals in other disciplines. Opportunity to meet and establish concrete collaborations in which the advantages of disciplinary diversity can be showcased.	A good symposium or talk at a professional society may not translate into collaboration or action.
Teaching and hiring at schools of public health	Improves the likelihood that theory and practice will become acquainted. Opportunity to work on diverse programs and grants. Opportunity to train professionals going into the area of public health.	Funding and critical mass issues. Few IDE practitioners are currently employed in traditional departments of biostatistics, epidemiology and environmental health sciences.
Teaching and hiring at veterinary schools	Desirable, but first the schools of veterinary medicine must be convinced of the benefits of committing funds to such programs.	Requires convincing veterinary schools that such a curriculum is a valuable asset. Also, funding and critical mass issues exist.

NCEAS	Opportunity to bring the different communities together to draft collaborative studies and papers. The NCEAS approach to metadata collection could function to collect available IDE models, simulations, and the like and link them to their primary creator. Possibility to disperse intellectual property or contact owner as a consultant in tailoring programs to individual needs.	Follow-up products to meeting are not assured, although the NCEAS record of productivity is extraordinary. Compiling a catalogue of models and simulations doesn't ensure the nonecological community will be interested without prior attempts to reach out.
Consortiums	Good way to bring together interested parties with different talents, but a shared interest. Conservation biology has provided examples of the effectiveness of this approach.	Requires identifying specific problems attractive to, or already being investigated by, multiple disciplinary groups. Requires an active coordinator.
Integrative training grants	Chance to cross-foster ideas and knowledge. Academic institutions might collaborate with a veterinary school, a school of public heath, or federal agencies involved directly with heath-related issues.	Often requires interpersonal relationships among key scientists drafting the proposal. Requires specific examples of how integration would occur and how it would benefit student training.

Note: Many of the ideas in the table came from the audience and panelists at the 2005 Cary Conference. The phrasing of the suggested paths for action and the assessment of the strengths and limitations for different action plans is solely the responsibility of the author.

there is a gap in communication and mutual appreciation for what each community brings to bear on disease control issues.

Bridging the Gap

In my opinion, the onus for making the first overtures to bridge communication between these two communities rests with the IDE community. The rationale for this conclusion is twofold. First, the IDE community has the most to lose by continuation of the status quo. Only by reaching out will IDE scientists realize the potential of their intellectual creations to make a difference in the real world of human and animal health. Second, if IDE scientists don't make these efforts, it is uncertain that others will. Because there is no profit-driven middleman or identifiable governmental agency or subsidy to aid in the transition of IDE's intellectual property into PVHP applications, that responsibility can only be assumed by the IDE community.

The IDE community must create the sustainable demand for IDE products within the standards of current public and veterinary health practices. A number of venues have been suggested to open a dialogue and to increase the visibility and influence IDE scientists may have in future health practice and future training of heath practitioners (e.g., DVMs, MDs, and MPHs)(table 21.1). Virtually all the avenues suggested in the table will require individuals in the IDE community to dedicate time and effort to this cause. However, the benefit that could arise from such efforts is substantial: the opportunity for IDE scientists to play a significant role in human and veterinary health practice.

The optimal solution will result in collaborations between the IDE and PVHP communities when the IDE community's input would be most valuable to devising policy, planning interventions, and developing study designs. Perhaps the best routes to this sort of collaborative process are white papers drafted by representatives of both communities that outline means of solving problems stemming from miscommunication or misunderstanding, working consortiums, NCEAS working groups, identifying areas for active collaboration, or the use of integrative training grants that allow graduate students to be intermediaries or ambassadors bridging communities. The proactive response should not only increase the flow of information and build understanding between communities, it should also reduce post hoc assessments of the applications and effort of the IDE community. When IDE practitioners are denied an initial opportunity to plan and determine the most appropriate means of incorporating their efforts into responses to an immediate disease threat, the potential for successful application of their work is diminished and the likelihood of perceived or actual failure is enhanced.

LITERATURE CITED

Allen, L. J., and P. J. Cormier. 1996. Environmentally driven epizootics. Mathematical Biosciences 131:51–80.

Anderson, R. M., and R. M. May. 1991. Infectious Diseases of Humans. Oxford: Oxford University Press.

Brownstein, J. S., T. R. Holford, and D. Fish. 2003. A climate-based model predicts the spatial distribution of the Lyme disease vector *Ixodes scapularis* in the United States. Environmental Health Perspective 111:1152–57.

Coyne, M. J., G. Smith, and F. E. McAllister. 1989. Mathematic model for the population biology of rabies in raccoons in the mid-Atlantic states. American Journal of Veterinary Research 50:2148–54.

Ferguson, N. M., C. A. Donnelly, and R. M. Anderson. 2001. The foot-and-mouth epidemic in Great Britain: Pattern of spread and impact of interventions. Science 292:1155–60.

Glass, G. E., T. L. Yates, J. B. Fine, T. M. Shields, J. B. Kendall, A. G. Hope, C. A. Parmenter, C. J. Peters, T. G. Ksiazek, et al. 2002. Satellite imagery characterizes local animal reservoir populations of Sin Nombre virus in the southwestern United States. Proceedings of the National Academy of Sciences of the United States of America 99:16817–22.

Jones, C. G., R. S. Ostfeld, M. P. Richard, E. M. Schauber, and J. O. Wolff. 1998. Chain reactions linking acorns to gypsy moth outbreaks and Lyme disease risk. Science 279:1023–26.

Keeling, M. J., M. E. Woolhouse, D. J. Shaw, L. Matthews, M. Chase-Topping, D. T. Haydon, S. J. Cornell, J. Kappey, J. Wilesmith, and B. T. Grenfell. 2001. Dynamics of the 2001 UK foot and mouth epidemic: Stochastic dispersal in a heterogeneous landscape. Science 294:813–17.

Nowak, M. A., and R. M. May. 2000. Virus Dynamics: Mathematical Principles of Immunology and Virology. Oxford: Oxford University Press.

Russell, C. A., D. L. Smith, J. E. Childs, and L. A. Real. 2005. Predictive spatial dynamics and strategic planning for raccoon rabies emergence in Ohio. Public Library of Science Biology 3:1–7.

Salman, M. D. 2004. Controlling emerging diseases in the 21st century. Preventive Veterinary Medicine 62:177–84.

Smith, K. F., A. P. Dobson, F. E. McKenzie, L. A. Real, D. L. Smith, and M. L. Wilson. 2005. Ecological theory to enhance infectious disease control and public health policy. Frontiers in Ecology 3:29–37.

Waller, L. A., and C. A. Gotway. 2005. Applied Spatial Statistics for Public Health Data. New York: Wiley.

CHAPTER TWENTY-TWO
Educating about Infectious Disease Ecology

*Carol A. Brewer, Alan R. Berkowitz, Patricia A. Conrad,
James Porter, and Margaret Waterman*

Summary

HUMAN HEALTH DEPENDS ON HEALTHY ECOSYSTEMS, and ecosystem health depends on human choices. Because people play key roles in moving diseases around and modifying the environment in ways that determine the prevalence, spread, and severity of diseases, every citizen needs to understand the basic linkages between ecosystems, diseases, and health. In this chapter we address the thinking skills and essential habits of mind needed to understand the ecology of infectious diseases (IDE) and to develop the self-confidence, motivation, and skills to use their understanding in their professional and personal lives. Today, there are few examples of IDE or of effective teaching materials evident in K–12, undergraduate, or graduate and professional school curricula. Key elements of an educational agenda to integrate IDE knowledge into curricula from K–12 through graduate and professional schools are (1) the development of a compelling vision of what citizens, students, and professionals need to know, (2) synthesizing current knowledge about people's understandings, (3) identifying how IDE is taught and what resources are currently available, (4) articulating a research program to provide the intellectual foundation and to guide future practice in teaching and learning, and (5) promoting the implementation, assessment, documentation, and dissemination of innovative and effective teaching practices. Implementing the IDE education agenda presents new opportunities for transdisciplinary teaching, learning, and teamwork connecting knowledge bases and logic systems across disciplines, promoting an understanding of ecological complexity, and integrating social and environmental dimensions of understanding.

The Need for IDE Education

IDE is an emerging field of local, regional, and global importance. There are two reasons why it is imperative to teach all citizens about IDE. First, people's health and the health of the plants, animals, and microbes on which humans depend are inextricably linked to the ecosystems in which we live, and the health of these ecosystems is in turn significantly affected by disease. Other chapters in this book provide eloquent testimony to and abundant evidence of the pervasive, diverse, and deep effects diseases have on critical ecosystem functions. As an essential part of ecological literacy, people need to know about diseases, how they affect ecosystems, and how ecosystems affect them.

Second, teaching students and the general public about IDE is imperative because it is one of the best and most effective vehicles for developing key understandings, skills, and attitudes that are highly valued in our education system and society. People learn best when they are highly motivated by a sense of wonder and a recognition of the importance and utility of their subject. Disease ecology examples and case studies are highly engaging to general audiences, presenting life-and-death relevance for people, their companion animals, their domesticated plants and animals, and the other living organisms on which they depend. Disease detection stories are wonderful mysteries, exhibiting great detective work, intriguing types of evidence, and dramatic heroes and villains. They also embody the many dimensions of complexity that fields as diverse as ecology, epidemiology, and economics are anxious to embrace and champion, requiring us to synthesize across the social, biological, physical domains.

Not only should all citizens have a basic understanding of IDE inculcated, so also should the numerous and diverse scientists and practitioners invested by society with the responsibility for developing scientific explanations, predictions, interventions, solutions and management regimes for preventing and coping with diseases.

In the enterprise to understand and manage IDE, humans' knowledge, values, and ability to use information play a central role in the system itself (see Chapin et al., chapter 3, this volume). People, whether as disease victims or as caretakers of other organisms that are victims, could use knowledge to limit exposure to disease or to improve health and thus resistance. People also play key roles in directly moving diseases around locally, regionally, and globally, and in modifying the environment in ways that determine the prevalence, spread, and severity of diseases. Thus, education plays a crucial role, not just as an add-on or afterthought but as an essential tool of intervention and a topic of investigation. Education shows up both as an explanatory variable in epidemiological studies and as part of the solution to managing diseases.

Understanding the ecology of infectious diseases is crucial to achieving ecological literacy (Berkowitz et al. 2005). Take, for example, the "key systems" dimension of ecological literacy. For someone to have a rich and useful knowledge of his or her "ecological address"—the important ecosystems and populations in the immediate environment—that person must understand the prominent diseases in the home environment. For example, knowledge of Lyme disease is critical for people in the affected regions of the United States and Europe, as is knowledge of malaria in tropical and subtropical parts of the globe. Understanding IDE needs to be grounded in all the dimensions of ecological thinking that are the complementary components of ecological literacy (e.g., Berkowitz et al. 2005). The disease organism must be understood using scientific and systems thinking as both a population in its own right, and in its ecosystem context. Critical thinking about the forms of evidence that are available about the disease, its spatial and temporal patterns, its potential for harm and the alternatives for its management and treatment all comprise what we would call ecological literacy about the disease.

Status of IDE in the Curriculum

Is the dynamic and important topic of IDE well articulated in the K–12, undergraduate, and graduate curriculum? A quick search of the National Science Education Standards (NRC 1996) reveals the term *disease* appears in five places in the Science Content Standards chapter. Three of these appear in the personal health content standards under the Science in Personal and Social Perspectives section for each of the grades K–4, 5–8 and 9–12. None of these references alludes to any connection between disease, ecology, or features of ecosystems. The other two places where *disease* appears are in the Life Science standards for grades 5–8; one reference is in the structure and function of living systems standards and the other is in the populations and ecosystems standards. These standards place some emphasis on the ecological role of disease as one of several possible factors limiting populations, and there is one mention that some diseases "are the result of damage by infection by other organisms." However, the dynamic two-way interaction between diseases and ecosystems is not well addressed. Whether a similar result would be obtained in a more thorough analysis of state or local standards, or of the actual curricula delivered in K-12 schools is beyond the scope of this chapter, but developing a clearer picture of the place IDE does or does not play in the K–12 curriculum is an important task for the future. Some excellent resources already exist for K–12 teachers in IDE education (e.g., Preserving Health Through Biological Diversity on AAAS's SciNetLinks Web site: http://www.sciencenetlinks.com/lessons.cfm?DocID=438), with prospects for more resources as interest in IDE continues to build.

All Citizens Should Know

- Human health and the health of all organisms is inextricably linked to the ecosystems we live in, and ecosystem health depends on human choices.
- Diseases can be important in affecting socioecological systems, including populations, communities, ecosystems, and landscape processes.
- Disease effects on ecosystems can affect people; people's effects on ecosystems can affect diseases.
- The effects of disease can be hard to see and hard to anticipate.
- We can understand diseases as having either fast or slow impacts, especially if we understand what species are affected and how the ecosystem functions.

At the other end of the formal education spectrum, undergraduate and graduate training, we observe that IDE provides an excellent venue to train advanced students. Despite the clear relevance, however, there is little evidence that IDE is well covered at the undergraduate level. An Internet search revealed no course syllabi at the introductory college level for courses such as Ecology of Infectious Disease, Conservation Medicine, Invasion Ecology, or other likely IDE titles. However, the topic is tangentially found in introductory biology courses, and numerous syllabi on EID were located for upper-level biology, graduate, public health, veterinary, medical, and professional continuing education. Indeed, some courses are entirely online and freely available.

EID and Introductory Biology

Most of today's 15,000,000 undergraduates take an introductory course in the life sciences as part of general education requirements. For many, it is the only and last science course they will take as undergraduates. For the 68% of students who do not major in science, technology, engineering, or mathematics (STEM) fields (NSF 2002), an introduction to IDE is essential to their preparation as knowledgeable citizens and future decision and policy makers. For some of these students, the IDE component of an introductory biology course may be fundamental to showing them how to use an understanding of IDE to better manage and sustain ecosystems for health. Some of these students are people on whom the future of teaching and research in IDE will depend. For students majoring in a STEM field, introductory science courses provide a conceptual framework for the discipline and engage students in doing science, as well as foreshadow potential careers related to the field.

IDE is not a topic found in the typical college biology textbook. We decided to examine more closely a text widely used in introductory college biology courses (Campbell and Reese 2005). Topics related to disease, emerging diseases, pathogenic organisms, and antibiotic resistance were present, but not in an ecological context. A few ecological topics incorporated effects of disease (e.g., American chestnut disease, disease in density-dependent population regulation). Despite the modest coverage in such texts, there are many opportunities for faculty to introduce IDE examples and concepts. Topics such as disturbances, species diversity and composition, disruptions by humans of global chemical cycles, disruption of interaction networks, the need for biodiversity for human welfare, and ecological restoration provide opportunities to link to current research and understanding of IDE.

There is a clear need for examples of excellent teaching materials in IDE, as well as particularly engaging pedagogical approaches. An important venue for integrating IDE into courses is through the use of supplemental case studies and data sets. For example, "Teaching Issues and Experiments in Ecology" contains curriculum modules on IDE issues such as the decline of coral reefs and the ecological impact of high densities of deer, along with images, data sets, instructional ideas, and background information. In addition, there are many interesting case studies and investigative tools that can be readily adapted to IDE teaching. Some excellent examples are listed in the accompanying box.

**Example Sources for Materials, Cases, and Data Suitable
for IDE Teaching**

Teaching Issues and Experiments in Ecology: http://tiee.ecoed.net/vol/
toc_all.html

Biological Inquiry: A Workbook of Investigative Cases: Waterman and
Stanley (2005)

Investigative Case Based Learning: http://www.bioquest.org/lifelines

(continued)

(*continued*)

> National Center for Case Studies in Science: http://ublib.buffalo.edu/
> libraries/projects/cases/case.html
> Problem Based Learning Clearinghouse: www.mis4.udel.edu/Pbl/
> Bioinformatics Education Project *BEDROCK*: www.bioquest.org/
> bedrock/problem_spaces/index.php
> Ecohealth 101: Environmental Change and Our Health. (Johns Hopkins
> and UW-Madison): http://ecohealth101.org

Professional Education of Teachers

Teacher education relies on university science departments to provide content, tools of inquiry, scientific dispositions, and research skills. This content needs to include IDE. Further professional development in IDE should be made available for in-service teachers through the usual channels of grant-supported research opportunities, workshops at professional meetings, and short courses for teachers.

Education courses and professional development opportunities can prepare teachers for IDE teaching by emphasizing methods of teaching that support IDE and by having teachers learn to locate available teaching materials. For example, K–12 teachers need to be ready to use case studies and problem-based learning, to engage students in inquiry, and to relate all science content to students' prior knowledge and the societal contexts in which they live.

Current Trends in Veterinary Schools

A variety of approaches to teaching about infectious diseases are utilized in the curricula of North American veterinary colleges and schools; most are oriented by discipline (e.g., pathology, microbiology, parasitology) or the type of host species affected (e.g., companion animals, food animals, wildlife). In addition, some veterinary colleges employ a "problem-based" learning approach that incorporates infectious diseases in the first two years of preclinical training in their curriculum. Only a small number of veterinary schools are involved in teaching infectious disease from an ecological perspective (e.g., Gilardi et al. 2004; Kaufman et al. 2004; Waltner-Toews et al. 2004).

Environmental influences on disease dynamics in animals has long been a consideration, though not always a specific focus, in veterinary medicine. The diagnosis and treatment of individual animals, particularly pets, is an important part of the professional training for veterinary students. However, there is an increasing consideration of animal populations and their interactions with the environment, other animals, and humans. In training food animal veterinarians, the herd health approach to prevention and control of infectious disease has generally replaced individual animal medicine. Similarly, shelter medicine and wildlife health programs are becoming more common in veterinary schools, and courses in these areas emphasize the importance of maintaining healthy environments for disease control and prevention.

In recent years there has been an increasing need for competent veterinarians to help deal with zoonotic pathogens of importance to public health and environmental management, as well as infectious disease threats to wildlife and domestic animals. Teaching disease ecology to veterinary students is needed to understand pathogen transmission, disease interactions and the role of animals in ecosystem health. Proponents of transdisciplinary ecosystem health and conservation medicine training contend that this training promotes a more global, generational, and preventative approach to health care.

Current Trends in Schools of Medicine and Public Health

In schools of medicine and public health, there has been progress toward more integrated learning by following a more "systems-based" approach to medical teaching. For example, most schools of medicine now combine modules spanning the disciplines of anatomy, physiology, pathology, and medical technology rather than teaching each in isolation. Similarly, biostatistics is routinely taught in conjunction with epidemiology and an integrating capstone practicum in many schools of public health. Yet, with the exception of a few schools, there remains a large gap between the disciplines of ecology and health.

Why do most schools fail to adequately integrate the fields of ecology and health, even as the linkages are becoming more and more apparent (Patz et al. 2004)? A century ago, Western approaches relied primarily on either microbiology or epidemiology to explain and control infectious diseases, and social factors were known to have a strong influence on infections. Later discoveries of vector-borne diseases revealed the importance of transmission ecology. Research from these periods led to the agent-host-environment model of infectious disease, which still serves as the central model for examining emergence of specific diseases. However, this model fails to shed light on the large-scale social and ecological settings that underlie the processes leading to increased IDE emergence (Bienen et al. 2005). More recently, an ecological context for teaching has lost further ground in the wake of successes from highly specific drug treatments and vector control programs. As a direct result, many epidemiologists have shifted their efforts from understanding epidemics of infectious diseases to studying individual risk factors in chronic disease.

In the future, promoting transdisciplinary approaches in medical schools is key to improving our response to the challenges posed by IDE · that stem from interlinked social and ecological systems, and across scales from the village to the globe. New forms of knowledge that integrate the natural and human sciences, attend to dynamics at multiple scales—both spatial and temporal—and engage diverse ways of understanding and intervening are needed in training programs for health professionals.

Essential Habits of Mind for Learning About IDE in All Educational Settings

Critical Thinking

First and foremost, we must teach critical thinking. Everyone must understand that the delicate balance between human health and disease is intimately connected to the health of the natural world. As scientists interested in the ecology of infectious diseases, we are in a unique position to explain this delicate balance, and to do so using inherently interesting and, therefore, memorable case studies and examples.

Scientific Thinking

Scientific thinking is about how we integrate and synthesize knowledge at the same time as we develop the processes, skills, and attitudes that are necessary for citizenship, or as professionals, to make substantive contributions in science. Furthermore, in our curricula, we acknowledge that habits of mind and discipline-specific understanding are tightly linked with modes of thinking. The curricular challenge is to construct learning experiences that build a conceptual framework and worldview for understanding, managing, and participating in complex natural systems embedded in the biosphere.

Dealing with Complexity

One of the hardest things to teach in any discipline is complexity, but IDE incorporates this naturally into its instructional requirements. Mikkelson (2001) points out that ecological systems are very complex in the sheer number of causal factors affecting them. The ecology of infectious disease is an especially good example of this kind of complexity. Having the habit of mind that accommodates complexity has important implications for dealing with uncertainly. Disease outbreaks can produce complex impacts on ecosystems (Chapin et al., chapter 3, this volume). Consequently, IDE can be used to teach the importance of incorporating uncertainty into ecological models. Policy makers often are uncomfortable with uncertainty and frequently want to ignore advice that is given with many qualifiers and alternative outcomes (Brewer and Gross 2003; Clark et al. 2001; Michener et al. 2002). Because people are so familiar with disease, however, some complexity and uncertainty can be introduced into the policy dialogue without automatically losing decision makers. If the results of this dialogue are perceived by the public as credible, then we will have succeeded in introducing ecological principles into the decision-making process as well as into society's response (Gross 1994; Pielke and Conant 2003).

Using IDE to develop the habit of mind of dealing with complexity may also have other indirect benefits. If, for example, we can capture the public's attention on this complex ecological subject, then it may also be possible to introduce other difficult environmental concepts that are needed to anticipate the consequences of global change (Brewer and Gross 2003). For instance, while it is true that newly emerging diseases are likely to come from new contacts between humans and reservoirs or vectors from the wild, it is also likely that cures will come from bioprospecting in the natural world. Exploring complex cases can link ecology with fields such as economics and ethics, and may lead to strong arguments favoring preservation of the natural world.

Transdisciplinary Thinking

Developing the capacity for transdisciplinary thinking requires that students possess a working framework for when and how to reach out into surrounding disciplines, integrate the information, and apply it to the IDE questions under study. Students must develop the facility for thinking beyond ecology to consider geology, geography, social sciences, physics, atmospheric science, chemistry, and mathematics, and to make effective intellectual leaps between these fields and ecological phenomena and questions. Having the dispositions and personal traits needed to work critically in this way requires enthusiasm for going beyond one's field, confidence, the use of evidence-based and systems thinking, and the ability and motivation to collaborate. Successful collaboration relies on expertise from many disciplines and openness to different perspectives and ways of knowing (e.g., natural science vs. social science).

Team Work and Interpersonal Skills

Current and future students need to learn to work as effective members of collaborative teams and collaboration skills must be part of the curriculum. Although no one member of a research team can be expected to have expertise in all the relevant areas, it can and should be expected that all members of the team have a good foundation in the general areas of interest. Members of collaborative teams need in-depth expertise in one critical area related to IDE, such as ecology, social science, modeling environmental change, or visualizing large data sets. Beyond that, collaborative team members need to be able to understand each other and to teach each other using familiar terms. In preparation for such teamwork, students must learn to communicate clearly and succinctly in writing and presentations, so that the results of their work can be accessible not only to their peers but to nonspecialists as well.

Basic interpersonal skills and management skills should be included in IDE curricula to address differences in discourse and language among disciplines. Formal training in collaborative management or team building and facilitating team dynamics (or key leadership skills) as well as in base skills such as active listening and conflict resolution will facilitate more productive collaborations in interdisciplinary research. While problem-based approaches are important to promoting interdisciplinary learning, professional skills can be taught and practiced alongside scientific content.

AN AGENDA FOR TEACHING AND LEARNING IN IDE

In the coming few years, two fundamental questions to discuss and answer about IDE are (1) what do disease ecologists and the general public in the twenty-first century need to know about IDE, and (2) how can we successfully translate new knowledge about IDE to scientists, managers, teachers, and the general public. These questions about teaching and learning, frequently underappreciated by the academic scientific community, are critical to the field of IDE.

Because of a high level of relevance and familiarity, most people have well-developed, if not scientifically valid, ideas about the causes and consequences of diseases. Future research work in teaching and learning about IDE will benefit from careful consideration of the sources and diversity of these ideas. People's worldviews and conceptions of nature shape how they perceive the mechanisms and outcomes of ecological interactions and their expectations about the future. We need to identify these worldviews and pervasive attitudes and misconceptions.

Concrete Recommendations to Guide an
Education Initiative about IDE

1. Develop a compelling vision for what people need to know and be able to do, focusing on three audiences: all citizens, undergraduate students, and professionals.

2. Synthesize current knowledge about people's understandings of the ecology of infectious diseases.
   a. Learn from work in education and cognitive research, anthropology and psychology.
   b. Look at cognitive understanding of linkages between ecology and disease, people's perceptions of disease and health with respect to the environment, prevailing and pervasive attitudes and misconceptions, and worldviews as they shape and constrain understanding of ecology and disease.

3. Describe current practices and resources for teaching about the ecology of infectious diseases.
   a. How is IDE addressed in curricula, education standards, educational programs, and institutions?
   b. What resources are available?

4. Develop a research agenda to provide the intellectual foundation and guide future practice in teaching and learning about IDE.
   a. How do people learn about complex, multifaceted systems and phenomena such as IDE?
   b. How do transdisciplinary teaching, research, and learning take place?
   c. What institutional and societal changes?
   d. How are information and understanding of EID used in decision making at the individual and collective levels?

5. Foster the implementation, assessment, documentation, and dissemination of innovative and effective practices for teaching about EID.
   a. Audience-specific recommendations.
   b. Common strategies including cases, inquiry and student-centered teaching, transdisciplinary teams, restructuring curricula and departments, partnering with professional associations, and working to incorporate IDE into the required curriculum and assessments.

How do people learn about complex, multifaceted systems and phenomena such as IDE? This needs to be considered at many levels: teachers—who they are, what they know, what curricula and materials they use, and how they teach; resources in the media and other non-school information transfer venues; the K–16 system—what works, what are the appropriate scope and sequences, and how accountability can be achieved.

Since IDE is a new component of introductory biology, little is known about teaching and learning IDE. At the introductory level it will be important to look at the prior knowledge students bring to the classroom. What are their concepts of ecology, of disease? Do students already connect the notions of ecology and disease together? How might their understanding of evolution help or hinder subsequent learning of IDE concepts? New IDE teaching materials will need to be field tested for their efficacy in implementation in the classroom as well as for their effectiveness in encouraging the knowledge, skills, and dispositions of EID.

Restructuring Academies for Teaching and Learning about EID

THE K–12 SYSTEM

In an exchange of views on the No Child Left Behind Act (NCLBA), Blank and Brewer (2003) acknowledged the fact "that students learn what they are taught, and that teachers teach what they are held accountable for (Shavelson et al. 1990)". These authors argue for a profound change in what is taught to include rudimentary concepts of environmental sustainability. The ecology of infectious diseases might be one subject superbly suited to achieve this goal because the Malthusian concept of disease as a population control is widely accepted and understood, even among the lay public. Bybee (2003) further points out that accepting at face value the presidential challenge contained in the NCLBA of "global economic leadership and security in the twenty-first century" would actually require teaching ecology as a primary subject in every grade. Nevertheless, this articulated goal potentially creates common ground between the top-down mandates of the NCLBA and the necessity to create an environmentally literate public.

COLLEGES AND UNIVERSITIES

Anyone with even rudimentary training in biology will already have an appreciation of the complexity inherent in parasitoid biology, parasite life history characteristics, and pathogen life cycles. To advance IDE, we need a new emphasis on courses that ecology undergraduate and graduate students often shy away from: physiology, microbiology, epidemiology, cell and molecular biology, genetics, virology, protozoology, and parasitology. Moreover, it is critical to provide undergraduates in biology, mathematics, and computer science with opportunities to (1) learn about the ecology of infectious disease, (2) develop their abilities to communicate and work closely with their peers from different disciplines, and (3) experience the challenges and scientific benefits of interdisciplinary and transdisciplinary learning.

How do you teach the next generation of ecologists to undertake interdisciplinary and transdisciplinary research? Although it is well known that major discoveries often occur at the interface between traditionally separate disciplines, such as between biology and chemistry, increasingly it is understood that major advances will come from investigations informed by experts from many disciplines. Specialization is no longer how major discoveries, those with broad societal implications, are made. With its emphasis on and inclusion of researchers with very different skill sets, IDE is an ideal field to illustrate why and how complex questions are now being answered.

To accomplish the kind of transdisciplinary teaching and research needed, there are challenges inherent in the way our colleges and universities are structured (Brewer and Maki 2005). Relevant disciplines are segregated into different departments, and developing interdisciplinary courses can be remarkably difficult, especially when they require listing in multiple departments. Despite these challenges, teaching and learning about IDE will require innovative courses that span disciplinary boundaries. Clearly, administrative structures are needed to make such courses not only possible, but routine.

What are some possible solutions to the challenges of transdisciplinary teaching? Brewer and Maki (2005) highlight four. First, faculty need to view their teaching as more closely aligned with their research practices. Although many faculty members may collaborate across disciplines, it is the rare set of individuals who translate that same kind of interdisciplinary environment into their classrooms. Classrooms, like research, must be transdisciplinary (or at least open to other disciplines and perspectives) for students to understand the depth and breadth of IDE in its real-world context. Furthermore, teaching IDE at the introductory level means not only that new materials need to be made available but that faculty must be prepared to teach these materials. Faculty may need to develop expertise in teaching approaches supportive of IDE goals through professional development opportunities supported by agencies like the National Science Foundation as well as by professional societies.

Second, faculty should highlight the unique relationships between their disciplines, and encourage students to collaborate by welcoming students and researchers from other disciplines and setting up opportunities for them to learn from one another. Clearly, to advance the field of IDE, faculty must move beyond the way they were taught in undergraduate and graduate courses.

Third, department heads and other administrators need to develop new ways to reward faculty so that those who participate in transdisciplinary courses and become more involved in their teaching are properly acknowledged in the institution's reward structure. The ways in which course loads and student credit hour production are determined need to be modified to better support the inclusion of multiple faculty and multiple departments in teaching a single course. Moreover, faculty need time (e.g., reduced course loads, travel to relevant conferences) for this work and to engage in transdisciplinary discussions and meetings across departments. This is essential for younger faculty who could sustain these kinds of student-centered changes.

Fourth, these kinds of changes require changes at the national funding level, where it is important that research support move beyond the principal investigator model. The National Science Foundation, the National Institutes of Health, and other large funding agencies should continue to develop innovative programs that facilitate participation by multiple investigators representing diverse fields. While funding agencies are funding more interdisciplinary research, researchers need to know that this is a serious and long-term commitment that extends to their teaching, too.

VETERINARY SCHOOLS

The integration of IDE into the professional veterinary curriculum is not without challenges. Paramount may be the need to demonstrate the importance and applications of this training for veterinary graduates. Ultimately, the success and sustainability of any curricular changes will depend on the enthusiasm of the faculty leading the educational effort and the support they receive from their peers and administration.

Few veterinary faculty members have sufficient knowledge or experience to teach the ecology of infectious diseases. To be effective these courses require an integrated transdisciplinary educational effort that breaks from traditional teaching styles applied in disciplines within a departmental structure. Creation of cross-disciplinary programs that might best support courses and research in infectious disease ecology requires cooperation and communication amongst the faculty involved. Integration of new courses, a change in focus, or an infusion of more disease ecology into existing core courses also will require careful planning and coordination so there is a logical flow in course progression and content. This may require faculty training and the shared development of cases and teaching materials, but the end result may well be a better coordinated and effective program delivered by enthusiastic instructors.

In addition, veterinary students may pose some challenges to curricular change. Although students complain about the mind-numbing demands for excessive rote memorization in their courses, they may resist moving out of this comfort zone when courses require synthesis and application of knowledge from multiple disciplines to solve complex problems relating to IDE. This resistance might be lessened by introducing an ecological perspective and presenting cases for consideration in the first year of the curriculum, and continuing its integration throughout the discussion of infectious diseases in the curriculum.

Several avenues are available for the introduction and expansion of IDE in the professional veterinary curriculum. Courses can be introduced either in the core curriculum, which all students are required to take, or in elective courses that students with a special interest can select for more extensive and detailed coverage of a topic. In addition, existing core courses in infectious disease could be modified to allow for the integration of more ecological considerations. Conceptual knowledge in disease ecology could be augmented with real or simulated cases that would require students to apply their knowledge to solve problems relating to infectious disease. These case scenarios could be developed in collaboration with faculty that have relevant experience in ecology, biology, public health, or environmental sciences, as well as with faculty from other disciplines within the veterinary school.

Experiential learning, a hallmark of veterinary medical education, is widely viewed as the most effective method for building clinical knowledge and skills. Unfortunately, in most professional training programs there is limited opportunity for students to acquire much hands-on experience with infectious disease problems within the preclinical curriculum of the first two to three years. During this period, teaching cases presented by instructors in the classroom or via interactive computer programs are the next best thing to a real experience. Cases focused on the ecology of infectious disease could be particularly challenging for students because they require the synthesis of knowledge from multiple disciplines and promote the exploration of complex problems. They could also be used to demonstrate how to pursue scientific investigations and critically analyze data that might be relevant to solving the case. And cases may stimulate an interest in some students to pursue postgraduate research training in disease ecology.

Research opportunities for veterinary students and graduates are critical for the improved involvement of veterinarians in disease ecology. Postgraduate research training will allow students to deepen their understanding of infectious disease ecology, contribute more effectively to disease investigation teams, and prepare them to be future leaders in this field. Ideally, research projects should involve transdisciplinary collaborations that strengthen relations with environmental, public health, and medical programs. Veterinary students should have the opportunity to work with medical students, biologists, ecologists, and other environmental scientists in elective courses and research projects.

The majority of students entering veterinary programs in North America and Europe are focused on a future in clinical veterinary medicine, most often in small animal practice. Extracurricular seminars and training programs, such as EnvironVet, should be expanded to show veterinary and graduate students potential career opportunities in disease ecology. Even students who remain on track to practice clinical veterinary medicine will be better prepared to consider the impact of environmental factors on their patients and their clients' impact on the environment.

SCHOOLS OF MEDICINE AND PUBLIC HEALTH

Recent and ongoing outbreaks of avian flu, monkeypox, SARS, mad cow disease, Ebola hemorrhagic fever, and numerous other diseases (Lashley 2004) have drawn public and scientific attention to the fact that problems of emerging diseases are not easily solved by "magic bullet" interventions from medical technology. In this millennium, IDE reveals two converging realities: the causes of social and ecological change (including global poverty and inequity, security, loss of ecosystem services, and others) underlie IDE emergence, and a fragmented or overspecialized approach to disease control is often inadequate, and in some cases can exacerbate the original problem (Bienen et al. 2005). Drawing on knowledge, both across disciplines as well as across knowledge sources (e.g., from academics to indigenous communities), will enhance understanding and subsequent response to complex societal and ecological problems contributing to IDEs (Sapiro 2004; Somerville and Rapport 2000), and warrants close examination as an option to improve understanding, prevention and control of future IDE practitioners.

Ecosystem approaches to health explicitly advocate the engagement of local communities to understand the ecological and social systems within which health is created or threatened (Lebel 2003). And while these approaches provide some common principles in conducting participatory-based community assessments to expand our knowledge base, we must also consider interdependent systems from which diseases emerge, maintain themselves, and evolve on a multitude of levels, from molecular and organismal to communal, national, and global (Wilcox 2005). Unfortunately, much of the mainstream biomedical community is biased against complex social, political, and environmental research that incorporates multilayered approaches with results that parallel the uncertainty of real life and may not lend themselves to rapidly implemented policies or interventions (Ravetz 2005). Recent developments, however, are making gains to show the value of transdisciplinary approaches, and, if accompanied by appropriate uncertainty analysis and transparent risk communication, should improve decision making around sustainable health and environment.

CONCLUSIONS

The challenges facing teaching IDE are quite similar to those facing instruction in the area of conservation biology. In both, the difficulties inherent in creating a sustainable future require that we do more than accept a trickle-down approach (Brewer 2001) to teaching these subjects. They are simply too important. Both face the same problem described by Wilson (2002) to bring together professionals long separated by academic and practical tradition. Neither discipline has defined the skill sets and aptitudes required for one to be a successful practitioner in the field, and neither has determined the most effective way to include the mix of professionals, citizens, legislators, and policy makers who ultimately will determine the success or failure of initiatives in these fields. Berkowitz et al. (1997) argue that a new transdisciplinary approach that incorporates both the knowledge bases and the logic systems of several disciplines may be more effective in creating future professionals, and an environmentally literate public, than narrow disciplinary approaches to teaching and learning. And we must also do a much better job of communicating beyond our scientist peers.

What tangible actions are needed to move the field of the ecology of infectious disease forward, and to ensure that the current and next generation of students will be ready to step into this field? First, there is a clear need for a more comprehensive assessment of the status of IDE in K–12, undergraduate, and graduate-level curricula, and of resources and institutional needs currently available to support teaching of IDE. Second, educational research is called for to help us better understand areas such as teaching complexity; teaching and fostering transdisciplinary thinking, research, and teamwork; people's worldviews and how they influence their learning; and people's use of knowledge and how it influences decisions and actions. Third, efforts are needed to include medical and public health perspectives in education about IDE, as well as to integrate IDE into the training of medical professionals. And finally, bringing researchers and educators together in more research-oriented efforts, such as the Cary Conference on IDE, is essential to build on the links between pedagogical and epistemological inquiry, and to create effective education programs and resources.

ACKNOWLEDGMENTS

We thank Yasmin Rubio-Palis and Felicia Keesing for helping us develop many of the ideas presented in this chapter.

LITERATURE CITED

Berkowitz, A. R., M. Arcade, and D. Simmons. 1997. Defining environmental literacy: A call to action. Bulletin of the Ecological Society of America 78:170–72.
Berkowitz, A., M. Ford, and C. A. Brewer. 2005. A framework for integrating ecological literacy, civics literacy and environmental citizenship in environmental education. *In* Environmental Education or Advocacy: Perspectives of Ecology and Education in Environmental Education, ed. E. A. Johnson and M. J. Mappin, 227–66. New York: Cambridge University Press.
Bienen L., M. Parkes, J. Breil, L. Hsu, M. McDonald, J. A. Patz, J. Rosenthal, M. Sahani, A. Sleigh, et al. 2005. All hands on deck: Transdisciplinary approaches to emerging infectious disease. EcoHealth 2:258–72.
Blank, L., and C. A. Brewer. 2003. Ecology education when no child is left behind. Frontiers in Ecology and Environment 1:383–84.
Brewer, C. A. 2001. Cultivating conservation Literacy: Trickle-down education is not enough. Conservation Biology 15:1203–5.
Brewer, C. A., and L. J. Gross. 2003. Training ecologists to think with uncertainty in mind. Ecology 84:1412–14.
Brewer, C. A. and D. Maki. 2005. Building the Renaissance team. *In* Math & Bio 2010: Linking Undergraduate Disciplines, ed. L. A. Steen, 45–50. Washington, DC: The Mathematical Association of America.
Bybee, R. W. 2003. Ecology education when no child is left behind. Frontiers in Ecology and Environment 1:383–84.
Campbell, N., and J. Reece. 2005. Biology, 7th ed. San Francisco: Benjamin Cummings.
Clark, J. S., S. R. Carpenter, M. Barber, S. Collins, A. Dobson, J. A. Foley, D. M. Lodge, M. Pascual, R. A. Pielke, Jr., et al. 2001. Ecological forecasts: An emerging imperative. Science 293:657–60.
Gilardi, K. V. K., J. G. Else, and V. R. Beasley. 2004. Envirovet summer institute: Integrating veterinary medicine into ecosystem health practice. EcoHealth 1:50–55.
Gross, L. J. 1994. Limitations of reductionist approaches in ecological modeling: Model evaluation, model complexity, and environmental policy. *In* Wildlife Toxicology and Population Modeling: Integrated Studies of Agroecosystems, ed. R. J. Kendall and T. E. Lacher, 509–18. Boca Raton, FL: CRC Press.

Kaufman, G. E., J. G. Else, K. Bowen, M. Anderson, and J. Epstein. 2004. Bringing conservation medicine in to the veterinary curriculum: the Tufts example. EcoHealth 1:43–49.

Lashley, F. R. 2004. Emerging infectious diseases: Vulnerabilities, contributing factors and approaches. Review of Anti-Infective Therapy 2:299–316.

Lebel, J. 2003. Health: An Ecosystem Approach. Ottawa: International Development Research Centre.

Michener, W. K., T. J. Baerwald, P. Firth, M. A. Palmer, J. L. Rosenberg, E. A. Sandlin, and H. Zimmerman. 2002. Defining and unraveling biocomplexity. BioScience 51:1018–23.

Mikkelson, G. M. 2001. Complexity and verisimilitude: Realism for ecology. Biology and Philosophy 16:533–46.

National Research Council (NRC). 1996. The National Science Education Standards. Washington, DC: National Academy Press.

National Science Foundation, 2002. Science and Engineering Degrees: 1966–2000, NSF 02-327. Arlington, VA: NSF.

Patz, J. A., P. Daszak, G. M. Tabor, A. A. Aguirre, M. Pearl, J. Epstein, N. D. Wolfe, A. M. Kilpatrick, J. Foufopoulos, et al. 2004. Unhealthy landscapes: Policy recommendations on land use change and infectious disease emergence. Environmental Health Perspectives 112:1092–98.

Pielke, R. A., and R. T. Conant. 2003. Best practices in prediction for decision-making: Lessons from the atmospheric and earth sciences. Ecology 84:1351–58.

Ravetz, J. 2005. The No-Nonsense Guide to Science. Toronto: New Internationalist Publications.

Sapiro, V. 2004. Interdisciplinary and collaborative teaching at the University of Wisconsin–Madison: Overcoming barriers to vitality in teaching and learning. Available: http://www.polisci.wisc.edu/users/sapiro/papers/interdisciplinary.pdf (accessed August 10, 2005).

Shavelson, R. J., N. B. Carey, and N. M. Webb. 1990. Indicators of science achievement: Options for a powerful policy instrument. Phi Delta Kappan 71:692–97.

Somerville, M. A. and D. Rapport. (Eds.). 2000. Transdisciplinarity: Recreating integrated knowledge. Oxford: EOLSS Publishers.

Waltner-Toews, D., J. A. Van Leeuwen, B. Hunter, N. Larivuère, D. Bèlanger, and J. Smits. 2004. Four veterinary colleges and a common vision: ecosystem health in Canadian veterinary colleges. EcoHealth 1:56–62.

Waterman, M., and E. Stanley. 2005. Biological Inquiry: A Workbook of Investigative Cases. San Francisco: Benjamin Cummings.

Wilcox, B. A. 2005. Emerging infectious diseases: Bridging the divide between biomedical and bioecological science. EcoHealth 2:167–168.

Wilson, E.O. 2002. The Future of Life. New York: Vantage Books.

PART IV

Concluding Comments: Frontiers in the Ecology of Infectious Diseases

The Ecology of Infectious Diseases: Progress, Challenges, and Frontiers

Richard S. Ostfeld, Felicia Keesing, and Valerie T. Eviner

RECENT YEARS HAVE SEEN EXPLOSIVE GROWTH in research activity devoted to understanding the ecology of pathogens and the diseases they cause. The importance of pathogens and diseases to humans, wild and domesticated animals, and plants has been recognized since antiquity. Only recently, however, has the science of ecology assumed a major role in understanding disease dynamics. Ecologists are beginning to routinely include pathogens as focal study organisms and to incorporate them and their consequences into the communities and ecosystems under study. The incorporation of ecological perspectives into the forecasting, prevention, and management of human diseases, however, remains rare; it is much more common in agricultural and silvicultural systems (Garrett and Cox, chapter 17, this volume). The degree to which ecological approaches to disease become integrated into the various health sciences seems to depend on the ability of ecologists to develop either conceptual-theoretical models that can be applied directly to specific disease systems or empirical understandings of specific systems that provide concrete direction to health specialists (Peters, chapter 20, this volume, Childs, chapter 21, this volume).

Conversely, the full potential of the various health sciences in contributing to ecology and environmental management has yet to be reached. Pathogens often have profound effects on the ways that ecosystems function, including their ability to provide services to people. For example, in agroecosystems, diseases strongly impact the production of crops, livestock, and seafoods (Chapin et al. chapter 13, this volume, Garrett and Cox, chapter 17, this volume). In more natural systems, diseases alter forest, grassland, and marine productivity, as well as contribute to erosion and other disturbance regimes (Eviner and Likens, chapter 17, this volume, Middelboe, chapter 11, this volume). The ability to detect pathogens, understand the modes and rates of transmission, and assess the consequences of infection in individual hosts is fundamental to the health sciences. Specialists in these disciplines, working together with ecologists, are beginning to make great strides in understanding the impacts of diseases on ecosystem services and consequently on human health and well-being.

This book provides a view of many of the cutting edges in the community and ecosystem ecology of disease. As in any rapidly moving field, we expect these cutting edges to require repeated sharpening as they penetrate the tough material of the unknown. We think, though, that even if the empirical and theoretical details given in the various chapters become outdated, the basic conceptual issues will continue to provide a foundation for many years of subsequent research. In this wrap-up chapter, we provide a brief synthesis and integration of the major contributions of each chapter in an attempt to anticipate future directions for the field of disease ecology.

EFFECTS OF ECOSYSTEMS ON DISEASE

Many components of ecosystems can affect disease dynamics.* Among these ecosystem components are host diversity, vector diversity, pathogen diversity, landscape structure, and eutrophication, all of which are explored in the first section of the book. Begon's chapter 1 on the effects of host diversity on disease dynamics explores the ways in which increasing species diversity can potentially decrease or increase rates of pathogen transmission and disease risk. He finds generally greater plausibility to the hypothesis that high host diversity will reduce rather than increase pathogen transmission, but he also argues that empirical support for a buffering effect of high host diversity (dilution effect) is not strong. His discussion suggests several important frontiers in the exploration of diversity-disease interactions. First, how do the effects of host diversity interact with diversity in other components of the broader disease system? For example, diversity of species that do not function as hosts for a focal pathogen might play an important role in disease dynamics if these species regulate host abundance, affect encounter rates between hosts and pathogens, or affect the nutritional or other physiological states of hosts (Keesing et al. 2006). A second frontier is distinguishing the effects of host diversity per se (e.g., species richness and evenness) from those of species composition in the host community. Host communities of equal species richness or evenness can be composed of different members or different relative abundances of the same members, with potentially strong impacts on pathogen dynamics. Finally, the shape of the relationship between host diversity and disease risk is a crucial frontier. For some disease systems, a minimal host

---

* Note that we define ecosystems broadly to include communities of interacting species, the abiotic components of ecological systems, and the landscapes within which species, energy, and nutrients interact.

diversity is required for the pathogen to persist, invoking an amplification effect (Keesing et al. 2006) as diversity increases from very low to low levels. But above a threshold level of diversity required for pathogen persistence, further increases in diversity could result in a dilution effect.

The impacts of host diversity on disease dynamics are only part of the story. Many pathogens require arthropod vectors to reach their hosts, setting the stage for vector diversity and community composition, in addition to host diversity, to play important roles in disease dynamics. Power and Flecker in chapter 2 point out just how little is known about the impact of vector diversity on pathogen dynamics. For diseases in which only one vector transmits the pathogen, vector diversity (in the strict sense) is a non sequitur. However, in many cases more than one arthropod is capable of transmitting a particular pathogen, leading to the potential for strong indirect interactions among vectors. For example, if different vector species compete for pathogens—that is, if some vector species deplete the pool of pathogens available to others—the high diversity could contribute to a dilution effect. In systems characterized by "bridge vectors," such as when one vector species maintains an enzootic cycle of transmission and a second vector species is responsible for transmission to humans or other "spillover" hosts, a minimum threshold diversity of vectors might be necessary for the occurrence of a zoonosis. However, this same threshold of diversity could, in principle, reduce pathogen transmission within reservoir hosts by deflecting transmission events to other host species that act as dead ends. As for the situation with host diversity, the shape of the curve relating vector diversity to disease transmission becomes important. Moreover, broadening the definition of vector diversity to include taxonomic relatives of the focal species (insects, fungi, nematodes, mites) suggests that species diversity could impact pathogen dynamics via regulation of the primary vector by its competitors and predators.

The ways in which the presence of more than one pathogen in a host might affect diseases caused by any of them have received surprisingly little attention. Superficially, a greater diversity of pathogens might be expected to produce a greater total burden of disease in a host population. Therefore, high diversity among this group should pose a threat to health, all else equal. In chapter 3, Rohani et al. show how this expectation may in fact be wrong. The presence of pathogens within a host can influence both the probability of infection with a new pathogen and the population growth rate and virulence of other pathogens that are able to establish. Such interpathogen effects are usually indirect, mediated by cross-reactivity by the host immune system to multiple pathogens, by competition for cellular attachment sites, or by behavioral responses to infection (reduced activity, quarantine) that reduce contact rates with new pathogens. In such cases, high pathogen diversity can serve a protective role. In other cases, however, the reverse is true. As examples, sequential exposures to multiple genetic strains of dengue virus dramatically increase the severity of disease, and infection with HIV vastly increases host susceptibility to many other pathogens and the severity of the consequent disease. A major frontier is the exploration of the conditions under which multiple pathogens engage in mutual interference versus facilitation.

In addition to variation in species diversity within hosts, vectors, and pathogens, anthropogenic disturbances such as pollution, climate change, and habitat destruction or fragmentation can strongly affect disease dynamics. Johnson and Carpenter in chapter 4 address the effects of ecosystems on disease from a more traditional bottom-up framework by asking, what are the consequences of nutrient superabundance on pathogens, parasites, and disease? They address this question specifically for aquatic eutrophication, finding multiple pathways by which excessive nutrient inputs into aquatic systems can influence pathogens, parasites, and hosts directly, as well as through interactions between pathogen and host. In an interesting twist on the interactions between pathogen diversity and disease outcomes, they find two important nonlinear relationships. The first is the effect of eutrophication on the species richness of pathogens and parasites, with richness initially increasing with low to moderate eutrophication but decreasing when nutrient loading is extreme. The second is the dramatic increase in disease severity in highly eutrophic systems, which occurs despite reduced pathogen diversity. Eutrophication of aquatic systems would seem to offer disease ecologists the opportunity to integrate the impacts of top-down with bottom-up forces, as has recently taken place in other branches of ecology.

Diseases occur within landscapes, but the role of landscape attributes in transmission dynamics tends to be ignored in favor of assessing highly localized influences on pathogens and disease. The nature of the localized interactions of pathogens with communities and environmental factors is likely to change across the landscape, as McCallum points out in chapter 5. Because landscape structure influences abundances and movements of hosts, vectors, and pathogens, it seems self-evident that landscapes should strongly influence disease dynamics. However, as with any process studied in an explicit landscape framework, the extent to which local dynamics are influenced solely by local factors (such as habitat quality) versus landscape features (such as patch size or degree of isolation) is rarely determined. Even when it is established that pathogen dynamics vary as a function of their position in a landscape, it is often poorly understood whether the composition of the landscape (e.g., relative abundances of different land cover types) or the configuration of those land cover types is more important. A major frontier in landscape epidemiology or spatial epidemiology (Ostfeld et al. 2005) is determining whether pathogen and disease dynamics can be understood sufficiently on the basis of localized interactions, or whether position in the landscape fundamentally alters the nature of localized interactions (Power 2006).

In summary, considerable progress has been made recently in addressing the impacts of several key aspects of ecological community structure and dynamics on epidemiological processes (including those in nonhuman animals and plants). The integration of effects of diversity among the various groups (hosts, vectors, pathogens, landscape elements) and of top-down and bottom-up effects remains to be attempted.

Effects of Disease on Ecosystems

Although a large number of case studies demonstrate that pathogens can have dramatic impacts on their hosts, substantially fewer studies have explored the ecological consequences of these pathogen-induced changes in hosts. Collectively, the chapters in the book's second section demonstrate that pathogens play an integral role in populations, communities, ecosystems, landscapes, and socioeconomic systems, and that our understanding of every level of ecological organization, and the interactions between these levels, is greatly enhanced by the explicit inclusion of pathogens in our conceptual ecological models. For example, the conceptual model of marine biogeochemical cycling was transformed by the inclusion of bacterial lysis by viruses, which accounted for the previously unexplained high ratio of bacteria to net primary production in oligotrophic waters (reviewed in Middelboe, chapter 11, this volume). In some cases, disease is the major mechanism underlying population dynamics and community interactions such as competition, succession, invasion, trophic interactions, maintenance of diversity, and self-thinning (reviewed in Collinge et al., chapter 6, Clay et al., chapter 7, Perkins et al., chapter 8, Lafferty, chapter 9, Hall et al., chapter 10, Middelboe, chapter 11, and Eviner and Likens, chapter 12, this volume). The response of these population and community processes to environmental changes would be hard to predict without explicitly understanding the role of pathogens. This is partly because interactions of host species with other community members can change dramatically if these hosts experience environmentally induced changes in the impacts of their pathogens (reviewed in Eviner and Likens, chapter 12, this volume) and partly because host organisms face important trade-offs between disease resistance and growth, competition, or herbivory (reviewed in Clay et al., chapter 7, and Perkins et al., chapter 8, this volume). These changes in host populations and communities can lead to large changes in ecosystem processes, disturbance regimes, local to global biogeochemistry, and human activities (reviewed in Lafferty, chapter 9, Middelboe, chapter 11, Eviner Likens, chapter 12, and Chapin et al., chapter 13, this volume), with potential long-term legacies. As a result, an understanding of current ecosystem structure and function must be considered incomplete without some recognition of the historical impacts of pathogens (examples from Collinge et al., chapter 6, Eviner and Likens, chapter 12, and Chapin et al., chapter 13, this volume).

Collinge et al. in chapter 6, describe how disease can profoundly affect the structure and function of ecosystems when the host is a key—keystone or dominant—species. Pathogens might be among the only species that strongly regulate abundance or cause population crashes of top predators, whereas for dominant species, pathogens might be among the only species able to regulate these hosts. As Collinge et al. point out, pathogens that are dynamically coupled with their hosts are likely to act in a regulating fashion, with trickle-down effects on other aspects of community structure, whereas those that are not tightly coupled with hosts (e.g., the pathogen causing chestnut blight, which persists in the environment even when its host does not) might tend to cause local or broader extinctions. A crucial frontier is determining the nature of the feedback structure within systems in which key species are under pathogen attack. What attributes of the system might predispose it toward stabilizing negative feedback, with disease outbreaks reducing the likelihood of subsequent disease, versus positive feedback, with outbreaks facilitating subsequent pathogen attacks?

Clay et al. in chapter 7 postulate that pathogens can constitute a potent force for increasing species diversity within ecological communities, with potential knock-on effects on ecosystem functions facilitated by high species diversity. They apply tenets of the Red Queen hypothesis, which include the argument that pathogens preferentially attack the most common genotypes within populations, thereby providing an advantage to rare genotypes, to entire communities. For pathogens to enhance community diversity, they must (1) be at least moderately host-specific and (2) preferentially attack the most common species. In addition, the hosts must (3) be subject to a trade-off between resistance to particular pathogens and other components of fitness, and the host-pathogen system must (4) be subject to dynamic fluctuations. The review of the relevant literature on plants and their pathogens by Clay et al. lends strong support to all four tenets, leading to the conclusion that pathogens can and do foster high species diversity in plant communities. Given the established relationship between high species diversity in plant communities and the performance of several important ecosystem functions, Clay et al. posit that pathogens have strong indirect effects on ecosystem functioning. The nature and strength of feedback loops linking pathogen effects on community diversity to diversity effects on ecosystem functioning, to community and ecosystem effects on pathogen communities, should be fruitful foci for future research.

The review by Perkins et al. in chapter 8 helps establish a research frontier at the interface between invasion biology and disease ecology. They ask how strong a role parasites and pathogens play in regulating host abundance and controlling their population dynamics in general, and apply this question specifically to hosts that are exotic organisms. The answers to these questions will ultimately depend not only on the degree to which exotic species lose their parasites when invading a new ecosystem but also on the rate at which they accumulate new parasites. Parasite accumulation can occur either through immigration, or through translocation of once distant parasites, or through evolutionary changes in parasites that facilitate their adaptation to new hosts. Perkins et al. argue that a major research frontier is determining the indirect effects of parasite gain or loss on the ecosystems occupied by exotic species.

Lafferty in chapter 9 further integrates parasites and pathogens into food webs, asking how they are involved in all categories of ecological interactions, both directly and indirectly. Lafferty advocates an approach that has been used to ask how consumers affect the structure and dynamics of components of food webs: theoretical exploration of how varying topologies of food web modules (subsets of entire food webs) affect processes such as stability or biomass accumulation. The sheer number of parasites and their potential to regulate hosts suggests these species need to be incorporated explicitly into food web studies, particularly those that focus on the outcomes of interactions that span more than two trophic levels. Hall et al. in chapter 10 extend the coverage of this topic to ask whether parasitism as a process is qualitatively different from other consumer-driven processes (e.g., predation, herbivory), and if not, whether it should be subsumed under the broader category of predation. If parasitism is simply a special case of predation, then one would expect the well-developed predation theory to provide a strong foundation for future explorations of parasitism. On the other hand, if parasitism is a fundamentally different process in some critical aspects, then new theory will be required. It seems that the choice of whether to subsume parasitism under predation depends on the research questions being addressed. Certainly, in many cases parasite-host interactions differ both quantitatively and qualitatively from predator-prey interactions.

Even though most disease ecologists focus on microbial pathogens of higher organisms such as vertebrates or vascular plants, infectious diseases can impact any organism. In chapter 11, Middelboe describes the magnitude and consequences of bacterial death by lysis, inflicted by viruses in the ocean. The sheer number and the biomass of bacteria and cyanobacteria infected by viruses are staggering. Perhaps more important, the virus-induced conversion of particulate organic matter to dissolved organic matter profoundly affects marine nutrient cycling at all scales, from the local to the global. Although the focus of Middelboe's explorations is on biogeochemical cycling, viral lysis has the potential to affect virtually every ecological process from the population dynamics of phytoplankton and bacterioplankton species to community diversity and organization to food web structure and biomass. Research frontiers in this area include how population and community processes are influenced by viral attack, spatial and temporal scales of variation in the magnitude of viral diseases and their consequences, and the degree to which viral lysis of prokaryotes and single-celled eukaryotes affects ecological processes in terrestrial and freshwater ecosystems.

Eviner and Likens in chapter 12 address some of these questions, and others, as they apply to plant pathogens in terrestrial ecosystems. In an exploration of principles that might underlie the effects of plant pathogens on terrestrial ecosystem processes, they identify seven factors that are expected to influence the ecosystem consequences of pathogen infection: (1) pathogen effects on host fitness, (2) the predominant life stages attacked, (3) host infection prevalence, (4) spatial extent of infection, (5) the rate of pathogen effects on hosts relative to the recovery rate of the host population, (6) the degree of functional similarity between the victim population and its replacement following a disease outbreak, and (7) the frequency and duration of the pathogen impact. Eviner and Likens ask how the impacts of pathogens on ecosystem processes are likely to change through time, and how variation in their seven factors will affect the nature, strength, and duration of these effects. Both general and specific predictions arising from these explorations are important areas for future research.

In a continuing expansion of the scope of disease studies, Chapin et al. in chapter 13 expand the focus to pursue how pathogens and disease affect both landscape structure and socioecological systems. Chapin et al. start where several other authors end their discussions, with the impacts of pathogens on keystone or dominant animals, plants, and microbes, but then pursue the consequences of these impacts for the larger landscapes within which these biotic interactions occur. They argue that pathogen-induced changes in disturbance regimes, microclimate, and nutrient availability that result from alterations in the abundance or dynamics of animal, plant, and microbe hosts can penetrate into regional and even global processes. Diseases of humans and of some nonhuman animals and plants change public health and infrastructure in myriad ways, with knock-on consequences for policies, business, and use of natural resources. These socioecological effects of disease potentially can result in feedbacks into the ecological processes controlling the spread of disease in the first place. Chapin et al. use resilience theory to catalogue features of socioecological systems that enhance the ability of systems that experience disease-caused "shocks" to recover, illustrating their arguments with descriptions of bark beetle outbreaks, rinderpest, malaria, schistosomiasis, and HIV/AIDS. Deem et al. in chapter 14 place similar arguments in a historical context, asking whether ecosystem-level impacts of wildlife diseases have been expanding in an increasingly human-dominated world and whether human activities are increasing the historical rate of emergence and the impacts of infectious diseases in natural ecosystems. They pose an ambitious research agenda to address these questions.

The various chapters in part II strongly suggest that a complex and sophisticated model of the impacts of pathogens and diseases on ecosystems is needed. One major challenge to erecting such a model is integrating the effects of disease on ecosystems with the effects of ecosystems on disease. For example, the African rinderpest epizootic of the twentieth century had profound and widespread impacts on savanna ecosystems by decimating populations of some grazing ungulates, such as wildebeest (*Connochaetes taurinus*). In response to the sharp declines in herbivory, woody plants invaded large areas that had been dominated by grasses. One of the many consequences of this dramatic change in ecosystem structure and function was growth in the abundance and distribution of tsetse flies (*Glossina* spp.), which transmit the trypanosome that causes African sleeping sickness (L. Talbot, personal communication). More such examples of feedback loops between disease and ecosystems should be expected.

MANAGEMENT AND APPLICATIONS

Infectious disease ecology as a discipline is more than emerging; it is exploding. Have we progressed sufficiently to allow disease outbreaks to be anticipated, consequences to be predicted, interventions to be imposed? How sophisticated must our understanding of the ecological underpinnings of disease be to provide the basis for action? Can outbreaks be predicted simply by identifying the reservoir for the pathogen and tracking its population dynamics? Can the dynamics and consequences of an epidemic be predicted simply by estimating R_0 for a specific pathogen? Or, conversely, must we know quite a bit about the communities and food webs within which (often multiple) reservoirs are embedded, about the heterogeneities among individuals within populations, and about human behaviors associated with landscapes of varying risk? The chapters making up part III indicate that simple ecological models rarely suffice, that significant challenges remain in creating, testing, and adapting more complex models to allow appropriate intervention, but that the seeds for joining ecology and epidemiology have begun to sprout.

Predators, and especially apex predators, are disproportionately reduced by pervasive human activities such as habitat destruction, fragmentation, and direct exploitation of wildlife. In chapter 15, Holt uses the concept of food webs and a series of mathematical models to address the conditions under which predators (or the loss of predators) are expected to alter the dynamics of infectious disease either quantitatively or qualitatively. He finds that, for a host population that is not regulated by either a pathogen or a predator (but instead by territoriality or resources), loss of the predator can result in pathogen invasion that would not be possible otherwise. But details of the host-pathogen interaction are important in determining the outcome of predator loss. For example, if hosts acquire immunity after recovering from the infection, predator loss will reduce the prevalence of infection. When hosts are regulated by the pathogen, predator loss is likely to increase the density of infected hosts, potentially leading to epizootics and spillover events. Upsurges in pathogen prevalence in hosts that accompany predator loss are especially likely for pathogens that can infect multiple hosts, particularly if the hosts compete. Holt's mathematical models, of course, are abstractions of nature, but ones in which assumptions are made explicit, systematically varied, and outcomes determined quantitatively. Although these models are sophisticated in that they incorporate multiple trophic levels, direct and indirect interactions, and both behavioral and physiological (immunological) complexities, they still represent highly simplified versions of nature. However, the combination of mathematical simplicity, ecological sophistication, and clarity of assumptions means that empirical tests should be tractable.

Hudson et al. in chapter 13 address the question of how emerging infectious diseases of wildlife can be controlled, either by preventing the initial case of spillover that might initiate an outbreak or by disrupting an initiated outbreak through either culling or vaccination of individuals. Before proceeding to this step, however, they emphasize two important points: (1) the assumption that a high initial rate of pathogen spread in an entirely susceptible host population (R_0) leads directly to the persistence of an infection in that host population is generally incorrect, and (2) transmission dynamics for a particular pathogen can be governed by interindividual heterogeneity within the host population, with superspreaders sometimes profoundly changing disease dynamics. Hudson et al. provide lively illustrations of how culling can either curtail or exacerbate a disease outbreak, depending on whether culling simply reduces the density of reservoir hosts or instead has more complex effects, such as disrupting social structure and regulation. In addition, they show how mathematical models can provide clear recommendations on what percentages of both reservoir and victim (e.g., endangered species) populations need to be vaccinated to prevent spillover events with dire outcomes. Finally, they describe the importance of determining the life history or other correlates of superspreaders that might be used to identify targets for control. Clearly, vaccination or culling of superspreaders could often prevent epidemics before major morbidity and mortality ensue, although identifying superspreaders before they exert their impact is a major challenge. Altogether, the chapter by Hudson et al. provides the basis for some optimism about the notion that wildlife diseases can be managed under realistic scenarios of knowledge of host-pathogen interactions and tried-and-true mechanisms of intervention.

Garrett and Cox in chapter 17 provide many more reasons for optimism that community and ecosystem ecology can be and are being applied to disease management and prevention. They document consistent inhibitory effects of a high diversity of either plant species or cultivars on the spread of diseases caused by several taxa of pathogens. Such a dilution effect has been demonstrated in both natural and agricultural systems. Interestingly, the mechanisms that underlie this dilution effect include the direct effects of reducing host plant frequency or density on pathogen transmission aboveground, but also the indirect effects of plant diversity on soil microbial diversity, which can suppress soil-borne pathogens quite strongly. Garrett and Cox further describe the ways in which this knowledge has been used to increase yield and profit while also listing impediments, such as artificial economic incentives, to its broader application. From both a theoretical and applied economic perspective, a major frontier is to determine what levels of diversity (genomic, cultivar, species, landscape) are most effective in reducing the impacts of disease on both natural and agricultural plant systems.

Porter et al. in chapter 18 tell a success story in which the ecology of an infectious disease of corals in the Florida Keys entered the realm of mass media, politics, and policy. In their case study, the ecosystem is best defined as one in which coral reefs, adjacent lands that generate and export the pathogen, ecologists, media, politicians, and voting citizens are identifiable components. The scientists documented the abrupt, massive decline of elkhorn corals and identified a likely etiological agent, the white pox bacterium, *Serratia marcescens*, a common fecal enteric bacterium found in the guts of humans and other terrestrial animals. By interacting judiciously with media and politicians, the ecologists helped to identify the likely sources of the pathogen and actions that might mitigate the problem. Local, state, and federal officials, with public support, then took steps to reduce wastewater and storm runoff, even though the costs of such action were high. Whether the pathogen loads are reduced and the corals rebound remains to be determined. Assuming that these anticipated outcomes materialize, a major frontier will be to determine whether systems of greater complexity, such as those with multiple hosts and multiple reservoirs, can be successfully managed on the basis of ecological understanding.

Ecological sleuthing to determine the cause of mysterious illness is not new. In chapter 19, Johnson recounts his efforts to determine the cause of an outbreak of severe hemorrhagic fever among the human inhabitants of San Joaquin, Bolivia, in the early 1960s. With the material support of the U.S. military and the ingenuity to invent a portable biosafety level 4 laboratory, Johnson and colleagues discovered a new, highly lethal virus, Machupo virus, and its main reservoir, *Calomys* mice. Mouse eradication, whether by direct human action or competitive exclusion courtesy of larger, more aggressive rodents, eliminated cases of human disease. Even though Johnson and his team had no formal training in ecology, they were able to determine, for instance, that mouse diet had a profound impact on the rate of virus shedding in urine, with wet diets (bananas) elevating shedding rates by a factor of 10 over dry diets (dried corn). Perhaps even more impressive was their discovery that two genotypes of *Calomys* exist, with one (type A) unable to mount an immune response to virus and therefore typically excreting massive quantities of virus and another (type B) that partially cleared infection via immune response and shed much less virus. Due to an apparent trade-off between immunity to Machupo virus and reproductive output, Johnson and colleagues surmised that, at low population density when the mice typically occur in isolated social groups, Type A genotypes should have higher fitness and population growth should be rapid. But dense populations dominated by this genotype are highly susceptible to Machupo virus epizootics, which cause female sterility only in type A mice, with the result that type B genotypes are strongly favored during and after peak densities. This hypothesis, only superficially described here, can account for maintenance the pathogen, dramatic population fluctuations in the host and virus, and strong temporal changes in human exposure rates. The ecology of infectious diseases was alive and well in the 1960s.

Not all observers of the role of ecology in controlling human infectious diseases are equally optimistic. Peters in chapter 20 describes ten emerging viral diseases of humans in the context of factors that have been proposed to cause disease emergence. Although in each case, ecological factors clearly are among the most important causes of emergence, Peters is not convinced that knowledge of how these factors act will increase our ability to prevent or control disease. Key obstacles identified by Peters include the notion that massive manipulations of the environment might be necessary to reduce disease transmission, and such acts generally are infeasible. Similarly, for diseases driven by regional or global climatic changes, acting on ecological knowledge might not be possible. Peters grants, however, that a better understanding of the factors that govern virus circulation outside human hosts could lead to actionable ecological understanding. In chapter 21, Childs provides a somewhat different perspective on the role of ecology in disease control, including a prescription for increasing positive interactions between practitioners of infectious disease ecology (IDE) and public and veterinary health providers (PVHP). Using the 2001 foot-and-mouth disease outbreak in the United Kingdom as a focal point, Childs characterizes the perspective of the IDE community as arguing that its theoretical models have direct utility and that its expertise should have been tapped sooner by the public health community in order to curtail the outbreak. In stark contrast, Childs describes the skepticism of the PVHP community toward the IDE community and its models, arguing that IDE models were slow to be developed, insufficiently validated, and impractical to apply. Childs exhorts the IDE community to redress these differences by initiating efforts to reach out to the PVHP community. He argues that the IDE community has more to lose from the status quo, although it seems one could equally argue that the onus for reaching out to cross-disciplinary counterparts is on the PVHP community, whose mandate is to protect human and nonhuman animal health. Clearly, even when ecological studies provide clear prescriptions for disease control, sociological, linguistic, and cultural barriers continue to divide allies in the fight against disease.

Education in disease ecology might be one means of bridging cultural divides between disciplines involved in disease research. Brewer et al. in chapter 22 describe the potential of education, spanning K–12, undergraduate, postgraduate, and informal realms, for uniting students of both ecosystems and infectious diseases and creating a new generation of broadly trained scientists and practitioners. They point out that disease is a great motivator and can represent many key concepts that span the subdisciplines of ecology. Despite the inherent interest in and pedagogical usefulness of infectious disease, Brewer et al. describe how poorly represented this topic is in K–12, undergraduate, and graduate curricula, at least in the United States. They provide some compelling reasons why, despite increasingly broad recognition of the importance of ecology in understanding disease dynamics, medical schools typically fail to teach appropriate ecological principles or case studies. Some veterinary schools appear to be setting an example with postgraduate, professional training that transcends boundaries between traditionally biomedical and traditionally ecological realms. After describing five different habits of mind needed across all educational settings for understanding the ecology of infectious disease, Brewer et al. propose a bold agenda for teaching and learning in this area.

Conclusions

In the introduction to this book we described two pressing needs that should be met to improve our ability to predict the occurrence, dynamics, and consequences of infectious diseases. One was the integration of empirical and theoretical case studies into broader conceptual frameworks that can inform and accelerate studies of new disease systems; the second was to forge more and stronger alliances between ecologists and the more traditional suite of infectious disease specialists. We think the pages of this book demonstrate dramatic progress on both fronts. The early twenty-first century is an exciting time for the transdisciplinary study of infectious diseases. The power of modern molecular techniques to identify pathogens and track their transmission, combined with conceptual, theoretical, and technological developments in ecology, biogeography, and climate science, provide a new alchemy with extraordinary potential. The beneficiary of a deeper understanding of disease ecology will be humankind and the ecosystems in which we are embedded.

Literature Cited

Keesing, F., R. D. Holt, and R. S. Ostfeld. 2006. Effects of species diversity on disease risk. Ecology Letters 9:485–98.

Ostfeld, R. S., G. Glass, and F. Keesing. 2005. Spatial epidemiology: An emerging (or re-emerging) discipline. Trends in Ecology & Evolution 20:323–31.

Power, M. E. 2006. Environmental controls on food web regimes: A fluvial perspective. Progress in Oceanography 68:125–33.

INDEX

Lightning Source UK Ltd.
Milton Keynes UK
UKHW022211171122
412377UK00005B/229